T0231191

SECOND EDITION

Protocols for Secure Electronic Commerce

Advanced and Emerging Communications Technologies Series

Series Editor-in-Chief: Saba Zamir

The Telecommunications Illustrated Dictionary, Second Edition,
Julie K. Petersen

Handbook of Emerging Communications Technologies: The Next Decade,
Rafael Osso

ADSL: Standards, Implementation, and Architecture, Charles K. Summers

Protocols for Secure Electronic Commerce, Mostafa Hashem Sherif

Protocols for Secure Electronic Commerce, Second Edition,
Mostafa Hashem Sherif

After the Y2K Fireworks: Business and Technology Strategies,
Bhuvan Unhelkar

Web-Based Systems and Network Management, Kornel Terplan

Intranet Performance Management, Kornel Terplan

Multi-Domain Communication Management Systems, Alex Galis

Fiber Optics Illustrated Dictionary, Julie K. Petersen

Electronic Bill Presentment and Payment, Kornel Terplan

SECOND EDITION

Protocols for Secure Electronic Commerce

Mostafa Hashem Sherif, Ph.D.

AT&T Laboratories, New Jersey

Series Editor-in-Chief
Saba Zamir

CRC PRESS

Boca Raton   London   New York   Washington, D.C.

Library of Congress Cataloging-in-Publication Data

Sherif, Mostafa Hashem.
    [Monnaie électronique. English]
    Protocols for secure electronic commerce / Mostafa Hashem Sherif. — 2nd ed.
       p. cm. (The CRC Press advanced and emerging technologies series)
    Includes bibliographical references and index.
    ISBN 0-8493-1509-3 (alk. paper)
    1. Electronic commerce. 2. Bank credit cards. 3. Computer networks—Security
measures. I. Title. II. Series.

HF5548.32.S5213 2003
658.8′72′028558—dc22   2003061098

Visit the CRC Press Web site at www.crcpress.com

© 2004 by CRC Press LLC

No claim to original U.S. Government works
International Standard Book Number 0-8493-1509-3
Library of Congress Card Number 2003061098
Printed in the United States of America 1 2 3 4 5 6 7 8 9 0
Printed on acid-free paper

Preface

The presence and influence of online commerce are growing steadily, despite, if not because of, the burst of the dot.com frenzy. With the speculators gone and in the absence of unsubstantiated claims, it is now possible to face the real problems of the information society in a rational and systematic manner. As more virtual services are offered to the general public or among businesses, security of the networked economy will be entangled with many other considerations. Potential solutions can go along so many directions as additional parties with different priorities and requirements are brought online. The interconnection and fusion of local spaces can only mean that electronic commerce (e-commerce) security will require global actions, including global technical standards and organizational agreements. These activities, however, do not occur in vacuum; compromises will have to be made to cope with existing infrastructures, processes, laws, or social organizations that were not designed for online activity.

The aim of this book is to help the reader address these challenges. Its intended audience ranges from readers of the periodic IT-Review of the *Financial Times*, who may want to understand the technical reasons behind the analysis, to graduate students in technical and informational domains, who would like to understand the context in which technology operates. In updating the text, I strove to maintain the goals of the first edition of providing a comprehensive, though readable, compendium to the protocols for securing e-commerce and electronic payments. I tried to provide enough technical details so that readers could gain a good grasp of the concepts, while leaving the rest to more specialized works as indicated in the bibliography. Chapters were revised or completely rewritten to reflect technical advances and continuous developments as well as to include new areas, such as mobile commerce (m-commerce). In doing so, I benefited from the experience gained in teaching the material to improve the presentation and correct errors. In some cases, such as for secure electronic transaction (SET), I decided to maintain topics that did not correspond to market successes because of the many innovative ideas that were involved.

For academic use, I followed the suggestions of several instructors and added review questions at the end of each chapter. In addition, contains PowerPoint® presentations will be available from the CRC Web site: www.crcpress.com on the topics discussed in each of the book's chapters.

My French editor, Mr. Eric Sulpice, generously supplied me with information on the development of smart cards in Europe. Mr. Kazuo Imai, vice president and general manager, network laboratories of NTT DoCoMo, provided me with technical information on i-mode®. Professors Manu Malek, of the Stevens Institute of Technology (Hoboken, New Jersey), and Mehmet Ulema, from Manhattan College, New York, gave me useful comments on the content and its presentation.

Once again, I must thank CRC Press LLC. In particular, Dr. Saba Zamir, editor-in-chief of the series, for her confidence, the editorial team of Nora Konopka, Samar Haddad, and Jamie Sigal for their assistance, and Lori Eby for her excellent copyediting skills.

Finally, the trust and encouragement of relatives and friends were, as usual, indispensable.

Tinton Falls, New Jersey, July 2002–September 2003

Preface to the First Edition

The purpose of this book is to present a synthesis of the protocols currently used to secure electronic commerce. The book addresses several categories of readers: engineers, computer scientists, consultants, managers, and bankers. Students interested in computer applications in the area of payment will find this volume a useful introduction that will guide them toward more detailed references.

The book is divided into three parts. The first consists of Chapters 1 through 3 and is a general introduction to the multiple aspects of electronic commerce. The second part is formed by Chapters 4 through 12 and details the various aspects of electronic money: Electronic Data Interchange (EDI), payments with bank cards, micropayments with electronic purses, digital money, and virtual checks. The final section comprises Chapters 13 through 15 and presents smart cards, efforts for converging heterogeneous payment systems, and some thoughts on the future of electronic commerce.

Because the field of electronic commerce covers several topics that are evolving continuously, it is not possible to cover all aspects in this first presentation. We would be grateful to readers to indicate errors, omissions, or additional material for consideration.

This book appears in a French version co-authored with Professor Ahmed Sehrouchni, for the École Nationale Supérieure des Télécommunications (ENST), Paris, France and published by Eyrolles under the title *La Monnaie Électronique: Systèmes de Paiement Sécurisé*.

The discussions that the author had with participants in the project PECUNIA of the now-defunct AT&T Unisource helped clarify many details concerning the payment systems. I would like to thank in particular Maria Christensen, Greger S. Isaksson, and Lennart E. Isaksson, all three from the research unit of the Swedish operator, Telia. I would also like to thank Philip Andreae (consultant) and Patrick Scherrer who led the project. Aimé Fay, my former colleague at AT&T France and author of the dictionary on banking technology, *Dico Banque*, graciously guided my first steps in the field of payment systems. The research conducted with Luis Lucena while he was a graduate student at ENST-Paris as well as with my colleagues at the National Technical University of Athens, Greece — Maria Markakis, Georges Mamais, and Georges Stassinoupoulos — helped me evaluate the effect of computer telephony integration (CTI) on electronic commerce. Chapters 6 and 7 were influenced profoundly by the contributions of A. Yassin Gaid and Farshid Farazmandnia during the course of their internship at AT&T France in 1997 as part of their ENST-Paris graduation project. The results of their work have been published in French and in English.

CRC Press has been patient throughout the long gestation of this book. The project would not have started without Saba Zamir, Editor-in-Chief of the series, "Advanced and Emerging Communications Technologies," and Gerald T. Papke, Senior Editor at CRC Press.

My thanks also extend to Donna Coggshall who reviewed and edited the first English version of the manuscript. Fred Burg, my colleague at AT&T, reviewed the first two chapters and suggested some stylistic improvements. Andrea Tarr introduced me to Bert V. Burke, the founder and CEO of Every Penny Counts, Inc. (EPC), who provided information included in Chapter 14.

Finally, I am grateful to friends and relatives who generously gave me their support throughout the time needed to research and write this book.

Neuilly-sur-Seine, France, October 1997
Tinton Falls, New Jersey, October 1999

Author

Mostafa Hashem Sherif, Ph.D., is a Principal Member of the technical staff at AT&T. He earned degrees from Cairo University, Egypt, the University of California, Los Angeles, and Stevens Institute of Technology, Hoboken, NJ.

He is a senior member of the Institute of Electrical and Electronics Engineers (IEEE), a standards editor for the *IEEE Communications Magazine* and is a certified project manager at the Project Management Institute (PMI).

Table of Contents

1   Overview of Electronic Commerce

Abstract 1

1.1   What Is Electronic Commerce? 1

1.2   Categories of Electronic Commerce 3

      1.2.1   Examples of Business-to-Business Commerce 4

      1.2.2   Examples of Business-to-Consumer Commerce 5

      1.2.3   Examples of Neighborhood Commerce and Payments
              to Automatic Machines 7

      1.2.4   Examples of Peer-to-Peer Commerce 8

1.3   The Influence of the Internet...8

      1.3.1   Some Leading Examples 8

      1.3.2   Internet and Transactional Security 9

      1.3.3   Putting the Internet in Perspective 11

1.4   Infrastructure for Electronic Commerce 13

1.5   Network Access 15

      1.5.1   Wireline Access 16

      1.5.2   Wireless Access 16

      1.5.3   Traffic Multiplexing 17

1.6   Consequences of E-Commerce 21

      1.6.1   Clients 21

      1.6.2   Suppliers 22

      1.6.3   Substitutes 22

      1.6.4   New Entrants 23

      1.6.5   Banks 24

      1.6.6   Role of Governments 24

1.7   Summary 25

Questions 25

2   Money and Payment Systems

Abstract 27

2.1   The Mechanisms of Classical Money 27

2.2   Instruments of Payment 29

      2.2.1   Cash 31

      2.2.2   Checks 33

      2.2.3   Credit Transfers 37

      2.2.4   Direct Debit..40

      2.2.5   Interbank Transfers 41

      2.2.6   Bills of Exchange 42

      2.2.7   Payment Cards 42

2.3 Types of Dematerialized Monies.. 46
    2.3.1 Electronic Money 46
    2.3.2 Virtual Money 47
    2.3.3 Digital Money 48
2.4 Purses and Holders 49
    2.4.1 Electronic Purses and Electronic Token (Jeton) Holders..........49
    2.4.2 Virtual Purses and Virtual Jeton Holders 50
    2.4.3 Diffusion of Electronic Purses 51
2.5 Transactional Properties of Dematerialized Currencies 53
    2.5.1 Anonymity 54
    2.5.2 Traceability 55
2.6 Overall Comparison of the Means of Payment 55
2.7 The Practice of Dematerialized Money 57
    2.7.1 Protocols of Systems of Dematerialized Money 57
    2.7.2 Direct Payments to the Merchant 62
    2.7.3 Payment via an Intermediary 62
2.8 Banking Clearance and Settlement 65
    2.8.1 United States 66
    2.8.2 United Kingdom 67
    2.8.3 France 68
2.9 Summary 69
Question 70

3   Algorithms and Architectures for Security
Abstract 71
3.1 Security of Commercial Transactions 71
3.2 Security of Open Financial Networks 72
3.3 Security Objectives 73
3.4 OSI Model for Cryptographic Security 75
    3.4.1 OSI Reference Model 75
    3.4.2 Security Services: Definitions and Locations 75
3.5 Security Services at the Link Layer 78
3.6 Security Services at the Network Layer 79
3.7 Security Services at the Application Layer 82
3.8 Message Confidentiality 83
    3.8.1 Symmetric Cryptography 83
    3.8.2 Public Key Cryptography 84
3.9 Data Integrity 86
    3.9.1 Verification of the Integrity with a One-Way Hash
            Function 87
    3.9.2 Verification of the Integrity with Public Key
            Cryptography 88
    3.9.3 Blind Signature 91
    3.9.4 Verification of the Integrity with Symmetric
            Cryptography 91

3.10 Identification of the Participants 94

 3.10.1 Biometric Identification 94

   3.10.1.1 Voice Recognition 95

   3.10.1.2 Handwritten Recognition 96

   3.10.1.3 Keystroke Recognition 96

   3.10.1.4 Retinal Recognition 97

   3.10.1.5 Iris Recognition 97

   3.10.1.6 Face Recognition 98

   3.10.1.7 Fingerprint Recognition 99

   3.10.1.8 Recognition of Hand Geometry 100

 3.10.2 Summary and Evaluation 100

3.11 Authentication of the Participants 102

3.12 Access Control 104

3.13 Denial of Service 106

3.14 Nonrepudiation 108

 3.14.1 Time-Stamping and Sequence Numbers 109

3.15 Secure Management of Cryptographic Keys 110

 3.15.1 Production and Storage 110

 3.15.2 Distribution 111

 3.15.3 Utilization, Withdrawal, and Replacement 111

 3.15.4 Key Revocation 112

 3.15.5 Deletion, Backup, and Archiving 112

 3.15.6 Comparison between Symmetric and Public Key
     Cryptography 112

3.16 Exchange of Secret Keys: Kerberos 113

 3.16.1 Message (1) — Request of a Session Ticket 114

 3.16.2 Message (2) — Acquisition of a Session Ticket 114

 3.16.3 Message (3) — Request of a Service Ticket 115

 3.16.4 Message (4) — Acquisition of the Service Ticket 115

 3.16.5 Message (5) — Service Request 116

 3.16.6 Message (6) — Optional Response of the Server 117

3.17 Public Key Kerberos 117

 3.17.1 Where To Find Kerberos? 118

3.18 Exchange of Public Keys 118

 3.18.1 Diffie–Hellman Exchange 118

3.19 ISAKMP (Internet Security Association and Key Management
   Protocol) 119

3.20 SKIP (Simple Key Management for Internet Protocols) 121

3.21 Key Exchange Algorithm 121

3.22 Certificate Management 122

 3.22.1 Basic Operation 125

 3.22.2 Description of an X.509 Certificate 126

 3.22.3 Certification Path 128

 3.22.4 Hierarchical Certification Path 128

 3.22.5 Nonhierarchical Certification Path 131

 3.22.6 Cross-Certification 131

3.22.7   Online Management of Certificates 133

3.22.8   Banking Applications 133

3.22.9   Example: VeriSign 134

        3.22.9.1   Certificate Classes 135

        3.22.9.2   Operational Life 136

        3.22.9.3   Revocation 136

        3.22.9.4   Archival 137

        3.22.9.5   Recovery 137

        3.22.9.6   Liability 137

3.22.10   Procedures for Strong Authentication 138

        3.22.10.1 One-Way Authentication 138

        3.22.10.2 Two-Way Authentication 139

        3.22.10.3 Three-Way Authentication 139

3.22.11   Certificate Revocation 140

3.22.12   Attribute Certificates 141

3.22.13   Audits 143

3.23   Encryption Cracks 143

3.24   Summary 146

3.25   Appendix I: Principles of Symmetric Encryption 147

    3.25.1   Modes of Algorithm Utilization for Block Encryption 147

    3.25.2   Examples of Symmetric Block Encryption Algorithms 153

        3.25.2.1   Advanced Encryption Standard (AES) 153

        3.25.2.2   Data Encryption Standard (DES) 154

        3.25.2.3   Triple DES 154

        3.25.2.4   IDEA 154

        3.25.2.5   SKIPJACK 154

3.26   Appendix II: Principles of Public Key Encryption 155

    3.26.1   RSA 156

        3.26.1.1   Practical Considerations 157

    3.26.2   Public Key Cryptography Standards (PKCS) 157

    3.26.3   Pretty Good Privacy (PGP) 159

    3.26.4   Elliptic Curve Cryptography (ECC) 159

3.27   Appendix III: Principles of the Digital Signature Algorithm

    (DSA) 161

3.28   Appendix IV: Comparative Data 162

    3.28.1   Performance Data for JSAFE 1.1 163

    3.28.2   Performance for S/WAN 164

    3.28.3   Performance for BSAFE™ 3.0 165

    3.28.4   Performance for BSAFE™ 4.1 166

Questions 166

4   Business-to-Business Commerce

Abstract 173

4.1 Overview of Business-to-Business Commerce 174

4.2 Examples of Business-to-Business Electronic Commerce 177

	4.2.1	A Short History of Business-to-Business Electronic Commerce	177
	4.2.2	Banking Applications	178
	4.2.3	Aeronautical Applications	178
	4.2.4	Applications in the Automotive Industry	179
	4.2.5	Other Examples	180
	4.2.6	Effect of the Internet	180
4.3		Business-to-Business Electronic Commerce Platforms	181
4.4		Obstacles Facing Business-to-Business Electronic Commerce	182
4.5		Business-to-Business Electronic Commerce Systems	184
	4.5.1	Generation and Reception of Structured Data	185
	4.5.2	Management of the Distribution	187
	4.5.3	Management of Security	187
4.6		Structured Alphanumeric Data	187
	4.6.1	Definitions	188
	4.6.2	ANSI X12	189
	4.6.3	EDIFACT	190
		4.6.3.1 UNB/UNZ and UIB/UIZ Segments	191
		4.6.3.2 UNH/UNT Segments	192
		4.6.3.3 UNS Segment	193
		4.6.3.4 UNG/UNE Segments	193
		4.6.3.5 UNO/UNP Segments	193
		4.6.3.6 Structure of an Interchange	194
		4.6.3.7 Partial List of EDIFACT Messages	194
		4.6.3.8 Interactive EDIFACT	195
	4.6.4	Structural Comparison between X12 and EDIFACT	195
4.7		Structured Documents or Forms	195
	4.7.1	SGML	197
	4.7.2	XML	198
	4.7.3	Integration of XML with Alphanumeric EDI	198
		4.7.3.1 BizTalk®	200
		4.7.3.2 Commerce XML (cXML)	200
		4.7.3.3 Electronic Business XML (ebXML)	201
		4.7.3.4 SAML (Security Assertion Markup Language)	201
		4.7.3.5 SOAP (Simple Object Access Protocol)	202
		4.7.3.6 UDDI (Universal Description, Discovery, and Integration)	202
		4.7.3.7 WSDL (Web Services Description Language)	203
4.8		EDI Messaging	203
	4.8.1	X.400	203
	4.8.2	Internet (SMTP/MIME)	204
4.9		Security of EDI	206
	4.9.1	X12 Security	207
	4.9.2	EDIFACT Security	208
		4.9.2.1 Security of EDIFACT Documents Using In-Band Segments	209

   4.9.2.2 Security of EDIFACT Documents with Out-of-Band
       Segments: The AUTACK Message 213
  4.9.3 IETF Proposals 216
   4.9.3.1 PGP/MIME Encrypted and Signed 217
   4.9.3.2 S/MIME Message Encrypted and Signed 219
  4.9.4 Protocol Stacks for EDI Messaging 220
  4.9.5 Interoperability of Secured EDI and S/MIME 221
  4.9.6 Security of XML Exchanges 223
 4.10 Relation of EDI with Electronic Funds Transfer 223
  4.10.1 Funds Transfer with EDIFACT 226
  4.10.2 Funds Transfer with X12 228
 4.11 Electronic Billing 228
 4.12 EDI Integration with Business Processes....................................229
 4.13 Standardization of the Exchanges of Business-to-Business
  Electronic Commerce 230
  4.13.1 EDI/EDIFACT 230
  4.13.2 XML/EDI Integration 234
   4.13.2.1 CEFACT 234
   4.13.2.2 CommerceNet 234
   4.13.2.3 IETF (Internet Engineering Task Force) 234
   4.13.2.4 Open Buying on the Internet (OBI) 234
   4.13.2.5 Open Trading Protocol (OTP) Consortium 234
   4.13.2.6 Organization for the Advancement of Structured
       Information Standards (OASIS) 235
   4.13.2.7 RosettaNet 235
  4.13.3 XML 235
 4.14 Summary 236
 Questions 236

5 SSL (Secure Sockets Layer)
Abstract 239
5.1 General Presentation of the SSL Protocol 239
  5.1.1 Functional Architecture 240
  5.1.2 SSL Security Services 241
   5.1.2.1 Authentication 242
   5.1.2.2 Confidentiality 242
   5.1.2.3 Integrity 243
5.2 SSL Subprotocols 243
  5.2.1 SSL Exchanges 244
   5.2.1.1 State Variables of an SSL Session 245
   5.2.1.2 State Variables of an SSL Connection 246
  5.2.2 Synopsis of Parameters Computation 247
  5.2.3 The Handshake Protocol 249
   5.2.3.1 General Operation 249
   5.2.3.2 Opening of a New Session 249

		5.2.3.3	Identification of the Cipher Suites	249
		5.2.3.4	Authentication of the Server	252
		5.2.3.5	Exchange of Secrets	253
		5.2.3.6	Verification and Confirmation by the Server	255
		5.2.3.7	Summary: Session Establishment	255
		5.2.3.8	Connection Establishment	255
	5.2.4	The ChangeCipherSpec Protocol		258
	5.2.5	The Record Protocol		258
	5.2.6	The Alert Protocol		259
	5.2.7	Summary		261
5.3	Example of SSL Processing			261
	5.3.1	Assumptions		262
	5.3.2	Establishment of a New Session		263
		5.3.2.1	Message Size	263
		5.3.2.2	ClientHello Message	263
		5.3.2.3	ServerHello Message	264
		5.3.2.4	Certificate Message	264
		5.3.2.5	ClientKeyExchange Message	265
		5.3.2.6	Calculation of the Cipher Suite	265
		5.3.2.7	ServerHelloDone Message	267
		5.3.2.8	Finished Message	267
		5.3.2.9	Processing at the Record Layer	268
	5.3.3	Processing of Application Data		270
		5.3.3.1	MAC Computation and Encryption	270
		5.3.3.2	Decryption and Verification of the Data	270
	5.3.4	Connection Establishment		271
		5.3.4.1	Connection Establishment in an Existing Session	271
		5.3.4.2	Session Refresh	272
		5.3.4.3	Summary	273
5.4	Performance Acceleration			274
5.5	Implementations			276
5.6	Summary			277
Questions				278

Appendix 5.1   Structures of the Handshake Messages

A5.1	Messages of the Handshake		279
	A5.1.1	Header	279
	A5.1.2	HelloRequest	280
	A5.1.3	ClientHello	280
	A5.1.4	ServerHello	281
	A5.1.5	Certificate	281
	A5.1.6	ServerKeyExchange	281
	A5.1.7	CertificateRequest	282
	A5.1.8	ServerHelloDone	283

	A5.1.9	ClientKeyExchange	283
	A5.1.10	CertificateVerify	284
	A5.1.11	Finished	284

6 TLS (Transport Layer Security) and WTLS (Wireless Transport Layer Security)

Abstract			285
6.1	From SSL to TLS		285
	6.1.1	Start of the Encryption of Transmitted Data	286
	6.1.2	The Available Cipher Suite	286
	6.1.3	Computation of MasterSecret and the Derivation of Keys	286
	6.1.4	Alert Messages	288
	6.1.5	Responses to Record Blocks of Unknown Type	289
6.2	WTLS		290
	6.2.1	Architecture	290
	6.2.2	From TLS to WTLS	292
		6.2.2.1 The Formats of Identifiers and Certificates	293
		6.2.2.2 Cryptographic Algorithms	294
		6.2.2.3 The Content of Some Handshake Messages	295
		6.2.2.4 The Exchange Protocol during the Handshake	296
		6.2.2.5 Calculation of Secrets	297
		6.2.2.6 Parameter Sizes	299
		6.2.2.7 Alert Messages	299
		6.2.2.8 Record	299
	6.2.3	Service Constraints	299
		6.2.3.1 Possible Location of the WAP/Web Gateway	300
		6.2.3.2 ITLS	301
		6.2.3.3 NAETEA	302
6.3	Summary		305
Questions			306

7 The SET Protocol

Abstract			307
7.1	SET Architecture		308
7.2	Security Services of SET		311
	7.2.1	Cryptographic Algorithms	312
	7.2.2	The Method of the Dual Signature	314
7.3	Certification		316
	7.3.1	Certificate Management	316
		7.3.1.1 Cardholder Certificate	318
		7.3.1.2 Merchant Certificates	319
		7.3.1.3 Certificate of Financial Agents	319
		7.3.1.4 Certificates of the Root Authority	319
		7.3.1.5 Certificate Durations	320

    7.3.2   Registration of the Participants 320
            7.3.2.1 Cardholder Registration 320
            7.3.2.2 Merchant Registration 325
7.4 Purchasing Transaction 326
    7.4.1   SET Payment Messages 327
    7.4.2   Transaction Progress 329
            7.4.2.1 Initialization 329
            7.4.2.2 Order Information and Payment Instruction 330
            7.4.2.3 Authorization Request 334
            7.4.2.4 Granting Authorization 334
            7.4.2.5 Capture 336
7.5 Optional Procedures in SET 337
7.6 SET Implementations 338
7.7 Evaluation...339
7.8 Summary 341
Questions 341

8 **Composite Solutions**
Abstract 343
8.1 C-SET and Cyber-COMM 343
    8.1.1   General Architecture of C-SET 344
    8.1.2   Cardholder Registration 346
    8.1.3   Distribution of the Payment Software 348
    8.1.4   Purchase and Payment 348
    8.1.5   Encryption Algorithms 351
    8.1.6   Interoperability of SET and C-SET 352
            8.1.6.1 Case 1: Cardholder Is C-SET Certified and Merchant
                     Is SET Certified 352
            8.1.6.2 Case 2: SET-Certified Cardholder and C-SET-Certified
                     Merchant 352
8.2 Hybrid SSL/SET Architecture 353
    8.2.1   Hybrid Operation SET/SSL 356
    8.2.2   Transaction Flows 358
            8.2.2.1 SSL Session between the Client and
                     the Intermediary 358
            8.2.2.2 Payment Authorization 360
            8.2.2.3 Notification of the Merchant and the Client 360
            8.2.2.4 Financial Settlement 360
    8.2.3   Evaluation of the Hybrid Mode SET/SSL 361
8.3 3-D Secure 362
    8.3.1   Enrollment 364
    8.3.2   Purchase and Payment Protocol 365
    8.3.3   Clearance and Settlement 367
    8.3.4   Security 368
8.4 Payments with CD-ROM 369

8.5 Summary 370
Questions 370

9   Micropayments and Face-to-Face Commerce
Abstract 371
9.1 Characteristics of Micropayment Systems 372
9.2 Potential Applications 373
9.3 Chipper® 374
9.4 GeldKarte 376
      9.4.1   Registration and Loading of Value 377
      9.4.2   Payment 377
      9.4.3   Security 380
9.5 Mondex 381
      9.5.1   Loading of Value 382
      9.5.2   Payment 382
      9.5.3   Security 383
      9.5.4   Pilot Experiments 384
9.6 Proton 384
      9.6.1   Loading of Value 385
      9.6.2   Payment 385
      9.6.3   International Applications 386
9.7 Harmonization of Electronic Purses... 386
      9.7.1   Authentication of the Purse by the Issuer 387
      9.7.2   Loading of Value 388
      9.7.3   Point-of-Sales Payments 388
9.8 Summary 389
Questions 389

10 Remote Micropayments
Abstract 391
10.1 Security without Encryption: First Virtual 392
      10.1.1 Buyer's Subscription 392
      10.1.2 Purchasing Protocol 392
      10.1.3 Acquisition and Financial Settlement 394
      10.1.4 Security 394
      10.1.5 Evaluation 395
10.2 NetBill 395
      10.2.1 Registration and Loading of Value 395
      10.2.2 Purchase 396
            10.2.2.1 Negotiation 398
            10.2.2.2 Order 398
            10.2.2.3 Delivery 398
            10.2.2.4 Payment 399
      10.2.3 Financial Settlement 401
      10.2.4 Evaluation 401

10.3   KLELine 402
         10.3.1   Registration 403
         10.3.2   Purchase and Payment 403
         10.3.3   Financial Settlement 406
         10.3.4   Evaluation 406
         10.3.5   Evaluation and Evolution 407
10.4   Millicent 408
         10.4.1   Secrets 409
         10.4.2   Description of the Scrip 409
         10.4.3   Registration and Loading of Value 411
         10.4.4   Purchase 412
         10.4.5   Evaluation 414
10.5   PayWord 415
         10.5.1   Registration and the Loading of Value 416
         10.5.2   Purchase 417
                   10.5.2.1   Commitment 417
                   10.5.2.2   Delivery 418
         10.5.3   Financial Settlement 419
         10.5.4   Computational Load 419
                   10.5.4.1   Load on the Broker 419
                   10.5.4.2   Load on the User 420
                   10.5.4.3   Load on the Vendor 420
         10.5.5   Evaluation 421
10.6   MicroMint 421
         10.6.1   Registration and Loading of Value 422
         10.6.2   Purchase 422
         10.6.3   Financial Settlement 422
         10.6.4   Security 422
                   10.6.4.1   Protection against Forgery 423
                   10.6.4.2   Protection against Coin Theft 423
                   10.6.4.3   Protection against Double Spending 424
         10.6.5   Evaluation 424
10.7   eCoin 424
10.8   Comparison of the Different First-Generation Remote
         Micropayment Systems 425
10.9   Second-Generation Systems 427
         10.9.1   Prepaid Cards Systems 427
         10.9.2   Systems Based on Electronic Mail 427
                   10.9.2.1   PayPal 428
         10.9.3   Minitel-like Systems 430
Questions 431

11   Digital Money

Abstract 433
11.1   Building Blocks 434

11.1.1 Case of Debtor Untraceability 434
    11.1.1.1 Loading of Value 435
    11.1.1.2 Purchase 436
    11.1.1.3 Deposit and Settlement 436
    11.1.1.4 Improvement of Protection 436
11.1.2 Case of Creditor Untraceability 438
11.1.3 Mutual Untraceablity 438
11.1.4 Description of Digital Denominations 439
11.1.5 Detection of Counterfeit (Multiple Spending) 442
    11.1.5.1 Loading of Value 443
    11.1.5.2 Purchasing 444
    11.1.5.3 Financial Settlement and Verification 444
    11.1.5.4 Proof of Double Spending 444
11.2 DigiCash (Ecash) 445
11.2.1 Registration 446
11.2.2 Loading of Value 446
11.2.3 Purchase 447
11.2.4 Financial Settlement 448
11.2.5 Delivery 448
11.2.6 Evaluation 449
11.3 NetCash 449
11.3.1 Registration and Value Purchase 450
11.3.2 Purchase 450
11.3.3 Extensions of NetCash 451
11.3.4 Evaluation 454
11.4 Summary 455
Questions 456

12 Dematerialized Checks
Abstract 457
12.1 Classical Processing of Paper Checks 458
12.1.1 Checkbook Delivery 458
12.1.2 Check Processing 458
12.2 Dematerialized Processing of Paper-Based Checks 459
12.2.1 Electronic Check Presentment 460
12.2.2 Point-of-Sale Check Approval 461
12.2.3 Check Imaging 461
12.3 NetCheque 462
12.3.1 Registration 463
12.3.2 Payment and Financial Settlement 464
12.4 Bank Internet Payment System (BIPS) 466
12.4.1 Types of Transactions 466
12.4.2 BIPS Service Architecture 467
12.5 eCheck 470
12.5.1 Payment and Settlement 470

	12.5.2	Representation of eChecks 473
12.6	Comparison of Virtual Checks with Bankcards 474

12.7	Summary 476
Questions	 477

13 Security of Integrated Circuit Cards

Abstract	 479
13.1	Overview	 479
	13.1.1	Classification of Smart Cards and Their Applications 480
	13.1.2	Integrated-Circuit Cards with Contacts 482
	13.1.3	Contactless Integrated-Circuit Cards 482
13.2	Description of Integrated-Circuit Cards 484
	13.2.1	Memory Types 484
	13.2.2	Operating Systems 485
13.3	Standards for Integrated-Circuit Cards 486
	13.3.1	ISO Standards 486
	13.3.2	EMV (EuroPay, MasterCard, Visa) 487
		13.3.2.1 Properties of Encryption Keys 488
		13.3.2.2 Migration to EMV 488
13.4	Security of Microprocessor Cards 489
	13.4.1	Security during Production 489
	13.4.2	Physical Security of the Card during Usage 492
	13.4.3	Logical Security of the Card during Usage 493
		13.4.3.1 Authentication with Symmetric Encryption 493
		13.4.3.2 Authentication with Public-Key Encryption 494
	13.4.4	Examples of Authentication 496
		13.4.4.1 Memory Card Reader for the Minitel 496
		13.4.4.2 Smart Card of French Banks 497
		13.4.4.3 EMV Card 498
	13.4.5	Evaluation 503
13.5	Multiapplication Smart Cards 504
	13.5.1	File System of ISO/IEC 7816-4 504
	13.5.2	The Swedish Electronic Identity Card 506
	13.5.3	Management of Applications in Multiapplication Cards 506
		13.5.3.1 Secondary Applications Controlled by the Primary Application 507
		13.5.3.2 Federation of Several Applications under a Central Authority 507
		13.5.3.3 Independent Multiapplications on the Same Card 508

13.6   Integration of Smart Cards with Computer Systems 509
      13.6.1   OpenCard Framework 510
      13.6.2   PC/SC 511
13.7   Limits on Security 512
      13.7.1   Logical (Noninvasive) Attacks 512
      13.7.2   Physical (Destructive) Attacks 513
      13.7.3   Attacks due to Negligence in the Implementation 513
      13.7.4   Attacks against the Chip-Reader Communication
                Channel 514
13.8   Summary 515
Questions 517

14   Systems of Electronic Commerce
Abstract 519
14.1   SEMPER 519
      14.1.1   SEMPER Architecture 520
      14.1.2   Payment Terminology in SEMPER 522
      14.1.3   The Payment Manager 523
14.2   CAFE 523
14.3   JEPI 526
14.4   PICS and P3P 526
14.5   Analysis of User Behavior 527
14.6   Fidelity Cards 528
14.7   Quality of Service Considerations 529
14.8   Summary 530
Questions 531

15   Electronic Commerce in Society
Abstract 533
15.1   Communication Infrastructure 534
15.2   Harmonization and Standardization 536
15.3   Issuance of Electronic Money 537
15.4   Protection of Intellectual Property 538
15.5   Electronic Surveillance and Privacy 539
15.6   Filtering and Censorship 543
15.7   Taxation of Electronic Commerce 544
15.8   Fraud Prevention 545
15.9   Archives Dematerialization 545
Questions 547

Web Sites
General 549
Standards 549
Encryption 550
Kerberos 550

Certification.. 551
Biometrics 551
    General 551
    Standards Organizations 552
    Products 552
        Face Recognition 552
        Fingerprints 552
        Iris Scan 553
        Hand Geometry 553
        Keyboard Recognition 553
        Retinal Scan 553
        Speech Recognition 553
EDIFACT 553
XML 554
Integration XML/EDIFACT 554
SSL/TLS/WTLS 555
SET 555
Purses 555
Micropayments 556
Smart (Microprocessor) Cards 556
Electronic and Virtual Checks 557
SEMPER 558
Labeling Organizations 558
Organizations 559

Acronyms.. **561**

References **575**

Index **597**

1

Overview of Electronic Commerce

ABSTRACT

Electronic commerce (or e-commerce) is a multidisciplinary activity that influences the behavior of the participants and the relations that they establish among themselves. In practice, it can take several forms, and this may cause some confusion. To clarify these multiple meanings before going to the heart of the subject, this chapter presents a general introduction to the principal aspects of e-commerce: its framework, types, and changes that it may cause in the banking and financial domains.

1.1 What Is Electronic Commerce?

In this book, we will adopt the definition of the French Association for Commerce and Electronic Interchange,[1] a nonprofit industry association created in 1996 to promote e-commerce: electronic commerce is "the set of relations totally dematerialized that economic agents have with respect to each other." Thus, e-commerce can be equally about physical or virtual goods (software, information, music, books, etc.) or about users' profiles, because some operators build their business models around the systematic exploitation of demographic and behavioral data collected during online transactions. The transactions can occur on Minitel, the Internet, or through Electronic Data Interchange (EDI), and the means of payment can be classic or emerging, such as electronic or virtual purses (whether they store legal or token values), electronic or virtual checks, and digital monies. It seems to us that this definition has the advantage of covering the gamut of dematerialized transactions and avoids the drawbacks of an excessive concentration on transactions over the Internet, as many authors have unfortunately done (Cho, 1999; Industry Canada, 1998; McCarthy, 1999; MENTIS, 1998; Lacoste et al., 2000).

---

[1] Association Française pour le Commerce et les Échanges Électroniques (AFCEE).

In fact, many aspects of the e-commerce infrastructure have been in place for two or three decades, thanks to progress in microelectronics, information processing, and telecommunications. These advances modified the role of computers in the enterprise tremendously. From a tool of computation and production control, the computer became essential to the tasks of analysis, data management, and text and transaction processing. In the 1980s, financial applications became commonplace for transactions processing and electronic fund transfers (through cash-withdrawal cards, bank cards, etc.). Money became guaranteed data in the form of bits moving around the world in the digital networks tying financial institutions. This decade saw the emergence of Minitel, the French Télétel system, which was used for business-to-business as well as business-to-consumer exchanges and payments. The usage of e-commerce spread such that in 1998, for example, 39% of French enterprises with more than 10 workers were exchanging information by electronic means, either by Minitel or by EDI (*Télécommunications*, 1998).

It should also be noted that a significant portion of the Internet economy is still nonmonetary, founded on mutual trust and the concept of community good. Free software comes in three forms (Chavanne and Paris, 1998):

1. Software with source code that is freely available — The authors, mostly students, do not produce to earn a living, but as part of professional activities, whether paid or voluntary. The work output is shared to solicit comments, contributions, modifications, or improvements to be included in future revisions.

2. Shareware — This includes programs distributed freely on the condition that users, after a trial-and-evaluation period, pay a symbolic fee to the developers.

3. Freeware — This includes programs that are free for use, but their code source is not available.

The Free Software Foundation, founded by Richard Stallman, introduced a new type of software licensing, called "general public license," to protect free software from commercial takeovers and from technical or legal prevention of their diffusion, utilization, or modification (Lang, 1998). The widespread availability of free software, even in industrial applications, has forced major commercial companies to modify their distribution policies, for example, to make some versions of their software available free of charge. This economy can be considered as the cybernetic form of nonmonetary exchange systems, such as LETS (Local Exchange Trading System), SEL (Systèmes d'Échange Locaux — Local Exchange Systems), and RRES (Réseaux Réciproques d'Échange de Savoirs — Mutual Networks for the Exchange of Knowledge) (Plassard, 1998).

Clearly then, e-commerce covers a wider area than the Internet and the applications usually associated with it. Its exchanges are not concerned with the selling of merchandise; the values exchanged can be nonmonetary; and the parties involved can belong to the same organization, to enterprises, to governments, or to the general public.

1.2   Categories of Electronic Commerce

The movement toward e-commerce has foundations that are at the same time commercial, socioeconomic, and industrial. This interest appears in a context where ambitious growth rates are not physically sustainable due to market saturation in rich countries, the progressive depletion of natural resources, and the risks of pollution (Haesler, 1995). It is linked to the prospect of a "virtual" economy, free of physical constraints whatever their origin (temporal, geographical, functional, or organizational) (Lefebvre and Lefebvre, 1998). On the one side, the reorganization of work in industrialized countries, the flattening of pyramidal structures, and the decentralization of decision centers augmented the need for exchange and communication. On the other side, the evolution toward a service society produced a virtual and speculative economy, where electronic monies flow without state control, which gives the illusion of an immediate abundance that is without problems. As a consequence, the theme of e-commerce gives a meaning, a blueprint, and a collective goal for an economy that, seemingly, has been liberated from the constraints of the reality.

From an operational viewpoint, the evolution toward e-commerce can be explained by several objectives. The first ambition is to increase productivity and reduce costs by improving the reliability and speed of communications with business partners. Less dependence on paper reduces the amount of data reentry and, hence, errors, while efficient communication reduces exposure to inventory risks. The second drive is the need to increase revenues of existing products and services by enhancing the supply network or by establishing additional distribution channels. In addition, electronic data collection of market data facilitates analyses of customers and channels information for better prediction of market conditions and scheduling of production. Finally, e-commerce opens doors to new services, such as online distribution of virtual goods.

Depending on the natures of the economic agents and the types of relations among them, the applications of e-commerce fall within one of four main categories of business relations:

1. *Business-to-business relations*, where the customer is another enterprise or another department within the same enterprise. A characteristic of these types of relations is their long-term stability. This stability justifies the use of costly data-processing systems, the installations of which are major projects. This is particularly true in information technology systems linking the major financial institutions.

2. *Business-to-consumer relations* allow an individual to act at a distance through a telecommunications network.

3. *Neighborhood or contact commerce* includes face-to-face interactions between the buyer and the seller, as in supermarkets, drugstores, coffee shops, etc.

4. *Peer-to-peer (P2P) commerce* takes place without intermediaries. This category may also include the transfer of money from one individual to another.

1.2.1 Examples of Business-to-Business Commerce

We give some examples among the electronic networks used for business-to-business e-commerce that were established before the Internet era:

1. The SITA (Société Internationale de Télécommunications Aéronautiques — International Society for Aeronautical Telecommunications) today links 350 airline companies and around 100 companies that are tied to them. This network allows the exchange of data regarding airline reservations, tariffs, departures and arrivals, etc.

2. SABRE, the airline reservation system SABRE, formerly of American Airlines, and Amadeus, created in 1987 by Air France, Iberia, and Lufthansa to link travel agents, airline companies, hotel chains, and car rental companies.

3. The SWIFT (Society for Worldwide Interbank Financial Telecommunications) network, established in 1977 to exchange standardized messages that control the international transfer of funds among banks.

4. The BSP (Bank Settlement Payment) network is dedicated to the settlement of travel tickets among airline companies.

5. The SAGITTAIRE (Système Automatique de Gestion Intégrée par Télétransmission de Transactions avec Imputation de Règlements Étrangers — Automatic System for Integrated Management with Teletransmission of Foreign Settlement Transactions with Charging) network used for the settlement of international transactions in France.

6. The bank settlement systems used to transport interbank instructions such as: NACHA (National Automated Clearing House Association) and ACH (Automated Clearing House) in the United States; BACS (Banker's Automated Clearing Service) in the United Kingdom; the SIT (Système Interbancaire de Télécompensation — Interbank Settlement System) in France; the Swiss Interbank Clearing (SIC) in Switzerland, etc.

Most of these networks are still governed by proprietary protocols, and this translates into greater dependency on suppliers. The first attempts to overcome these obstacles by standardizing the transport mechanisms as well as the messages associated with them led to the X12 standard in North America and EDIFACT (Electronic Data Interchange for Administration, Commerce and Transport) in Europe. The European Commission issued a model EDI contract to guide the European organizations and businesses using electronic exchanges in the course of their commercial activities (European Commission, 1994). To simplify and expedite the procedures, various European customs authorities currently recognize declarations submitted by electronic means (Granet, 1997). Finally, the United Nations Commission on International Trade Law (UNCITRAL) proposed a model law for the commercial use of international contracts in e-commerce that national legislation could use as a reference (UNCITRAL, 1996).

In the U.S., the CALS (Continuous Acquisition and Life-cycle Support) was started in the early 1980s to improve the flow of information between the Department of Defense (DOD) and its suppliers. In 1993, President Clinton extended the use of commercial and technical data in electronic form to all branches of the federal government (Presidential Executive Memorandum, 1993). The Federal Acquisition Streamlining Act of October 1994 required the use of EDI in all federal acquisitions. A taxonomy was later developed to describe various entities and assign them a unique identifier within the Universal Data Element Framework (UDEF). With the installation of the Federal Acquisition Computer Network (FACNET) in July 1997, federal transactions can be completed through electronic means from the initial request for proposal to the final payment to the supplier.

Today, the adoption of the Internet as the worldwide network for data exchange is encouraging the migration toward open protocols and the production of a series of standards, some of which will be presented in Chapter 4.

1.2.2   Examples of Business-to-Consumer Commerce

Interest in business-to-consumer e-commerce started to grow in the 1980s, although to different degrees in different countries. In Germany, and before the Internet took off, most banks offered their clients the possibility of managing their accounts remotely through the BTX (*Bildschirmtext*) system. In BTX, security was achieved using a personal identification code and a six-digit transaction number (Turner, 1998).

Minitel is undoubtedly one of the largest successes of business-to-consumer e-commerce systems. In this system, access is through a special terminal connected through the Public Switched Telephone Network (PSTN) to an X.25 data network called Transpac. Until 1994, the rate of penetration of the Minitel in French homes exceeded that of personal computers in the U.S. (Hill, 1996), thereby assuring more uniform access to services by all socioeconomic classes and ethnic groups. In 1994, approximately 10,000 providers offered about 25,000 Télétel services, which were hosted on around 4,000 servers. The business turnover was approximately $1.7 billion (9.2 billion French francs), mostly (about 70%) related to "value-added services," i.e., services that are outside simple directory services, such as information queries or remote tax filing (France Télécom, 1995). Compare these numbers to the global turnover of e-commerce using the Internet which grew from $0.7 billion in 1996 to $2.6 billion in 1997, reaching about $5.6 billion in 1998 (Jupiter Communications, 1998). As late as 2000, 16 million users relied on the Minitel regularly to manage bank accounts, query government administrations and local authorities, file taxes, and conduct mail-order purchases or other transactions regarding travel, tourism, and entertainment (*Le Canard*, 2001; France Télécom, 1997).

The importance of the kiosk model of the Minitel is that it shows how a nonbank, in this case the telephone operator France Télécom, can be a payment intermediary for information services sold to the public. According to this model, the service provider delegates the billing and the collection to the telephone operator. If the payment is made by a bank card, the user sends the payment information (on a credit or debit card) in the clear to the intermediary, who collects the amount through its bank, retains a percentage of the amount, and then forwards the rest to the service provider. The user's telephone bill reflects the connect time to the various servers in terms of telephone units. After collection of the bill, the operator compensates the content providers according to an established payment grid.

The intermediation functions consist of the following:

1. Authentication of the service providers to the users and guarantee of their good faith according to a code of conduct defined for telematic services
2. Identification of users through their telephone numbers
3. Certification of the telephone subscribers, because the telephone companies know the addresses of their residences (see Chapter 3 regarding certification)
4. Measurement of the duration of the communication using the telephone impulse
5. Summation of all transactions
6. Billing and recovery of the amounts for a set percentage

Notice that the telephone unit plays the role of an instrument of payment for the purchase of information services.

The financial intermediation of the telephone operator implies the collection of payments on behalf of Minitel service providers for a given percentage of the revenues. This infringement on the prerogatives of financial institutions can be justified because it is very difficult for banks to propose, alone, an economical alternative for billing and collecting sums that are individually marginal. At the same time, financial institutions benefit from having a unique interlocutor that accumulates for them the amounts for each individual transaction. In addition, the Minitel also proved that the sense of security is not merely a question of sophisticated technical means because business-to-consumer transactions are sent without encryption, but of a "trust" between the user and the operator.

This model kept its attractiveness, which explains its update of use in the Internet and in mobile networks. Thus, in some systems, such as ClickShare, WISP, or iPIN, which will be studied in Chapter 10, the access service provider records the client's transactions, bills the subscribers, and collects the amounts. Then it reimburses the merchants after withholding its commission and pays a usage fee to the supplier of the management or the payment software.

In particular, it is the Japanese mobile telephony operator NTT DoCoMo that reutilized the Minitel model in the design of its i-mode® service so as to mask the complexity of the Internet for the user (Enoki, 1999; Matsunaga, 1999). This service allows the mobile subscriber to consult information providers to exchange messages, or to participate in networked games. Just like for the Minitel, the operator bills the subscriber according to the schedule set by the content supplier plus a surcharge for the use of the mobile network. Having collected collecting the bills, the operator compensates the service providers after deducting a commission.

1.2.3 Examples of Neighborhood Commerce and Payments to Automatic Machines

Prepaid cards form another aspect of business-to-consumer e-commerce, particularly in neighborhood commerce.

In Japan, about 90% of the population utilizes this means of payment. In addition to telephony applications, prepaid cards are commonly used to play *pachinko*, a form of gambling that involves a pinball machine that propels balls, with the objective of producing a winning combination of numbers.

In France, telephone cards are widely used, and many local municipalities issue prepaid cards to access municipal services. Telephone cards represent about 76% of the market for smart cards in Europe and 93% of the cards issued in 1997 (Adams, 1998). In Australia, in 1998, the telephone operator Telstra started a project to equip all telephone booths with smart-card readers to replace magnetic-strip cards. South Africa pioneered the use of prepayment meter systems in electrification projects. This reduced operational costs to less than 5% of the turnover by ensuring collection and eliminating the need for meter reading in rural and remote areas (Anderson and Bezuidenhoudt, 1996). Although the reception of prepaid cards has been less enthusiastic in the U.S., they are nevertheless used in closed communities, such as on university campuses and military bases, and are slowly gaining acceptance in public transportation and telephony.

All of these experiences demonstrate that, in some cases, it is possible to replace cash with prepaid cards. Thus, banks as well as financial and political authorities are pondering the future role of prepaid cards in the ensemble of monetary operations and the implications of their use in the financial system.

1.2.4   Examples of Peer-to-Peer Commerce

This category of transactions was practically unknown a few years ago. The growth of the mechanisms for peer-to-peer (P2P) exchanges is a tribute to Napster, a software used to exchange pop music files over the Internet without passing by the publishers. The Mondex electronic purse has a function that allows for the transfer of value between two purses without the intermediation of a financial institution. In both cases, by resistance or through judicial threats, the large enterprises succeeded in derailing or stopping the technical evolution.

1.3   The Influence of the Internet

The arguments in favor of the Internet sound, at a decade interval, like echoes of those previously formulated in praise of the Minitel (de Lacy, 1987). From a technical viewpoint, the major advantage of the Internet over the Minitel is that its protocols are standardized, which means that the programs and applications of the Internet are independent of the physical platforms. Traffic from several applications or users can take on distinct infrastructures without worrying about interoperability. This technical advantage translates into economies of scale in installing and administering networks, provided that the challenges of security are met.

1.3.1   Some Leading Examples

The auction site eBay® illustrates a successful innovation of the Internet era, having contributed to the creation of a virtual marketplace. The eBay site supplies a space for exhibiting merchandise and for negotiating selling conditions, and, in particular, it provides a platform that links participants in return for a commission on the selling price. The setup is characterized by the following properties:

- Participants can join from anyplace they may be, and the site is open to all categories of merchandise or services. The market is thus fragmented geographically or according to the commercial offer.
- Buyers have to subscribe and establish accounts at eBay to obtain logins and define their passwords.
- The operator depends on the evaluations of each participant by its correspondents to assign the participant a grade. The operator preserves the right to eliminate those who do not meet their obligations.
- The operator does not intervene in the payment and does not keep records of the account information of the buyers.

These conditions allowed eBay to be profitable, which is exceptional in consumer-oriented sites. Amazon.com®, despite its notoriety, remained more than 6 years in the red, and its first profitable quarter was the last trimester of 2001, for a total yearly loss of $567 million.

Targeting individual consumers and home workers, the systems for electronic Stamps.com™ or Neopost allow the printing of postal fees with a simple printer instead of postage meters, thereby avoiding going to the post office. A two-dimensional bar code contains, in addition to the stamp, the destination address and a unique number that allows tracking of the letter. Stamps.com operates online and requires the intervention of an authorization center each time a stamp needs to be printed. In contrast, Neopost is a semionline system, where stamping of envelopes continues without central intervention, as long as the total value of the stamps does not exceed the amount authorized by the authorization server.

The operational difficulties arise from the precise specifications of postal authorities for the positioning of the impressions, which are, in turn, a consequence of the requirements of automatic mail sorters. There is also a need to adapt to users' software and to all printer models. Users must pay a surcharge of about 10% to the operator. The total cost of the operator includes that of running a call center to assist users in debugging their problems.

1.3.2 Internet and Transactional Security

Although the Internet was able to achieve in the 1990s an international or even a global dimension that the Minitel was never able to attain, it was not originally intended for conducting commercial transactions. As an experimental network subsidized by public funds in the U.S. as well as by the large telecommunications companies, the Internet was used to encourage the free distribution of information and the sharing of research efforts. An informal honor code shunned commercial uses, and utilization of the Internet for profit was prohibited. The Internet allowed collaboration without geographic proximity or financial compensation. The birth of a new community spirit was translated into a nonmonetary social interaction and an economy of donations and exchanges, in the form of free advice or software freely shared. Even today, despite the domination of financial interests, the growth of Internet technologies depends to a large extent on volunteers who put their efforts at the disposal of everyone. Which Internet user has not benefited from the information freely given on newsgroups or distribution lists? In this manner, developers and users form electronic communities with common objectives, and the sharing of knowledge worldwide allows for the rapid evolution of products through fast fault detection and correction. Free information and free software have other consequences as well because they increase the available services on the Internet and attract more participants to the network.

The U.S. decision to privatize the backbone of the Internet starting in 1991 encouraged the authorities directing the Internet to review their line of conduct and to consider for the first time the market economy. This started a campaign to establish the Internet as a way to realize the project of the

Clinton–Gore Administration for an information highway, without incurring the prohibitive costs of installing the infrastructure of broadband networks (Sherif, 1997). Furthermore, the invention of the World Wide Web, with its visual and user-friendly interface, stimulated the development of virtual storefronts. Similarly, the introduction of XML (Extensible Markup Language) and its specialized derivatives, improved the ease with which business data are exchanged.

Nevertheless, the transformation of the "county fair" into a "supermarket" is taking longer than originally anticipated. For one, the utilization of the Internet for economic exchanges clashes with the culture of availability of information free of charge, a culture that the music industry, for example, is currently confronting. Other impediments include the absence of a central authority and the legitimate concerns regarding the security of information on the network. Security on the public Internet is an afterthought. As a consequence, in 1999, half of card payment disputes and frauds in the European Union were related to Internet transactions, even though they represented only 1% of the turnover.[1] In the U.S., the fraud rate on e-commerce transactions was around 2% (1.8% in 2002, down from 1.92% in 2000), which is about 20 times the fraud rate for offline transactions (Richmond, 2003; Waters, 2003). It should be noted that users may have legitimate concerns regarding the collection and the reuse of their personal data from the Web. The consolidation of information tying buyers and products, which allows the constitution of individualized portfolios corresponding to consumer profiles, could be a threat to individual privacy. Another plague poisoning the life of many users is unsolicited electronic advertisement or spam.

From an operational viewpoint, the lack of integration and the nonharmonization of various software programs or payment mechanisms remain a handicap for a merchant aiming for worldwide operation, as is the problem of currency for the individual consumer. In this regard, the non-localization of the participants in a commercial transaction introduces completely new aspects, such as the conflict of jurisdictions on the validity of contracts, the standing of electronic signatures, consumer protection, the taxation of "virtual" products, etc. Finally, new approaches are needed to address virtual products, such as information, images, or software — products that pose major challenges to the concepts of intellectual property and copyrights. In the late 1990s, the contradictory predictions of market research firms or specialized magazines were telltale signs of market immaturity that many neglected to their chagrin.[2]

---

[1] *Financial Times*, April 12, 1999.
[2] On page 14 of the April issue of *Banking Technology*, the article "UK business slow on e-commerce" stated that 90% of small-business banking was still conducted by visits or telephone calls to local branches, because "many customers like to have regular face to face contact with people who know them and understand their business." Yet, on page 48 of the same issue, T. George reported in the article "On a virtual roll" that "suppliers involved in the internet banking business are in a buoyant mood" (*Banking Technology*, 1998; George, 1998).

1.3.3 Putting the Internet in Perspective

We see that many forms of e-commerce predate the Internet. Furthermore, the growth of e-commerce needs a legal framework in order for the "Information Society" to protect the rights of its citizenry, such as safeguards for the protection of participants' private information, prevention and repression of fraud or abusive use, warranties on merchandise, etc. In this regard, use of the Internet is a social activity, thus influenced by the cultural environment.

Figure 1.1 depicts the rate of penetration of the Internet within the population and within households in Western European countries. These results are consistent with many other surveys and confirm that the Internet is more popular in the Nordic countries, Germany, and England. For example, in November 1998, the countries with the highest PC usage were Finland and Denmark, respectively, 9% and 5.5%, compared with 3.5% in the U.S. and 0.9% in France (Catinat, 1999). A comparison of the volume of e-mail exchanges among companies and their customers in France and in the U.S. shows that the telephone remains the preferred means of communication in France, with the exception of companies with activities that revolve around the Internet (Internet Professionel, 2002). In a poll conducted at the beginning of 2002, out of the 69% of French people that did not have a connection to the Internet, 73% had no desire to be connected (Froissard, 2002). We remind the reader that the data compiled in the first edition of this book based on the information available on the site *http://www.nic.fr* supported the same conclusions.

These numbers can be explained by taking into account the classification of societies into "low-context" and "high-context" societies (Hall and Hall, 1990). In high-context societies, interpersonal relations and oral networks have a much more important place than in low-context societies, where communication takes explicit and direct means, such as that written. This explains why the Internet has been well received in low-context societies such as the U.S., of course, but also, the U.K., Germany, and the Nordic countries. In contrast, high-context societies, in particular those of Southern Europe (France, Italy, and Spain), are less receptive, particularly because the Internet has to compete with other social networks. Even if the dominance of cyber-English has a role in the observed difference, the success of the Internet in Finland, for example, cannot be understood without considering social information networks. This is consistent with the fact that the written press is a more important source of information in Northern Europe (Finland, Germany, the Netherlands, Sweden, the U.K.) than in Southern Europe (Belgium, France, Greece, Italy, and Spain), where the high-context culture in Northern Europe favors radio and TV (*Futuribles*, 1999). Also, in 2002, a ranking of European countries depending on the ease of locating financial information on companies from Web sites found Sweden to be the first, followed by Finland and Norway (FT-IT Review, 2002).

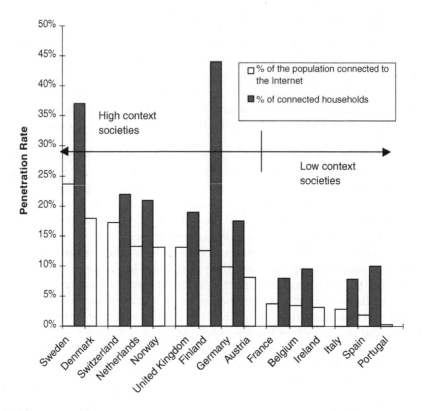

FIGURE 1.1
Penetration of the Internet in Western Europe. (From *Internet Professionel,* June 2000, 43, 16; ITU *Year Book of Statistics,* 2001.)

We will see in Chapter 13 that smart cards remain a European specialty, even though the market looks promising in Asia, and that its diffusion in the U.S. remains relatively weak. This is why e-commerce applications using smart cards are rarely discussed in the U.S.

Finally, looking at the geographic distribution of mobile commerce (m-commerce), i.e., transactions from mobile terminals (telephones or pocket organizers), highlights another aspect of diversity. Forecasts of turnover between the years 2000 and 2004 are illustrated in Figure 1.2. They show that the U.S. is expected to lag behind in this area, with respect to Europe and Asia. One main cause of this phenomenon is the fragmentation of the U.S. markets among several transmission standards (Nakamoto, 2002; Norton, 2001).

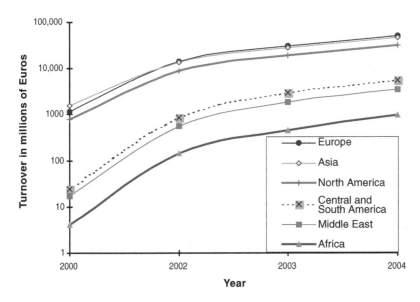

FIGURE 1.2
Geographic distribution of the turnover forecast for mobile commerce (logarithmic scale). (Adapted from Ovum, cited in Boucher, X., *Internet Professionel*, 52, 48–51, April 2001.)

1.4 Infrastructure for Electronic Commerce

To get an overall look at the necessary infrastructure for e-commerce, it is sufficient to consider a simple transaction between a merchant and a buyer. This transaction includes four types of exchanges:

1. Documentation
2. Agreement on the terms and conditions of the sale and payment
3. Payment instructions
4. Shipment and delivery of the items acquired

The documentation relates to the descriptions of the goods and services offered for sale, the terms and conditions of their acquisitions, the guarantees that each party offers, etc. These details can be presented online or offline or in catalogs recorded on paper or on electronic media.

The agreement between the client and the merchant is generally translated into an order defining the required object, the price, the required date of delivery and acceptable delays, and the means and conditions for payment. The exchanges of this phase comprise the transmission of a command from the client to the supplier, the response of the supplier, the issuance of an invoice, and the recording of the order and the invoice.

The payment method in a commercial transaction depends on several factors, such as the amount in question; the distance or proximity of the merchant and the client; and the cultural and historical specificity of the country. However, regardless of the method used, payment instructions have a different path than that for the exchange of financial value. For example, the check can be handed in person or sent by mail, but the exchange of monetary value flows through specific interbanking networks.

Finally, the means of delivery depends on the nature of the purchase object and the terms of the sale; it can precede, follow, or accompany the payment. The delivery of electronic or digital objects such as files, images, or software can be achieved through telecommunications. In contrast, the processing, the delivery, and the guarantees on physical goods or services require detailed knowledge of insurance procedures and, in international trade, of customs regulations.

Figure 1.3 illustrates the various exchanges that come into play in the acquisition of a physical good and its delivery to the purchaser.

Partial or complete dematerialization of commercial transactions introduces new requirements. These requirements relate to the authentication of both parties in the transaction, to the personalization of the presentation to display only the data that correspond to the user's profile, to guarantees for the integrity of the exchanges, to the collection of proofs in case of disagreements, and to the security of remote payments. These functions are generally carried out by distinct software (from numerous suppliers) with heterogeneous

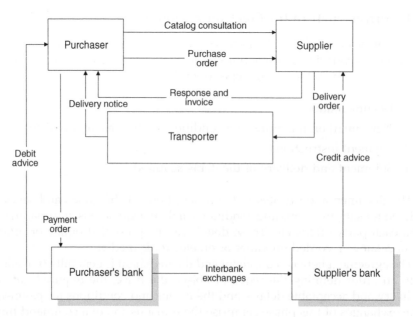

FIGURE 1.3
Typical exchanges and actors in an acquisition transaction.

interfaces. Interconnecting elements or *middleware* mask this heterogeneity through protocol conversion, mapping, and translation of messages among applications, management of database systems, etc.

Finally, for the large-scale use of smart cards as electronic means of payment, an infrastructure is needed with the following components:

- Low-cost card readers that resist physical intrusions and include security modules
- For rechargeable cards, a network of recharging points that can verify the identity of cardholders and, by remote queries to the banking system, their creditworthiness
- A secure telecommunication network to protect the financial exchanges

Let us have a closer look at the informatics infrastructure of e-commerce.

Portals form a single point of entry to a Web site from a workstation equipped with a browser. They provide an easy way for communication by aggregating data from multiple sources, whether unstructured data or databases. An efficient taxonomy is necessary to organize the catalogs that will be searched by search engines.

Low-end catalog HTML (HyperText Markup Language) pages have commands scripted in Perl or in Visual Basic. More sophisticated catalogs can communicate with existing databases through appropriate interfaces. High-end catalogs can adapt the classification scheme depending on usage statistics.

Payment servers are often hosted by a financial institution; their role is to convert purchase orders into financial instructions to banks. The handling of micropayments can be left to a telecommunications operator or to an Internet service provider, according to a contract between the vendor and the operator that takes into account the amount of data, the nature of the articles, the duration, etc.

Back-office processing relates to accounting, inventory management, client relations, supplier management, logistical support, analysis of customer's profiles, marketing, as well as relations with government entities, such as with the online submission of tax reports.

1.5 Network Access

Network access can be through fixed lines or through radio links for mobile users. The quality of access to the telecommunications network is characterized by the capacity of the link (i.e., the bandwidth) in bits per second (bits/sec), its reliability in terms of downtime or time to repair, as well as the blocking probability of a call for lack of resources in the network.

1.5.1   Wireline Access

The physical transmission medium can be copper cables, optical fibers, or radio or satellite links. The bit rates depend on the access technology. With DSL (Digital Subscriber Line) techniques, twisted-pair copper lines can achieve high bit rates in one or two directions. ADSL (Asymmetrical Digital Subscriber Line) establishes a downstream channel with a bit rate of 1.5 to 8 Mbit/sec, respectively, at 3.4 and 1.7 miles from the central office, and an upstream channel from 64 to 640 kbit/sec. Variations of ADSL include RADSL (Rate Adaptive Digital Subscriber Line) and VDSL (Very High Bit Rate Digital Subscriber Line) (Goralski, 1998).

In the case of large enterprises, access can be through ISDN (Integrated Services Digital Network) lines, which are channels with bandwidths in multiples of 64 kbit/sec, usually 128 kbit/sec.

1.5.2   Wireless Access

Several wireless access protocols allow for the exchange of data. On GSM (*Groupe Spécial Mobile* — Global System for Mobile Communication), the bit rate that can be obtained with SMS (Short Message Service) does not exceed 9.6 kbit/sec. To reach 28 or 56 kbit/sec (with a maximum bit rate of 114 kbit/sec), the use of GPRS (General Packet Radio Service) is essential.

Access to e-commerce applications can be seen as an extension of the Internet to mobile terminals or a consultation of Web sites through a mobile phone. The first approach was the starting point for WAP (Wireless Application Protocol), while the Japanese operator NTT DoCoMo selected the second path for its i-mode service. This arrangement retains the simple interface with which the general public is already familiar. The telecommunications operator guarantees all participant subscribers, merchants, and intermediaries that it identifies and authenticates. In addition, the operator plays the role of a payment intermediary by billing for the consumed services and collecting the payment on behalf of the provider, for a commission. This business model, which is in many ways reminiscent of that of Minitel, proved to be judicious: in about 18 months, there were 12 million Japanese subscribers to i-mode services, as indicated in Figure 1.4.

Finally, wireless local area networks can offer access points, in particular, IEEE 802.11b and IEEE 802.11a/g technologies. These operate, respectively, at the frequencies of 2.4 GHz and 5 GHz with theoretical bit rates of 11 Mbit/sec or 54 Mbit/sec. Nevertheless, the actual bit rates depend on the local topology and the number of users. Thus, at 5 m from the access point and without obstacles, the best bit rate that a single user can obtain will not exceed 5 Mbit/sec. Similarly, an 802.11g link reaches, in practice, a bit rate of 20 Mbit/sec under optimal conditions. These bit rates will certainly be reduced once security procedures are taken into account.

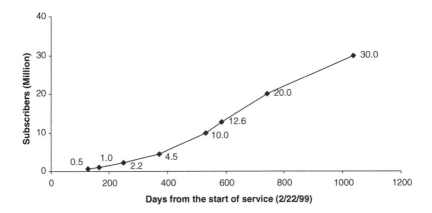

FIGURE 1.4
Growth of Japanese subscribers to i-mode® services. (Adapted from NTT DoCoMo, *Financial Times*, October 7–19, 2002, and Pimont, T., *Décision Micro & Réseaux*, 444, 20, 13, November 2000.)

1.5.3 Traffic Multiplexing

Multiplexing of voice and data channels is inextricably tied to complex commercial transactions. This is particularly true when the possibilities of choice differ from the standardized tracks of a robot or an "intelligent agent," such as during the negotiation of a trip with several stops, which would require human intervention (Billaut, 1997). Some systems for payment by bank cards are designed to invoke the intervention of a human operator for verification of the transaction when the amount exceeds a specific limit. In such cases, two communication channels are needed: one for the exchange of data (search of a virtual catalog, transmission of card information, etc.) and the other for oral communication. Figure 1.5 depicts the connections to be established.

These connections are readily made when an ISDN connection is available. For analog lines, multiplexing at the customer's premises is possible using the adaptor defined by ETSI (European Telecommunications Standards Institute) specifications ETS 301 141-1 for Narrowband Multiservice Delivery Systems (NMDSs), shown in the block diagram of Figure 1.6.

In this service configuration, the analog port of the UNI (User Network Interface) is connected to the analog telephone set, while the ISDN user port is connected to a computer equipped with an ISDN card.

A similar service can, in theory, be offered through the PSTN (Public Switched Telephone Network). Figure 1.7 depicts the various connections involved are depicted, irrespective of the technique used at the physical layer.

In this case, the first connection between the PC (Personal Computer) and the Web server for e-commerce is established through the IP (Internet Protocol) network (Connection 1). When the user initiates a vocal contact by

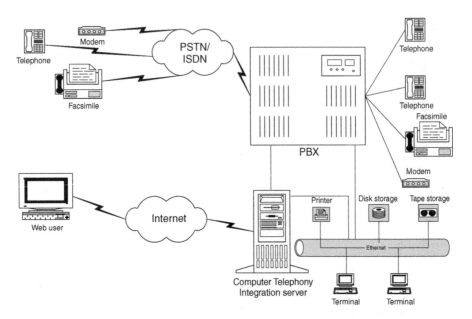

FIGURE 1.5
An access multiplexing architecture.

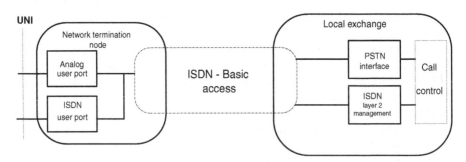

FIGURE 1.6
A narrowband multiservice delivery system (NMDS).

clicking on the appropriate key, the Web server contacts the call center via the IP network (Connection 2). The call center, in turn, sets up a PSTN call to the user (Connection 3). Thus, the telephone conversation and the Internet connection can progress simultaneously.

The transport of voice traffic in IP packets can improve the service, provided that a gateway is placed between the PSTN and the IP network, as shown in Figure 1.8.

Voice, coded between 6 and 8 kbit/sec, is packeted using the protocol stack RTP (Real-Time Protocol)/UDP (User Datagram Protocol)/IP. This choice means that there is substantial overhead, because to transport a payload of 20 octets, which corresponds to voice samples collected during

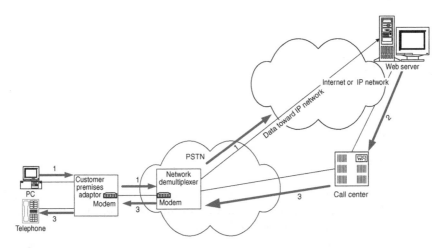

FIGURE 1.7
Connection of a user to an e-commerce server through a call center.

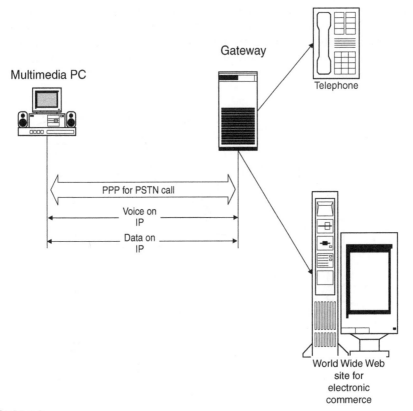

FIGURE 1.8
Use of voice on an IP for e-commerce. (From Yamada et al., *Proc. Int. Symp. Services Local Access*, 259–264, ©1998 IEEE. With permission.)

20 ms at the bit rate of 8 kbit/sec, a 40-octet header will be added. To avoid this drawback, it is possible to add a new protocol layer above the IP layer to compress the header and reduce the overhead to about 2 to 4 octets. While increasing bandwidth usage, the algorithm is capable of producing a burst of lost packets, in case some errors are encountered during transmission, which undoubtedly will degrade the quality of the transmitted voice (Mamais et al., 1998). This degradation is particularly noticeable if, to reduce cost, most of the trajectory is on the IP network and the separation of the joint flow into its constituents is as close as possible to the destination.

To avoid these problems, the joint flow can be limited to the local loop between the user PC and the router of the Internet service provider. After that point, the traffic will be separated and routed differently: voice on the PSTN or the ISDN, and data on the IP network. The router can be managed by the Internet service provider (Case I), the telephone operator (Case II), or a value-added network (Case III) (Yamada et al., 1998). These three possibilities are depicted in Figure 1.9.

In the first case, the gateway sets up the telephone call in lieu of the user. The second configuration requires that the gateway send signaling messages conforming to the Signaling System No. 7 (SS7) protocol to exploit the intelligence of the PSTN. Finally, if the gateway is in a third-party network, tight coordination is needed to ensure a smooth integration of the various networks.

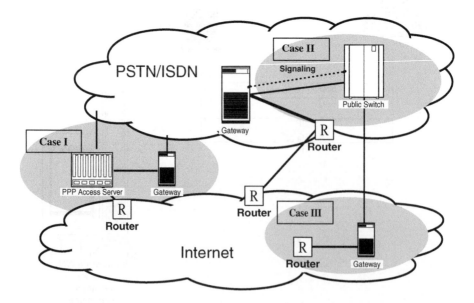

FIGURE 1.9
Alternative locations of routers in voice/data multiplexing on the link layer using point-to-point protocol (PPP). (From Yamada et al., *Proc. Int. Symp. Services Local Access*, 259–264, ©1998 IEEE. With permission.)

1.6 Consequences of E-Commerce

By increasing the speed and the quantity as well as the quality of business exchanges, e-commerce rearranges the internal organizations of enterprises and modifies the configurations of the various players. Innovative ways of operation eventually emerge, with new intermediaries, suppliers, or marketplaces. In the long run, the whole financial and banking environment could be modified. Porter's model, shown in Figure 1.10, allows us to appreciate the effects of the pressure from players (customers, suppliers, or competitors), the role of regulation, and the threats of substitution.

1.6.1 Clients

Whether the client is an individual or an enterprise, a technological innovation cannot be embraced voluntarily without adaptation to the ambient culture. The main criteria that the new means of payment should satisfy seem to be the simplicity of implementation and utilization, the level of

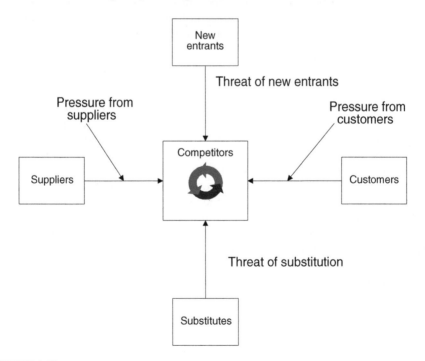

FIGURE 1.10
The competitive environment of e-commerce. (Adapted with the permission of The Pree Press, a Division of Simon & Shuster Adult Publishing Group, from *Competitive Strategy* by Porter, M.E., Copyright ©1980, 1998 by The Free Press.)

security, and the control of payment schedule. The acceptance of business-to-consumer e-commerce would probably be slower than expected, as long as the issues regarding the protection of personal data are not resolved. This means that the confidentiality of the transaction must be "guaranteed," and that privacy is protected, not only against surveillance by the state, but also against the abuses of unethical merchants. Another factor that is at least as important is the necessity of buying the necessary equipment, for example, a secured card reader, or of installing an infrastructure.

1.6.2 Suppliers

The operation, maintenance, and update of merchant sites is a full-fledged service, from simple hosting of the computation platform to an ensemble of services, including network engineering, equipment configuration, data storage, security management, payment processing, integration with legacy systems, etc. This complexity increases the dependency on computer manufacturers and software suppliers and their capabilities of worldwide support of their product. International standards are therefore needed, not only to reduce efforts of the suppliers, but to facilitate interconnections, to ensure a minimum quality of service, and to avoid lock-ins into proprietary solutions that would prevent price or quality competition.

1.6.3 Substitutes

Let us consider the substitutes in terms of new payment instruments as well as intermediaries.

The progressive substitution of paper money with electronic representations of monetary values could lead, in the long run, to the management of money by a "virtual" bank that is totally dematerialized and without any tangible means of payment. If the conditions of security are taken into consideration, the production and the distribution of this money can be completely automated through telecommunications networks.

To replace the physical fiduciary money, this electronic money must meet the following conditions (Fay 1997, pp. 113, 115):

1. It must be issued by a source that has the confidence of those that will hold that money.
2. Each monetary unit must have a unique number and must be unfalsifiable.
3. Clearly identified signs must guarantee the quantity represented.

The unsuccessful experiment of DigiCash (see Chapter 11) demonstrates that these conditions are not easy to meet.

Concerning new intermediaries, aggregators are companies that specialize in the collection, integration, synthesis, and online presentation of consumer data obtained from several sources. The goal is to save end customers the headache of managing multiple passwords of all the Web sites that have their financial accounts by replacing them with a single password to a site from which they can recover all their statements at once: bank statements, fidelity programs, investment accounts, money market accounts, mortgage accounts, etc. Ultimately, these aggregators may be able to perform some banking functions. On the other hand, traditional forms of music distribution are currently under question, due to free online distribution of music and the ability to burn CDs.

Security of payments covers many aspects: certification of merchants and clients, the production and escrow of keys, fabrication and issuance of smart cards, and constitution and management of electronic and virtual purses. Other activities include the detection of fraud, the recording of exchanges to ensure nonrepudiation, the formation and distribution of revocation lists, etc. These functions should lead to the birth of new legal roles, such as electronic notaries, trusted third parties, and certification authorities (Lorentz, 1998), with exact responsibilities that remain to be defined.

1.6.4 New Entrants

The introduction of such virtual banks requires substantial investments from all the actors (banks, merchants, individuals, and enterprises) as well interdisciplinary collaboration. Past experiences show that new means of payments require a long gestation period (on the order of 10 years). Despite the increase in deposits, most strictly virtual banks have not reached the threshold of profitability, even in countries where Internet penetration is high. Banks that thought they could save on the cost of "bricks and mortar" by closing branches had to change course (George, 2001).

In theory, telecommunications network operators may also compete with the banks (without violating their monopoly on the management of money supply) by capturing the cash flow of enterprises and investing it. These nonbanks are already responsible for managing the transport networks, for hosting merchant sites, and in some cases, in detecting and managing fraud. Furthermore, telephone operators have a core competence that the banks lack, namely, billing for small amounts or micropayments, particularly if these amounts are expressed in terms of telephone impulses, such as for the Minitel or for i-mode. In France, in particular, there are about 180,000 public phones that, with slight modifications, could function as terminals to recharge smart cards with monetary value or to utilize to make remote payments.

1.6.5 Banks

In most countries, central banks have the monopoly for issuing legal money; they define the framework for the administration of means of payment as well as govern the supply and demand of capital. As long as this exclusivity lasts and states are able to extract financial benefits from this monopoly, banks will remain the unshakable pedestal for the edifice of e-commerce. Under this hypothesis, one can assume that central banks will keep the responsibilities of administering and tracking monetary transactions, even if the traditional instruments (coins, bills, and checks) are replaced, in part or in total, by new electronic means of payment. Thus, as long as e-commerce substitutes one form of scriptural money with another (see Chapter 2 for the definition of scriptural money), the consequences on monetary policy and banking institutions would seem to be limited. This is particularly true if, as some central banks request, the value of the purchasing power contained in the various electronic purses is taken into account in the various money aggregates. In 1994, in fact, the European Monetary Institute stated that the "funds representing the purchasing power charge in an electronic purse must be considered as bank deposits that only credit institutions should be allowed to hold" (Sabatier, 1997, p. 35).

If these funds are not taken into account, the substitution of coins and bills with money that is not accounted for will reduce the importance of central banks' accounts and their capabilities to affect interest rates. The demateri-alization of money and the emergence of e-commerce may thus stimulate a redistribution of power among the existing economical, political, and social forces. In the extreme case, the privilege to mint money could be privatized, thereby turning the situation upside down.

Even without going to such an extreme, online financial services may threaten some traditional banking functions.

1.6.6 Role of Governments

It should be apparent that the development of e-commerce, if not of the information society, requires the definition of new global rules, such as the legal recognition of electronic signatures, the uniform protection of individ-ual and consumer rights, as well as the protections given to intellectual properties. This is why the role of governments and intergovernmental orga-nizations in the progress of electronic levels is undeniable. For sometime already, governments have encouraged, through legislation, the circulation of documents in electronic form. For example, in Italy, laws governing the legal and fiscal aspects of "electronic invoices" as well as the transmission and storage of electronic documents were introduced in 1990 (Pasini and Chaloux, 1995). Since 1992, the French DGI (Direction Générale des Impôts — General Taxation Directorate) allowed companies to file their tax docu-ments by electronic means, and a more general law was promulgated in 1994 to expand the scope of file exchanges between businesses and government authorities.

The security of payments is not sufficient to protect users. It is legislation that must prevent fraud and breaches of trust and protect the right to privacy. Public authorities are thus directly concerned by e-commerce and not just because of its potential effects on employment in the banking sector. Laws need to be written for monetary transactions and the purchase of nonmaterial goods online, especially on a worldwide basis. Most of the examples mentioned in this regard relate to taxation and the exploitation of personal data collected during transactions. This subject will be discussed in Chapter 15.

1.7 Summary

The initial applications of e-commerce in the 1980s were stimulated by the desire of the economic agents, such as banks and merchants, to reduce the cost of data processing. With the Internet and mobile networks in place, e-commerce targets a wider audience, at least in some countries. One condition for the acceptance of e-commerce is that the security of transactions and the protection of private information be improved. However, the commercialization of cryptography, which a few years back was strictly a military application, may contribute to overcoming many justified hesitations.

The obstacles that e-commerce has to face are technical, cultural, and financial. A performing telecommunications infrastructure is essential, and the security of the whole system — not only of the transactions — requires solid computer expertise. The diffusion of various techniques of e-commerce depends on the cultural context and the encouragement of public authorities. Finally, the cost of switching to e-commerce has to include, in addition to equipment, software, or network access, the cost of training, reorganizing work, and maintaining and managing the back-office systems during the operational life of the system.

Questions

1. Comment on the following definitions of e-commerce, which are adapted from the September 1999 issue of the *IEEE Communications Magazine*:

   a. It is the trading of goods and services, where the final order is placed over the Internet (John C. McCarthy).

b. It is the sharing and maintaining of business information and conducting of business transactions by means of a telecommunications network (Vladimir Zwass).

c. It consists of Web-based applications that enable online transactions with business partners, customers, and distribution channels (Sephen Cho).

2. How can e-commerce reduce operating costs?

3. What is sold in e-commerce?

4. Compare the characteristics of online and offline electronic payments.

2

Money and Payment Systems

ABSTRACT

In this chapter, we describe the financial context within which the dematerialization of means of payment is taking place. The first part of the chapter is dedicated to the "classical" forms of money and the means of payment in some developed countries. The second half corresponds to "emerging" monies, either in "electronic" or "virtual" forms.

2.1 The Mechanisms of Classical Money

The term *money* designates a medium that can be used to certify the value of the items exchanged with respect to a reference system common to all parties of the transaction (Berget and Icard, 1997; Dragon et al., 1997, p. 17; Fay, 1997, p. 112; Mayer, 1997, p. 37). Thus, money represents the purchasing power for goods and services and has three functions:

- It serves as a standard of value to compare different goods and services. These values are subjective and are affected, among other things, by currency fluctuations.

- It serves as a medium of exchange, as an intermediary in the process of selling one good for money, thereby replacing barter.

- It serves as a store of value and of purchasing power. Money permits postponement of the utilization of the product of the sales of goods or services. This saving function is maintained on the condition that the general level of prices remains stable or increases only slightly.

The practical terms of money depend on theoretical considerations on its nature and its intrinsic value. Primitive forms of money corresponded to needs for storage and exchange on the basis of valued objects. Accordingly, money first took a materialistic nature, in the form of a coin with a specific weight and minted from a precious metal. Today, the value of money corresponds to a denomination that is independent of the material support medium.

A monetary unit is a sign with a real discharging power that an economic agent would accept as payment in a specific geographic region. This discharging power is based on a legal notion (i.e., a decision of the political power) accompanied by a social phenomenon (acceptance by the public). This sign must satisfy specific conditions:

- It must be divisible to cover a wide range of small, medium, and large amounts.
- It must be convertible to other means of payment.
- It must be recognized in an open community of users. This is because money exists only inasmuch as its issuer enjoys the trust of other economic agents.
- It must be protected by the coercive power of a state.

As a consequence, the only monetary sign that has real discharging power is the set of notes issued by a central bank or the coins minted by a government mint. This set, which is called *fiduciary money*, is total and immediate legal tender within a specific territory, usually a national boundary, with two important exceptions. On one side, 10 countries "dollarized" their economy by adopting the U.S. dollar as currency, while 34 others indexed their currency to its value. On the other, the European Union adopted the Euro as currency without a political union. Note, however, that payment by coins can be restricted by legislation.

While the nominal power corresponds to the face value imprinted on the note or the coin, the real value resides in the trust in the issuer. This is the same for the money that a bank, or generally a credit institution, creates by making available to a nonfinancial agent a certain quantity of means of payment to be used, in exchange for an interest proportionate to the risks and the duration of the operation. This money is called *scriptural money* and is a monetary sign tied to the trust that the issuer enjoys in the economic sphere. For example, when Bank A creates scriptural money, the discharging power of that scriptural money depends on the confidence that this bank enjoys, and on the system of guarantees that surround its utilization, under the supervision of political authorities (for example, a central bank).

It should be noted that a merchant is free to accept or reject payments with scriptural money but not with fiduciary money. Note also that scriptural money is traceable, while fiduciary money is not.

To ensure its practical utility, the material support of classical money must meet the following requirements (Camp et al., 1995; Kelly, 1997):

- Be easily recognizable
- Have a relatively stable value across transactions
- Be durable
- Be easy to transport and use
- Have negligible production cost compared with the values exchanged in the transactions

The power of money can be transferred from one economic agent to another with the help of a means of payment or an *instrument of payment*. Let us briefly review these instruments.

2.2 Instruments of Payment

Instruments of payment facilitate the exchange of goods and services and respond to specific needs. Each instrument has its own social and techno- logical history that orients its usage in specific areas. Today, banks offer a large number of means tied to the automatic processing of transactions and to the progressive dematerialization of monetary supports. The means uti- lized vary from one country to another. A general inventory of the means of payment takes the following forms:

- Cash (in the form of metallic coins or paper notes)
- Checks
- Credit transfers
- Direct debits
- Interbank transfers
- Bills of exchange or negotiable instruments
- Payment cards (debit or credit)

The emerging means of payment are based on dematerialized money stored in smart (chip) cards or in electronic or virtual purses.

Note that some of the instruments are merely banking inventions with no corresponding legal status. For example, in France, credit transfers and the Interbank Payment Title (Tip) are regulated only by the CFONB (Comité Français d'Organisation et de Normalisation Bancaires — French Center for Banking Organization and Standardization) and interbank organizations. Similarly, in the U.S., electronic funds transfer (EFT) was developed without a strict legal status under the auspices of NACHA (National Automated Clearing House Association), which is a private entity.

Reproduced in Table 2.1 are data from the Bank for International Settle- ments (BIS) regarding the use of various instruments of payment in selected countries in 2000[1] (Bank for International Settlements, 2002).

---

[1] The BIS is owned by 50 of the world's biggest central banks.

TABLE 2.1

Utilization of Scriptural Money in Selected Countries in 2000

			Millions of Transactions				
Country	Checks	Debit Cards	Credit Cards	Credit Transfers	Direct Debit	Electronic Purse	Total
Belgium	70.7	408.2	53.8	656.8	166.2	51.3	1,407
Canada	1,658.2	1,960.1	1,270.8	569.5	444.4	—	5,903
France	4,493.7	3,292	—	2,093.6	1,968.6	—	11,847.9
Germany	436.6	1,037.1	351.6	7,132.9	5,532.1	26.6	14,516.9
Hong Kong	138.6	—	—	16.6	35.2	—	190.4
Italy	602	317.5	272.3	1,018.6	319.6	—	2,530
Japan	225.9	3.2	1,641	1,215.4	—	—	3,085.5
The Netherlands	14.2	801.5	57.1	1,140.4	836.2	25	2,874.4
Singapore	91.7	85.5	—	14.6	17.3	100.1	309.2
Sweden	2	254	66	715	91	2.9	1,130.9
Switzerland	11.2	172	71.5	545.2	46.1	18	864
U.K.	2,698	2,337	1,452	1,848	2,010	—	10,345
U.S.	49,604	9,550.1	20,485.1	3,486.1	1,947.3	—	85,072.6

			Percentage Utilization				
Country	Checks	Debit Cards	Credit Cards	Credit Transfers	Direct Debit	Electronic Purse	Total (%)
Belgium	5.02	29.01	3.82	46.68	11.81	3.65	100
Canada	28.09	33.21	21.53	9.65	7.53	—	100
France	37.93	27.79	—	17.67	16.62	—	100
Germany	3.01	7.14	2.42	49.14	38.11	0.18	100
Hong Kong	72.79	—	—	8.72	18.49	—	100
Italy	23.79	12.55	10.76	40.26	12.63	—	100
Japan	7.32	0.10	53.18	39.39	—	—	100
The Netherlands	0.49	27.88	1.99	39.67	29.09	0.87	100
Singapore	29.66	27.64	0.00	4.72	5.60	32.38	100
Sweden	0.18	22.46	5.84	63.22	8.05	0.26	100
Switzerland	1.30	19.91	8.28	63.10	5.34	2.08	100
U.K.	26.08	22.59	14.04	17.86	19.43	—	100
U.S.	58.31	11.23	24.08	4.10	2.29	—	100

These data show that, in 2000, checks were still the mostly used scriptural money in Hong Kong (72.8%), in the U.S. (58.3%), and in France (37.9%). In contrast, bank card transactions were dominant in Canada (54.74%), Japan (53.28%), and in the U.K. (36.63%). Note, however, that three-fifths of these transactions in Canada and in the U.K. were by debit card, while in Japan, the overwhelming majority of the transactions were by credit card. Finally, thanks to the Postal Gyro system, in which debtors authorize their banks to debit their account regularly to pay their creditors, credit transfers were the most important scriptural money instrument in Sweden (63.22%), Switzerland (63.10%), Germany (49.14%), Belgium (46.68%), Italy (40.26%), and in the Netherlands (39.67%). In most of these countries, checks have almost disappeared; the volume of check transactions has fallen to 0.18% in Sweden, 0.49% in the Netherlands, 1.3% in Switzerland, 3% in Germany, and 5% in Belgium. Finally, electronic purses transactions formed about a third of the volume of transactions in Singapore.

Clearly, the techniques used for electronic commerce (e-commerce) will have to take into account the differences in behavior and the current trends in the different societies. For example, systems of electronic or virtual checks would be difficult to introduce in Sweden or Switzerland, whereas they may be of interest in the U.S. and France. In fact, it is in these last countries that research is being actively conducted to replace the check with other electronic means. A closer examination of the various types of monies follows.

2.2.1   Cash

In each country, cash constitutes the fiduciary money that the central bank and the public treasury issue in the form of notes and coins (Fay, 1997, p. 83). This instrument of payment is available free of charge to the individuals. Banks cover the costs for managing the payments, withdrawals from branches or teller machines, as well as the costs of locking up the money. In retail commerce, banks usually charge their customers for their services if they have to process large amounts of notes or coins and perform the counting and the sorting of bills and coins.

Cash is the preferred means of payment for face-to-face commerce. The current trend in Western countries is to use cash for relatively small amounts, while medium and large amounts are handled with scriptural instruments. On the basis of this suggestion, the French Comité des Usagers (Users Committee) defined micropayment as a "payment, particularly in the case of a face-to-face payment, where, given the absence of any specific constraint, cash is the preferred instrument" (Sabatier, 1997, p. 22).

Depicted in Table 2.2 is the part of cash in the narrow money for selected countries between 1985 and 2000 (Dragon et al., 1997; Bank for International Settlements, 1996, 1997, 2000, 2001, 2002). In most of the countries, narrow money is measured using the M1[1] monetary, with the exceptions of Sweden, which uses the M3 aggregate, and the U.K., which has been using the M2 aggregate since 1989. The data are presented in graphical form in Figure 2.1.

Clearly, the contribution of fiduciary money varies tremendously among countries. However, with the exception of the U.S., the general trend is a

[1] M1 is the total amount of currency in circulation as well as monies in checking accounts. M2 is M1 plus monies in saving accounts and money market funds. M3 is M2 plus bank certificates of deposits and other institutional accounts, such as accounts in foreign currencies and, for the U.S., Eurodollar deposits in foreign branches of U.S. banks.

TABLE 2.2

Percentage of Cash in the Narrow Money in Selected Countries (1985–2000)

	1985	1986	1987	1988	1989	1990	1991	1992	1993	1994	1995
Belgium	36.6	35.3	34	34.8	32.5	31.3	31.5	31.5	29.6	27.1	27.2
Canada	44.5	45.7	43.6	44.6	43.8	46.1	47	47.1	44	44.2	42.8
France	16	15.3	15.2	15.2	15.2	15.1	15.8	15.9	15.3	15.1	14.2
Germany	31.1	31.3	32.2	33.4	32.6	27.1	29.9	29.9	29.6	29.6	29.1
Italy	14.2	14.1	13.8	14.3	15	14.4	14.2	15.7	15.5	16	16.3
Japan	28.6	27.4	28.3	31.5	35.3	36	33.1	31.2	31.1	30.7	29.2
The Netherlands	32.1	31.2	32.4	31.3	30.3	29.5[a]	28.6[a]	27.2	25.1	25	22.1
Sweden[a]	—	—	—	—	—	—	11.5	10.8	10.7	10.7	10.5
Switzerland	—	—	—	—	—	—	21.8	21.6	19.7	30.6	18
U.K.[b]	19.5	17	14.5	14.1	6.5	6	5.6	4.8	4.5	4.6	4.6
U.S.	31	29.4	28.7	27.2	28.2	29.2	28.5	28.5	28.5	30.7	32.9

	1996	1997	1998	1999	2000
Belgium	27.5	26.5	23.8	20.4	19.3
Canada	14.3	14.2	14.5	15.6	13.7
France	13.3	13.1	11.0	12.6	11.8
Germany	27.6	27.2	24.1	23.4	21.8
Hong Kong	38.6	42.8	45.5	48.5	45.0
Italy	16.1	16.1	16.1	14.4	14.3
Japan	26.1	25.8	25.3	24.8	25.0
The Netherlands	18.0	15.7	14.1	12.8	11.4
Singapore	38.1	38.9	37.3	36.4	33.9
Sweden[a]	9.9	10.0	10.2	10.6	—
Switzerland	17.3	15.6	15.5	15.3	15.8
U.K.[b]	4.9	5.0	5.0	5.0	5.0
U.S.	36.0	39.0	41.4	45.4	48.1

[a] As a percentage of the M3 monetary aggregate.
[b] As a percentage of the M2 aggregate starting from 1989.

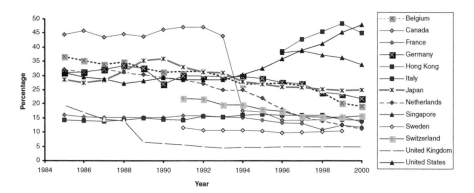

FIGURE 2.1
Percentage of cash in the narrow money for selected countries.

decreasing or constant contribution. The U.S. situation can be explained using socioeconomic factors, because the cost of banking services put them out of reach for an increasing sector of the population (about 25 to 30%) (Hawke, 1997; Mayer, 1997, p. 451). If we exclude Sweden and the U.K. because they use their own definition of narrow money, it is seen that the ratio of cash to narrow money is lowest in France. The countries with the highest ratio are, respectively, the U.S., Hong Kong, and Singapore. One conclusion is that the extensive use of checks and bank cards does not automatically reduce the importance of cash transactions.

Because counterfeit money cannot be exchanged with legitimate money, the use of cash relies on the reciprocal trust of the various parties. To sustain this trust, the authorities multiply various security measures. The protection of bank notes relies on the utilization of special paper that cannot be easily reproduced, on the protection of supplies to the banks, on detecting counterfeit money, and on a guarantee of replenishing the stocks with new notes and pieces. The protection must cover the whole life cycle of the money, from the components used in the fabrication, until the recall and destruction of worn-out notes or coins as well as counterfeit ones. The rate of counterfeit varies from 0.002% for the old French francs to 1% for U.S. dollar bills. Added security has a price; the unit cost of the franc bills amounted to 1.1 to 1.4 francs, which was almost double the unit cost for the German mark or the British pound (which were around 0.7 francs) (Dragon et al., 1997, pp. 90–91). To this cost, one must add the cost of fraud-detection equipment at merchants and banks.

2.2.2 Checks

Table 2.3 shows the relative importance of checks in the total volume of scriptural transactions in selected countries from 1991 to 2000 using data from the BIS.

These data are represented in graphical form in Figure 2.2. They show that the contribution of checks is decreasing in most countries, even though the

TABLE 2.3

Percentage of Checks in the Volume of Scriptural Payments in Selected Countries (1991–2000)

	1991	1992	1993	1994	1995	1996	1997	1998	1999	2000
Belgium	21.6	18.8	16	11.7	10.6	9.4	8	7	5.8	5
Canada	64.8	62.4	58.7	52.7	52.1	45.5	39.4	34.7	31.6	28.1
France	52.2	50.6	49.1	46.9	44.8	48.6	46.6	44	40.1	37.9
Germany	9.6	8.8	8.1	7.9	7	6.4	5.7	4.8	3.3	3
Hong Kong	—	—	—	—	—	100	76.9	74.1	73.2	72.8
Italy	41.6	40	37.2	34.9	33.7	34.5	31.3	29.6	27.3	23.8
Japan	—	—	—	—	—	12.1	10.8	9.5	8	7.3
The Netherlands	14.3	12.3	8.1	8.5	5.8	4	2.8	1.9	1	0.5
Singapore	—	—	—	—	—	47.9	45.5	39.2	33.5	29.7
Sweden	—	—	—	—	—	4.6	2	0.4	0.4	0.2
Switzerland	5.4	4.4	3.3	2.6	2	1.6	1.3	10	0.8	1.3
U.K.	48.5	45.4	43	40.2	36.7	37.5	34.5	31.8	28.8	26.1
U.S.	81.6	81.1	80.1	77.9	75.4	74.1	72.2	70	68.6	58.3

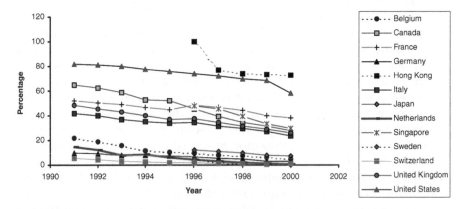

FIGURE 2.2
Check contribution in the volume of scriptural transactions in selected countries.

total volume of scriptural transactions increased. A detailed examination of the patterns reveals three large classes of use:

- The largest use of checks is essentially in Hong Kong and the U.S. where the volume of transactions by checks in 2000 was between 50 and 70% of the total number of scriptural transactions.

- In Canada, France, Italy, Singapore, and the U.K., the contribution made from checks varied between 30 and 50% of the volume of scriptural transactions. In this category of countries, check use, while decreasing continuously, seemed to tend to a stable plateau.

- The countries where checks play an insignificant role (less than 10% of the volume of transactions) are Belgium, Germany, Japan, the Netherlands, and Switzerland. It is even possible to envision the total disappearance of the check in these countries within a few years.

The relative importance of the values exchanged in the same countries in the period from 1991 to 2000 is given in Table 2.4 and presented in graphical form in Figure 2.3.

TABLE 2.4

Percentage Contribution of Checks in the Value Exchange with Scriptural Money in Selected Countries (1991–2000)

	1991	1992	1993	1994	1995	1996	1997	1998	1999	2000
Belgium	5.4	6.2	5.4	4.6	4.3	2.9	2.9	3.2	0.6	0.5
Canada	99	98.8	98.8	98.8	98.1	97.3	97	96.5	21.5	14.6
France	7.3	6.4	4.6	4.4	4.7	4	3.9	2.6	2.9	2.9
Germany	2.8	2.4	2.3	2.3	2.1	8.8	7.9	7.2	3.9	3.4
Hong Kong	—	—	—	—	—	100	97.3	94.2	93.2	93.1
Italy	9.1	7.1	5.4	4.5	4.5	3.7	3.3	2.9	3.7	3.2
Japan	—	—	—	—	—	3	2.4	2.7	2.7	2.7
The Netherlands	0.2	0.2	0.1	0.1	0.1	0.3	0.2	0.1	0.1	0.1
Singapore	—	—	—	—	—	7.1	5.9	4.9	5.1	4.7
Sweden	—	—	—	—	—	—	—	0.5	0.4	0.3
Switzerland	5.4	0.1	0.1	0.1	2	0.1	0	0	0	0
U.K.	16.1	11.6	9.4	7.6	5.3	5.9	5	4.4	2.8	2.5
U.S.	13.7	13.1	12.6	12.2	11.9	11.2	10.5	10.3	11.2	6.5

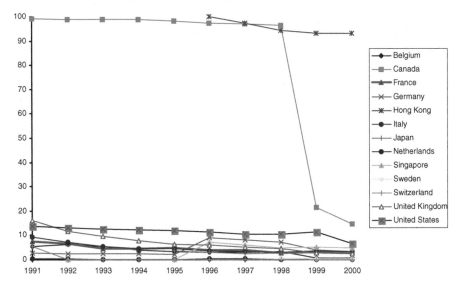

FIGURE 2.3
Check contribution to the value of scriptural transactions in selected countries.

The data show that in Hong Kong, payment by checks is responsible for more than 90% of the value exchanged. This was also the case in Canada in the early 1990s, before the contribution of checks dropped dramatically to 14.6% by the end of the decade. In the Netherlands, Sweden, and Switzerland, the amounts exchanged by checks are almost insignificant. In other countries, the contribution of checks is between 2 and 6%. In the U.S., the decline from 11.2 to 6.5% is the direct consequence of the federal law that took effect on January 1, 1999, mandating the use of credit transfers for all payments by the federal government, with the exception of tax refunds. The driving reason is that credit transfers are 20 times less expensive than checks, and postal costs are avoided. However, social inequality in the U.S. is impeding total implementation of these mandates. The cost of opening a bank account discourages more than 40 million people whose annual income is less than $25,000 from having a checking account, which puts them outside the financial circuits (Hawke, 1997; Mayer, 1997, p. 451). In addition, about 10,000 banks do not have all the capabilities of EDI and are not capable of translating the incoming transfers into a form that can be read by the small- and medium-sized enterprises that are dealing with the federal government.

The total cost of processing an individual check hovers around 50 cents to a dollar (Dragon et al., 1997, pp. 110–126). This cost includes fabrication, security, distribution, return (sorting, identifying the signature, capturing written data, rejecting, etc.), archiving, and the price for stolen checks, in addition to the cost of bad checks, which is about 1% of the total amount of checks. In addition, about 800,000 checks are lost, stolen, delayed, or damaged before arrival, and more than 75,000 checks are counterfeit or fraudulent. Finally, paperless transfers avoid postal costs. This means that checks are the most expensive instrument of payments, not only for banks but also for heavy users.

The volume of business-to-business payments in the total payments by checks in the U.S. was 42.7% in 2001, while 15.5% of the volume is related to business-to-consumer payments, and 10.3% is related to transactions from consumers to businesses. The remaining 32.5% was for other transactions, such as refunds, government checks, payroll checks, social security, etc. (Schneider, 2002). In France, likewise, business-to-business payments by checks stabilized around 44% of the total payments by checks (Dragon et al., 1997, pp. 105–110). One possible explanation for this commonality is that businesses attempt to take advantage of the cash flow due to the float, i.e., the calendar difference between when the check is issued and when the funds are actually withdrawn. However, as will be seen in Chapter 12, schemes for electronic check presentment (ECP) will do away with float and one advantage of checks for businesses. This could eventually drive small- and medium-sized companies to other means of payments, leaving checks to payments by individual consumers. However, consumer-to-business payments differ from business-to-business payments in that they are more frequent but have a much lower monetary value (less than $50). To prevent a costly instrument from being used more frequently for decreasing values, banks will have to induce customers to switch to new scriptural instruments, such as payment cards or direct debits. An interesting experiment is going on currently, because, in July 2002, French banks generalized the use of ECP; at the same time, checks are free for account holders in France, but this is not the case for other instruments, such as bank cards.

Note that check processing, which is essentially manual, provides direct and indirect employment to tens of thousands in the public and private sectors. As a consequence, any cost reduction through the dematerialization of checks or their replacement by electronic means of payment will have important social repercussions.

2.2.3 Credit Transfers

Credit transfers are a means by which to transfer funds between accounts at the initiative of the debtor. This instrument requires the debtor to know the beneficiary's bank and bank accounts. This is the reason it is usually used in bulk transfers, such as for salaries and pensions. The data available from the BIS, reproduced in part in Table 2.5 and represented in Figure 2.4, underline the evolution of the contribution of credit transfers in the volume of scriptural transactions between 1991 and 2000.

The same groupings of the various countries as related to check usage exists but in reverse order:

- Countries where the role of checks is not significant are those where credit transfers are used the most, particularly in Sweden and Switzerland (more than 60%). The percentage is between 40 and 60% in Belgium, Germany, Japan, Italy, and the Netherlands.
- The use of credit transfers in France and in the U.K. stabilized around 20% of the total number of transactions.
- In Hong Kong and the U.S., where check usage is the highest, credit transfers constitute less than 10% of the total number of transactions. Credit transfers in Canada are also infrequent (less than 10%).

We note the saturation if not a small decline in the proportion of credit transfers almost everywhere. For a better analysis of the situation, we study the data reproduced in Table 2.6 and represented by the curves of Figure 2.5. These data illustrate that between 1991 and 2000, and in almost all countries, the values exchanged by credit transfers remained constant or increased slightly. In Canada, the increase was dramatic. Thus, even if the number of transfers diminishes, the values exchanged remain constant. In reality, a small percentage of the transfers, less than 5% in volume, are related to large movements of capital, particularly to interbank operations (lending or borrowing from markets, settling foreign-exchange operations) as well as financial operations (transfers among different accounts of a single entity or group). The differences among countries seem to stem mostly from large

TABLE 2.5

Percentage of Credit Transfers in the Volume of Scriptural Payments in Selected Countries (1991–2000)

	1991	1992	1993	1994	1995	1996	1997	1998	1999	2000
Belgium	57.0	56.9	58.5	60.9	60.2	59.4	58.0	54.0	51.9	46.7
Canada	3.9	4.4	5.2	7.1	8.2	8.9	8.9	9.5	9.3	9.6
France	15.2	15.4	15.4	15.7	15.6	17.5	17.7	17.8	18.4	17.7
Germany	51.3	49.8	45.6	48.7	48.8	49.1	48.2	50.5	52.5	49.1
Hong Kong	—	—	—	—	—	—	6.9	8.1	8.5	8.7
Italy	40.9	42.1	44.6	46.8	45.0	47.9	46.5	44.3	41.2	40.3
Japan	—	—	—	—	—	42.8	42.3	41.7	39.0	39.4
The Netherlands	61.3	61.3	66.0	59.8	52.7	48.6	45.9	42.9	40.6	39.7
Singapore	—	—	—	—	—	7.9	7.8	7.0	5.0	4.7
Sweden	76.9	77.6	84.5	82.3	79.4	73.8	72.3	68.6	67.2	63.2
Switzerland	82.7	81.3	80.1	78.1	76.3	74.4	72.5	72.1	68.4	63.1
U.K.	20.9	20.6	20.4	20.1	18.2	18.6	18.5	18.3	18.1	17.9
U.S.	1.6	1.8	1.9	2.5	2.4	2.6	2.7	3.1	3.2	4.1

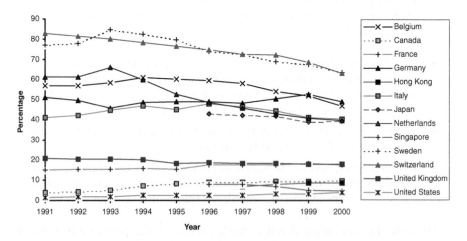

FIGURE 2.4

Evolution of credit transfers in selected countries (in volume).

payments from social organizations and enterprises (salaries or retirement benefits). As already indicated, U.S. federal law mandates that all government payments be made with credit transfers, with the exception of tax returns. This law became effective on January 1, 1999, which explains the increase in the values exchanged by credit transfers from 83 to 86.4% between 1999 and 2000.

TABLE 2.6

Percentage of Credit Transfers in the Exchange of Value in Selected Countries
(1991–2000)

	1991	1992	1993	1994	1995	1996	1997	1998	1999	2000
Belgium	94.3	93.4	94.2	95.8	96.3	96.7	96.7	96.3	98.9	99.0
Canada	0.6	0.7	0.7	0.7	1.0	1.5	1.8	2.1	77.5	84.8
France	89.9	91.2	93.5	93.5	93.0	95.0	95.1	96.6	96.2	96.2
Germany	95.4	95.5	95.7	95.7	95.8	79.2	79.3	79.0	83.9	86.3
Hong Kong	—	—	—	—	—	—	2.3	5.0	6.0	6.1
Italy	88.6	91.1	93.2	94.2	94.1	96.0	96.3	96.8	95.7	96.2
Japan	—	—	—	—	—	96.4	96.9	97.5	97.2	97.3
The Netherlands	98.4	98.6	98.8	98.8	98.9	93.7	93.8	93.8	93.4	93.5
Singapore	—	—	—	—	—	92.6	93.7	94.8	94.6	95.0
Sweden	84.9	86.3	95.8	96.2	95.7	95.7	95.8	94.9	94.3	93.9
Switzerland	99.8	99.9	99.9	99.8	99.8	99.8	99.8	99.8	99.8	99.7
U.K.	82.5	87.1	89.5	91.2	92.4	92.7	93.8	94.4	96.3	96.6
U.S.	85.4	85.8	86.4	86.7	87.0	75.3	76.0	82.0	83.0	86.4

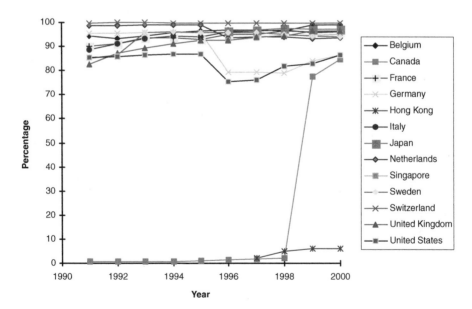

FIGURE 2.5
Evolution of the exchange of value with credit transfers in selected countries.

2.2.4   Direct Debit

Direct debit is a means of payment used for recurrent payments (for example, electricity consumption, subscription renewals, etc.).To start a direct debit, the debtor signs on paper an agreement to pay future amounts. This agreement can also be signed by electronic means, for example, the Tep (Titre Électronique de Paiement — Electronic Payment Title), in France. Large bill producers, such as the utilities and telecommunications companies, find this instrument to be convenient, but its progress is impeded by the suspicion of the debtors. The contribution of direct debits to the volume of scriptural payments in some countries is shown in Table 2.7. These data are also presented graphically in Figure 2.6.

TABLE 2.7

Percentage of Direct Debit in the Volume of Scriptural Payments in Selected Countries (1991–2000)

	1991	1992	1993	1994	1995	1996	1997	1998	1999	2000
Belgium	8.2	8.8	9	9.4	9.5	9.7	9.8	9.4	10.2	11.8
Canada	3.5	4.3	5	6.1	6.4	6.8	7.2	7.4	7.5	7.5
France	9.3	10.6	10.6	11.2	11.3	13.2	13.6	14.4	15.5	16.6
Germany	37.3	39.3	43.7	40.3	40.6	40.3	42.1	37.3	35.9	38.1
Hong Kong	—	—	—	—	—	—	16.2	17.8	18.3	18.5
Italy	3.8	4.1	4.4	4.7	5.4	8.3	9.6	10.3	12	12.6
The Netherlands	22.6	23.9	21.7	25.6	28.1	27.9	28	28.8	29.4	29.1
Singapore	—	—	—	—	—	11.7	12	9.2	6.3	5.6
Sweden	4.4	4.9	5.7	6.1	6.4	6.3	7.1	8	8	8
Switzerland	2.3	2.5	2.8	3.1	3.3	3.3	3.6	3.5	3.5	5.3
U.K.	14.2	15.1	15.6	16.5	16.4	16.9	17.7	18.4	18.8	19.4
U.S.	0.8	1	1.1	1.3	1.3	1.4	1.5	1.6	1.7	2.3

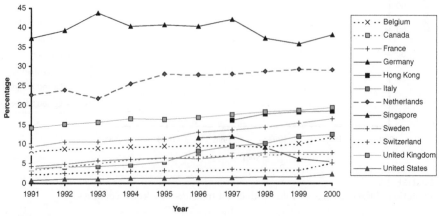

FIGURE 2.6
Contribution of direct debits to the volume of scriptural payments in selected countries.

The consequence of the gyro systems appears one more time — the countries where direct debit is most popular are Germany and the Netherlands. However, it is used essentially for small amounts, as indicated in Table 2.8 and depicted in Figure 2.7.

2.2.5 Interbank Transfers

The Tip (Titre Interbancaire de Paiement — Interbank Payment Title) is a specific instrument introduced in France in 1988. It is different from typical debit transfers in that a signature is required for each payment on a special form that the creditor supplies. Its main advantage is that it can be easily integrated into an architecture of electronic payment using telephone or computer services to allow remote payments. In this way, the creditor still

TABLE 2.8

Percentage of Direct Debit in the Transfer of Value by Scriptural Payments in Selected Countries (1991–1996)

	1991	1992	1993	1994	1995	1996
Belgium	0.2	0.2	0.3	0.5	0.3	0.4
Canada	0.1	0.2	0.2	0.2	0.3	0.5
France	0.7	0.6	0.7	0.8	0.9	1.0
Germany	1.8	2.1	2.0	2.0	2.1	2.5
Italy	0.3	0.2	0.2	0.2	0.2	0.2
The Netherlands	1.4	1.2	1.1	1.1	1.2	1.2
Switzerland	—[a]	—[a]	—[a]	0.1	0.1	0.1
U.K.	1.2	1.1	1.0	1.0	1.0	1.1
U.S.	0.8	1.0	0.9	0.9	0.9	0.9

[a]Less than 0.1%.

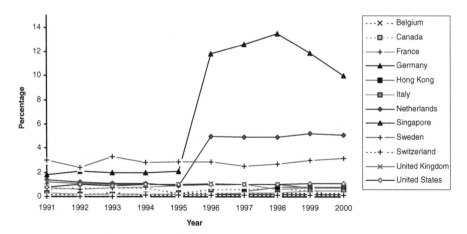

FIGURE 2.7
Contribution of direct debits to the value of scriptural payments in selected countries.

sends the Tip by postal mail, while the client signs electronically. Currently, two solutions are available: (1) the télé-Tip, where the signature is sent on the Minitel; and (2) the audio-Tip, where the signature is sent by entering a special code over the phone.

2.2.6   Bills of Exchange

A bill of exchange (or a negotiated instrument) is a remote payment reserved for professional relations, giving either the debtor or the creditor the initiative of the payment. If the debtor is the initiator, the instrument is called a "promissory note," whereas if the creditor has the initiative, it is a "bill of exchange" proper. In either case, creditors give the documents they possess to their banks that then send the bill of exchange to the debtor banks. The promissory note resembles a check drawn on a checking account, with the assurance of payment and the possibility of a discount fee for the beneficiary.

2.2.7   Payment Cards

Depending on the service offered, there are several types of payment cards:

- Check guarantee cards.
- Cash withdrawal cards.
- Bank payment cards:
  - Immediate debit cards, i.e., the withdrawal from the debtor accounts occurs immediately at the conclusion of the transaction.
  - Deferred debit cards, i.e., the debtor account is debited at a fixed date, such as the end of the month.
  - Credit cards.
- Restricted usage cards, which have limited applications, in distinction of bank cards that are universally applicable.
- Charge cards, such as American Express or Diner's Card, that can be defined as "international deferred debit cards." They differ from bank cards by the nature of the issuing financial institutions that control the network of affiliates.
- Private fidelity cards are issued by merchants to retain their customers and offer credit facilities (with the help of credit institutions). One of the uses of these cards is to construct customers' profiles of their consuming habits to focus marketing and sales campaigns.
- Cards that are focused on business usages, such as the following:
  - Corporate cards, which allow a company to optimize the expenses incurred by the employees during the course of their work-related activities.

- Purchasing or procurement cards, which are deferred debit cards used to cover the payments made for nonrecurrent charges and small amounts. While the cardholder represents the enterprise in making the purchases, it is the enterprise account that will be debited for the sales incurred. The processing of the data relative to these cards includes the generation of management reports and accounting and fiscal reports on all operations used with this card.

The protocols for bank card purchases require the intervention of several actors in addition to the buyer and the seller, in particular, the banks of each of the parties and the credit card scheme, for example, Visa or MasterCard. The merchant's bank is called the acquiring bank because it acquires the credits, and the buyer's bank is called the issuing bank because it issued the cards to its members that it authenticated. The bank card schemes call for the intervention of authorization servers connected to call centers whose role would be to filter out abusive transactions. The filtering process utilizes preestablished criteria, for example, whether a spending ceiling was reached, or if a large number of transactions took place in a specific interval, etc. Finally, the transaction is cleared, and settlements are made among the banks by using national and international circuits for interbank exchanges. Depicted in Figure 2.8 are these exchanges.

Systems for bank card payments on open networks of the Internet type must be integrated within this framework. The adaptation efforts attempt to take advantage of the storage capacity and the computation capabilities of the new generation of integrated circuits cards, called microprocessor or smart cards. These processing capabilities make the cards suitable for securing e-commerce in addition to other nonbanking telematics applications. The architecture of multiapplication cards will be presented in Chapter 12.

Table 2.9 provides the proportion (in volume) of scriptural payments made by bank cards in selected countries is provided. The data are depicted in graphical form in Figure 2.9.

The use of bank cards is increasing in all countries, except in Singapore, where it is decreasing. One possible reason is the rapid development of electronic purses that seem to have replaced bank cards as an instrument of payment, as will be seen later. The greatest bank card use is in Japan and Canada, while the lowest is in Germany. In most countries, the percentage in volume of scriptural payments made by bank cards is between 23 and 36%, making this instrument the second most popular. The tremendous increase in the volume of transactions by bank cards can be explained by several factors, such as plans for the diffusion of cards in the population as well as good geographic coverage by automatic teller machines (ATMs).

Table 2.10 gives the proportion of value exchanged by bank cards in the same countries from 1991 to 2000 is given. The data are depicted in graphical form in Figure 2.10. These results confirm that this instrument is actually a mass instrument to be used for small amounts (less than 1% of the total

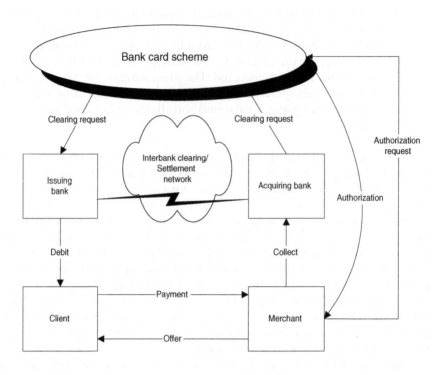

FIGURE 2.8
Message exchanges during a bank card transaction.

TABLE 2.9

Percentage in Volume of Payments by Bank Cards in Selected Countries (1991–2000)

	1991	1992	1993	1994	1995	1996	1997	1998	1999	2000
Belgium	13.3	15.6	16.5	18	19.7	21.4	23.4	27.4	28.9	32.8
Canada	27.8	28.9	31.1	28.1	33.3	38.8	44.5	48.4	51.6	54.7
France	14.5	15	15.7	16.2	17.6	20.6	22.2	18.2	26.1	27.8
Germany	1.8	2.1	2.6	3.1	3.6	4.2	4.1	7.3	8.2	9.6
Italy	3.1	3.7	4.1	5.2	6.6	9.5	12.6	15.8	19.6	23.3
Japan	—	—	—	—	—	45.1	46.9	48.8	53	53.3
The Netherlands	1.8	2.6	4.2	6.1	13.4	19.5	23.2	25.7	28.2	29.9
Singapore	—	—	—	—	—	32.6	34.3	32.8	27.4	27.7
Sweden	8.8	8.2	9.8	11.6	14.2	15.3	18.6	23	24	28.3
Switzerland	9.7	11.8	13.8	16.2	18.4	20.7	22.6	22.9	26	28.2
U.K.	16.4	18.8	21	23.3	24.1	26.9	29.3	31.5	34.3	36.6
U.S.	16	16.8	17.5	18.6	20	21.4	22.9	24.6	26.6	35.3

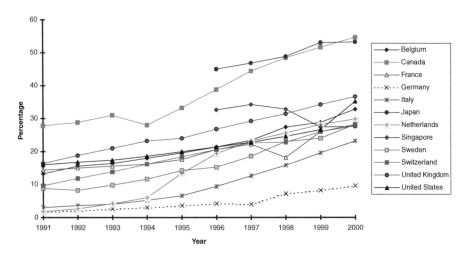

FIGURE 2.9
Percentage in volume of scriptural payments by bank cards in selected countries.

TABLE 2.10

Percentage in Value of Payments by Bank Cards in Selected Countries (1991–2000)

	1991	1992	1993	1994	1995	1996	1997	1998	1999	2000
Belgium	0.1	0.2	0.1	0.1	0.1	0.1	0.2	0.2	0.1	0.2
Canada	0.3	0.3	0.3	0.3	0.4	0.7	0.7	0.8	0.5	0.6
France	0.2	0.2	0.2	0.2	0.2	0.2	0.2	0.1	0.2	0.2
Germany	—a	—a	—a	—a	—a	0.2	0.2	0.3	0.3	0.4
Italy	—	—	—	—	0.1	0.0	0.0	0.0	0.1	0.1
Japan	—	—	—	—	—	—	—	—	—	0.1
The Netherlands	—	—	—	0.1	0.1	1.0	1.1	1.2	1.4	1.4
Singapore	—	—	—	—	—	0.2	0.1	0.2	0.2	0.2
Sweden	0.7	0.7	0.9	1.0	1.4	1.4	1.6	1.9	2.3	2.6
Switzerland	—a	—a	—a	—a	—a	0.1	0.0	0.0	0.1	0.1
U.K.	0.2	0.2	0.2	0.2	0.2	0.3	0.3	0.3	0.2	0.2
U.S.	0.1	0.1	0.1	0.1	0.2	0.2	0.2	0.2	0.2	0.3

aLess than 0.1%.

value exchanged by all scriptural instruments). Sweden and the Netherlands are the only countries where the bank card's share exceeded this threshold. This may explain why the electronic purse has had difficulties starting in Switzerland (*Le Matin*, 1998), because it was competing for the same niche as the bank card.

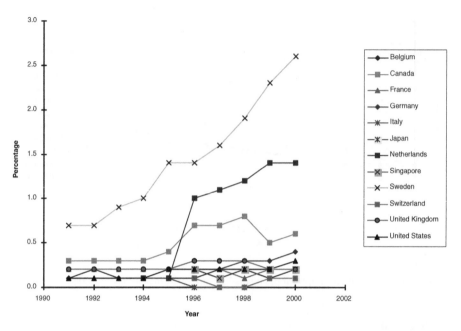

FIGURE 2.10
Percentage in value of scriptural payments by bank cards in selected countries.

2.3   Types of Dematerialized Monies

Several forms of dematerialized currencies appeared in the 1980s, with the increased use of prepaid cards, such as telephone cards, and the success of the Minitel in France (Martres and Sabatier, 1987, pp. 85–87). To clarify the presentation, three types of "emerging" monies, all of which are dematerialized, will be distinguished: electronic money, virtual money, and digital money.

2.3.1   Electronic Money

According to the Bank for International Settlements, "electronic money can be defined as a monetary value measured in fiduciary units that is stored in an electronic device owned or available to the consumer" (Bank for International Settlements, 1996, p. 13). It is thus a movable scriptural means of payment that carries the values in units of payment in an electronic store. This definition corresponds to a binary form of scriptural money, stored on portable support, such as a smart card. The scriptural character of the electronic money is related to the status of the issuer (because it is not issued by the central bank) and to the traceability of the transactions and the movement of money.

The units of payment contained in the cards or in the software are bought either with fiduciary money or by charging to a bank account. The discharging power of these units is restricted to those merchants who accept them. This is the reason certain experts consider that electronic money does not exist in a strict sense, because it is neither legal tender nor does it have discharging power (Fay, 1997, p. 113).

2.3.2 Virtual Money

Virtual money differs from electronic money in that its support, its representation, and its mode of payment do not take tangible forms. Virtual money can be contained in software programs that allow payments to be carried out on open networks, such as the Internet.

Starting with the definition of the BIS for electronic money, one can consider virtual money as a referent (or a pointer) to a bank account. The scriptural character of the virtual money is also tied to the status of the issuer (it is not issued by the central bank) and to the traceability of the transactions.

In the limiting case, virtual money may also be a virtual token (or *jeton*) issued by a trusted issuer for unique usage in a closed circuit.[1] These jetons are different from the electronic versions of legal tenders because they are recognized only in a restricted commercial circuit. This contrasts with electronic money, which is a multipurpose payment mechanism recognized in general commercial circuits. Millicent, for example, is one system that proposes a method for micropayments with a virtual jeton, the "scrip." A service provider issues a scrip, which does not have any direct relationship with the banking system but is a promise for future service. By generalizing this notion, service providers can issue their jetons and tie them with banking accounts that they maintain. They will remain within the perimeter defined by the law as long as these units are ascribed to a specific purchase within a well-defined circuit.

Telephone cards are a particular case of virtual purses issued by telephone companies. These cards are prepaid, and the values they store are dedicated to the settlement of telephone communications at a given service provider. The purchasing power is described in terms of "telephone jetons" that correspond to the impulse counts in the telephone networks. The experiences of the telephone card and of the Minitel kiosk show that, when the amounts under consideration are individually marginal, a telephone operator, although a "nonbank," can effectively help in collecting amounts that are

---

[1] The dictionary (*Webster's New Collegiate Dictionary*, 1975) definition of token is "a piece resembling a coin issued as money by some person or a body other than a de jure government." Despite this clear statement, there has been a tendency to mix legal coins with tokens (see, for example, Camp et al., 1995). To avoid the potential confusion, this book will use the French word *jeton* to mean a coin issued by a nongovernmental body.

individually marginal by attaching them to the telephone bill. By extending this role, the telephone unit could play the role of virtual money between the supplier and the purchaser in the case of micropayments. This advantage may even be extrapolated to the case where the two parties are not located in the same country. This is because telephone companies, over the years, have developed the ability to handle small payments in an efficient manner and have defined efficient procedures for settling accounts among themselves, even across currencies. In fact, the use of phone-ticks for micropayments was considered within the European project CAFE (Conditional Access for Europe), which ran from 1992 to 1996 (Pedersen, 1997). The value of the telephone unit is relatively more neutral than the legal tenders and is regulated by agreements within the ITU (International Telecommunication Union). It fluctuates less than currency and could thus be the standard of measure for micropayments on the international scene. Another proposal is to consider the transmission capacity (i.e., the available bandwidth) as the support for the electronic money.

Interbanking networks are strictly regulated and monitored by the monetary authorities in each, given that only the central banks have the monopoly to print money. The dispensation given to telephone cards was justified on the basis that telephone tokens represent future service consumption, paid with the legal money. Furthermore, it is difficult for banks to propose an economic alternative to the billing and collection of amounts that are individually marginal.

The example of the telephone card, whether discardable or rechargeable, could encourage telephone companies to aspire to an intermediary role in e-commerce, especially for micropayments. However, this ambition requires passage from the "virtual purse" mode to the "electronic purse" mode. In other words, the value stored in the telephone card (i.e., the billing impulses) must be recognized as new scriptural and universal monies, expressed in binary form. This poses the problem of how the financial authorities can regulate this new money supply, which must be resolved before that bridge can be crossed.

2.3.3 Digital Money

Digital money is an ambitious solution to the problem of online payment that will be further described in Chapter 11. Like regular money, each piece has a unique serial number. However, the support for this money is "virtual," the value being stored in the form of algorithms in the memory of the user's computer, on a hard disk, or in a smart card.

As will be shown later, one of the most salient characteristics of the digital money of DigiCash is that it is minted by the client but sealed by the bank. The creditor that receives the digital money in exchange of a product or a service verifies the authenticity using the public key of the issuer bank. Anonymity is thus guaranteed, but it is not easy to transfer the value among two individuals without the intervention of the bank of the issuer. Furthermore, as each algorithmic step is associated with a fixed value, the problem of change causes some complications.

As a new step in the dematerialization of money, the digital unit of money will be a monetary sign, with a real discharging power that the economic agents in as large an area as possible would be able to accept in return for payments. The exchange of value takes place in real time via the network using coded digital coins, but the clearance and settlement may be in real time or in nonreal time. The digital money can be exchanged with physical money at banking institutions after verification with an authentication database. This database can be centralized or distributed.

One of the characteristics of digital money compared with other electronic payment systems is the possibility of making the transactions completely anonymous, i.e., of dissociating the instrument of payment from the identity of the holder, just as in the case of fiduciary money.

One of the destabilizing aspects of this digital money is that it could lead to formation of new universal monies independent of the current monetary system. This is the reason attempts at creating digital money have encountered technical and legal difficulties. A digital currency that is international would collide with the various regional and local currencies and would disturb the existing economies. The question is no longer exclusively technological, as it touches upon aspects of national sovereignty and foreign intervention. The economic and political stakes of such a proposition are enormous and may lurk behind the screen of juridical disputes.

2.4 Purses and Holders

2.4.1 Electronic Purses and Electronic Token (Jeton) Holders

According to the BIS, an *electronic purse* is a "a reloadable multipurpose prepaid card that can be used for retail or other face-to-face payments." This means of payment can substitute, if the holder wishes, for other forms of monies. Electronic purses are thus portable electronic registers for the funds that the cardholder possesses physically. These registers contain a precharged value that can serve as an instrument of exchange in open monetary circuits. The protection afforded the stored value of money is based on the difficulty (if not the impossibility) of fabricating a fake card or manipulating the registers. Here the notion of "open networks" describes the final utilization of the means of payment to make purchases without any a priori restrictions and independent of the issuer. This notion of openness is different from that in telecommunications networks, where a network can be "open" or "closed" depending on whether the access and transmission protocols are standardized or proprietary.

Where an electronic purse is used depends on the identity of the issuer (merchant, bank, merchants association, etc.) and its prerogatives under the law. Banking networks are, by definition, open wherever the electronic money corresponds to a legal currency. In contrast, a purse that is issued by a nonbank is restricted because it can only contain jetons and can only be used in closed circles and for predefined transactions involving the issuer.

Jeton holders are analogous to private means of payments, such as restaurant or manufacturer's coupons. The jeton holder that is mostly used is found in the form of telephone cards, where the units of payment give the right to establish prepaid telephone connections.

Electronic purses are attractive to banks because they permit a reduction in the transaction cost and can replace coins, notes, or checks for small amounts. They can be considered a cybernetic form of the traveler checks that were first introduced by American Express in 1890.

Electronic purses and electronic jeton holders have already proved their economic effectiveness in face-to-face commerce and in payment through automatic machines. They have an advantage over traditional payment cards, which are not suited to micropayments, and even to face-to-face commerce, because the transaction cost may exceed the amounts involved. It is possible, however, to combine electronic purses and jeton holders in a multiapplication card. A merchant may be associated with a bank to issue a fidelity card while offering credit facilities (as managed by the bank). Table 2.11 summarizes the financial and legal differences between electronic purses and electronic jeton holders.

2.4.2   Virtual Purses and Virtual Jeton Holders

A *virtual purse* is an account precharged with units of legal money and stored in the collection system of a nonbank (for example, a virtual mall) (Remery, 1989; Bresse et al., 1997, p. 26; Sabatier, 1997, p. 94). Online access to this virtual purse is achieved with software installed in the personal computer of the client to effect online micropayments.

The system functions as follows. Operators open in their banks and under their own accounts several subaccounts. These subaccounts are then allocated to subscribers of their systems, whether buyers or merchants. The client's subaccount is called a virtual purse, while the merchant subaccount is denoted as the virtual cash register. The purse is called "virtual" because the value stored is not physically touchable, yet the units of payment correspond to legal tenders.

The client's purchasing power is indicated in the virtual purse and refers to the subaccount under the operator's account. What clients have on the hard disk of their personal computer is a copy of the balance of this subaccount. In addition, the hard disk may contain various files that are needed for the cryptographic security of the operation. This approach has an additional advantage in that it protects the clients' assets, even when their computers fail.

TABLE 2.11

Comparison between Electronic Purses and Electronic Jeton Holders

Characteristic	Electronic Purse	Electronic Jeton Holder
Expression of purchasing power	Legal tender	Consumption unit
Unit of payment	Universal: can settle any payment in a defined territory	Specific to transactions involving the issuer
Guarantor of purchasing power	Bank	Service provider
Charging of value	By a bank or its agent	Unregulated
Circuit of financial services	Open	Closed

Each purchase debits the client's virtual purse and credits the merchant's virtual purse with the amount of the transaction minus the operator's commission. At predefined intervals, the operator makes an overall payment to each merchant, corresponding to the amounts that have accumulated in their respective virtual cash registers. The grouping of payments before initiating the compensation makes this approach economical for micropayments.

In principle, virtual jeton holders could help settle informatics purchases with micropayments, in particular, information or other virtual products sold over the Internet. The purchasing power would be expressed in units of promises for service or for consumption at specific vendors. This value represented in jetons would be stored in memory and would have a limited scope of application. It would only be used in transactions with suppliers that the operator of the payment system registered. However, because the interest of the operator is to attract the participation of the largest number of merchants, an aggressive recruitment may put the operator in an ambiguous position with respect to credit institutions, which are the only institutions legally allowed to operate in the general sphere. From the examples currently proposed, Millicent and Payword will be discussed.

2.4.3 Diffusion of Electronic Purses

Table 2.12 gives the portion of the volume of scriptural payments performed using electronic purses in several countries. These data are depicted in the graph of Figure 2.11. Singapore is distinct from all other countries as being the first where payments from electronic purses form an important part of the volume of transactions (about one third). Belgium is in a distant second place where the proportion of transactions using electronic purses does not even reach 4% of the total volume.

The growth of the electronic purse in Singapore from 1996 is the fruit of a planned effort to replace coins with a contactless electronic purse, Cash-Card, which was introduced for small amounts (tolls, parking fees, etc.). Furthermore, starting from 2008, Singapore will accord to electronic purses

TABLE 2.12

Percentage of Transactions with Electronic Purses in
the Volume of Scriptural Payments in Selected
Countries (1996–2000)

	1996	1997	1998	1999	2000
Belgium	0.13	0.79	2.25	3.26	3.64
Germany	0.00	0.03	0.10	0.14	0.18
The Netherlands	—	—	0.68	0.82	0.87
Singapore	0.02	0.35	11.77	27.82	32.38
Sweden	—	—	0.22	0.40	0.26
Switzerland	—	—	0.53	1.25	2.08

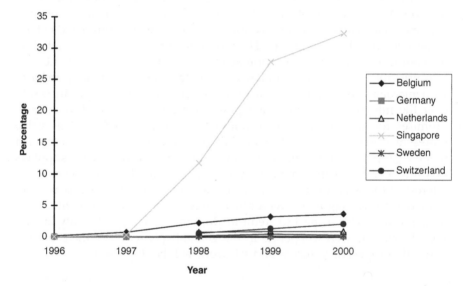

FIGURE 2.11
Percentage of payments by electronic purses in the volume of scriptural payments in selected
countries.

the same legal status as cash. The numbers in Table 2.13 confirm that, as
expected, the values settled with electronic purses form a tiny proportion of
the total values exchanged. However, for an equal value, the volumes of
transactions vary considerably. Thus, the value exchanged with electronic
purses in Sweden exceeds that of Singapore, with a more modest volume of
transactions.

TABLE 2.13

Pecentage of Contribution of Electronic Purses to the Exchange of Values by Scriptural Payments in Several Countries (1998–2000)

	1998	1999	2000
Belgium	0.0009	0.0013	0.0012
Germany	0.0004	0.0004	0.0003
The Netherlands	0.0039	0.0037	0.0031
Singapore	0.0000	0.0010	0.0020
Sweden	0.0025	0.0040	0.0025
Switzerland	0	0	0.0002

2.5 Transactional Properties of Dematerialized Currencies

From an information technology viewpoint, computer monetary transactions must satisfy certain conditions that can be expressed in terms of the following properties (Camp et al., 1995):

- *Atomicity*: This is an all-or-none property. A transaction has to occur completely for its consequences to take place. Otherwise, the state anterior to the transaction must be restored.
- *Consistency*: All parties must agree on the critical fact of the exchange.
- *Isolation*: The absence of interference among transactions so that the final result of a set of transactions that may overlap will be equivalent to the result when the transactions are executed in a nonconcurrent serial order.
- *Durability*: This is the property where the system returns to its previous state following a breakdown during operation.

From an end-user viewpoint, the reliability of the system as a whole depends on the atomicity of the transactions; i.e., a transaction must occur in its entirety or not at all. No buyer should be forced to pay for an interrupted transaction. Atomicity is the property of payments made by cash, by checks, by credit transfers, or by debit cards. In contrast, transactions by credit cards or by deferred credit are not always atomic if the client can revoke the transaction during the time interval between the end of the transaction and the instant at which the amount is debited to the client's account. Although cash payments are isolated, check transactions do not have this characteristic, because an overdraft may occur depending on the order of successive deposits and withdrawals.

2.5.1   Anonymity

Anonymity means that the identity of the buyer is not explicitly utilized to settle the obligations. Personalization, in contrast, establishes a direct or indirect relationship between the debtor and the means of payment. Cash in the form of notes and metallic coins is anonymous because it has no links to the nominal identities of holders and their banking references. In the case of remote financial transactions, anonymity raises two questions: the ability to communicate anonymously and the ability to make anonymous payments. Clearly, an anonymous communication is a necessary condition for anonymous payments, because once the source of a call is identified, the most sophisticated strategies for masking the exchanges would not be able to hide the identity of the caller.

For bank cards and electronic or virtual purses and holders, there are four types of anonymity (Sabatier, 1997, pp. 52–61, 99):

- The plastic support is anonymous if it does not contain any identifier that can establish a link with the holder. This is the case with telephone cards. On the other side, the support of a bank card is not anonymous because it carries the card number as well as the cardholder's name and account.

- Recharging an electronic purse with value is an anonymous transaction if it does not establish a link with the identity of the holder, for example, charging a smart card with the aid of cash. The transaction loses its anonymity temporarily if it is protected by a personal identification number (PIN), because the identity is taken into consideration. However, anonymity can be restored if the transaction is not archived.

- A transaction is partially anonymous if the information collected during its progress does not establish a link with the holder's bank account. One such example is when payment transactions are grouped by accumulating the total sum of the transactions within a given period. In this case, however, it is possible to discover the identity of the cardholder, because the grouped transactions must be tied with a bank account for clearance and settlement.

- Anonymity for face-to-face transactions is different from anonymity for remote transactions. In face-to-face commerce, the utilization of a smart card with offline verification can protect the identity of the holder and the subject of the transaction. This is because the algorithms for authentication and identification, which are stored within the memory of the card, will operate independently of any management center. The case of remote commercial transactions, whose smooth operation requires that both parties identify themselves without ambiguity to prevent any future contest of the authenticity of the exchange is different. In this case, complete anonymity is incompatible with nonrepudiation. The maximum that can be achieved is partial anonymity; for example, merchants would not have access to the references of the holder, and this information would be collected and stored by a trusted third party that will be an arbiter if a dispute arises.

2.5.2 Traceability

Scriptural money is tied to the status of the issuer and the user, which allows for the monitoring of a transaction in all its steps; it is thus personalized and traceable. Nontraceability means that the buyer would not only be anonymous, but also that two payments made by the same person could not be linked to each other, no matter what (Sabatier, 1997, p. 99). In smart cards, for example, a "protected zone" preserves an audit trail of the various operations executed. However, by ensuring total confidentiality of the exchanges with the help of a powerful cryptographic algorithm, third parties external to the system would not be able to trace the payments or link two different payments made with the same card.

Any guarantee for merchandise delivery as well as ambitions to arbitrate disputes run counter to nontraceability of transactions. The question of proof quickly becomes complicated, because the laws on "guarantees" and "confidentiality" vary widely among countries.

Table 2.14 compares the different means of payments on the bases of the previous properties.

2.6   Overall Comparison of the Means of Payment

The multiplicity of instruments for payment suggests that they are not all adapted to the same types of applications. As a consequence, the success of emerging payment instruments will depend on socioeconomic factors of a given society.

Among the classical means, the choice for face-to-face commerce is limited to cash, checks, and bank cards. The choice is much larger for remote payments, which indicates that the requirements differ according to applications, and that there is not a uniformly optimal solution. Three means are more suitable for remote payments in business applications: credit transfers, direct debit, and, when available, various types of interbank exchanges.

While the main strength of cash is in the area of retail commerce, it is not suitable for remote payments or for business-to-business payments. The check is the only means of payment that is adapted to most cases, which explains its resistance to electronic innovations in many countries. However, the cost of transactions by checks or by bank cards does not make them suitable for micropayments. Stored-value systems, such as electronic or vir-

TABLE 2.14

Transactional Properties of Different Methods of Payment

	Atomicity	Consistency	Isolation	Durability	Anonymity	Traceability
Cash	Yes	Yes	Yes	Yes	Yes	No
Checks	Yes	Yes	No	Yes	No	Yes
Credit transfer	Yes	Yes	Yes	Yes	No	Yes
Direct debit	Yes	Yes	Yes	Yes	No	Yes
Debit card	Yes	Yes	Yes	Yes	No	Yes
Credit card	No	Yes	Yes	Yes	No	Yes
Electronic purse	Yes	Yes	Yes	Yes	Maybe	Maybe
Virtual purse	Yes	Yes	Yes	Yes	Maybe	Maybe

tual purses, may be able to displace cash and checks in this area because they can satisfy, more or less, the same need, while offering the possibility of making small payments in an economic manner.

It is worth noting that checks are often used to obtain cash, and that cash can be used to feed a checking account. Currently, not all proposed electronic purses retain the bidirectionality property. In fact, the electronic purse can be charged with cash, from a checking account, or even through a bank card; however, the money is not discharged in one of these forms. Figure 2.12 represents the circuit of monetary flow for a unidirectional electronic purse

Summarized in Table 2.15 is the previous discussion on the domain of utilization of the various means of payment.

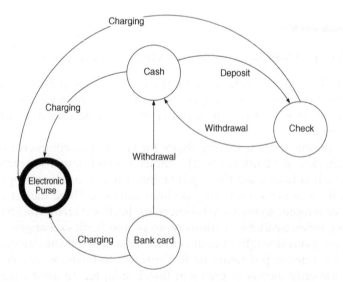

FIGURE 2.12

Monetary flow among different means of payments for a unidirectional electronic purse.

TABLE 2.15

Domains of Utilization of Means of Payment

Means of Payment	Face-to-Face Payment	Remote Payment	Business-to-Business Payment
Cash	Yes	—	—
Check	Yes	Yes	Yes
Credit transfer	—	Yes	Yes
Direct debit	—	Yes	—
Interbank transfer	—	Yes	—
Bank card	Yes, with a reader	—	Yes
Electronic or virtual purse	Yes, with a reader	Possible	Possible

A worldwide solution for e-commerce will have to integrate easily and without distinction the various operational systems of payment.

Summarized in Table 2.16 are the various properties of money in terms of six criteria:.

- The nature of money
- The support of money (the container)
- The location of the value store
- The representation of the value
- The mode of payment
- The means or instruments of payment

2.7 The Practice of Dematerialized Money

2.7.1 Protocols of Systems of Dematerialized Money

Depicted in the block diagram of Figure 2.13 are the financial and control flows among participants in a system of dematerialized money (Sabatier, 1997, pp. 46–47):

- *Relation 1* defines the interface between the client (the purse holder) and the operator responsible for charging the purse with electronic monetary values. This operator verifies the financial solvency of the holder or the validity of the payment that the holder makes with the classical instruments of payment. After verification, the operator updates the value stored in the electronic or virtual purse.
- *Relation 2* controls the junction between the charging operator and the issuing bank, if the operator is a nonbank.

TABLE 2.16
Properties of Money

Type of Money	Nature of Money	Support (the Container)	Value Store	Value Representation	Mode of Payment	Means of Payment (Instrument)
Fiduciary	Concrete, material	Paper, piece of metal	Safe, wallet, purse	Bank notes, coins	Face-to-face transaction	Bank notes, coins
Scriptural	Immaterial (an account maintained by a credit institution)	Magnetic, optical, electronic	Account maintained by a credit institution	Numerical value	Remote, face-to-face (retail automatic machines)	Check, debit card, credit card, credit transfer
		Integrated circuit card	Electronic purse			
		Computer	Virtual purse (memory allocated by an intermediary)			Electronic fund transfer

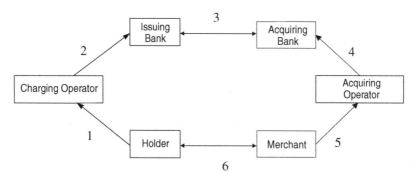

FIGURE 2.13
The flow in a transaction by dematerialized money.

- *Relation 3* relates to the interbanking relations between the issuing bank and the acquiring bank (the bank of the merchant) and depends on the regulations at hand.

- *Relation 4* defines the interface of the acquiring operator and the acquiring bank to acquire the credits owed to the merchant. These two entities are generally the same.

- *Relation 5* describes the procedures for collection and compensation to credit the merchant's account with the values corresponding to the electronic values exchanged.

- *Relation 6* represents the purchase transaction and the transfer of electronic value from the client to the merchant, simultaneously.

The *charging protocol* of a system of dematerialized money specifies the procedures for requesting authorization and transfer of electronic value toward the holder's purse in exchange for a payment acceptable to the charging operator (for example, cash, bank card, checks, or even another electronic purse). The protocol relates to Relations 1 and 2 when the charging operator is a nonbank, otherwise to Relation 1 only. In this latter case, Relation 2 falls within the realm of interbank relationships. Feeding an electronic or virtual purse is considered a collection of funds from the public, which in most countries is a banking monopoly. Only a credit institution is allowed to credit a purse with units that can be utilized for the purchase of products that were not previously defined. With the help of a system for point-of-sale activation and recharge (POSA/R of cards), the reloading of value can be done from points connected to the banking networks.

Relation 6 includes two distinct protocols: a *purchasing protocol* during the negotiation of the price and the purchase conditions and a *payment protocol*. The payment can be made directly to the merchant or through an intermediary. The corresponding architectures will be discussed below. In general, the means used for the security of payments do not extend to the purchase protocol, even though the simple fact of knowing that a communication between the partners is taking place can be interesting information. An attempt to protect the negotiation that precedes the payment was considered within the JEPI (Joint Electronic Payment Initiative) presented in Chapter 14.

The interrogation of the authorization server can be the responsibility of the merchant or supplier who directly queries the financial circuits. However, an intermediary can relieve merchants of this job and collect, in their stead, the necessary authorization, in return for a negotiated fee.

In systems where the verification is online, interrogation of the authorization server is systematic for each purchase, irrespective of the amounts. These systems are predominant in the U.S. for credit cards, because the cost of telephone communication is negligible. Online verification was retained by Visa and MasterCard in the SET (Secure Electronic Transaction) protocol for remote payments by bank cards on the Internet.

Systems with semionline verification interrogate the authorization server only for certain situations, for example, when the amount of the transaction exceeds a critical threshold or when the transaction takes place with merchants who are more exposed to risk because of the nature of their activity (such as gas stations, etc.). An automatic connection is set up periodically to transmit the details of the transactions and to update the security parameters (blacklisted cards, authorization thresholds, etc.). The French system for bank cards is semionline.

Finally, the whole verification is done locally, in the case of offline architectures based on secure payment modules incorporated in the merchant cash registers. Remote collection and update of the security parameters take place once every 24 hours, usually at night.

The terminals used for electronic payment in semionline or offline payment systems are computationally more powerful than those for the online systems. Intelligent terminals have the following responsibilities: (1) reading and validating the parameters of the means of payment; (2) authenticating the holders; (3) controlling the ceiling expenditures allowed to the holder (calculating the proof of payment, generating the sales ticket, and recording the acceptance parameters); and (4) periodically exchanging data and files with the collection and authorization centers. These terminals must therefore be equipped with an adequate Security Application Module (SAM) to perform the operations of authentication and verification according to the protocols of the payment system used.

The security of online systems is theoretically higher, because they allow for continuous monitoring of the operating conditions and real-time evaluation of the risks. This assumes that the telecommunications network is reliable and is available at all times. The choice of a semionline system can be justified if the cost of connection to the telecommunications network is important or if the cost of the computational load is too high for the amounts involved.

The protocols used must be able to resist attacks from outside the system as well as from any misappropriation by one of the participants (Zaba, 1996). Thus, a third party that is not a participant must not be able to intercept the messages, to manipulate the content, to modify the order of the exchanges, or to resend valid but old messages (this type of attack is called the man-in-the-middle attack). Similarly, the protocols must resist false charges, for example:

- Attributing the recharge to a different purse than the one identified
- Debiting a purse by a false server
- Attributing a different amount than the amount requested
- Replaying a previously authenticated charge
- Repudiating a charge that was correctly executed or revocating a payment that was made

In general, the protocols must be sufficiently robust to return to the previous state following a transmission error, particularly if the recharging is done through the Internet.

Finally, the protocol for collection, acquisition, clearance, and settlement, which Relations 4 and 5 describe, varies depending on whether the acquiring operator is a bank. The purpose is to collect in a secure manner the electronic values stored in the merchants' terminals, to group these values according to the identity of each acquiring bank, and to inform the respective bank of the acquired amount. In the case where the acquiring operator is a bank, which is the most common situation envisaged, Relation 4 falls within the domain of the interbanking relations defined by the law.

It should be noted that the functioning of the system must include other protocols that are not represented in Figure 2.13:

- An initialization protocol to allow the purse holder to subscribe to an account at the operator of the system of e-commerce
- A peer-to-peer transfer protocol to allow the transfer of the electronic monetary value from one purse to another, among holders equipped with compatible readers, and without the intervention of a third party
- A discharging protocol to control the inverse transfer of the electronic money in the purse to a bank account
- A shopping protocol, which is not treated in this book

Some systems of dematerialized money seem to be able to accept peer-to-peer transfers and discharging operations. For example, the suppliers of the electronic purse Mondex indicate that the transfer of value among two purses is possible, just as the exchange of currency notes is possible from one person to another. However, because the technical specifications of Mondex are still proprietary, it is not possible to give more details on this mode of operation.

2.7.2   Direct Payments to the Merchant

In systems where the payment is directly given to the merchant, clients transmit the coordinates of their accounts to the merchants. In a classical configuration, the merchant may use one of the well-tested mechanisms, such as direct debit or credit transfers.

To make a payment from the client computer using a purse or a bank card, a payment gateway must intervene to guarantee the isolation of the banking network from the Internet traffic. It is the gateway that will receive the client's request before contacting the authentication and authorization servers, to make the function completely transparent to the banking circuits. In this manner, the gateway operator is called upon to become a trusted third party and a notary.

The gateway operator cannot assume the role of charging operator unless it is certified by a credit establishment. In this case, the gateway can take on a supplementary role as a change agent and can accept payments in the currency of the client and pay the merchants the amount that is due in the currency of choice. An example of such an operation is KLEline, which will be discussed in Chapter 10.

The location of the payment gateway with the payment architecture is illustrated in Figure 2.14. Although the diagram shows access to the authorization server through the Internet, an alternative configuration is to have the server connected directly to the secure financial network.

The proliferation of projects for electronic purses throughout the world has led to incompatible products. The Electronic Commerce Modeling Language (ECML) is one step toward a unique payment interface. This language, described in the IETF RFC 3106 (2001), defines the exchanges between applications and the merchant sites. A software piece called *digital wallet* manages the various fields of an online order (buyer's name, address, banking coordinates, delivery address, etc.), thereby ensuring that the online forms can be automatically filled using data stored once in the buyer's computer.

The disadvantage of direct payments is that the cardholder and the merchant will have to agree on all the details of the protocol beforehand, which impedes open or spontaneous exchanges. The merchant site will have to be able to manage all payment schemes that could potentially be used. Finally, the buyer would have to own a purse for each currency that may be used, which, due to cost of inconvenience, may hinder acceptance of the scheme.

Payments mediated by intermediaries can overcome some of these drawbacks.

2.7.3   Payment via an Intermediary

Figure 2.15 shows the position of a payment intermediary in the circuits of e-commerce. The function of the intermediary is to hide from the participants the differences among the various purse schemes. This allows participants to avoid the hassle of having specific software for the various systems of payment.

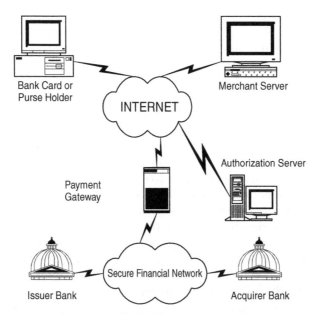

FIGURE 2.14
Position of the payment gateway in e-commerce.

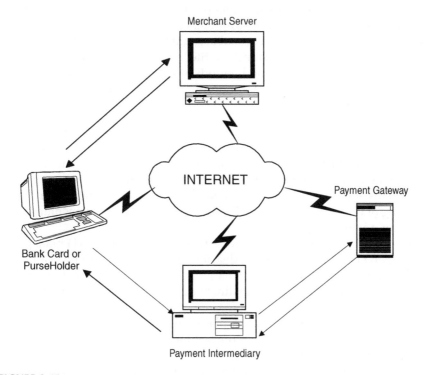

FIGURE 2.15
Position of the payment intermediary in e-commerce.

The function depends on prior subscription by the clients to the intermediary to give the proxy to make the payment. There are two possibilities, depending on whether the payment will be done by bank card or electronic purse in one case, or by a virtual purse in the other:

- For payments by bank card or by electronic purse, the intermediary usually will know the client's payment coordinates because they were previously sent through a secure channel. The intermediary uses this information to instruct the banking authorities to debit the buyer's account for the purchases made and to credit the suppliers with the amounts due to them. To establish a connection, the holder utilizes an identifier (that could be encrypted with a secret key) as an indication to the intermediary. Settlements can be made for each transaction individually or, in the case of small payments, by a periodic global invoice grouping the individual amounts. In the particular case of the Minitel kiosk, the identifier is the telephone line number, and the Internet is replaced by a combination of the PSTN and the X.25 network, Transpac.

- For payments with a virtual purse, as was previously explained, the intermediary opens in its own bank subaccounts for the various users and merchants that subscribe to the intermediation service. Users prepay their subaccounts by direct credit or by a bank card or any other established instrument. Following each transaction, the intermediary debits the user subaccount to the benefit of the "virtual cash register" (subaccount) of the merchant. The intermediary groups the transactions and periodically sends requests to the banking network to settle the accounts after withdrawing commission on the turnover.

The same operator can add to the function of the intermediary other roles, such as management of a virtual mall, billing and collection for the suppliers, management of the payment instruments for the merchant, or management of the cross-borders commerce (exchange rates, import and export taxes, shipping of physical goods, etc.). These roles are often complementary, especially for a worldwide operation.

This trilateral architecture calls for a trusted third party to (1) manage the encryption keys, their generation, distribution, archiving, and revocation; (2) manage the subscriptions of merchants and clients, their certification and authentication; and (3) update the directories and the blacklists or revocation lists.

The electronic notary can put in place a nonrepudiation service to time-stamp the exchanges, archive the transactions, etc. Depending on the legislation, the intermediary may also act as a small-claims judge to settle differences between the merchant and the client on faulty deliveries, defective or nondelivered items, incorrect deciphering keys, etc.

If telephone companies act as the intermediaries, the telephone unit could play the role of virtual money between the supplier and the purchaser in the case of micropayments. This advantage may even be extrapolated to the case where the two parties are not located in the same country.

Other administrative functions may be added as well. For example, the DGI (Direction Générale des Impôts — General Taxation Directorate), the French Internal Revenue Service, certifies some service providers as "relaying organizations" that can transmit supporting fiscal data according to the TDFC (Transfer de Données Fiscales et Comptables — Transfer of Fiscal and Accounting Data) procedures available since 1992 (Granet, 1997). Each intermediary identifies and certifies its clients and gives them the possibility of signing electronic documents. The digest of the document and the symmetric key are encrypted with the public key of the DGI using the RSA algorithm. With the generalization of EDIFACT (Electronic Data Interchange for Administration, Commerce and Transport) starting from 2000, only EDI partners that signed an agreement with the fiscal authorities can send fiscal declarations or represent the taxpayers. It should be noted, however, that adding all of these functions to intermediaries could augment the computational charge that they have to support, particularly if encryption is extensively used.

2.8   Banking Clearance and Settlement

"Clearance and settlement" among financial institutions was alluded to earlier, and it seems useful to present the main outlines to facilitate understanding of the way scriptural payment systems function. Historically, clearance and settlement took place when all bank representatives would meet every working day in a special house to compare their respective credits in the various financial instruments and then settle their accounts by exchanging money. Today, that system has been replaced by a computer network. However, the unique evolution of the financial circuits of each country, the differences in the notions of security, and the diversity of formats used means that several different models exist in Europe. Similarly, the clearance and settlement systems in the U.S. are different from those in Europe.

From a technical viewpoint, the European architectures vary: regional and national systems in France, Italy, and Spain; competing and incompatible bilateral and multilateral systems in Germany; and centralized systems in Belgium, Portugal, and the U.K.

Classification of the settlement networks can be based on several criteria, for example:

- The nature of the processing:
  - Large-value systems
  - Mass systems that process many daily transactions of relatively small values
- The ownership and management of the network:
  - Public network owned by the central bank
  - Private network owned by members of a group of banks
  - Private network leased to the banks on a use basis
- The way the settlement is done:
  - Real-time gross settlement occurs the same day and in real time
  - Netting involves the consolidation of various transactions
  - Grouping is used when the transfer occurs among different entities of the same group of companies to avoid paying settlement charges repeatedly

The following sections contain additional clarifications on the clearance and settlement systems in the U.S., the U.K., and France.

2.8.1   United States

There are two large-value settlement systems in the U.S., Fedwire and CHIPS (Clearing House Interbank Payment System). Fedwire is the network of the Federal Reserve and is for real-time settlements. It is used for a few interbank transactions that exchange large values. In contrast, CHIPS is a private system managed by the New York Clearing House (NYCH), and it first consolidates the operations of its member institutions before starting a settlement action.

For large-scale settlements, a private system, under the surveillance of the Federal Reserve, coexists with the federal system. The private system comprises 32 regional clearinghouses under the administration of the NACHA (National Automated Clearing House Association) located in Washington, D.C. The Automated Clearing House (ACH), which was developed by NACHA as a private institution, is involved in EFT.

Table 2.17 summarizes the contributions from each of these settlement systems in 1995 and 2000 (Bank for International Settlements, 2001, 2002). This table also includes transactions among customers of the same bank ("on us").

These entries reveal the downwards tendency of the amounts of settlement in number of transaction and in value, as well as the increased relative weight of the compensations done through the Federal Reserve.

The information exchanged is coded using one of several formats: CCD (Cash Concentration and Disbursement); CTP (Corporate Trade Payments); and CTX (Corporate Trade Exchange) (Thierauf, 1990, pp. 170–172; Emmelhainz, 1993).

TABLE 2.17

Contribution of Various Settlement Systems in the U.S. in 1995 and 2000

	1995		2000	
Nature of the Contribution	Volume (in millions of transactions)	Value (in billions of U.S. $)	Volume (in millions of transactions)	Value (in billions of U.S. $)
Large-value systems				
CHIPS	51.0	310,021.2	59.8	191,147.1
Fedwire	75.9	222,954.1	108.3	379,756.4
Checks				
Private clearinghouses	28,145.0	—	17,413.0	—
Federal Reserve	16,128.0	12,083.0	17,486.0	14,161.9
"On-us" checks	18,690.0		14,705.0	
Automated clearinghouses				
Private (ACH)	249.7	1,095.2	613.8	2,417.3
Federal Reserve	2,645.0	8,934.8	4,650.5	14,0424.4
"On-us" ACH	595.0	2,201.5	1,674.7	4,966.8
Total	66,579.6		56,712.1	

CCD is the mandatory format that all institutional members of NACHA use, whereas the support of CTP and CTX is optional. CCD is used for transfer and direct debit and does not require that the information systems of the various institutions be interoperable. The check is presented using fields of 94 characters; among these, a field of up to 34 characters is reserved for annexes and notes. These addenda are not standardized, which makes it difficult to automate the processing of the messages.

In CTP and CTX, the messages are formed of units of 99 octets of length, which can be concatenated up to 4999 units. CTX was influenced by ANSI ASC (American National Standards Institute Accredited Standards Committee) X12 and allows variable-length fields. This format accommodates automated processing and is used for EFT.

2.8.2   United Kingdom

The systems for clearance and settlement in the U.K. depend on the payment instrument and the value exchanged. The Clearing House Automated Payment System (CHAPS) is for large-value transfers (credit and direct debit). The Town Clearing Company Ltd. was responsible for same-day settlement of transactions of very large values (£500,000 or more) until it ceased operation on February 24, 1995 (Eaglen, 1988; Tyson-Davies, 1994). The services of the Cheque & Credit Clearing Company Ltd. include checks and paper instruments. Finally, BACS (Banker's Automated Clearing Service), founded in 1968, is the world's oldest and largest system dealing with credit transfers and direct debits (Fallon and Welch, 1994).

TABLE 2.18

Contribution of Various Settlement Systems in the U.K. in 1995 and 2000

	1995		2000	
	Volume[a]	**Value[b]**	**Volume[a]**	**Value[b]**
CHAPS				
CHAPS Sterling	13	26,719	21.7	49,146
CHAPS Euro			3.3	25,316
Town[c]		59		
Check and Credit Clearings				
Checks	2,202	1,237	1,869	1,365
Paper-based credit transfer	171	99	164	88
BACS				
Credit transfers	969	742	1,307	1,405
Direct debits	1,299	312	2,010	517
Total	4,654	29,168	5,375	77,837

[a]In millions of transactions.
[b]In billions of pounds sterling.
[c]Ceased operation on February 24, 1995.

Table 2.18 gives a breakdown of the transactions cleared through each of these systems in 1995 and 2002 (Bank of International Settlements, 2001, 2002).

2.8.3   France

The structure of the French clearance system changed profoundly between 1995 and 2000 as a consequence of efforts to streamline the process and to follow technological evolution, in addition to the adoption of the Euro as a single European currency. Currently, settlement of small amounts relies on the following systems:

- The SIT (Système Interbancaire de Télécompensation — Interbanking Clearance System) whose mission is to allow continuous settlement. Introduced in 1995, it is gradually replacing all other systems. Thus, it absorbed the functions of the network of bank cards in 1996, and in July 2002, it replaced clearinghouses following the generalization of the exchange of check images.
- The Creic (Centre Régionaux d'Échanges d'Images-Chèques — Regional Centers for the Exchange of Check Images) will, in time, be replaced by the SIT.

The SIT utilizes Transpac, an X.25 network. The ETEBAC5 (Échange Télématique Entre les Banques et leurs Clients — Telematic Exchange among Banks and Their Clients) protocol is used to secure the point-to-point file transfer, while the settlement dialogues follow the PESIT (Protocole de Transfert de Fichier pour le Système Interbancaire de Télécompensation — File

Transfer Protocol for the Interbanking System for Remote Clearance and Settlement) protocol. As standardized by the CFONB in 1988, ETEBAC5 ensures integrity, confidentiality, mutual authentication of the parties, and nonrepudiation of the exchanged messages.

Settlements for large amounts utilize:

- The TBF (Transferts Banque de France) is the system management by the French Central Bank and constitutes the French component of TARGET (Trans-European Automated Real-Time Gross Settlement Express Transfer system). This is a European settlement system of Euro transactions in real time (less than 2 minutes after debiting the issuer account).

- The Paris Net Settlement (PNS) replaced, in April 1999, the Système Net Protégé (SNP) that started functioning in 1997. The PNS is technically managed by the CRI (Centrale des Règlements Interbancaires — Union of Interbanking Payments), a society jointly owned by the French central bank and other credit institutions. PNS absorbed a large portion of the exchanges that were executed before in clearinghouses.

- The Paris clearinghouse is also used.

The SAGITTAIRE network, which was established in 1984 to route international transactions of large amounts, was shut down in 1998. Similarly, the old credit transfers from the Banque de France were retired in 1998.

The SWIFT (Society for Worldwide Interbank Financial Telecommunications) network provides the access to PNS or TBF.

Depicted in Table 2.19 are the contributions of each of these systems in 1995 and 2000 (Bank for International Settlements, 2001, 2002).

2.9 Summary

The acceptance of payment systems depends on many technical, political, and social factors. A worldwide solution for e-commerce will have to fit easily with the existing structure of the payment systems in place. Intermediaries may be needed for cost-effective billing and collection of moneys, particularly in the case of micropayments. One possible role for any of these intermediaries is to be able to offer clients and merchants a single interface independent of the underlying system of payment. There is another problem that must be resolved for micropayments, and this is the differences in currencies and the fluctuations in the exchange rates, which adds significant financial risks to individuals, merchants, and operators.

TABLE 2.19

Clearance and Settlement Transactions in France in 1995 and 2000

	1995		2000	
	Volume[a]	Value[b]	Volume[a]	Value[b]
Large-value systems				
TBF	—	—	3.0	52,804.7
PNS	—	—	5.5	21,844.9
Transfers through the Banque de France	29.4	5,616.0	—	—
SAGITTAIRE	4.5	15,941.1	—	—
Small-value systems				
Creic	281.8	155	303.0	25.9
Clearinghouses	3,588.4	137,412.8	—	—
Automatic clearance	4,744.7	10,375.4	—	—
Network of bank cards[c]	1,872.6	589.7	—	—
SIT	2,590.3	9,625.6	6,485.3	2,458.4

[a]In millions of transactions.
[b]In billions of Euros.
[c]Replaced by the SIT in 1996.

Question

The technology *S*-curve is used to evaluate incremental and discontinuous progress in technology, while the value chain can be used to evaluate the effects of market changes.

Evaluate the position of the various payment instruments (cash, bank cards, checks, electronic bill presentment) on any performance criterion (e.g., cost, security, user's convenience). Estimate the core competencies used for each technology and what would cause a change in the view that each actor has of its core competencies.

An introduction of the technology *S*-curve is available in Betz, F., *Strategic Technology Management*, McGraw-Hill, New York, 1993. The value chain is described by Christensen, C.M., *The Innovator's Dilemma: When New Technologies Cause Great Firms to Fail*, Harvard Business School Press, Boston, MA, 1997. Finally, an example that combines both approaches is available in Sherif, M.H., When standardization is slow?, *Int. J. IT Stand. & Stand. Res.*, 1, 1, 19–32, January–March, 2003.

3

Algorithms and Architectures for Security

ABSTRACT

In this chapter, a brief review of the state of the art in the application of security systems for electronic commerce is presented. In particular, the chapter deals with the following themes: definition of security services in open networks; security functions and their possible locations in various layers of the distribution network; mechanisms to implement security services; certification of the participants; and the management of the encryption keys. Some potential threats to security are highlighted, particularly as they relate to cracks in the protection walls of cryptography.

The chapter has four appendices. Appendices I and II contain a general overview of the symmetric and public key encryption algorithms, respectively. Described in Appendix III are the main operations of the Digital Signature Algorithm (DSA) of ANSI X9.30:1 (1997). Appendix IV contains comparative data on the performance of various security algorithms.

3.1 Security of Commercial Transactions

Commercial transactions depend on the participants' trust in their mutual integrity, trust in the quality of the exchanged goods, as well as trust in the reliability of the systems for payment transfer or for purchase delivery. Because the exchanges associated with electronic commerce (e-commerce) take place mostly at a distance, it is indispensable to establish a climate of trust that is conducive to business, even if the participants do not meet in person or if they use dematerialized monies.

Security functions for e-commerce have three aspects, at least: protection of the communication networks between the merchant and the buyer on the one side, and the merchant and its banks on the other; protection of the financial exchanges; and whenever necessary, protection of the merchandise

(Girolle and Guerin, 1997). It should be noted that telecommunication services are built simultaneously on network elements and their management systems, on the operations support systems (for provisioning, billing, etc.), and on the policies for maintenance and operation. The availability of the telecommunication network relies on the quality of operations of these three components. Thus, the network architecture must be capable of withstanding potential faults without important service degradation, and the physical protection of the network must be insured against fires, earthquakes, floodings, vandalism, or terrorism. This protection will primarily cover the network equipment (switches, trunks, information systems) but can be extended to user end-terminals as well. Procedures to ensure such protection are beyond the scope of this chapter. [Note that in a technical report from ISO (ISO/IEC TR 13335-5, 2001), several measures to ensure information security are suggested. Part 5, in particular, relates to the means for physical protection of network equipment.]

At the level of the transaction, security of e-commerce covers service access; the correct identification and authentication of participants (so as to provide them the services they subscribed to); the integrity of the exchanges; and, if needed, their confidentiality. It may be necessary to preserve the evidence that can help to resolve disputes and litigation. These aspects are the subject of this chapter. Nevertheless, protective measures taken by a network operator may counter users' expectations regarding anonymity and nontraceability of transactions.

3.2   Security of Open Financial Networks

A full-fledged security infrastructure with encryption is not always necessary. Neither the French Minitel nor its Japanese counterpart for wireless services (i-mode) have, at any time, raised users' misgivings because of the absence of encryption. This may be attributed to the fact that a single operator is responsible for running the telecommunications network. One can thus assume that it is the openness of the network that generated the feeling of insecurity. Furthermore, the risks of dysfunction increase with the number of operators and the multiplication of equipment. In 1996, Bank of America conducted with the Lawrence Livermore National Laboratory a pilot experiment on the use of the Internet for electronic fund transfer. The data showed that 49% of the difficulties could be attributed to systems going down or being offline, 24% to document delivery problems (duplication, delays, or loss), 17% to applications and operating systems incompatibilities, 5% to message truncations, and only 5% to decryption problems (Segev et al., 1996). These figures confirm the results obtained for other services on public networks.

In the contemporary context, network fragmentation and the compartmentalization of end-to-end connection management continue relentlessly for three main reasons:

1. The worldwide phenomenon of deregulation of telecommunications prevents a single operator from getting all the traffic, even in a restricted zone.
2. The emergence of new players in niche markets poses new problems of interconnectivity.
3. The Internet covers the main business sites worldwide, even though it is administered by a multiplicity of federated authorities without central organization.

3.3 Security Objectives

Several types of information exposures in an open network affect user data and applications as well as the network elements or the network infrastucture. Recommendations X.509 (2000) and X.800 (1991) of the ITU-T identify several types of information threats that can be classified in two categories, as follows:

1. Passive attacks:
   a. Interception of the identity of one or more of the participants by a third party with mischievous intent
   b. Data interception through clandestine monitoring of the exchanges during a communication by an outsider or an unauthorized user
2. Active attacks:
   a. Replay of a previous message, in its entirety or in part, after its recording
   b. Accidental or criminal manipulation of the content of an exchange by substitution, insertion, deletion, or reorganization of a user's data exchanged in a communication by a nonauthorized third party
   c. Users' repudiation or denial of their participation in part or in all of a communication exchange
   d. Misrouting of messages from one user to another (the objective of the security service would be to mitigate the consequences of such an error)

e. Analysis of the traffic and examination of the parameters related to a communication among users (i.e., absence or presence, frequency, direction, sequence, type, volume, etc.); this analysis would be made more difficult by producing unintelligible additional traffic (by a fill-in traffic) and by using encrypted or random data

f. Masquerade, whereby one entity pretends to be another entity

g. Denial of service and the impossibility of accessing the resources usually available to authorized users following the breakdown of communication, link congestion, or the delay imposed on time-critical operations

Based on the preceding threats, the objectives of security measures are as follows:

- Prevent an outsider other than the participants from reading or manipulating the contents or the sequences of the exchanged messages without being detected. In particular, this third party must not be allowed to play back old messages, replace blocks of information, or insert messages from multiple distinct exchanges without detection.

- Impede the falsification of payment instructions or the generation of spurious messages by users with dubious intentions. For example, dishonest merchants or processing centers must not be capable of reutilizing information about their clients' bank accounts to generate fraudulent orders. They should not be able to initiate the processing of payment instructions without expediting the corresponding purchases. At the same time, the merchants will be protected from excessive revocation of payments or malicious denials of orders.

- Satisfy the legal requirements on, for example, payment revocation, conflict resolution, consumer protection, privacy protection, and the exploitation of data collected on clients for commercial purposes.

- Ensure access to service, according to terms of the contract.

- Give the same level of service to all customers, irrespective of their location and the variations in climate, temperature, humidity, erosion, etc.

The International Organization for Standardization (ISO) standard ISO 7498 (1994) Part 2 (ITU-T Recommendation X.800, 1991) describes a reference model for the service securities in open networks. This model, which was used in Recommendation X.509, will be the framework for the discussion here. It should be noted that the latter recommendation, which was approved for the first time in 1988, was subsequently revised in 1993, in 1996, and in 2000, without modifying the basic premises. Recommendation X.509 is also a joint standard of ISO and the International Electrotechnical Commission (IEC), known as ISO/IEC 9594-8. ANSI (American National Standards Institute) also ratified a corresponding standard known as ANSI X9.57 (1997). A list of security standards is available in Menezes et al. (1997).

3.4 OSI Model for Cryptographic Security

3.4.1 OSI Reference Model

It is well known that the OSI (Open Systems Interconnection) reference model of data networks establishes a structure for exchanges in seven layers:

1. The *physical layer* is where the electrical, mechanical, and functional properties of the interfaces are defined (signal levels, rates, structures, etc.).

2. The *link layer* defines the methods for orderly and error-free transmission between two network nodes.

3. The *network layer* is where the functions for routing, multiplexing of packets, flow control, and network supervision are defined.

4. The *transport layer* is responsible for the reliable transport of the traffic between the two network end points as well as the assembly and disassembly of the messages.

5. The *session layer* handles the conversation between the processes at the two end points.

6. The *presentation layer* manages the differences in syntax among the various representations of information at both end points by putting the data into a standardized format.

7. The *application layer* ensures that two application processes cooperate to carry out the desired information processing at the two end points.

To each layer was assigned some cryptographic security functions that are detailed in the following section.

3.4.2 Security Services: Definitions and Locations

Security services for exchanges used in e-commerce employ mathematical functions to reshuffle the original message into an unreadable form before it is transmitted. After the message is received, the authenticated recipient must restore the text to its original status. The security consists of six services (Baldwin and Chang, 1997):

- *Confidentiality,* i.e., ensuring that the exchanged messages are not divulged to a nonauthorized third party. In some applications, the confidentiality of addresses may be needed as well to prevent the analysis of traffic patterns and the derivation of side information that could be used.

- *Integrity* of the data, i.e., proof that the message was not altered after it was expedited and before the moment it was received. This service guarantees that the received data are exactly what were transmitted by the sender and that they were not corrupted, either intentionally or by error, in transit in the network. Data integrity is also needed for network management data, such as configuration files, accounting, and audit information.

- *Identification* of the participants, i.e., the verification of a preestablished relation between a characteristic (for example, a password or cryptographic key) and an entity. This allows for control of access to the network resources or to the offered services based on the privileges associated with a given identity. One entity may possess several distinct identifiers. Furthermore, some protection against denial-of-service attacks can be achieved using access control.

- *Authentication* of the participants (users, network elements, and network element systems), i.e., the corroboration of the identity that an entity claims, with the guarantee of a trusted third party. Authentication is necessary to ensure nonrepudiation of users as well of network elements.

- *Access control,* i.e., ensuring that only the authorized participants, whose identities were duly authenticated, can gain access to the protected resources.

- *Nonrepudiation* is the service that offers an irrefutable proof of the integrity of the data and of their origin in a way that can be verified by a third party, for example, the nonrepudiation that the sender sent the message or that a receiver received the message. This service may also be called authentication of the origin of the data.

Unfortunately, not all of the services offered on the Internet can be easily protected. The case of mobile IP illustrates this point. According to this protocol, a mobile node outside the zone that its home agent serves must register with the foreign agent in whose region it is currently located. Yet, the protocol does not provide the means with which to authenticate the foreign agent by initiating the exchange of the secret key that will be used to protect the resubscription data (Perkins, 1998, pp. 134–139, 189–192).

The implementation of the security services can be made over one or more of the layers of the OSI model (Ford and O'Higgins, 1992; Rolin, 1995). The choice of the layer depends on the following criteria:

1. If the protection has to be accorded to all the traffic flow in a uniform manner, the intervention has to be at the physical or the link layers. The only cryptographic service available at this level is confidentiality, by encrypting the data or by similar means (frequency hopping, spread spectrum, etc.). The protection of the traffic at the physical layer covers all the flow, not only user data but also the information related to network administration: alarms, synchronization, updates of routing table, etc. The disadvantage of the protection at this level is that a successful attack will destabilize the whole security structure, because the same key is utilized for all transmissions. At the link layer, encryption can be end-to-end, based on the source/destination, provided that the same technology is used all the way through.

2. For a selective bulk protection that covers all the communications associated with a particular subnetwork from one end system to another end system, network layer encipherment will be chosen. Security at the network layer is also needed to secure the communication among the network elements, particularly for link state protocols, such as OSPF (Open Short Path First) or PNNI (Private Network-to-Network Interface), where updates to the routing tables are automatically generated based on received information and are then flooded to the rest of the network.

3. For a protection with recovery after a fault, or if the network is not reliable, the security services will be at the transport layer. The services of this layer apply end-to-end, either singly or in combination. These services are authentication (whether *simple* by passwords or *strong* by signature mechanisms or certificates), access control, confidentiality, and integrity.

4. If a high granularity of protection is required, or if the nonrepudiation service has to be assured, the encryption will be at the application layer. It is at this level that most of the security protocols for commercial systems operate, which frees them from a dependency on the lower layers. All security services are available.

It should be noted that there are no services at the session layer. In contrast, the services offered at the presentation layer are confidentiality, which can be selective, such as by a given data field, authentication, integrity (in whole or in part), and nonrepudiation with a proof of origin or proof of delivery.

The Secure Sockets Layer (SSL)/Transport Layer Security (TLS) protocols are widely used to secure the connection between a client and a server (Freier et al., 1996; IETF RFC 2246, 1999). With respect to the OSI reference model, SSL/TLS lie between the transport layer and the application layer.

Nevertheless, it may be sufficient for an attacker to discover that a communication is taking place among partners and then attempt to guess, for example:

- The characteristics of the goods or services exchanged
- The conditions for acquisition: delivery intervals, conditions, and means of settlement
- The financial settlement

The establishment of an enciphered channel or "tunnel" between two points at the network layer can constitute a shield against such types of attack. It should be noticed, however, that other clues, such as the relative time to execute the cryptographic operations, or the variations in the electric consumption or the electromagnetic radiation, can permit an analysis of the encrypted traffic and ultimately lead to breaking of the encryption algorithms (Messerges et al., 1999).

3.5   Security Services at the Link Layer

IETF RFC 1661 (1994) defines the link-layer protocol PPP (Point-to-Point Protocol) to carry traffic between two entities identified with their respective (Internet Protocol) IP addresses. The Layer 2 Tunneling Protocol (L2TP) defined in IETF RFC 2661 (1999) extends the PPP operation by separating the processing of IP packets within the PPP frames from that of the traffic flowing between the two ends at the link layer. This distinction allows a remote client to connect to a network access server (NAS) in a private (corporate) network through the public Internet, as follows. The client encapsulates PPP frames in an L2TP tunnel, prepenses the appropriate L2TP header, and then transports the new IP packet using the User Datagram Protocol (UDP). The IP addresses in the new IP header are assigned by the local Internet Service Provider (ISP) at the local access point. Figure 3.1 illustrates the arrangement where the size of the additional header ranges from 8 to 16 octets: 1 to 2 octets for PPP, and 8 to 16 octets for L2TP. Given that the overhead for UDP is 8 octets and that the IP header is 20 octets, the total additional overhead ranges from 37 to 46 octets.

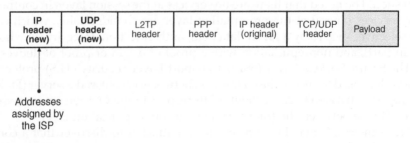

FIGURE 3.1
Layer 2 tunneling with L2TP.

Although L2TP does not provide security services, it is possible to use Internet Protocol Security (IPSEC) to secure the Layer 2 tunnel, because L2TP runs over IP. This is shown in the following section.

3.6 Security Services at the Network Layer

The security services at this layer are offered from one end of the network to the other. They include network access control, authentication of the users and hosts, and authentication and integrity of the exchanges. These services are transparent to applications and end users, and their responsibilities fall on the administrators of network elements.

The purpose of network access control is to limit the actions and the privileges of an entity based on the network addresses of both end points (e.g., IP addresses). As explained earlier, this is important in link-state protocols, such as OSPF or PNNI, to protect the routing tables of the various network elements.

Authentication at the network layer can be simple or strong. *Simple* authentication uses a name and password pair (the password may be a one-time password), while *strong* authentication utilizes digital signatures or the exchange of certificates issued by a recognized certification authority. The use of strong authentication requires the presence of encryption keys at all network nodes, which imposes the physical protection of all these nodes.

IPSec is a protocol suite defined in IETF RFCs 2401 to 2412 (1998) to secure communications at the network layer between two peers. The overall security architecture is described in IETF RFC 2401, while a road map to the IPSEC documentation is in IETF RFC 2411.

IPSec offers authentication, confidentiality, and key management. The Authentication Header (AH) protocol defined in IETF RFC 2402 provides the cryptographic services to authenticate and verify the integrity of the payload as well as the routing information in the original IP header. The Encapsulating Security Payload (ESP) protocol is described in IETF RFC 2406, and it gives the means to assure the confidentiality of the original payload and to authenticate the encrypted data as well as the ESP header. Both IPSec protocols provide some protection against replay attacks, with the help of a monotonically increasing sequence number that is 32 bits long. Although these two mechanisms are available in the IP Version 6 (IPv6) protocol (Huitema, 1996), IPSec makes them available with the current IP Version 4. The key exchange is performed with the IKE (Internet Key Exchange) protocol defined in IETF RFC 2409. [Note that a new ESP draft uses 64-bit sequence numbers and takes into consideration the new symmetric encryption algorithm Advance Encryption Standard (AES).]

IPSec operates in one of two modes: the transport mode and the tunnel mode. In the transport mode, the protection covers the payload and the transport header only, while the tunnel mode protects the whole packet, including the IP addresses. The transport mode secures the communication between two hosts, while the tunnel mode is useful when one or both ends of the connection is a trusted entity, such as a firewall, which provides the security services to an originating device. The tunnel mode is also employed when a router provides the security services to the traffic that it is forwarding (Doraswamy and Harkins, 1999). Both modes are used to secure virtual private networks with IPSec, as shown in Figure 3.2. Typically, the AH protocol can be used for the transport mode, while the ESP is applicable to both modes. This explains why there is a decreasing tendency to use the AH protocol.

Illustrated in Figure 3.3 is the encapsulation in both cases. In this figure, the IPSec header represents either the ESP or both the ESP and the AH headers. Thus, routing information associated with the private or corporate network can be encrypted after establishment of a TCP tunnel between the firewall at the originating side and the one at the destination side. [Note that ESP with no encryption (i.e., with a NULL algorithm) is equivalent to the AH protocol, which is another reason usage of the latter is limited.]

In verifying the integrity, the contents of fields in the IP header that change in transit (e.g., the "time to live") are considered to be zero. With respect to transmission overheads, the length of the AH is at least 12 octets (a multiple of 4 octets for IPv4 and of 6 octets for IPv6). Similarly, the length of the ESP header is 8 octets. However, the overhead includes 4 octets for the initial-

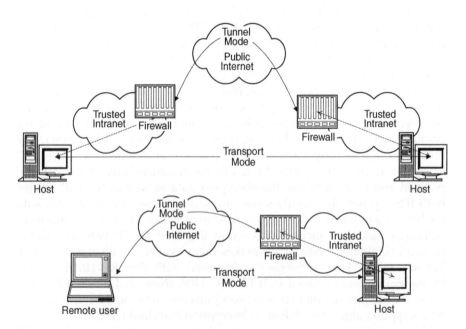

FIGURE 3.2
Securing virtual private networks with IPSec.

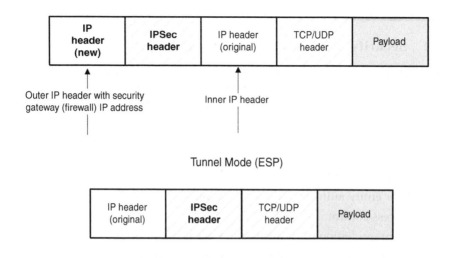

Tunnel Mode (ESP)

Transport Mode (AH and/or ESP)

FIGURE 3.3
Encapsulation for IPSec modes.

ization vector (if it is included in the payload field) as well as an ESP trailer of at least 6 octets that comprise a padding and authentication data.

Let us return to the protection of L2TP (control data or user information) traffic with the IPSec protocol suite as described in IETF RFC 3193 (2001). When IPSec and L2TP are used together, the various headers are organized as shown in Figure 3.4. [Note that in the 1996–1998 time frame, RSA Data Security, Inc., and the Secure Wide Area Network (S/WAN) consortium were actively promoting a specific implementation of IPSec to ensure interoperability among firewalls and Transmission Control Protocol (TCP)/IP products. However, the free-software advocates cooperated under the umbrella of FreeS/WAN to distribute an open source implementation of IPSec and its default exchange protocol IKE, written for Linux. As a consequence, S/WAN is no longer an active initiative. Details on ongoing projects for Linux are available at *http://www.freeswan.org.*]

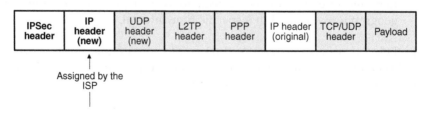

FIGURE 3.4
Encapsulation for secure network access with L2TP and IPSec.

3.7   Security Services at the Application Layer

The majority of security protocols for e-commerce operate at the application layer, which makes them independent of the lower layers. The whole gamut of security services is now available, namely, the following:

1. Confidentiality, total or selective by field or by traffic flow
2. Data integrity
3. Peer entity authentication
4. Peer entity authentication of the origin
5. Access control
6. Nonrepudiation of transmission with proof of origin
7. Nonrepudiation of reception with proof of reception

To illustrate, the Secure Shell (SSH[®1]) provides security at the application layer — it allows a user to log on, execute commands, and transfer files securely. Thus, it can replace other applications, such as telnet, rlogin, rsh, and rcp (Carasik, 2001; Ylönen, 1995, 1996). In reality, there are two distinct protocols: SSH1 and SSH2. Both bind to the same TCP port. One important difference is that SSH2 has an explicit capability to secure ftp as well. Both are freely available specifications with freeware and commercial implementations. Guidelines for management of security with SSH are available (AF-SEC-0179.000, 2002).

Additional security mechanisms are specific to a particular usage or to the end-user application at hand. For example, several additional parameters are considered to secure electronic payments, such as the ceiling of allowed expenses or withdrawals within a predefined time interval. Fraud detection and management depend on the surveillance of the following (Sabatier, 1997, p. 85):

- Activities at the points of sale (merchant terminals, vending machines, etc.)
- Short-term events
- Long-term trends, such as the behavior of a subpopulation within a geographical area and in a specific time interval, etc.

In these cases, audit management takes into account the choice of events to collect and register, the validation of an audit trail, definition of the alarm thresholds for suspected security violations, etc.

---

[1] Secure Shell and SSH are registered trademarks of SSH Communications Security, Ltd. of Finland.

The rights of intellectual property to dematerialized articles sold online pose an intellectual and technical challenge. The aim is to prevent the illegal reproduction of what is easily reproducible using "watermarks" incorporated in the product (Anderson et al., 1998). The means used differ depending on whether the products protected are ephemeral (such as news), consumer-oriented (such as films, music, books, articles, or images), or for production (such as enterprise software). While the technical aspects are not treated in this book, we will briefly go over the legislative efforts in Chapter 15.

In the rest of this chapter, we give an overview of the mechanisms used to implement security service. The objective is to present sufficient background for understanding the applications and not to give an exhaustive review. For a comprehensive discussion of the mathematics of cryptography and its applications, the reader is invited to consult the literature for more detailed descriptions (Schneier, 1996a; Menezes et al., 1997).

3.8 Message Confidentiality

Confidentiality guarantees that information will be communicated solely to the parties authorized for its reception. Concealment is achieved with the help of encryption algorithms. There are two types of encryption: symmetric encryption, where the operations of message obfuscation and revelation use the same secret key, and public key encryption, where the encryption key is secret, and the revelation key is public.

3.8.1 Symmetric Cryptography

Symmetric cryptography is the tool employed in classical systems. The key that the sender of a secret message utilizes to encrypt the message is the same as the one that the legitimate receiver uses to decrypt the message. Obviously, key exchange among the partners has to occur before the communication, and this exchange takes place through other secured channels. The operation is illustrated in Figure 3.5.

Let **M** be the message to be encrypted, with a symmetric key **K** in the encryption process **E**. The result will be the ciphertext **C**, such that:

$$E[K(M)] = C$$

The decryption process **D** is the inverse function of **E** that restores the clear text:

$$D(C) = M$$

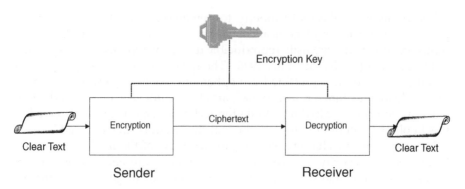

FIGURE 3.5
Symmetric encryption.

There are two main categories of symmetric encryption algorithms: block encryption algorithms and stream cipher algorithms. Block encryption acts by transforming a block of data of fixed size, generally 64 bits, in encrypted blocks of the same size. Stream ciphers convert the clear text one bit at a time by combining the stream of bits in the clear text with the stream of bits from the encryption key using an Exclusive OR (XOR).

Table 3.1 presents the algorithms for symmetric encryption that are often used in applications of e-commerce.

The main drawback of symmetric cryptography systems is that both parties must obtain, one way or another, the unique encryption key. This is possible without too much trouble within a closed organization; on open networks, however, the exchange can be intercepted. Public key cryptography, which was proposed in 1976 by Diffie and Hellman, is one solution to the problem of key exchange (Diffie and Hellman, 1976).

3.8.2 Public Key Cryptography

Algorithms of public key cryptography introduce a pair of keys for each participant, a private key SK and a public key PK. The keys are constructed in such a way that it is practically impossible to reconstitute the private key with the knowledge of the public key.

Consider two users, A and B, each having a pair of keys (PK_A, SK_A) and (PK_B, SK_B), respectively. Thus,

1. To send a secret message x to B, A encrypts it with B's public key and then transmits the encrypted message to B. This is represented by

$$e = PK_B\left(x\right)$$

TABLE 3.1

Symmetric Encryption Algorithms in E-Commerce

Algorithm	Name and Comments	Type of Encryption	Key Length (bits)	Standard
AES	Advanced encryption standard	Blocks of 128, 192, or 256 bits	128, 192, or 256	FIPS 197
DES	Data encryption standard	Blocks of 64 bits	56	FIPS 81; ANSI X3.92, X3.105, X3.106; ISO 8372, ISO/IEC 10116
IDEA (Lai and Massey, 1991a, 1991b)	International data encryption algorithm (apparently one of the best and most secure algorithms commercially available)	Blocks of 64 bits	128	—
RC2	Developed by Ronald Rivest (Schneier, 1996a, pp. 319–320)	Blocks of 64 bits	Variable, 40 bits for export from the U.S.	No, and proprietary
RC4	Developed by Ronald Rivest (Schneier, 1996a, pp. 397–398)	Stream	40 or 128	No, but posted on the Internet in 1994
RC5	Developed by Ronald Rivest (1995)	Blocks of 32, 64, or 128 bits	Variable up to 2048 bits	No, and proprietary
SKIPJACK	An algorithm developed in the U.S. by the National Security Agency (NSA) for applications with the PCMCIA card Fortezza[a]	Blocks of 64 bits	80	Declassified algorithm; version 2.0 is available at http://csrc.nist.gov/encryption/skipjack-kea.htm
Triple DES	Also called TDEA	Blocks of 64 bits	112	ANSI X9.52

[a] Fortezza is a Cryptographic Application Programming Interface (CAPI) that the NSA define for security applications on PCMCIA cards incorporating SKIPJACK.

2. B recovers the information using his or her private key SK_B. It should be noted that only B possesses SK_B, which can be used to identify B. The decryption operation can be represented by

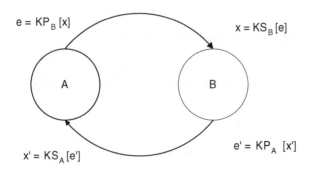

FIGURE 3.6
Confidentiality of messages with public key cryptography. [From ITUT-T. Recommendation X.509 (2001). With permission.]

$$x = SK_B(e) \text{ or } x = SK_B\left[PK_B(x)\right]$$

3. *B* can respond to *A* by sending a new secret message x' encrypted with the public key PK_A of *A*:

$$e' = PK_A \, x'$$

4. *A* obtains x' by decrypting e':

$$x' = SK_B \, e' \text{ or } x' = SK_A\left[PK_A \, x'\right]$$

The diagram in Figure 3.6 summarizes these exchanges.

It is worth noting that the preceding exchange can be used to verify the identity of each participant. More precisely, *A* and *B* are identified by the possession of the decryption key, SK_A or SK_B, respectively. *A* can determine if *B* possesses the private decryption key SK_B if the initial message x is included in the returned message x' that *B* sends. This indicates to *A* that the communication was established with the entity that possesses SK_B. *B* can also confirm the identity of *A* in a similar way.

The de facto standard for public key encryption is the algorithm RSA invented by Ronald Rivest, Adi Shamir, and Leonard Adleman in 1977.

3.9   Data Integrity

The objective of the integrity service is to eliminate all possibilities of non-authorized modification of messages during their transit from the sender to the receiver. The traditional form to achieve this securityis to stamp the letter envelope with the wax seal of the sender. Transposing this concept to electronic transactions, the seal will be a sequence of bits associated univocally with the document to be protected. This sequence of bits will constitute a unique and unfalsifiable "fingerprint" that will accompany the document sent to the destination. The receiver will then recalculate the value of the fingerprint from the received document and compare the value obtained with the value that was sent. Any difference will indicate that the message integrity was violated.

The fingerprint can be made to depend on the message content only by applying a *hash function*. A hash function converts a sequence of characters of any length into a chain of characters of a fixed length, L, usually smaller than the original length, called a hash value. However, if the hash algorithm is known, any entity can calculate the hash value from the message using the hash function. For security purposes, the hash value depends on the message content and the sender's private key in the case of a public key encryption algorithm, or a secret key that only the sender and the receiver know in the case of a symmetric encryption algorithm. In the first case, anyone knowing the hash function can calculate the fingerprint with the public key of the sender; in the second case, only the intended receiver will be able to verify the integrity. It should be noted that lack of integrity can be used to break confidentiality. For example, the confidentiality of some algorithms may be broken through attacks on the initialization vectors.

The hash value has many names: compression, contraction, message digest, fingerprint, cryptographic checksum, Message Integrity Check (MIC), etc. (Schneier, 1996a, p. 31).

3.9.1 Verification of the Integrity with a One-Way Hash Function

A *one-way hash function* is a function that can be calculated relatively easily in one direction but with considerable difficulty in the inverse direction. A one-way hash function is sometimes called a compression function or a contraction function.

To verify the integrity of a message with a fingerprint that was calculated with the hash function $H()$, this function should also be a one-way function, i.e., it should meet the following properties:

1. *Absence of collisions*: In other words, the probability of obtaining the same hash value with two different texts should be almost null. Thus, for a given message x_1, the probability of finding a different message x_2 such that $H(x_1) = H(x_2)$, is extremely small. For the collision probability to be negligible, the size of the hash value L should be sufficiently large.

2. *Impossibility of inversion*: Given the fingerprint h of a message x, it is practically impossible to calculate x such that $H(x) = h$.

3. *A wide spread among the output values*: This is so that a small difference between two messages should yield a large difference between their fingerprints. Thus, any slight modification in the original text should, on the average, affect half of the bits of the fingerprint.

Consider the message X. It will have been divided into n blocks, each consisting of B bits. If needed, padding bits would be appended to the message, according to a defined scheme, so that the length of each block reaches the necessary B bits. The operations for cryptographic hashing are described using a compression function $f()$ according to the following recursive relationship:

$$h_i = f(h_{i-1}, x_i), i = 1, ..., n$$

In this equation, h_0 is the vector that contains an initial value of L bits, and $x = \{x_1, x_2, ..., x_n\}$ is the message subdivided into n vectors of B bits each. The hash algorithms commonly used in e-commerce are listed in Table 3.2.

For MD5 and SHA-1, the message is divided into blocks of 512 bits. The padding consists in appending to the last block a binary "1," then as many "0" bits as necessary for the size of the last block with padding to be 448 bits. Next, a suffix of 8 octets is added to contain the length of the initial message (before padding) coded over 64 bits, which brings the total size of the last block to 512 bits of 64 octets.

In 1994, two researchers, van Oorschot and Wiener, were able to detect collisions in the output of MD5 (van Oorschot and Wiener, 1994), which explains its gradual replacement with SHA-1. (Note that many authors use SHA1, SHA-1, and SHA interchangeably.)

3.9.2   Verification of the Integrity with Public Key Cryptography

An encryption algorithm with a public key is called *permutable* if the decryption and encryption operations can be inverted, i.e., if

$$M = PK_X[SK_X(M)]$$

In the case of encryption with a permutable public key algorithm, an information element M that is encrypted by the private key SK_X of an entity X can be read by any user possessing the corresponding public key PK_X. A sender can, therefore, sign a document by encrypting it with a private key reserved for the signature operation to produce the seal that accompanies the message. Any person who knows the corresponding public key will be able to decipher the seal and verify that it corresponds to the received message.

Another way of producing the signature with public key cryptography is to encrypt the fingerprint of the document. This is because the encryption of a long document using a public key algorithm imposes substantial computations and introduces excessive delays. Therefore, it is beneficial to

TABLE 3.2

Hash Functions Utilized in E-Commerce Applications

Algorithm	Name	Length of the Finger-print (L) (bits)	Block Size (B) (bits)	Standardi-zation
AR/DFP	Hashing algorithms of German banks	—	—	German banking standards
DSMR	Digital signature scheme giving message recovery	—	—	ISO/IEC 9796
MCCP	Banking key management by means of public key algorithms; algorithms using the RSA cryptosystem; signature construction by means of a separate signature	—	—	ISO/IEC 1116-2
MD4	Message digest algorithm	128	512	No, but described in RFC 1320
MD5	Message digest algorithm	128	512	No, but described in RFC 1321
NVB7.1, NVBAK	Hashing functions used by Dutch banks	—	—	Dutch banking standard, published in 1992
RIPEMD	Extension of MD4, developed during the European project RIPE (Menezes et al., 1997, p. 380)	128	512	—
RIPEMD-128	Dedicated hash function #2	128	512	ISO/IEC 10118-3
RIPEMD-160	Improved version of RIPEMD (Dobbertin et al., 1996)	160	512	—
SHA	Secure hash algorithm (NIST, 1993) (replaced by SHA-1)	160	512	FIPS 180
SHA1 (SHA-1)	Dedicated hash function #3 (NIST, 1995) (revision and correction of the secure hash algorithm)	160	512	ISO/IEC 10118-3 FIPS 180-1

use a digest of the initial message before applying the encryption. This digest is produced by applying a one-way hash function to calculate the fingerprint that is then encrypted with the sender's private key. At the destination, the receiver recomputes the fingerprint. With the public key of the sender, the receiver will be able to decrypt the fingerprint to verify if the received hash value is identical to the computed hash value. If both are identical, the signature is valid.

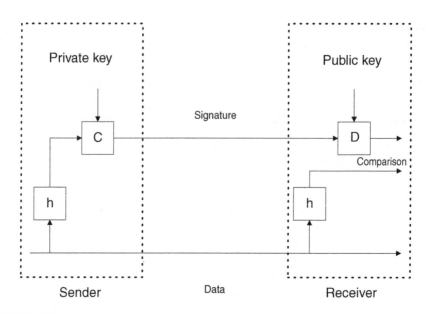

FIGURE 3.7
Computation of the digital signature using public key algorithms and hashing.

The block diagram in Figure 3.7 represents verification of the integrity with public key encryption. In this figure, h represents the hash function, C the encryption function, and D the decryption function.

The public key algorithms frequently used to calculate digital signatures are listed in Table 3.3.

TABLE 3.3

Public Key Algorithms Used to Compute Digital Signatures

Algorithm	Comments	Length of the Finger-print	Standard
DSA	Digital signature algorithm, which is a variant of the ElGamal algorithm; it is a part of the digital signature standard (DSS) that was proposed by NIST (National Institute of Standards and Technology) in 1994	512 to 1024 bits	FIPS 186I
ElGamal	Nondeterministic algorithm where a message corresponds to several signatures; it uses discrete logarithms (ElGamal, 1985)	Variable	—
RSA	This is the de facto standard algorithm for public key encryption; it can also be used to calculate signatures	512 to 1024 bits	ISO/IEC 9796

Note: The U.S. federal government mandates the use of the DSA for signing electronic procurements.

Even though this message allows for verification of the message integrity, it does not guarantee that the identity of the sender is authentic. In the case of public key encryption of the hash value, authentication requires the use of certificates, as will be explained later. [Note that a signature produced from a message with the signer's private key and then verified with the signer's corresponding public key is sometimes called a *signature scheme with appendix* (IETF RFC 2437, 1998).]

3.9.3   Blind Signature

A *blind signature* is a special procedure for a notary to sign a message using the RSA algorithm for public key cryptography without revealing the content (Chaum, 1983, 1989). One possible utilization of this technique is to time-stamp digital payments.

Consider a debtor who would like to have a payment blindly signed by a bank. The bank has a public key e, a private key d, and a public modulo N. The debtor chooses a random number k between 1 and N and keeps this number secret.

The payment p is "enveloped" by applying the following formula:

$$(p\ k^e)\ \text{mod}\ N$$

before sending the message to the bank. The bank signs it with its private key so that

$$(p\ k^e)^d\ \text{mod}\ N = p^d\ k\ \text{mod}\ N$$

and returns the payment to the debtor. The debtor can now extract the signed note by dividing the number by k. To verify that the note received from the bank is the one that was sent, the debtor can raise it to the e power, because (as will be shown in Appendix II):

$$(p^d)^e\ \text{mod}\ N \equiv p\ \text{mod}\ N$$

The various payment protocols for digital money take advantage of blind signatures to satisfy the conditions of anonymity.

3.9.4   Verification of the Integrity with Symmetric Cryptography

The Message Authentication Code (MAC) is the result of a one-way hash function that depends on a secret key. This mechanism guarantees, simultaneously, the integrity of the message content and the authentication of the sender. (As previously mentioned, some authors call the MAC the "integrity check value" or the "cryptographic checksum.")

The most obvious way to construct a MAC is to encrypt the hash value with a block symmetric encryption algorithm. The MAC is then affixed to the initial message, and the whole is sent to the receiver. The receiver recomputes the hash value by applying the same hash function on the received message and compares the result obtained with the decrypted MAC value. The equality of both results confirms the data integrity.

The block diagram in Figure 3.8 depicts the operations where h represents the hash function, C the encryption function, and D the decryption function.

Another variant of this method is to append the secret key to the message that will be condensed with the hash functions.

It is also possible to perform the computations with the compression function $f()$ and use as an initial value the vector of the secret key, k, of length L bits in the following recursion:

$$k_i = f(k_{i-1}, x_i), \ i = 1, \ldots, n$$

where $x = \{x_1, x_2, \ldots, x_n\}$ is the message subdivided into n vectors, each of B bits. The MAC is the value of the final output k_n.

The procedure that several U.S. and international standards advocates — for example, ANSI X9.9 (1986) for the authentication of banking messages, and ISO 8731-1 (1987) and ISO/IEC 9797-2 (2002) for implementing a one-

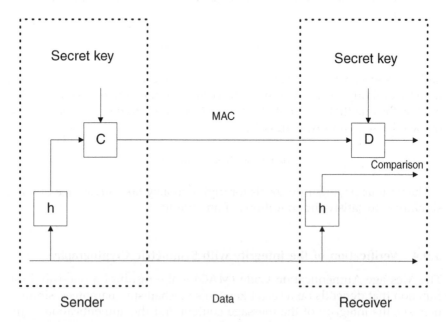

FIGURE 3.8
Digital signature with symmetric encryption algorithms.

way hash function — is to encrypt the message with a symmetric block encryption algorithm in the Cipher Block Chaining (CBC) or the Cipher Feedback (CFB) modes. The MAC is the last encrypted block, which is encrypted one more time in the same CBC or CFB mode.

The following key hashing method augments the speed of computation in software implementation and increases the protection, even when the one-way hash algorithm experiences some rare collisions (Bellare et al., 1996).

Consider the message X subdivided into n vectors of B bits each, and two keys (k_1 and k_2), each of L bits. The padding bits are added to the end of the initial message according to a determined pattern. The hashing operations can thus be described with the help of two compression functions $f_1()$ and $f_2()$:

$$k_i^1 = f_1(k_{i-1}^1, x_i), i = 1,...n$$

$$k_i^2 = f_2(k_{i-1}^2, k_i^1)$$

where k_0^1 and k_0^2 are the initial values of k_1 and k_2, respectively, and $x = x_1$, $x_2,..., x_n$.

The result that this method yields is denoted as the Nested Message Authentication Code (NMAC). It is, in effect, constructed by applying compression functions in sequence, the first on the padded initial message and the second on the product of the first operation after padding.

The disadvantage of this method is that it requires access to the source code of the compression functions to change the initial values. In addition, it requires the usage of two secret keys. This explains the current popularity of the Hashed Message Authentication Code (HMAC), which is described in IETF RFC 2104 (1997). This method uses one single key k of L bits.

Assuming that the function $H()$ represents the initial hash function, the value of the HMAC is computed in the following manner:

$$HMAC_k(x) = H\left[\bar{k} \oplus \text{opad} \| H(\bar{k} \oplus \text{ipad}, x)\right]$$

In this construction, $\bar{k}$ is the vector k of minimum length of L bits, which after padding with a series of 0 bits will reach a total length of B bits. The variables opad and ipad are constants for outer padding and inner padding, respectively. The variable opad is formed with the octet 0x36 repeated as many times as needed to constitute a block of B bits. The variable ipad is the octet 0x5C repeated as many times. For MD5 and SHA-1, the number of repetitions is 64. Finally, the symbols $\|$ and $\oplus$ in the previous equation denote, respectively, the concatenation and Exclusive OR operations.

It should be noted that with the following representation:

$$k^1 = f_1(\overline{k} \oplus \text{ipad})$$

$$k^2 = f_2(\overline{k} \oplus \text{opad})$$

the HMAC becomes the same as the nested MAC. [Note that it will be seen in Chapter 5 that for the SSL protocol, the HMAC is denoted as MAC.]

3.10 Identification of the Participants

Identification is the process of ascertaining the identity of a participant (whether a person or a machine) by relying on uniquely distinguishing features. This contrasts with authentication, which is the confirmation that the distinctive identifier corresponds to the declared user.

Authentication and identification of a communicating entity take place simultaneously when that party proposes to the verifier in private a secret that is only shared between them, for example, a password or a secret encryption key. Another possibility is to pose a series of challenges that only the legitimate user is supposed to be capable of answering.

Digital signature is the usual means of identification because it associates a party (a user or a machine) with a shared secret. Other methods of simultaneous identification and authentication of human users exploit biometric characteristics, such as fingerprints, voiceprints, the shape of the retina, the form of the hand, etc. This is elaborated in the following section.

3.10.1 Biometric Identification

Biometric identification techniques, reserved until recently for military uses and law enforcement agencies, are being considered for user identification in civilian applications. The use of biological attributes for identification and authentication bypasses some of the problems associated with cryptography (e.g., key management). This explains the interest in biometrics in large-scale civilian applications, such as in mobile telephony, e-commerce, or telework.

There are two main categories of biometric features. The first category relates to behavioral patterns and acquired skills, such as speech, handwriting, or keystroke patterns. In contrast, the second category comprises physiological characteristics, such as facial features, iris morphology, retinal texture, hand geometry, or fingerprints. Methods based on gait, odor, or genetic composition using DNA have limited applications for online systems.

The usage of biometric systems includes three steps: image acquisition during the registration phase, features extraction, and identification or verification. The digital image of the person under examination originates from a sensor in the computer peripheral (a microphone, for example). This image is processed to extract a compact profile that should be unique to that person.

This profile or signature is then archived in a reference database that can be centralized or distributed according to the architecture of the system. In most cases, registration cannot be done online; rather, the person has to be physically present in front of a registrar to record the necessary biometric template.

Biometric identification systems ascertain the identity of the end user by matching the biometric data with an entry in a database to supplement another identifier (password, badge, etc.). Verification systems, in contrast, match biometric data with what is stored in the user credential (e.g., a smart cart) to verify access privileges.

It should be noted that biometric systems are not foolproof. The accuracy of an identification system is measured in terms of the rate of mix-up of identities and the rate of rejects of authorized identities. In contrast, the performance of biometric verification systems is assessed in terms of rate of false rejects, i.e., the rejection of authorized identities and the rate of false acceptances. These rates are interdependent and are adjusted according to the required levels of security.

The choice of a particular systems depends on several factors:

1. Accuracy and reliability of the identification or verification: the result should not be affected by the environment or by aging
2. Cost of installation, maintenance, and operation
3. Scale of applicability of the technique; for example, handwriting recognition is not useful for illiterate people
4. Ease of use
5. Reproducibility of the results; in general, physiological characteristics are more reproducible than behavioral characteristics
6. Resistance to counterfeit and attacks

3.10.1.1 Voice Recognition

Identification techniques through voice recognition play one of two distinct functions:

- *Speaker identification*: The technology verifies the end user by comparing a digitized sample of a person's voice with a stored vocal print. Here, a vocal message is compared with a set of stored acoustic references to determine the person from his or her utterances.

- *Speaker verification*: This case consists in verifying that the voice imprint matches the acoustic references of the person that the speaker pretends to be.

These two types of identification can be carried out for the same application, such as the authentication of payment orders made by phone. In this case, the voice imprint that characterizes a subscriber is formed using one or several passwords that are recorded during registration. During the authentication, the user utters one or several of these passwords to allow the system to match the new sample with the previously recorded voice imprints before authorizing the financial transaction.

Depending on the compression algorithm and the duration of the record, the size of the voice imprints that characterize an individual varies between 1 to 70 K octets.

A bad sound quality can cause failures. In remote applications, this quality depends on several factors, such as the type of telephone handset, ambient noise (particularly in the case of hands-free telephony), the type of connection (wireline or wireless), etc. Using about 20 hours of professionally recorded material, some speech synthesis algorithms are perfectly capable of mimicking the speaker's voice characteristics. An easier method with which to defraud the system would be to play back recordings of authentic commands. This is why automatic speaker recognition systems must be supplemented with other means of identification.

3.10.1.2 Handwritten Recognition

The principle of handwritten recognition is to distinguish the permanent characteristics of an individual's handwriting from the changing characteristics to be able to identify the writer. The supposedly permanent characteristics are matched with a prerecorded sample of the handwriting of the person whose identity is to be verified

Handwritten recognition can be static or dynamic. In static verification, the signature is compared with an archived signature of the person to be authenticated. Systems of dynamic handwritten recognition use a special pen and a pressure-sensitive pad connected to a computer. The subject uses the pen to write on the pad, which captures the written text and transmits it to the analysis and verification system. The dynamic movement of the pen is described by tens of parameters, such as the pressure exercised on the pad, the speed and direction of the movement, the accelerations and decelerations, the angle of the letter, etc.

It goes without saying that handwritten recognition assumes that users have reached a certain level of education. Furthermore, the technique does not seem to reach the level of reliability needed for financial transactions (the rate of false rejects remains sufficiently high) (Nalwa, 1999). A current project in the U.S. initiated by the Financial Services Technology Consortium (FSTC) aims at improving the procedures for check processing using handwritten recognition.

3.10.1.3 Keystroke Recognition

Keystroke recognition is a technique based on an individual's typing patterns in terms of rhythm, speed, duration, and pressure of keystrokes, etc. This is because human behavior in repetitive and routine tasks is strictly individual. Keystroke measures are based on several repetitions of a known sequence of characters (for example, the login and the password) (Dowland et al., 2001; Obaidat and Sadoun, 1999).

Net Nanny Software International, Inc., developed software entitled *Bio-Password LogOn for NT* (*http://www.biopassword.com*) that uses keyboard recognition for stations using Windows NT. The sample used to form the reference pattern must contain at least eight characters and must be used eight times. The verification phase requires 15 successful entries.

3.10.1.4 Retinal Recognition

The retina is a special tissue of the eye that responds to light pulses by generating proportional electrical discharges to the optical nerve. It is supplied by a network of blood vessels according to a configuration that is characteristic of each individual and that is stable throughout life. The retina can even distinguish among twins. A retinal map can be drawn by recording the reflections of a low-intensity infrared beam with the help of a charge-coupled device (CCD) to form a descriptor of 35 octets

The necessary equipment has been commercialized since 1975 by EyeDentify, Inc., (*http://www.eye-dentify.com*) at the cost of about $5000 per unit. As a consequence, this technique is used for access control to high-security areas: military installations, nuclear plants, high-security prisons, bank vaults, network operation centers, etc. According to the manufacturer, the enrollment time is less than one minute, and the verification time for a library of about 1500 does not exceed 5 seconds. The rate of false acceptance is extremely low (one per million). However, the subject has to look directly into the infrared retinal probe through a special eyepiece, which may be inconvenient. Furthermore, the rate of false rejects seems to be relatively large. Currently, this technique is not suitable for remote payment systems or for large-scale deployment.

3.10.1.5 Iris Recognition

The iris is the colored area between the white of the eye and the pupil. Its texture is an individual characteristic that remains constant for many years. As a consequence, the description of the iris texture was made with a numeric code of 256 octets (2048 bits). The accuracy is very high, and the error probability is on the order of 1 for 1.2 million. It is even possible to distinguish among identical twins and to separate the two irises of the same person.

This technique was started and patented by Iridian Technologies — previously known as IriScan, Inc., — a company formed by two ophthalmologists and a computer scientist (*http://www.iriscan.com*). The inspection is less invasive than in the case for the retinal scan. The person to be identified needs merely to face a camera connected to a computer at a distance of about 1 m. The size of the initial image of the iris is 20 K octets, which is then processed to produce the corresponding digital code. The operation takes less than about 800 msec with a computer with a clock speed of 66 MHz (Daugman 1994, 1999; Flom and Safir, 1987; Wildes, 1997). In online verification systems, this code is used together with the subject's personal identification number and the number of the person's bankcard.

Some precautions need to be respected during image capture, particularly to avoid reflections by ensuring uniform lighting. Contact lenses are detected through the presence of a regular structure in the processed image.

Iris recognition is now being evaluated to speed passenger processing at airports. Other potential applications include the identification of users of automatic bank teller machines, the control of access either to a physical building or equipment, or control of access to network resources.

3.10.1.6 Face Recognition

Face recognition is done on the basis of a template with sizes that range from 100 to 800 octets, constructed on the basis of some parameters, such as the distance between the eyes, the gap between the nostrils, the dimensions of the mouth, etc. This method can detect a person from a set of 5,000 to 50,000 images. The duration of the verification can take from 3 to 20 seconds, according to the size of the image library. However, sunglasses, beards or mustaches, grins, or head tilts of even 15 degrees can cause recognition errors. Some algorithms are so sensitive to the adjustment of the optics that they require the use of the same equipment for the acquisition of the reference image and of the image used for identification/verification.

A detailed examination of the error rates took place in 1996 and 1997 (at the instigation of the U.S. Army Research Laboratory) (Pentland and Choudhury, 2000; Phillips et al., 2000). The study covered 1196 persons for different lighting conditions, time intervals between the acquisition of the reference image, and the image used for classification. The results underlined that the rate of false rejects increases with the interval that separates the two images, as shown in Table 3.4.

TABLE 3.4

Rate of False Rejects in Automatic Face Recognition as a Function of the Interval Separating the Image Acquisitions

Category	Percentage of False Alarms	Percentage of False Rejects
Same day, same illumination	2	0.4
Same day, different illumination	2	9
Different days	2	11
Over 1.5 years difference	2	43

Source: Phillips, P.J., Martin, A., Wilson, C.L., and Przybocki, M., *Computer*, 3, 2, 56–63, 2000.

TrueFace™ of Miros was the first product of face recognition to be certified by the International Computer Security Association (ICSA) in 1998. It is being evaluated for check cashing systems. Identrix (*http://www.identrix.com*) — formerly Visionics Corporation — commercializes the FaceIt® algorithm from the Rockfeller University (*http://www.Faceit.com*). Visage Technology uses the algorithm developed at the Massachusetts Institute of Technology Media Laboratory.

3.10.1.7 Fingerprint Recognition

It is common knowledge that fingerprints are permanent characteristics of each individual. The traditional method for collecting fingerprints is to swipe the fingertips (or the palm) in a special ink and then press them over paper to record a negative image. This image is processed to extract user-specific information or *minutiae*. New imaging methods allow the capture of the fingerprints with optical, optoelectronic, electric, or thermal transducers. These methods can easily be adapted to applications of online or mobile e-commerce.

Fingerprints can be collected electronically by measuring the fluctuations in the capacitance between the user's fingers and sensors on the surface of a special mouse. These fluctuations can help draw the contour of the fingerprint. Another technique relies on a low-tension alternating current injected into the finger pulps to measure the changes in the electric fields between a resin plate on which the finger rests and the derma. These variations in the electric field reproduce faithfully the details of the fingerprint. Thermal techniques rely on a transducer to measure the temperature gradient on the mouse's surface, thereby localizing points of friction. Finally, optoelectronic methods employ a layer of polymers to record the image of the fingerprint on a polymer layer that converts the image into a proportional electric current.

During the enrollment phase, the user's fingerprint is recorded and then processed to extract the features or minutiae. These minutiae form the reference signature during verification. Therefore, they must include a set of stable and reliable indices that are not sensitive to defects in the image that may be introduced by dirty fingers, wounds, or deformities. Each minutia takes about 16 octets on the average; therefore, the image size varies between 500 and 5,000 octets, depending on the number of minutiae preserved and the rate of compression used.

To verify the identity of a person, the minutiae extracted from the new imprint are compared with those extracted from the reference image. The algorithms used must be insensitive to potential translations, rotations, and distortions. The degree of similarity between the two images analyzed is described in terms of an index that varies from 0 to 100%. The percentage of false rejects in commercial systems reaches about 3%, and the rate of false acceptance is less than one per million. With some equipment, the image of the full length and not only that of the finger extremities is used (Takeda et al., 1990).

TABLE 3.5

List of Several Commercial Offers for Online Recognition of Fingerprints

Phenomenon Exploited	Firm	Product	URL
Capacitance	Infineon	Finger-print Security	http://www.infineon.com
	Secugen	EyeD Mouse	http://www.secugen.com
Electric field	Authentec	FingerLoc	http://www.authentec.com
	Veridicom	FPS110	http://www.veridicom.com
Optics	Identix	BioCard/Touchlock	http://www.identix.com
Optoelectronics	Who?Vision	TactileSense	http://www.whovision.com
Temperature	Thomson-CSF	FingerChip	http://www.tcs.thomson-csf.com

The Society for Worldwide Interbank Financial Telecommunications (SWIFT) sponsored the development of a mouse with a capacitance transducer. Nevertheless, Secugen was the first to offer a commercial product for online users. For illustrative purposes, Table 3.5 lists some commercial offers for different physical phenomena under consideration.

3.10.1.8 Recognition of Hand Geometry

In the last several years, hand geometry recognition has been used in large-scale commercial applications to control access to enterprises, customs, hospitals, military bases, prisons, etc. In the U.S., some airports (e.g., New York and Newark) are using it to accelerate the admission of frequent travelers (those with more than five entries per year).

The user positions the hand on a plate facing the lens of a digital camera by spreading the fingers and resting them against guiding pins soldered on the plate. This plate is surrounded by mirrors on three sides to capture the hand sideways and from the top with a digital camera. The time for taking one picture is about 1.2 sec. Several pictures (three to five) are taken, and the average is stored in memory as a reference to the individual. Using a three-dimensional model and 90 input parameters, the hand geometry is described using a 9-octet vector.

Among the companies active in this field, in alphabetical order: BioMet Partners Inc. (*http://www.biomet.ch*) and Recognition Systems (*http://www.recogsys.com*).

3.10.2 Summary and Evaluation

Given in Table 3.6 is the required memory for storing selected biometric identifiers (Sherman et al., 1994; Nanavati et al., 2002).

At this stage, and regardless of the biometric technology, there is little commonality among the various methods being proposed and their implementations. In the face of such a lack of standards, potential users hesitate

TABLE 3.6

Required Storage Memory for Biometric Identifiers

Identifier	Required Memory (octets)
Photo image	1000–1500
Voiceprint	1000–2000
Handwritten scan	500–1500
Face recognition	500–1000
Fingerprint	500–5000
Iris scan	256–512
Retinal scan	35
Hand geometry	9

to develop their particular solutions. There are no agreed-upon protocols for measuring and comparing total system performance in terms of processing speed, reliability, security, and vulnerability in an operational environment. Users are concerned about the long-term viability of any solution they may select, and the cost of switching methods or suppliers in the future. A related concern is that of being locked into a specific implementation or supplier. Software developers, in turn, are not sure as to what options to include in their programs. Application developers, also, are not sure what method deserves their full attention. Clearly, the lack of standards is hampering the wide-scale acceptance of biometric identification.

Awareness of these roadblocks spurred standardization activities to facilitate data exchanges among various implementations, irrespective of the biometric method. NIST and the Federal Bureau of Investigations (FBI) collaborated to produce a large database of fingerprints gathered from crime scenes, with their corresponding minutiae. This database will help train and evaluate new algorithms for automatic fingerprint recognition.

In 1995, the Committee on Security Policy Board established by President Clinton chartered the Biometric Consortium (BC) to be the focal point for the U.S. government on research, development, testing, evaluation, and application of biometric-based systems for personal identification and verification. The BC cosponsors activities at NIST and at San Jose State University in California.

The U.S. Department of Defense (DOD) initiated a program to develop a standard application interface called the *Human-Authentication Application Program Interface* (HA-API) to decouple the software of the applications from the technology used to capture the biometric data. After publishing, in April 1998, Version 2.0 of this API, activities merged with those of the BioAPI Consortium (*http://www.bioapi.org*). This consortium groups hardware and software companies as well as suppliers of biometric peripherals. In March 2000, the consortium published Version 1.0 of a BioAPI and reference realizations for Windows, Unix, Linux, and Java. All of these implementations are in the public domain. Despite Microsoft's withdrawal from the consortium, the BioAPI specification was the basis of the ANSI INCITS 358 (2002), a standard that the Technical Committee M1 on Biometrics for the InterNational Committee for Information Technology Standards (INCITS) developed as an ANSI standard.

In parallel, efforts within the ANSI X9.F4 working group resulted in a common format in which to exchange biometric data among various systems known as Common Biometric Exchange File Format (CBEFF). This is the format to be recognized by the BioAPI. It was agreed that the International Biometric Industry Association (IBIA), based in the U.S. (*http://www.ibia.org*), will act as the registration authority for the formats to be recognized. Finally, ANSI X9.84 (2001) defined a data object model that is compatible with CBEFF and is suitable for securing physical and remote access within the financial industry. The standard gives guidance on the proper controls and procedures for using biometrics for identification and authentication.

Other standardization initiatives are pursued by the Association for Biometrics (*http://www.afb.org.uk*) in the U.K. and the *Bundesamt für Sicherheit in der Informationtechnik* (BSI — Federal Information Security Agency) (*http://www.bsi.bund.de*) in Germany. Finally, joint work by ISO and IEC aims at a standard for personal verification through biometric methods with the use of integrated circuit cards (e.g., smart cards). Potential applications include driver licenses and travel documents. The standard will be issued as ISO/IEC 7816, Part 11.

3.11 Authentication of the Participants

The purpose of authentication of participants is to reduce, if not eliminate, the risk that intruders might masquerade under legitimate appearances to pursue unauthorized operations.

As previously stated, when the participants utilize a symmetric encryption algorithm, they are the only ones who share a secret key. As a consequence, the utilization of this algorithm guarantees, in theory, the confidentiality of the messages, the correct identification of the correspondents, and their authentication. The key distribution servers also act as authentication servers, and the good functioning of the system depends on the capability of all participants to protect the encryption key.

In contrast, when the participants utilize a public key algorithm, a user is considered authentic when that user can prove that he or she holds the private key that corresponds with the public key attributed to the user. A certificate issued by a certification authority indicates that it certifies the association of the public key (and therefore the corresponding private key) with the recognized identity. In this manner, identification and authentication proceed in two different ways, identity with the digital signature and authentication with a certificate. Without such a guarantee, a hostile user could create a pair of private/public keys and then distribute the public key as if it were that of the legitimate user.

Although the same public key of a participant could equally serve to encrypt the message that is addressed to that participant (confidentiality service) and to verify the electronic signature of the documents that the participant transmits (integrity and identification services), in practice, a different public key is used for each set of services.

According to the authentication framework defined by ITU-T Recommendations X.500 (2001) and X.811 (1995), simple authentication may be achieved by one of several means:

1. Name and password in the clear
2. Name, password, and a random number or a time stamp, with integrity verification through a hash function
3. Name, password, a random number, and a time-stamp, with integrity verification using a hash function

Strong authentication requires a certification infrastructure that includes the following entities:

1. *Certification authorities* to back the users' public keys with "sealed" certificates (i.e., signed with the private key of the certification authority) after verification of the physical identity of the owner of each public key.
2. A database of authentication data (*directory*) that contains all the data relative to the private encryption keys, such as their values, the duration of validity, and the identity of the owners. Any user should be able to query such a database to obtain the public key of the correspondent or to verify the validity of the certificate that the correspondent would present.
3. A *naming* or *registering authority* may be distinct from the certification authority, and its principal role is to define and assign unique *distinguished names* to the different participants.

The certificate guarantees correspondence between a given public key and the entity whose unique distinguished name is contained in the certificate. This certificate is sealed with the private key of the certification authority. When the certificate owner signs documents with the private signature key, the partners can verify the validity of the signature with the help of the corresponding public key contained in the certificate. Similarly, to send a confidential message to a certified entity, it is sufficient to query the directory for the public key of that entity and then use that key to encrypt messages that only the holder of the associated private key would be able to decipher.

3.12 Access Control

Access control is the process by which only authorized entities are allowed access to the resources as defined in the access control policy. It is used to counter the threat of unauthorized operations, such as unauthorized use, disclosure, modification, destruction of protected data, or denial of service to legitimate users. ITU-T Recommendation X.812 (1995) defines the framework for access control in open networks. Accordingly, access control can be exercised with the help of a supporting authentication mechanism at one or more of the following layers: the network layer, the transport layer, or the application layer. Depending on the layer, the corresponding authentication credentials may be X.509 certificates, Kerberos tickets, simple identity and password pairs, etc.

There are two types of access control mechanisms: identity-based and role-based. Identity-based access control uses the authenticated identity of an entity to determine and enforce its access rights. In contrast, for role-based access control, access privileges depend on the job function and its context. Thus, additional factors may be considered in the definition of the access policy, for example, the strength of the encryption algorithm, the type of operation requested, or the time of day. Role-based access control provides an indirect means of bestowing privileges through three distinct phases: the definition of roles, the assignment of privileges to roles, and the distribution of roles among users. This facilitates the maintenance of access control policies, because it is sufficient to change the definition of roles to allow global updates without revising the distribution from top to bottom.

At the network layer, access control in IP networks is based on packet filtering using the protocol information in the packet header, specifically, the source and destination IP addresses and the source and destination port numbers. Access control is achieved through "line interruption" by a certified intermediary or a firewall that intercepts and examines all exchanges before allowing them to proceed. The intermediary is thus located between the client and the server, as indicated in Figure 3.9. Furthermore, the firewall can be charged with other security services, such as encrypting the traffic for confidentiality at the network level or verifying integrity using digital signatures. It can also inspect incoming and outgoing exchanges before

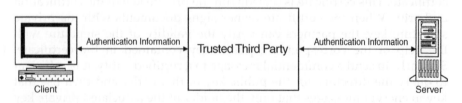

FIGURE 3.9
Authentication by line interruption at the network layer.

forwarding them to enforce the security policies of a given administrative domain. However, the intervention of the trusted third party must be transparent to the client.

The success of packet filtering is vulnerable to packet spoofing if the address information is not protected and if individual packets are treated independently of the other packets of the same flow. As a remedy, the firewall can include a proxy server or an application-level gateway that implements a subset of application-specific functions. The proxy is capable of inspecting all packets in light of previous exchanges of the same flow before allowing their passage in accordance with the security policy in place. Thus, by filtering incoming and outgoing electronic mail, file transfers, exchanges of Web applications, etc., application gateways can block nonauthorized operations and protect against malicious codes such as viruses. This is called a *stateful* inspection. The filter uses a list of keywords, the size and nature of the attachments, the message text, etc. Configuring the gateway is a delicate undertaking, because the intervention of the gateway should not prevent daily operation.

A third approach is to centralize the management of the access control for a large number of clients and users with different privileges with a dedicated server. Several protocols were defined to regulate the exchanges among network elements and access control servers. IETF RFC 2865 (2000) specifies Remote Authentication Dial-in User Service (RADIUS) for client authentication, client authorization, and collection of accounting information of the calls. In IETF RFC 1492 (1993), Cisco described a protocol called Terminal Access Controller Access System (TACACS) that was later updated in TACACS+. Both RADIUS and TACACS+ require a secret key between each network element and the server. Depicted in Figure 3.10 is the operation of RADIUS in terms of a client/server architecture. The RADIUS client resides

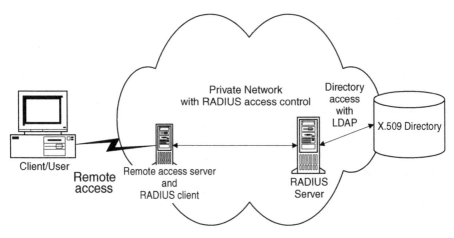

FIGURE 3.10
Remote access control with RADIUS.

within the access control server, while the server relies on an ITU-T X.509 directory through the protocol Lightweight Directory Access Protocol (LDAP). Both X.509 and LDAP will be presented later in this chapter.

Note that both server-to-client authentication and user-to-client authentication are outside the scope of RADIUS. Also, because RADIUS does not include provisions for congestion control, large networks may suffer degraded performance and data loss.

Commercial systems implement two basic approaches for end-user authentication: one-time password and challenge-response (Forrester et al., 1998). In a typical one-time password system, each user has a device that generates a number periodically (usually every minute) using the current time, the card serial number, and a secret key held in the device. The generated number is the user's one-time password. This procedure requires that the time reference of the access control server be synchronized with the card so that the server can regenerate an identical number.

In challenge-response systems, the user enters a personal identification number to activate handheld authenticators (HHA) and then to initiate a connection to an access control server. The access control server, in turn, provides the user with a random number (a challenge), and the user enters this number into a handheld device to generate a unique response. This response depends on both the challenge and some secret key shared between the user's device and the server. It is returned to the access control server to compare with the expected response and decide accordingly.

It should be noted that there are some known vulnerabilities in RADIUS or in its implementations (Hill, 2001).

3.13 Denial of Service

Denial of service attacks prevent normal network usage by blocking the access of legitimate users to the network resources they are entitled to, by overwhelming the hosts with additional or superfluous tasks to prevent them from responding to legitimate requests, or to slow their response times below satisfactory limits.

In a sense, denial of service results from the failure of access control. Nevertheless, these attacks are inherently associated with IP networks for two reasons: network control data and user data share the same physical and logical bandwidths; and IP is a connectionless protocol, where the concept of admission control does not apply. As a consequence, when the network size exceeds a few hundred nodes, network control traffic (due, for example, to the exchange of routing tables) may, under some circumstances, occupy a significant portion of the available bandwidth. Further, inopportune or ill-intentioned user packets may be able to bring down a network element (e.g., a router), thereby affecting not only all end points that rely on this network element for connectivity, but also all other network elements that depend on it to update their view of the network status. Finally, in distributed denial of service attacks (DDOS), a sufficient number of compromised hosts may send useless packets toward a victim at around the same time, thereby affecting the victim's resources or bandwidths or both (Chang, 2002; Moore et al., 2001).

As a point of comparison, the current public switched telephone network uses an architecture called Common Channel Signaling (CCS), whereby user data and network control data travel on totally separate networks and facilities. It is worth noting that CCS was introduced to protect against fraud. In the old architecture, called Channel-Associated Signaling (CAS), the network data and the user data used separate logical channels, on the same physical support. Similarly, experience has shown that Asynchronous Transfer Mode (ATM) can be exposed to the same risks of interruption, because the user traffic and the network control messages share the same facilities, even though they are virtually distinct (Sherif et al., 2001).

Let us illustrate the preceding discussion with a few examples of denial of service attacks using several protocols of the IP stack: TCP, Internet Control Message Protocol (ICMP), and HTTP (HyperText Transfer Protocol):

- The SYN flooding attack, one of the best-known mechanisms of denial of service, perturbs the functioning of the TCP protocol (Schuba et al., 1997). It is well known that the handshake in TCP is a three-way exchange: a connection request with the SYN packet, an acknowledgment of that request with the SYN/ACK packet, and finally a confirmation from the first party with the ACK packet (Comer, 1995, p. 216). Unfortunately, the handshake imposes asymmetric memory and computational loads on the two end points, the destination being required to allocate large amounts of memory without authenticating the initial request. Thus, an attacker can paralyze the target machine by exhausting its available resources by sending a massive number of fake SYN packets. These packets will have spoofed source addresses, so that the acknowledgments are sent to hosts that the victim cannot reach or that do not exist. Otherwise, the attack may fail, because unsolicited SYN/ACK packets at accessible hosts provoke the transmission of RST packets, which, upon arrival, would allow the victim to release the resources allocated for a connection attempt.

- ICMP is a protocol for any arbitrary machine to use to communicate control and error information back to the presumed source. This — an ICMP echo request, or "ping," with the victim's address falsely indicated as the source and sent to all the machines of a given network using the subnet broadcast address — can flood the victim with echo replies that will overwhelm its capacities.

- The Code Red worm exploits defects in the response of some Web servers to an HTTP GET request larger than the regular size (a payload of 62 octets instead of 60 octets). Under specific conditions, the buffer overflow causes an upsurge in HTTP traffic and the infection of neighboring machines, which increases network traffic, thereby causing a massive disruption (CERT/CC CA-2001-19, 2002).

Given that IP does not separate user traffic from that of the network, the best solution is to identify all with trusted certificates. However, authentication of all exchanges increases the computational load, which may be excessive in commercial applications, as the lack of success of the protocol for payments with bankcard Secure Electronic Transaction (SET) has shown. Short of this, defense mechanisms will be developed on a case-by-case basis to address specific problems as they arise. For example, resource exhaustion due to the SYN attack can be alleviated by limiting the number of concurrent pending TCP connections, by reducing the time out for the arrival of the ACK packet before calling off the connection establishment, and by blocking packets to the outside that have source addresses from outside.

Another approach is to reequilibrate the computational load between the two parties by asking the requesting client to solve a *puzzle* in the form of simple cryptographic problems before being granted the allocated resources needed to establish a connection. To avoid replay attacks, these problems are formulated using the current time, a server secret, and additional information from the client request (Juels and Brainard, 1999). This approach, however, requires programs for solving puzzles specific to each application that are incorporated in the client browser.

3.14 Nonrepudiation

Nonrepudiation is a service that prevents a person who accomplished an act from denying it later, in part or as a whole. Nonrepudiation is a legal concept to be defined through legislation. The role of informatics is to supply the necessary technical means to support the service offer according to the law. The building blocks of nonrepudiation include the electronic signature of documents, the intervention of a third party as a witness, time-stamping, and sequence numbers. Among the mechanisms for nonrepudiation are a security token sealed with the secret key of the verifier that accompanies the transaction record, time-stamping, and sequence numbers. Depending on the system design, the security token sealed with the verifier's secret key can be stored in a tamper-resistant cryptographic module. The generation and verification of the evidence often require the intervention of one or more entities external to parties to the transaction, such as a notary, a verifier, and an adjudicator of disputes.

ITU-T Recommendation X.813 (1996) defines a general framework for nonrepudiation in open systems. Accordingly, the service comprises the following measures:

- Generation of the evidence
- Recording of the evidence
- Verification of the evidence generated
- Retrieval and reverification of the evidence

There are two types of nonrepudiation services:

1. *Nonrepudiation at the origin*: This service protects the receiver by preventing the sender from denying having sent the message.
2. *Nonrepudiation at the destination*: This service plays the inverse role of the preceding function. It protects the sender by demonstrating that the addressee received the message.

Threats to nonrepudiation include compromise of keys or unauthorized modification or destruction of evidence. In public key cryptography, each user is the sole and unique owner of the private key. Thus, unless the whole system was penetrated, a given user cannot repudiate the messages that are accompanied by his or her electronic signature. In contrast, nonrepudiation is not readily achieved in systems that use symmetric cryptography. A user can deny having sent the message by alleging that the receiver compromised the shared secret or that the key distribution server was successfully attacked. A trusted third party would have to verify each transaction to be able to testify in cases of contention.

Nonrepudiation at the destination can be obtained using the same mechanisms but in the reverse direction.

3.14.1 Time-Stamping and Sequence Numbers

Time-stamping of messages establishes a link between each message and the date of its transmission. This permits the tracing of exchanges and prevents attacks by replaying old messages. If clock synchronization of both parties is difficult, a trusted third party can intervene as a notary and use its clock as reference.

The intervention of the "notary" can be in either of the following modes:

- Offline to fulfill functions such as certification, key distribution, and verification if required, without intervening in the transaction
- Online as an intermediary in the exchanges or as an observer collecting the proof that might be required to resolve contentions. This is a similar role to that of a trusted third party of the network layer (firewall) or at the application layer (proxy) but with a different set of responsibilities.

Let us assume that a trusted third party combines the functions of the notary, the verifier, and the adjudicator. Each entity encrypts its messages with the secret key that was established with the trusted third party before sending the message. The trusted third party decrypts the message with the help of this shared secret with the intervening party, time-stamps it, and then reencrypts it with the key shared with the other party. This approach requires establishment of a secret key between each entity and the trusted third party that acts as a delivery messenger. Notice, however, that the time-stamping procedures were not normalized, and each system has its own protocol.

Detection of duplication, replay, as well as the addition, suppression, or loss of messages is achieved with the use of a sequence number before encryption. Another mechanism is to add a random number to the message before encryption. All these means give the addressee the ability to verify that the exchanges genuinely took place during the time interval that the time-stamp defines.

3.15 Secure Management of Cryptographic Keys

Key management is a process that continues throughout the life cycle of the keys to thwart unauthorized disclosures, modifications, substitutions, reuse of revoked or expired keys, or unauthorized utilizations. Security at this level is a recursive problem, because the same security properties that are required in the cryptographic system must be satisfied, in turn, by the key management system.

The secure management of cryptographic keys relates to key production, storage, distribution, utilization, withdrawal from circulation, deletion, and archiving (Fumer and Landrock, 1993).

3.15.1 Production and Storage

Key production must be done in a random manner and at regular intervals, depending on the degree of security required.

Protection of the stored keys has a physical aspect and a logical aspect. Physical protection consists of storing the keys in safes or in secured buildings with controlled access, whereas logical protection is achieved with encryption.

In the case of symmetric encryption algorithms, only the secret key is stored. For public key algorithms, storage encompasses the user's private and public keys, the user's certificate, and a copy of the public key of the certification authority. The certificates and the keys may be stored on the hard disk of the certification authority, but there is some risk of possible attacks or of loss due to hardware failure. In cases of microprocessor cards, the information related to security, such as the certificate and the keys, are inserted during card personalization. Access to this information is then controlled with a confidential code.

3.15.2 Distribution

The security policy defines the manner in which keys are distributed to entitled entities. Manual distribution by mail or special dispatch (sealed envelopes, tamper-resistant module) is a slow and costly operation that should only be used for the distribution of the root key of the system. This is the key that the key distributor utilizes to send to each participant their keys.

An automatic key distribution system must satisfy all of the criteria of security, in particular:

- Confidentiality
- Identification of the participant
- Data integrity, by giving proof that the key was not altered during transmission or that it was not replaced by a fake key
- Authentication of the participants
- Nonrepudiation

Automatic distribution can be point-to-point or point-to-multipoint. The Diffie–Hellman key exchange method (Diffie and Hellman, 1976) allows the two partners to construct a master key with elements that were previously exchanged in the clear. A symmetric session key is formed next on the basis of the data encrypted with this master key or with a key derived from it and exchanged during the identification phase.

To distribute keys to several customers, an authentication server can also play the role of a trusted third party and distribute the secret keys to the different parties. These keys will be used to protect the confidentiality of the messages carrying the information on the key pairs.

3.15.3 Utilization, Withdrawal, and Replacement

The unauthorized duplication of a legitimate key is a threat to the security of key distribution. To prevent this type of attack, a unique parameter can be concatenated to the key, such as a time-stamp or a sequence number that increases monotonically (up to a certain module).

The risk that a key is compromised increases proportionately with time and with usage. Therefore, keys have to be replaced regularly without causing service interruption. A common solution that does not impose a significant load is to distribute the session keys on the same communication channels used for user data. For example, in the SSL protocol, the initial exchanges provide the necessary elements to form keys that would be valid throughout the session at hand. These elements flow encrypted with a secondary key, called a key encryption key, to keep their confidentiality.

Key distribution services have the authority to revoke a key before its date of expiration after a key loss or because of the user's misbehavior.

3.15.4 Key Revocation

If a user loses the right to employ a private key, if this key is accidentally revealed, or, more seriously, if the private key of a certification authority is broken, all the associated certificates must be revoked without delay. Furthermore, these revocations have to be communicated to all the verifying entities in the shortest possible time. Similarly, the use of the revoked key by a hostile user should not be allowed. Nevertheless, the user will not be able to repudiate all the documents already signed and sent before revocation of the key pair.

3.15.5 Deletion, Backup, and Archiving

Key deletion implies the destruction of all memory registers as well as magnetic or optical media that contain the key or the elements needed for its reconstruction.

Backup applies only to encryption keys and not to signature keys; otherwise, the entire structure for nonrepudiation would be put into question.

The keys utilized for nonrepudiation services must be preserved in secure archives to accommodate legal delays that may extend for up to 30 years. These keys must be easily recoverable in case of need, for example, in response to a court order. This means that the storage applications must include mechanisms to prevent unrecoverable errors from affecting the ciphertext.

3.15.6 Comparison between Symmetric and Public Key Cryptography

Systems based on symmetric key algorithms pose the problem of ensuring the confidentiality of key distribution. This translates into the use of a separate secure distribution channel that is preestablished between the participants. Furthermore, each entity must have as many keys as the number of participants with whom it will enter into contact. Clearly, management of symmetric keys increases exponentially with the number of participants.

Public key algorithms avoid such difficulties because each entity owns only one pair of private and public keys. Unfortunately, the computations for public key procedures are more intense than those for symmetric cryptography. The use of public key cryptography to ensure confidentiality is only possible when the messages are short, even though data compression before encryption with the public key often succeeds in speeding the computations. Thus, public key cryptography can complement symmetric cryptography to ensure the safe distribution of the secret key, particularly when safer means, such as direct encounter of the participants, or the intervention of a trusted third party, are not feasible. Thus, a new symmetric key could be distributed at the start of each new session and, in extreme cases, at the start of each new exchange.

3.16 Exchange of Secret Keys: Kerberos

Kerberos is the mostly widely known system for automatic exchange of keys using symmetric encryption. Its name is that of the three-headed dog that, according to Greek mythology, was guarding the gates of Hell.

Kerberos offers the services of online identification and authentication as well as access control using symmetric cryptography (Neuman and Ts'o, 1994). It allows management of access to the resources of an open network from nonsecure machines, such as management of student access to the resources of a university computing center (files, printers, etc.). Kerberos is now the default authentication option in Windows 2000.

The development of Kerberos started in 1978 within the Athena project at the Massachusetts Institute of Technology (MIT), financed by Digital Equipment Corporation (DEC) and IBM. Version 5 of Kerberos, which was published in 1994, is the version currently in use. This version is also included, with some modifications, in the micropayment system NetBill. This chapter presents the basic principles of Kerberos, leaving its adaptations to NetBill for Chapter 10

The system is built around a Kerberos key distribution center that enjoys the total trust of all participants with whom it has already established symmetric encryption keys. Symmetric keys are attributed to individual users for each of their accounts when they register in person.

The key distribution center consists of an authentication server (AS) and a ticket-granting server (TGS). The AS controls access to the TGS, which in turn, controls access to specific resources. Every server shares a secret key with every other server. The algorithm used for symmetric encryption is the Data Encryption Standard (DES). Finally, during the registration of the users in person, a secret key is established with the AS for each user's account. With this arrangement, a client has access to multiple resources during a session with one successful authentication, instead of repeating the authentication process for each resource. The operation is explained below.

After identifying the end user with the help of a log-in and password pair, the AS sends to the client a session symmetric encryption key to encrypt data exchanges between the client and the TGS. The session key is encrypted with the symmetric encryption key shared between the user and the AS. The key is also contained in the session ticket that is encrypted with the key preestablished between the TGS and the AS.

The session ticket, also called a ticket-granting ticket, is valid for a short period, typically a few hours. During this period, it can be used to request access to a specific service; this is why it is also called an initial ticket.

The client presents the TGS with two items of identification: the session ticket and an authentication title that is encrypted with the session key. The TGS compares the data in both items to verify client authenticity and its access privileges before granting access to the specific server requested.

Depicted in Figure 3.11 are the interactions among the four entities: the client, the AS, the TGS, and the desired merchant server or resource S.

The exchanges are now explained.

3.16.1 Message (1) — Request of a Session Ticket

A client C that desires to access a specific server S first requests an entrance ticket to the session from the Kerberos AS. To do so, the client sends a message consisting of an identifier (for example, a log-in and a password), the identifier of the server S to be addressed, a time-stamp H_1, as well as a random number Rnd, both to prevent replay attacks.

3.16.2 Message (2) — Acquisition of a Session Ticket

The Kerberos authentication server responds by sending a message formed of two parts: a session key K_{CTGS} and the number Rnd that was in the first

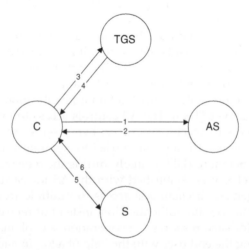

FIGURE 3.11
Authentication and access control in Kerberos.

message, both coded with the client's secret key K_C, and the session ticket T_{CTGS} destined for the TGS and encrypted by the latter's secret key between itself and the Kerberos AS.

The session (ticket-granting ticket) includes several pieces of information, such as the client name C, its network address Ad_C, the time-stamp H_1, the period of validity of the ticket *Val*, and the session key K_{CTGS}. All these items, with the exception of the server identity TGS, are encrypted with the long-term key K_{TGS} that the TGS shares with the AS. Thus,

$$T_{CTGS} = \text{TGS}, K_{TGS}\{C, Ad_C, H_1, Val, K_{CTGS}\}$$

and the message sent to the client is

$$K_C\{K_{CTGS}, Rnd\}, T_{CTGS}$$

where $K\{x\}$ indicates encryption of the message x with the shared secret key K. The client decrypts the message with its secret key K_C to recover the session key K_{CTGS} and the random number. The client verifies that the random number received is the same as was sent as a protection from replay attacks. The time stamp H_1 is also used to protect from replay attacks. Although the client will not be able to read the session ticket because it is encrypted with K_{TGS}, it can extract it and relay it to the server.

By default, the session ticket T_{CTGS} is valid for 8 hours. During this time, the client can obtain several service tickets to different services without needing new authentication.

3.16.3 Message (3) — Request of a Service Ticket

The client constructs an authentication title *Auth* that contains its identity C, its network address Ad_C, the service requested S, a new time-stamp H_2, and another random number Rnd_2, and then encrypts it with the session key K_{CTGS}. The encrypted authentication title can be represented in the following form:

$$Auth = K_{CTGS}\{C, Ad_C, S, H_2, Rnd_2\}$$

The request of the service ticket consists of the encrypted authentication title and the session ticket T_{CTGS}:

$$\text{Service Request} = Auth, T_{CTGS}$$

3.16.4 Message (4) — Acquisition of the Service Ticket

The TGS decrypts the ticket content with its secret key K_{TGS}, deduces the shared session key K_{CTGS}, and extracts the data related to the client's service request. With knowledge of the session key, the server can decrypt the authentication title and compare the data in it with that the client supplied. This comparison gives formal proof that the client is the entity that was given the session ticket by the server. The time-stamps confirm that the message was not an old message that was replayed. Next, the TGS returns a service ticket for accessing the specific server S.

The exchanges described by Messages (3) and (4) can be repeated for all other servers available to the user as long as the validity of the session ticket has not expired.

The message from the TGS has two parts: the first contains a service key K_{CS} between the client and the server S and the number Rnd_2, both coded with shared secret key K_{CTGS}; and the second includes the service ticket T_{CS} destined for the server S and encrypted by secret key, K_{STGS}, shared between the server S and the TGS.

As before, the service ticket destined for the server S includes several pieces of information, such as the identity of the server S, the client name C, its network address Ad_C, a time-stamp H_3, the period of validity of the ticket Val, and if confidentiality is desired, a service key K_{CS}. All these items, with the exception of the server identity S, are encrypted with the long-term key K_{STGS} that the TGS shares with the specific server. Thus,

$$T_{CS} = S, K_{STGS}\{C, Ad_C, H_3, Val, K_{CS}\}$$

and the message sent to the client is

$$K_{CTGS}\{K_{CS}, Rnd_2\}, T_{CS}$$

The client decrypts the message with the shared secret key K_{CTGS} to recover the service key K_{CS} and the random number. The client verifies that the random number received is the same as was sent as a protection from replay attacks.

3.16.5 Message (5) — Service Request

The client constructs a new authentication title $Auth_2$ that contains its identity C, its network address Ad_c, a new time-stamp H_3, and another random number Rnd_3 and then encrypts it with the service key K_{CS}. The encrypted authentication title can be represented as follows:

$$Auth_2 = K_{CS}\{C, Ad_C, H_4, Rnd_3\}$$

The request of the service consists of the encrypted new authentication title and the service ticket T_{CS}:

$$\text{Service Request} = Auth_2, T_{CS}$$

3.16.6 Message (6) — Optional Response of the Server

The server decrypts the content of the service ticket with the key K_{STGS} it shares with the TGS to derive the service key K_{CS} and the data related to the client. With knowledge of the service key, the server can verify the authenticity of the client. The time stamps confirm that the message is not a replay of old messages. If the client requested the server to authenticate itself, it will return the random number, Rnd_3, encrypted by the service key K_{CS}. Without knowledge of the secret key K_{CS}, the server would not be able to extract the service key K_{CS}.

The preceding description shows that Kerberos is mostly suitable for networks administered by a single administrative entity. In particular, the Kerberos key distribution center fulfills the following roles:

- It maintains a database of all secret keys (except of the key between the client and the server, K_{CS}). These keys have a long lifetime.

- It keeps a record of users' login identities, passwords, and access privileges. To fulfill this role, it may need access to an X.509 directory.

- It produces and distributes encryption keys and ticket-granting tickets to be used for a session.

3.17 Public Key Kerberos

The utilization of a central depot for all symmetric keys increases the potential of traffic congestion due to the simultaneous arrival of many requests. In addition, centralization threatens the whole security infrastructure, because a successful penetration of the storage could put all the keys in danger (Sirbu and Chuang, 1996). Finally, the management of the symmetric keys (distribution and update) becomes a formidable task when the number of users increases.

The public key version of Kerberos simplifies key management, because the server authenticates the client directly using the session ticket and the client's certificate sealed by the Kerberos certification authority. The session ticket is sealed with the client's private key and then encrypted with the server public key. Thus, the service request to the server can be described as follows:

$$\text{Service Request} = S, PK_S \{Tauth, Kr, Auth\}$$

with

$$Auth = C, certificate, [Kr, S, PK_C, Tauth]SK_C$$

where *Tauth* is the initial time for authentication, *Kr* is a one-time random number that the server will use as a symmetric key to encrypt its answer, {…} represents encryption with the server public key, PK_s, while […] represents the seal computed with the client's private key, SK_C. This architecture improves speed and security.

The operations of public key Kerberos are described in IETF RFC 1510 (1996).

3.17.1 Where To Find Kerberos?

The official Web page for Kerberos is located at *http://web.mit.edu/kerberos/www/index.html*. A FAQ (Frequently Asked Questions) file on Kerberos can be consulted at the following address: *ftp://athena-dist.mit.edu/pub/kerberos/KERBEROS.FAQ*. Tung (1999) offers a good compendium of information on Kerberos.

The Swedish Institute of Computer Science is distributing a free version of Kerberos, called Heidmal. This version was written by Johan Danielsson and Assar Westerlund and includes improvements in security protocols, such as the support of Triple DES. A commercial version is TrustBroker available from CyberSafe at *http://www.cybersafe.com*.

3.18 Exchange of Public Keys

3.18.1 Diffie–Hellman Exchange

The Diffie–Hellman algorithm, published in 1976, is the first algorithm for key exchange in public key algorithms. It exploits the difficulty in calculating discrete algorithms in a finite field, as compared with the calculation of exponentials in the same field.

The key exchange comprises the following steps:

1. The two parties agree on two random large integers, *n* and *g*, such that *g* is a prime with respect to *n*. These two numbers do not have to necessarily be hidden, but their choice can have a substantial impact on the strength of the security achieved.

2. *A* chooses a large random integer *x* and sends *B* the result of the computation:

$$X = g^x \bmod n$$

3. *B* chooses another large random integer *y* and sends to *A* the result of the computation:

$$Y = g^y \bmod n$$

4. *A* computes:

$$k = Y^x \bmod n = g^{xy} \bmod n$$

5. Similarly, *B* computes:

$$k = Y^x \bmod n = g^{xy} \bmod n$$

The value *k* is the secret key that both correspondents exchanged. Even by listening to all exchanges, it would be difficult to discover the key, unless there is a suitable way to calculate the discrete algorithm of *X* or of *Y* to rediscover the value of *x* or of *y*.

The SSL uses the method called ephemeral Diffie–Hellman, where the exchange is short-lived, thereby achieving *perfect forward secrecy*, i.e., that a key cannot be recovered after its deletion. The Diffie–Hellman parameters are signed with the algorithms RSA or the DSS to guarantee integrity.

It should be noted that on March 29, 1997, the technique for key exchange entered the public domain.

3.19 ISAKMP (Internet Security Association and Key Management Protocol)

IETF RFC 2408 (1998) defines ISAKMP (Internet Security Association and Key Management Protocol), a generic framework to negotiate point-to-point security associations and to exchange key and authentication data between two parties. In ISAKMP, the term security association has two meanings. It is used to describe the secure channel established between two communicating entities. It can also be used to define a specific instance of the secure channel, i.e., the services, mechanisms, protocol, and protocol-specific set of parameters associated with the encryption algorithms, the authentication mechanisms, the key establishment and exchange protocols, and the network addresses. In ISAKMP, a domain of interpretation (DOI) is the context of operation in terms of the relevant syntax and semantics. The IETF RFC 2407 (1998) defines the IP security DOI for security associations in IP networks within the ISAKMP framework.

ISAKMP specifies the formats of messages to be exchanged and their building blocks (payloads). A fixed header precedes a variable number of payloads chained together to form a message. This provides a uniform management layer for security at all layers of the ISO protocol stack, thereby reducing the amount of duplication within each security protocol. This centralization of the management of security associations has several advantages. It reduces connect setup time, improves reliability of software, and allows for future evolution when improved security mechanisms are developed, particularly if new attacks against current security associations are discovered.

To avoid subtle mistakes that can render a key exchange protocol vulnerable to attacks, ISAKMP includes five default exchange types. Each exchange specifies the content and the ordering of the messages during communications between the peers.

Although ISAKMP can run over TCP or UDP, many implementations use UDP on port 500. Because the transport with UDP is unreliable, reliability is built into ISAKMP.

The header includes, among other information, two 8-octet "cookies" (also called "syncookies") that constitute an *anticlogging* mechanism because of their role against TCP SYN flooding. Each side generates a cookie specific to the two parties and assigns it to the remote peer entity. The cookie is constructed, for example, by hashing the IP source and destination addresses, the UDP source and destination ports, and a locally generated secret random value. ISAKMP recommends including the data and the time in this secret value. The concatenation of the two cookies identifies the security association and gives some protection against the replay of old packets or SYN flooding attacks. The protection against SYN flooding assumes that the attacker will not intercept the SYN/ACK packets sent to the spoofed addresses used in the attack. As explained earlier, the arrival of unsolicited SYN/ACK packets at a host that is accessible to the victim will elicit the transmission of an RST packet, thereby telling the victim to free the allocated resources so that the host, whose address was spoofed, will respond by resetting the connection (Juels and Brainard, 1999; Simpson, 1999).

The negotiation in ISAKMP comprises two phases: the establishment of a secure channel between the two communicating entities and the negotiation of security associations on the secure channel. For example, in the case of IPSec, Phase I negotiation is to define a key exchange protocol, such as the IKE (Internet Key Exchange) and its attributes. Phase II negotiation concerns the cryptographic algorithms to achieve IPSec functionality.

IKE is an authenticated exchange of keys consistent with ISAKMP. It is a hybrid protocol that combines aspects of the Oakley Key Determination Protocol and of SKEME. Oakley utilizes the Diffie–Hellman key exchange mechanism with signed temporary keys to establish the session keys between the host machines and the network routers. SKEME is an authenticated key exchange that uses public key encryption for anonymity and nonrepudiation and provides a means for quick refreshment (Krawczyk, 1996). IKE is the default key exchange protocol for IPSec.

None of the data used for key generation is stored, and a key cannot be recovered after deletion, thereby achieving perfect forward secrecy. The price is a heavy cryptographic load, which becomes more important the shorter the duration of the exchanges. Therefore, to minimize the risks from denial of service attacks, ISAKMP postpones the computationally intensive steps until authentication is established.

Unfortunately, despite the complexity of IKE, the various documents that describe it do not use the best practices for protocol engineering. For example, there are no formal language descriptions or conformance test suites available. Nevertheless, IBM revealed some details on the architecture of its implementation (Cheng, 2001).

Although ISAKMP was designed in a modular fashion, implementations are often not modular for commercial or legal reasons. For example, to satisfy the restrictions against the export of cryptographic software, Version 5.0 of Microsoft Windows NT had to sacrifice the modularity of the implementation. Similarly, the version that Cisco produces, which is based on the cryptographic library of Cylink Corporation, is only available in North America (the U.S. and Canada). It should also be noted that the MIT distributes in North America the prototype of a version approved by the U.S. DOD. (Note that a new version of IKE is being prepared with the aim of removing problems that were uncovered. Some of these problems relate to hashing and to the protection cookies.)

3.20 SKIP (Simple Key Management for Internet Protocols)

Simple Key Management for Internet Protocols (SKIP) is an approach to key exchange that Sun Microsystems championed at one time. The principle is to exchange a master key according to the method of Diffie–Hellman, then store it in cache memory to construct the encryption key for subsequent sessions. In this manner, the protocol avoids the preliminary exchanges needed to define the secure channel before the message exchange. This may be useful in applications where efficient use of the transmission bandwidth available justifies reduced security.

SKIP operates at the network layer. The IP packets that contain the information used in SKIP have an IP AH, and their payloads are encapsulated according to the ESP procedures.

Although this method allows a reduction in the number of exchanges and alleviates the cryptographic loads, its success assumes that the master key is never compromised. Interest in SKIP seems to have subsided.

3.21 Key Exchange Algorithm

The Key Exchange Algorithm (KEA) is an algorithm from the U.S. National Security Agency (NSA). It is based on the Diffie–Hellman algorithm. All calculations in KEA are based on a prime modulus of 1024 bits generated as per the DSA specifications of FIPS 186. Thus, the key size is 1024 bits, and as in DSA, the size of the exponent is 160 bits.

KEA is used in the cryptographic PCMCIA card Fortezza and the SKIP-JACK encryption algorithm. The experimental specifications of IETF RFC 2773 (2000) describe its use for securing file transfers with ftp. Those of IETF RFC 2951 (2000) provide security to telnet sessions.

Consider its use with telnet. The aim is to replace the user-level authentication through its login and password being exchanged in the clear, with more secure measures and the ability to authenticate the server.

It is known that a telnet session is a series of exchanges on a character-by-character basis. With the combination of KEA and SKIPJACK, the encryption of the telnet bit stream can be with or without integrity protection. Without the integrity service, each character corresponds to a single octet online. Stream integrity uses the one-way hash function SHA-1 and requires the transmission of 4 octets for every character, i.e., it adds an overhead of 300%. (Note that Version 2.0 of KEA is available from NIST at *http://csrc.nist.gov/ encryption/skipjack-kea.htm.*)

3.22 Certificate Management

When a server receives a request signed with a public key algorithm, it must first authenticate the declared identity associated with the key. Next, it will verify if the authenticated entity is allowed to perform the requested action. Both verifications rely on a certificate that a certification authority signed. As a consequence, certification and certificate management are the cornerstones of e-commerce on open networks.

A Certification Practice Statement (CPS) describes the practices a certification authority employs in issuing certificates. It covers the obligations and liabilities of various entities, the requirements for physical and cryptographic security, the operational aspects for key management, as well as the life-cycle management of certificates. The IETF RFC 2527 (1999) gives guidance on how to write such a certification statement.

Certification can be decentralized or centralized. Decentralized certification utilizes PGP (pretty good privacy) and is popular among Internet users (Garfinkel, 1995). This model works by reference among users and, by obviating the need for a central authenticating authority, eliminates vulnerability to attacks on the central system and prevents potential for power abuse, which are the weak points of centralized certification. Each user, therefore, determines the credence accorded to a public key and assigns the confidence level in the certificate that the owner of this public key issued. Similarly, a user can recommend a new party to members of the same circle of trust. At one time, the World Wide Web Consortium (W3C) favored this approach in its Digital Signature Initiative. However, the absence of any collective structure forces users to manage the certificates by themselves (update, revocation, etc.). The load of this management increases exponentially with the number of participants, which makes this mode of operation impractical for large-scale operations such as commerce.

Centralized certification is denoted X.509 certification, using the name of the ITU-T recommendation that defines the framework for authentication in open systems. X.509 is identical to ISO/IEC 9594-1, a joint standard from the ISO and the IEC. It is this X.509 certification that is most often used in commercial applications.

The management of EDIFACT (Electronic Data Interchange for Administration, Commerce and Transport) certificates (which are used for EDI) is also centralized. The manner in which these certificates are administered, however, is distinct from that of X.509 certificates. The KEYMAN message defined in ISO standard 9735-9 (2002) is used; however, KEYMAN can include references to the certification path of X.509. This is discussed in Chapter 4.

In the following presentation, the focus will be on X.509 certificates, because these are usually used to secure payments. In some cases, parallel efforts will be cited for the sake of completeness, without aiming at exhaustiveness. The interested reader is invited to consult the literature on certification, for example, Ford and Baum (1997, pp. 357–404), whose first author, when the book was written, was a manager in VeriSign, a leading company in the area of certification.

The ITU-T and the ISO/IEC established a whole series of recommendations to describe the operation of a public key infrastructure (PKI). These are as follows:

- X.500 (ISO/IEC 9594-1) (2001) for a general view of the concepts, the models, and the services
- X.501 (ISO/IEC 9594-2) (2001) for the different models used in the directory
- X.509 (ISO/IEC 9594-8) (2000), which defines the framework for authentication through public key cryptography using identity certificates and attribute certificates
- X.511 (ISO/IEC 9594-3) (2001), which defines the abstract services of the directory (search, creation, deletion, error messages, etc.)
- X.520 (ISO/IEC 9594-6) (2001) and X.521 (ISO/IEC 9594-7) (2001), which, respectively, specify selected attributes (keywords) and selected object classes to ensure compatibility among implementations

These recommendations specify services, protocols, messages, and object classes to carry out the following functions:

- Retrieval of credentials stored in the directory by a directory user agent (DUA) at the client side and a directory system agent (DSA) at the server's side, with the Directory Access Protocol (DAP) defined in X.519 (ISO/IEC 9594-5) (2001)
- Distributed searches and referrals among directory system agents with the Directory System Protocol (DSP) of X.518 (ISO/IEC 9594-4) (2001)
- Information sharing among directory system agents through replication of the directory using the DISP (Directory Information Shadowing Protocol) of X.525 (ISO/IEC 9594-9) (2001)

The relationship among these different protocols is shown in Figure 3.12. In IP networks, a simplified version of DAP, the Lightweight Directory Access Protocol (LDAP), is often used for communication between user agents and system agents. The LDAP is the output of the Public Key Infrastructure (X.509) (PKIX) working group of the IETF. As defined in IETF RFC 2251 (1997), the main simplifications are as follows:

1. The LDAP is carried directly over the TCP/IP stack, thereby avoiding some of the OSI protocols at the application layer.

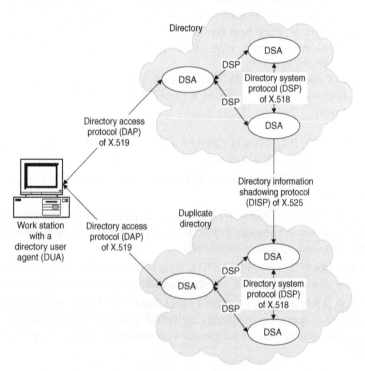

FIGURE 3.12
Communication protocols among the components of the X.500 directory system.

2. It uses simplified information models and object classes.

3. Being restricted to the client side, LDAP does not address what happens on the server side, for example, the duplication of the directory or the communication among servers.

4. Finally, Version 3 of LDAP (LDAPv3) does not mandate any strong authentication mechanism.

The latitude that LDAPv3 allowed developers with respect to strong authentication, however, resulted in some incompatibilities among different implementations of secure clients and servers. The IETF RFC 2829 (2000) specifies a minimum subset of security functions common to all implementations of LDAPv3 that use the SASL (Simple Authentication and Security Layer) mechanism defined in IETF RFC 2222 (1997). SASL adds authentication services and, optionally, integrity and confidentiality. Simple authentication is based on the name/password pair, concatenated with a random number or a time-stamp with integrity protection using MD5. Strong authentication is achieved on a session basis using the TLS protocol.

3.22.1 Basic Operation

After receiving over an open network a request encrypted using public key cryptography, a server has to accomplish the following tasks before answering the request:

1. Read the certificate presented

2. Verify the signature by the certification authority

3. Extract the requester public key from the certificate

4. Verify the requester signature on the request message

5. Verify the certificate validity by comparing with the certificate revocation lists (CRLs)

6. Establish a certification path between the certification authority of the requester and the authority that the server recognizes

7. Extract the name of the requester

8. Determine the privileges that the requester enjoys

The certificate permits the accomplishment of Tasks 1 through 7 of the preceding list. In the case of payments, the last step consists of verifying the financial data relating to the requester, in particular, whether the account mentioned has sufficient funds. In the general case, the problem is more complex, especially if the set of possible queries is large. The most direct method is to assign a key to each privilege, which increases the difficulties of key management. This topic is currently the subject of intense investigation.

3.22.2 Description of an X.509 Certificate

An X.509 certificate is a record of the information needed to verify the identity of an entity. This record includes the distinguished name of the user, which is a unique name that ties the certificate owner with its public key. The certificate contains additional fields with which to locate its owner's identity more precisely. Each version of X.509 introduces its allotment of supplementary information, although compatibility with previous versions is retained. The essential pieces of information can be found in the basic certificate (Version 1), whose content is illustrated in Table 3.7.

The certificate contains the digital signature using the private key of the certification authority. It is usually recommended that a distinct key be used for each security function (signature, identification, encryption, etc.). Accordingly, the same entity will have several certificates, and certificates that conform to Version 3 of X.509 may contain details on the security service for which they may be used, on the duration of their validity, on any restrictions on the use of the certificates, on cross-certifications with other certification authorities, etc.

In the initial version of X.509, the hierarchical arrangement of the distinguished names followed the rules for X.500. These rules were inspired by the worldwide assignment of telephone numbers in complete accordance with Recommendation X.400 for e-mail. The directory entries are described using the keywords defined in Recommendation X.520 (2000), a partial list of which is given in Table 3.8.

So, for the National University of Benin, in Cotonou, the corresponding clear entry is

<O = Université Nationale du Bénin; L = Cotounou; C = Bénin>

Figure 3.13 shows the hierarchical naming of this example according to the rules of X.400/X.500.

TABLE 3.7

Content of the Basic X.509 Certificate

Field Name	Description
version	Version of the X.509 certificate
serialNumber	Certificate serial number
signature	Identifier of the algorithm used to sign the certificate and the parameters used
issuer	Name of the certification authority
validity	Duration of the validity of the certificate
subject	User's references: distinguished name, unique identifier (optional), etc.
subjectPublicKeyInfo	Information concerning the public key algorithm of the sender, its parameters, and the public key

TABLE 3.8

Partial List of Keywords in X.520

Keyword	Meaning
C	Country
CN	Common name
L	Locality name
O	Organization name
OU	Organizational unit name

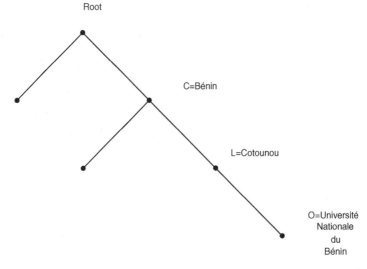

FIGURE 3.13
Example for the tree structure of an X.400/X.500 identifier.

The widespread use of the Internet spawned other models for hierarchical naming. Version 3 of X.509, which was approved in 1996, took this fact into account and authorized the use of a variety of distinguished names, such as the network addresses, passport or identity card numbers, social security numbers, Internet domain names, e-mail addresses, URL (Uniform Resource Locator) for Web applications, etc. The certificate can include additional pointers to the certified subject (physical name, postal address, electronic address) as well as identifiers related to specific applications, such as e-mail address, EDI identity, or even personal details, such as profession, photo ID, bank account number, etc. This additional flexibility requires a name registration system to ensure that any name used unambiguously identifies a certificate subject. Without this verification, automatic cross-checking of directory entries will be difficult, particularly on a worldwide basis.

Starting from Version 3 of X.509 (1996), the public key certificate may contain details on the security service for which the certified public key may be used, on the duration of its validity, on any restrictions on the use of the certificates, on cross-certifications with other certification authorities, etc. For example, X.509 now provides a way for a certificate issuer to indicate how the issuer's certificate policies can be considered equivalent to a different policy used by another certification authority [§8.2.2.7 of X.509 (2001) on policy mapping extension].

Version 4 of X.509 (2001) introduced several certificate extensions to improve the treatment of certificate revocation and to associate privileges with the identification public-key certificates or with attribute certificates.

3.22.3 Certification Path

The idea behind X.509 is to allow each user to retrieve the public key of certified correspondents so they can proceed with the necessary verifications. It is sufficient, therefore, to request the closest certification authority to send the public key of the communicating entity in a certificate sealed with the digital signature of that authority. This authority, in turn, relays the request to its own certifying authority, and this permits an escalation through the chain of authorities, or certification path, until reaching the top of the certification pyramid, where the Root Authority (RA) resides. Depicted in Figure 3.14 is this recursive verification.

Armed with the public key of the destination entity, the sender can include a secret encrypted with the public key of the correspondent and corroborate that the partner is the one whose identity is declared. This is because, without the private key associated with the key used in the encryption, the destination will not be able to extract the secret. Obviously, for the two parties to authenticate themselves mutually, both users have to construct the certification path back to a common certification authority.

Thus, a certification path is formed by a continuous series of certification authorities between two users. This series is constructed with the help of the information contained in the directory by going back to a common point of confidence. The tree structure of the certification path can be hierarchical or nonhierarchical. Similar to the system for telephone numbering, each country or region can have its own local root authority. However, to ensure worldwide communication, agreements for cross-certification among the various authorities would extend the zone of validity of their certification.

3.22.4 Hierarchical Certification Path

According to the notational convention used in X.509, a certificate is denoted by the following:

$$\text{authority}<<\text{user}>>$$

Thus,

$$X_1<< X_2>>$$

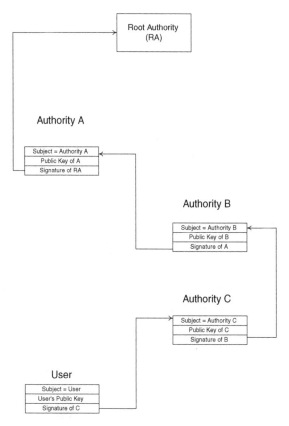

FIGURE 3.14
Recursive verification of certificates. (Adapted from Ford, W. and Baum, M.S., *Secure Electronic Commerce*, Pearson Education, Inc., Upper Saddle River, NJ, 1997.)

indicates the certificate for user X_2 that authority X_1 has issued, while

$$X_1 << X_2 >> X_2 << X_3 >> \ldots X_n << X_{n+1} >>$$

represents the certification path connecting user X_{n+1} to authority X_1. In other words, this notation is functionally equivalent to $X_1 << X_{n+1} >>$, which is the certificate that authority X_1 would have issued to user X_{n+1}. By constructing this path, another user would be able to retrieve the public key of user X_{n+1}, if that other user knows X_{1P}, the public key of authority X_1. This operation is called "unwrapping" and is represented by

$$X_{1P} \bullet X_1 << X_2 >>$$

where $\bullet$ is an infix operator, with a left operand that is the public key, X_{1P}, of authority X_1, and with a right operand that is the certificate $X_1 << X_2 >>$

delivered to X_2 by that same certification authority. This result is the public key of user X_2.

In the example depicted in Figure 3.15, assume that user A wants to construct the certification path toward another user B. A can retrieve the public key of authority W with the certificate signed by X. At the same time, with the help of the certificate of V that W issued, it is possible to extract the public key of V. In this manner, A would be able to obtain the chain of certificates:

$$X<<W>>, W<<V>>, V<<Y>>, Y<<Z>>, Z<<B>>$$

This itinerary, represented by $A \rightarrow B$, is the forward certification path that allows A to extract the public key B_P of B, by application of the operation • in the following manner:

$$B_P = X_P \bullet (A{\rightarrow}B) = X_P \bullet X<<W>> W<<V>> V<<Y>> Y<<Z>> Z<<B>>$$

In general, A also has to acquire the certificates for the return certification path $B \rightarrow A$, to send them to its partner:

$$Z<<Y>>, Y<<V>>, V<<W>>, W<<X>>, X<<A>>$$

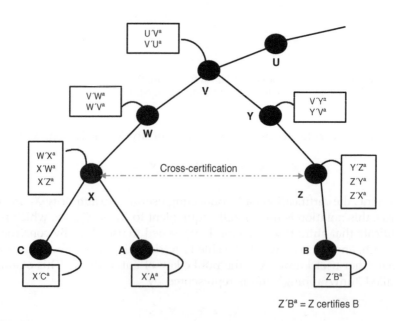

FIGURE 3.15

Hierarchical certification path according to X.509. (From ITUT-T. Recommendation X.509 (2001). With permission.)

When B receives these certificates from A, it can unwrap the certificates with its private key to extract the public key of A, A_p:

$$A_p = Z_p \bullet (B{\rightarrow}A) = Z_p \bullet Z{<<}Y{>>}\ Y{<<}V{>>}\ V{<<}W{>>}\ W{<<}X{>>}\ X{<<}A{>>}$$

As previously mentioned, such a system does not necessarily impose a unique hierarchy worldwide. In the case of electronic payments, two banks or the fiscal authorities of two countries can mutually certify each other. In the preceding example, assume that authorities X and Z have cross-certified their respective certificates. If A wants to verify the authenticity of B, it is sufficient to obtain:

$$X{<<}Z{>>}, Z{<<}B{>>}$$

to form the forward certification path, and

$$Z{<<}X{>>}$$

to construct the reverse certification path. This permits the clients of the two banks to be satisfied with the certificates supplied by their respective banks.

3.22.5 Nonhierarchical Certification Path

If certification authorities are not organized hierarchically, the users would have to construct the certification paths. In practice, the number of operations to be carried out can be reduced with various strategies, for example:

1. Two users served by the same certification authority have the same certification path, and the users can exchange their certificates directly. This is the case for entities C and A in Figure 3.15.
2. If one user is constantly in touch with users that a particular authority has certified, that user could store the forward and return certification paths in memory. This would reduce the effort expended for obtaining the other users' certificates to a query into the directory.
3. If two users know each other's certificates, they can mutually authenticate themselves without querying the directory. This reverse certification is based on the confidence that each user has in his or her certification authority.

3.22.6 Cross-Certification

As the number of electronic transactions on the Internet increases, the pressure to identify the correspondents augments. However, given the potential number of certification authorities, establishing a worldwide public key infrastructure faces a series of difficulties, such as the following:

1. Harmonize practices among different root authorities to ensure mutual recognition of their certificates
2. Establish criteria for measuring and evaluating the performance of the various certification authorities
3. Coordinate among the various naming authorities
4. Protect the confidential data stored in the certification warehouses

Clearly, this is a mixture of strictly technical problems with political and managerial problems.

In the U.S., NIST published NIST Special Publication 800-15 to ensure minimum interoperability among the components of a public key infrastructure. These specifications are called Minimum Interoperability Specification for PKI Components (MISPC) (Burr et al., 1997).

MISPC includes profiles for certificates and their certificate revocation lists and describes the sequence of exchanges and the message formats in basic transactions concerning a signature certificate. A reference implementation is also available. Several organizations have supported this initiative through trials, such as NASP (National Association of State Purchasing Officials) as well as state and federal governmental agencies. An updated specification is currently being developed to take into account the results of these trials.

In the same spirit and to facilitate electronic exchanges with suppliers of the federal agencies by replacing handwritten signatures as a means for authentication, the Canadian government launched the program GOCPKI (Government of Canada Public Key Infrastructure). The objective of GOCPKI is to establish uniform criteria to manage the keys and certificates among all Canadian federal agencies.

These efforts focus essentially on agreements about business-to-business exchanges that are defined in *interchange contracts*. In the case of consumer certification, the absence of a public service directory leaves room for enterprises to offer authentication services. However, the following observations should be made:

- The regulatory environment for business-to-consumer e-commerce, particularly on a worldwide level, is far from being defined. Little progress has been made on the responsibilities of the entities, the judicial competence in case of disagreements, the location of archived financial data, or the means for conserving electronic documents.
- Guarantees, particularly those that relate to the protection of private life or to consumer rights, are not universally recognized. In addition, some laws for the "war on terrorism" weakened this protection.

- Criteria to evaluate the request for certification are arbitrary, and there is no mandatory procedure to verify and correct the databases utilized.

- The implementation of a worldwide certification infrastructure would lead to the construction of international networks for data manipulation, automatic treatment of information, with neither a center nor a border, and which would not be responsible to any specific law.

- The lifetime of the support medium and its readability were not sufficiently addressed. This does not relate solely to the physical condition of the support but also to the availability of software and readers. A guarantee of 30 years may seem unattainable given the speed with which technological developments are taking place. Who today can easily access data stored on punched cards or on 8-in. disks?

3.22.7 Online Management of Certificates

The PKIX (X.509) working group of the IETF studied the support of X.509 certificates on the Internet. Among its proposals are the Certificate Management Protocol (CMP) of IETF RFC 2510 (1999) and the Online Certificate Status Protocol (OCSP) of IETF RFC 2560 (1999). The RFC 2585 (1999) describes the conventions for using the File Transfer Protocol (FTP) and HTTP to obtain certificates and certification revocation lists from their respective repositories.

3.22.8 Banking Applications

The difficulties related to the management and maintenance of a distributed database for certificates are neither mastered nor fully understood. Nevertheless, by limiting the scope of application, the task can be more easily accomplished. In particular, many banking applications are for internal use. Accordingly, a bank can certify its clients, knowing all their coordinates. The purpose of certification is to allow them access to their bank accounts across the Internet. Once access is given, the operation will continue as if the client was in front of an ATM. The interoperability of bank certificates can be achieved with interbank agreements, analogous to those that permitted the interoperability of bank cards. Each financial institution certifies its clients and is assured that the other institutions will honor that certificate. In this case, the certification directory will be the equivalent of the telephone directory.

To verify the feasibility of such an approach, the Internet Council of NACHA (National Automated Clearing House Association) conducted an experiment using electronic direct debits authenticated with certificates. The participants included Bank of America, Citibank, Mellon Bank, and Zion's Bank, as well as the certification authorities CertCo & Digital Trust Company, Entrust, GTE CyberTrust, IBM, and VeriSign. The experience has shown that the use of common (or interoperable) certificates among financial institutions enables large-scale electronic payments.

As the main victims of fraud, financial institutions established certification infrastructures. In the U.S., several banks, including Bank of America, Chase Manhattan, Citigroup, and Deutsche Bank, formed Identrus (*http://www.iden-trus.com*) in 2000 to distribute digital certificates to their partners and clients.

At the same time, about 800 entities, either singly, such as BNP or Société Générale, or through an association, such as *Groupement de cartes bancaires* (Association of Bank Cards), joined forces to form a Global Trust Authority (GTA) (*http://www.cartes-bancaires.com/html/grpmnt/comm1.html*). This is a nonprofit organization with a mission to put in place the necessary infra-structure for worldwide management of public key certificates stored in integrated circuits (smart) cards[1].

3.22.9 Example: VeriSign

VeriSign (*http://www.verisign.com*) established its VeriSign Trust Network[SM] for worldwide public key infrastructure for wired and wireless applications. VeriSign certificates can be issued to individuals, Web sites, or organizations. In an enterprise, the service provides certificates to individuals, such as employees, suppliers, customers, or partners, as well as devices, such as routers, servers, or firewalls.

VeriSign's certification practice statement went through several iterations, and the discussion below is based on Version 2.0 dated August 31, 2001, and available at *http://www.verisign.com/repository/CPS.*

VeriSign produces three types of certificates according to the strength of the link between the certificate and the owner's identity. For each class of certificates, a primary certification authority (PCA) certifies certification authorities or registration authorities. Registration relates to the approval and rejection of certificate applications and the request of revocation or renewal of certificates. Usually, certification authorities restrict themselves to the registration functions, leaving the key management and the certificate management to VeriSign. In enterprise applications, certification authorities may also generate the key pairs on behalf of the end users, whose certificate applications they approve, and transmit the key pairs to the end users via password-protected PKCS #12 files. Registration authorities perform a sim-ilar role, but their security module is implemented in software rather than in a specialized hardware module.

The certificates for the PCAs are available to end users by their inclusion in browser software. The key length and security level of the various

---

[1] For up-to-date news on public key certification, see, for example, *http://www.pkiforum.com.*

TABLE 3.9

Cryptographic Parameters for Various Entities

Entity	Key Size (bits)	FIPS 140–1 Security Level	Implementation
Primary certification authority	1024–2048	Level 3	Hardware
Other certification authorities	1024	Level 2	Hardware
Registration authority	1024	Level 1	Browser software
End user	1024	Level 1	Browser software

entities in VeriSign architecture are shown in Table 3.9. There are now three generations of PCAs; those of the third generation have 2048-bit RSA key pairs.

A naming authority defines the X.501 distinguished names used for the issuer of a certificate, and its user (subject) and a repository contain various documents on the certification policies as well as the data needed for operation of the PKI.

VeriSign implemented techniques that require the collaboration of multiple trusted operators to activate the private key of a certification authority. The activation data are split into m separate parts, and at least n of these parts are needed to activate the key stored in a hardware module. For normal operations, $m = 12$, and $n = 3$; while for disaster recovery, $m = 5$, and $n = 3$.

Signatures for wired communications use the SHA-1 and MD5 hash function and are encrypted with RSA. For wireless communications, the hash function used is SHA-1, and the encryption uses the Elliptic Curve Digital Signature Algorithm (ECDSA).

During online verification of the certificates, end users or registration authorities submit their public keys using the Certification Signing Request of PKCS #10 (IETF RFC 2986, 2000).

3.22.9.1 Certificate Classes

Class 1 certificates are for individuals only. The validation procedure confirms that the distinguished name the user presents is unique and unambiguous within the certification authority's domain, and that it corresponds to an e-mail address in the VeriSign repository. Class 1 certificates are used for modest enhancement of security through confidentiality and integrity verification. They cannot be used to verify an identity or to support nonrepudiation services.

Class 2 certificates are also restricted to individuals. They indicate that the information the user submitted during the registration process is consistent with information available in business records or in "well known" consumer databases. In the U.S. and Canada, one such reference database is maintained by Equifax, from Atlanta, Georgia.

Class 3 certificates are given to individuals and to organizations. To obtain a certificate of this class, an individual has to be physically present with their public key in possession before an authority to confirm the identity of the applicant with a formal proof of identity (passport, identity card, electricity or telephone bill, etc.) and the association of that identity with the given public key. If the individual is to be certified as a duly authorized representative of an organization, then the necessary verifications have to be made. Similarly, an enterprise will have to prove its legal existence. The authorities will have to verify these documents by querying the databases for enterprises and by confirming the collected data by telephone or by mail. Class 3 certificates have many business applications, for example, in EDI.

3.22.9.2 Operational Life

The maximum operational lives, in years, for VeriSign certificates are shown in Table 3.10.

3.22.9.3 Revocation

The CPS determines the circumstances under which certification of end users as well as various authorities can be revoked and defines who requested that revocation. To inform all the entities of the PKI, CRLs are published at regular intervals with the digital signature of the certification authority to ensure their integrity. Among other information, the CRL indicates the issuer's name, the date of issue, the date of the next scheduled CRL, the serial numbers of the revoked certificates, and the specific times and reasons for revocation. The CRL is published daily for authorities that certify end-user subscribers and quarterly or whenever a certificate of an authority is revoked for other certification authorities.

TABLE 3.10

Maximum Operational Life for VeriSign Certificates (years)

Certification Authority	Subject of Certification	Class 1	Class 2	Class 3
Primary certification authority	Self-signed (1024 bit key length)	—	30	—
Primary certification authority	Self-signed (2048 bit key length)	—	50	—
Primary certification authority	Certification authority	—	10	—
Certification authority	End user	2	—	2–5

TABLE 3.11

Archival Period per Certificate Class

Certificate Class	Duration (years)
1	5
2	10
3	30

3.22.9.4 Archival

Following certification expiration or revocation, the records associated with a certificate are retained for at least the time periods set forth in Table 3.11. Thus, archival of Class 1 certificates lasts for at least 5 years after expiration of the certificate or its revocation. The corresponding durations for Class 2 and 3 certificates are 10 and 30 years, respectively.

3.22.9.5 Recovery

VeriSign has implemented procedures to recover from computing failures, corruption of data, such as when a user's private key is compromised, as well as natural or man-made disasters.

A disaster recovery plan addresses the gradual restoration of information services and business functions. Minimal operations can be recovered within 24 hours. They include certificate issuance or revocation, publication of revocation information, and recovery of key information for enterprises' customers. If the disaster occurs at VeriSign's primary site, the time to full recovery extends to a week.

3.22.9.6 Liability

According to §1.3.4.3 of the CPS, VeriSign certificates cannot be used for fail-safe applications, where failure could lead directly to death, personal injuries, or severe environmental damages (such as the operation of nuclear facilities, aircraft navigation, communication systems, air traffic control systems, etc.). Nevertheless, the liabilities of certification authorities are capped for each class of certificates, according to Table 3.12. These liability caps limit damages recoverable outside a special protection plan. With such a plan, the liability caps range from US$1,000 to US$1,000,000.

TABLE 3.12

Certification Authority Limitations of Liability

Certificate Class	Liability ($)
1	100
2	5,000
3	100,000

3.22.10 Procedures for Strong Authentication

Having obtained the certification path and the other side's authenticated public key, X.509 defines three procedures for authentication, one-way or unidirectional authentication, two-way or bidirectional authentication, and three-way or tridirectional authentication.

3.22.10.1 One-Way Authentication

One-way authentication takes place through the transfer of information from User A to User B according to the following steps:

- A generates a random number R^A used to detect replay attacks.
- A constructs an authentication token $M = (T^A, R^A, I^B, d)$, where T^A represents the time stamp of A (date and time), and I_B is the identity of B. T^A comprises two chronological indications, for example, the generation time of the token and its expiration date, and d is arbitrary data. For additional security, the message can be encrypted with the public key of B.

A sends to B the message:

$$B \rightarrow A, A\{(T^A, R^A, I_B, d) \}$$

where $B \rightarrow A$ is the certification path, and $A\{M\}$ represents the message M encrypted with the private key of A.

B carries on the following operations:

- Obtain the public key of A, A_p, from $B \rightarrow A$, after verifying that the certificate of A did not expire
- Recover the signature by decrypting the message $A\{M\}$ with A_p. B then verifies that this signature is identical to the message hash, thereby simultaneously ascertaining the signature and the integrity of the signed message
- Verify that B is the intended recipient
- Verify that the time stamp is current
- Optionally, verify that R^A was not previously used

These exchanges prove the following:

- The authenticity of A, and that the authentication token was generated by A
- The authenticity of B, and that the authentication token was intended for B
- The integrity of the identification token

- The originality of the identification token, i.e., that it was not previously utilized

3.22.10.2 Two-Way Authentication

The procedure for two-way authentication adds similar exchanges to the previous unidirectional exchanges but in the reverse direction. Thus:

- B generates another random number R^B.
- B constructs the message $M' = (T^B, R^B, I_A, R^A, d)$, where T^B represents the time stamp of B (date and time), I_A is the identity of A, and R^A is the random number received from A. T^B consists of one or two chronological indications, as previously described. For security, the message can be encrypted with the public key of A.
- B sends to A the message:

$$B\{(T^B, R^B, I_A, R^A, d)\}$$

where $B\{M'\}$ represents the message M' encrypted with the private key of B. A carries out the following operations:

- Extracts the public key of B from the certification path, uses it to decrypt $B\{M'\}$, and recovers the signature of the message that B produced; A verifies next that the signature is the same as the hashed message, thereby ascertaining the integrity of the signed information
- Verifies that A is the intended recipient
- Checks the time stamp to verify that the message is current
- As an option, verifies that R^B was not previously used

3.22.10.3 Three-Way Authentication

Protocols for three-way authentication introduce a third exchange from A to B. The advantage is the avoidance of time-stamping and, as a consequence, of a trusted third party. The steps are the same as for two-way identification, but with $T^A = T^B = 0$. Then:

- A verifies that the value of the received R^A is the same that was sent to B.
- A sends to B the message:

$$A\{R^B, I_B\}$$

encrypted with the private key of A.

B performs the following operations:
- Verifies the signature and the integrity of the received information
- Verifies that the received value of R^B is the same as was sent

3.22.11 Certificate Revocation

Authentication establishes the correspondence between a public key and an identity only for a period of time. Therefore, certification authorities must refer to revocation lists that contain certificates that expired or were revoked. These lists are continuously updated. Table 3.13 shows the format of the revocation list that Version 1 of X.509 defined. In the third revision of X.509 were added other optional entries, such as the date of the certificate revocation and the reason for revocation.

In principle, each certification authority has to maintain at least two revocation lists: a dated list of the certificates it issued and revoked, and a dated list of all the certificates that the authority knows of and that it recognizes as having been revoked. The root certification authority and each of its delegate authorities must be able to access these lists to verify the instantaneous states of all the certificates to be treated within the authentication system.

Revocation can be periodic or exceptional. When a certificate expires, the certification authority withdraws it from the directory (but retains a copy in a special directory, to be able to arbitrate any conflict that might arise in the future). Replacement certificates have to be ready and supplied to the owner to ensure continuity of the service.

The root authority (or one of its delegated authorities) may cancel a certificate before its expiration date, for example, if the certificate owner's private key was compromised or if there was abuse in usage. In the case of secure payments, the notion of solvency, i.e., that the user has available the necessary funds, is obviously one of the essential considerations.

TABLE 3.13

Basic Format of the X.509 Revocation List

Field	Comment
signature	Identifier of the algorithm used to sign the certificates and the parameters used
Issuer	Name of the certification authority
thisUpdate	Date of the current update of the revocation list
nextUpdate	Date of the next update of the revocation list
revokedCertificates	References of the revoked certificates, including the revocation date

Processing of the revocation lists must be speedy to alert users and, in certain countries, the authorities, particularly if the revocation is before the expiration date. Perfect synchronization among the various authorities must be attained to avoid questioning the validity of documents signed or encrypted before withdrawal of the corresponding certificates.

Users must also be able to access the various revocation lists; this is not always possible, because current client programs do not query these lists.

In summary, when an entity has a certificate signed by a certification authority, this means that the entry for that entity in the directory that the certification authority maintains has the following properties:

1. It establishes a relationship between the entity and a pair of public and private cryptographic keys.

2. It associates a unique distinguished name in the directory with the entity.

3. It establishes that at a certain time, the authority was able to guarantee the correspondence between that unique distinguished name and the pair of keys.

3.22.12 Attribute Certificates

Some questioned the utility of an X.509-type directory for e-commerce applications. As a consequence, authentication structures modified so that access to a private key rather than the identity an entity would play, is the principal role. One example is the architecture of SDSI (Simple Distributed Security Infrastructure) that Ronald Rivest and Butler Lampson have proposed.

With Version 4 of X.509, a new type of public key certificate called an *attribute certificate* was introduced to link a subject to certain privileges separately from its authenticated identity. Attribute certificates allow for the verification of the rights or prerogatives of their subjects, such as access privileges (Feigenbaum, 1998). Thus, once an identity is authenticated with a public key certificate, the subject can use multiple attribute certificates associated with that public key certificate.

Although it is possible to use public key identity certificates to define what the holder of the certificate may be entitled to, a separate attribute certificate may be useful in some cases, for example:

1. If the authority for privilege assignment is distinct from the certification authority

2. If a variety of authorities will be defining access privileges to the same subject

3. If the same subject may have different access permissions depending on its role

4. If there is the possibility of delegation of privileges, in full or in part

5. If the duration of validity of the privilege is shorter than that of the public key certificate

Conversely, the public key identity certificate may suffice for assigning privileges whenever the following occur:

1. The same physical entity combines the roles of certification authority and of attribute authority
2. The expiration of the privileges coincides with that of the public key certificate
3. Delegation of privileges is not permitted, or if permitted, all privileges are delegated at once

The use of attribute certificates raises the need for a new infrastructure for their management. This is called Privilege Management Infrastructure (PMI). When a single entity acts as both a certification authority and an attribute authority, it is strongly recommended that different keys be used for each kind of certificate.

The Source of Authority (SOA) is the trusted entity responsible for assigning access privileges. It plays a role similar to the root certification authority; however, the root certification authority may control the entities that can act as SOAs. Thus, the SOA may authorize the holder of a set of privileges to further delegate these privileges, in part or in full, along a *delegation path*. There may be restrictions on the power of delegation capability, for example, the length of the delegation path can be bounded, and the scope of privileges allowed can be restricted downstream. To validate the delegation path, each attribute authority along the path must be checked to verify that it was authorized to delegate its privileges.

Attribute certification allows for modification of the privileges of a role without impacts on the public key identity certificates However, privilege verification requires an independent verification of the privileges attributed to a role. This can be done by prior agreement or through role-specification certificates. It is worth noting that hierarchical role-based access control allows role specifications to be more compact, because higher levels inherit the permissions accorded to subordinates.

X.509 supports role-based access control (RBAC), provided that role-specification certificates can be linked with the role assignments indicated in identity certificates or in attribute certificates. In addition, X.509 supports hierarchical RBAC through a "domination rule" that puts limits on the scope of delegated privileges. [An X.509 RBAC policy for privilege management using XML is available at *http://www.xml.org* and is based on work done at the University of Salford, U.K. (Chadwick and Ottenko, 2002).

3.22.13 Audits

The definition of accreditation criteria is beyond the scope of ITU-T Recommendation X.509, and a code of conduct for certification authorities is not yet available. For the time being, each operator defines its conduct, rights, and obligations in its own CPS.

Thus, it is the authority that defines the rigor with which it will verify the seriousness of the applications supplied for accreditation and certification, as well as the procedures to maintain the list of valid certificates. The authority operates at its own discretion and is not accountable to anyone for its decisions and is not obliged to justify its refusal to accredit an individual or an entity. Finally, no objective criterion today permits an evaluation of the quality of the services that the certification authorities are offering.

VeriSign "requires" its certification authorities to keep an audit trail of all exchanges and events, such as key generation, request for certification, validation, suspension, or revocation of certificates. A certified public accountant with "demonstrated expertise in computer security" or "an accredited computer security professional" is supposed to make security audits. These efforts are voluntary because the activity is not regulated, and a major complication is the lack of independent audits. Seeing a business opportunity, the American Institute of Certified Public Accountants and the Canadian Institute of Chartered Accountants announced the development of a procedure to evaluate the risks of conducting commerce through electronic means. The CPA WebTrust^SM is a seal that is supposed to indicate that a site is subject to quarterly audits on the procedures to protect the integrity of the transactions and the confidentiality of information.

It should be noted that nothing prevents a PKI operator from cashing in on the data collected on individuals and their purchasing habits by passing the information to all those who might be interested (merchants, secret services, political adversaries, etc.). If the certification authority produces the pair of keys and keeps them in escrow under its control, it will be able to decipher the messages to all the participants that it has certified and then extract intelligence it might be able to profit from. In an environment where the war against terrorism or on drugs is often invoked, rightly or wrongly, to justify secret accusations and in camera courts, the danger of misuse of the data collected within a PKI should not be minimized.

3.23 Encryption Cracks

While the role of encryption is to mask the messages, the objective of cryptanalysis is to uncover the flaws in the cryptographic algorithms to eavesdrop on the encrypted messages or at least to spread confusion.

Cryptanalysis consists in recovering the message without knowledge of the encryption key. Such an offensive penetrates the shield that encryption offers. The best-known cryptological attacks are of the following types:

1. Brute-force attacks, where the assailant systematically tries all possible encryption keys until getting the one that will reveal the plain text

2. Attacks on the encrypted text, assuming that the clear text has a known given structure, for example, the systematic presence of a header with a known format (this is the case of e-mail messages) or the repetition of known keywords

3. Attacks starting with chosen plaintexts that are encrypted with the unknown key, so as to deduce the key

4. Attacks by replaying old legitimate messages to evade defense mechanisms and to short-circuit the encryption

5. Attacks by interception of the messages (man-in-the-middle), where the interceptor eavesdrops at an intermediate point between the two parties; after intercepting, an exchange of a secret key, for example, the interceptor will be able to decipher the exchanged messages, while the participants think they are communicating in complete security; the attacker may also be able to inject fake messages that would be treated as legitimate by the two parties

6. Attacks by measuring the length of encryption times, of electromagnetic emissions, etc., to deduce the complexity of the operations, and hence their forms

Other techniques depend on the communication system. For example, corruption of the DNS can reorient packets to an address that the attacker chose. Among the recommended measures to fend off attacks are the following (Abadi and Needham, 1996):

1. The explicit indication of the identity of the participants, if this identity is essential for the semantic interpretation of the message

2. The choice of a sufficiently large key to discourage brute-force attacks, if the encryption algorithm is well designed; the key size needed grows with the computational power available to the adversaries

3. The addition of random elements, a time stamp, and other nonce values that make replay attacks more difficult

In some cases, the physical protection of the whole cryptographic system (cables, computers, smart cards, etc.) may be needed. For example, bending an optical fiber results in the dispersion of 1 to 10% of the signal power; therefore, well-placed acoustic-optic devices can capture the diffraction pattern for later analysis.

Thus, in the real world, there are easier ways than cryptanalysis to break cryptographic defenses. For example, when a program deletes a file, most commercial operating systems merely eliminate the corresponding entry in the index file. This allows recovery of the file, at least partially, with off-the-shelf software. The only means by which to guarantee total elimination of data is to systematically rewrite each of the bits that the deleted file was using. Similarly, the use of the virtual memory in commercial systems exposes vulnerability, because the secret document may be momentarily in the clear on the disk.

Errors in design, gaps in implementations, or operational deficiencies, particularly if the encryption is done in software, augment the vulnerability of the system. It is well known, for example, that GSM, IEEE 802.11b, IS-41, etc. have faulty or deliberately weakened protection schemes. A catalog of the causes of vulnerability includes the following (Fu et al., 2001; Schneier, 1996b, 1998a):

1. Nonverification of partial computations

2. Use of defective random-number generators, because the keys and the session variables depend on a good supply source for nonpredictable bits

3. Improper reutilization of random parameters

4. Misuse of a hash function in a way that increases the chances for collisions

5. Structural weakness of the telecommunications network

6. Nonsystematic destruction of the clear text after encryption as well as the keys used in encryption

7. Retention of the password or the keys in the virtual memory

8. No checking of correct range of operation; this is particularly the case when buffer overflows can cause security flaws (Recently, a problem with Kerberos was discovered through buffer overflow within a process that administers the database.)

9. Misuse of a protocol can lead to an authenticator traveling in plaintext [For example, IETF RFC 2109 (1997) specifies that when the authenticator is stored in a cookie, the server has to set the "secure" flag in the cookie header so that the client waits before returning the cookie until a secure connection is established with SSL/LS. Unfortunately, some Web servers neglect to set this flag, thereby negating that protection. The authenticator can also leak if the client software continues to be used even after the authentication is successful.]

Clearly, the resistance of a cryptographic system depends on the theoretical properties of the cryptographic algorithms used as well as the quality of the implementation. However, systems for e-commerce that are for the general public must be easily accessible and affordably priced. As a consequence, all the protective measures used in "top-secret" computers will not be used, and many compromises will be made to improve response time and ease of use. However, if one starts from the principle that, sooner or later, any system is susceptible to unexpected attacks with unanticipated consequences, it would be useful to design the system such that any possible attack will be detected. For example, by accumulating proof accepted by courts, the consequences would be alleviated and the possible damages reduced.

The starting point should be to correctly define the types of expected threats and the eventual attack plans. The model has to take into account users' practices and the way they will be using the system, as well as the motivations for possible attacks. Such a realistic evaluation of threats and risks permits a precise understanding of what should be protected, against whom, and for how long.

3.24 Summary

There are two types of attacks: passive and active. Protection can be achieved with suitable mechanisms and appropriate policies. Recently, security leaped to the forefront in priority because of changes in the regulatory environment and in technology. The fragmentations of operations that were once vertically integrated increased the number of participants in end-to-end information transfer. In virtual private networks, customers are allowed some control of their parts of the public infrastructure. Finally, security must be retrofitted in IP networks to protect systems from the inherent difficulties of having user traffic and network control traffic within the same pipe.

Security mechanisms can be implemented in one or more layers of the OSI model. The choice of the layer depends on the security services to be offered and the coverage of protection.

Confidentiality guarantees that only the authorized parties can read the information transmitted. This is achieved by cryptography, whether symmetric or asymmetric. Symmetric cryptography is faster than asymmetric cryptography but has a limitation in terms of the secure distribution of the shared secret. Asymmetric (or public key) cryptography overcomes this problem; this is why both can be combined. In online systems, public key cryptography is used for sending the shared secret that can be used later for symmetric encryption. Two public key schemes used for sharing the secrets are Diffie–Hellman and RSA. ISAKMP is a generic framework used to negotiate point-to-point security and to exchange key and authentication data among two parties.

Data integrity is the service for preventing nonauthorized changes to the message content during transmission. A one-way hash function is used to produce a signature of the message that can be verified to ascertain integrity.

Blind signature is a special procedure for signing a message without revealing its content.

The identification of participants depends on whether cryptography is symmetric or asymmetric. In asymmetric schemes, there is a need for authentication using certificates. In the case of human users, biometric features can be used for identification in specific situations. Kerberos is an example of a distributed system for online identification and authentication using symmetric cryptography.

Access control is used to counter the threats of unauthorized operations. There are two types of access control mechanisms: identity-based and role-based. Both can be managed through certificates defined by ITU-T Recommendation X.509. Denial of service is the consequence of failure of access control. These attacks are inherently associated with IP networks, where network control data and user data share the same physical and logical bandwidths. The best solution is to authenticate all communications by means of trusted certificates. Short of this, defense mechanisms will be specific to the problem at hand.

Nonrepudiation is a service that prevents a person who accomplished an act from denying it later. This is a legal concept defined through legislation. The service comprises the generation of evidence and its recording and subsequent verification. The technical means by which to ensure nonrepudiation include electronic signature of documents, the intervention of third parties as witnesses, time-stamping, and sequence numbering of the transactions.

3.25 Appendix I: Principles of Symmetric Encryption

3.25.1 Modes of Algorithm Utilization for Block Encryption

The principal modes for using symmetric algorithms of the block-cipher type are electronic code book (ECB) mode, cipher-block chaining (CBC) mode, cipher feedback (CFB) mode, and output feedback (OFB) mode.

The ECB mode is the most obvious, because each clear block is encrypted independently of the other blocks. However, this mode is susceptible to attacks by replay of blocks, which results in the perturbation of the messages, even without breaking the code. This is the reason this mode is only used to encrypt random data, such as the encryption of keys during authentication.

The other three modes have in common that they protect against such types of attacks with a feedback loop. They also have the additional property that they need an initialization vector to start the computations. These values can be revealed. The difference among the three feedback modes resides in the way the clear text is mixed, partially or in its entirety, with the preceding encrypted block.

In the CBC mode, input to the encryption module is the clear text mixed with the preceding encrypted block with an exclusive OR. This encryption operation is represented in Figure 3.16. Represented in Figure 3.17 is the decryption. In these figures, M_i represents the ith block of the clear message, while E_i is the corresponding encrypted block. Thus, the encrypted block E_i is given by

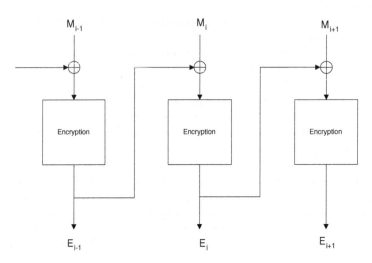

FIGURE 3.16
Encryption in the CBC mode.

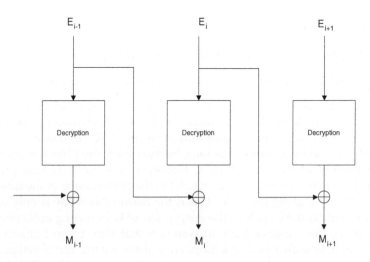

FIGURE 3.17
Decryption in the CBC mode.

$$E_i = E_K(M_i \oplus E_{i-1}), i = 0,1,\ldots$$

where $E_K()$ represents the encryption with the secret key K, and $\oplus$ is the exclusive OR operation. The starting value E_o is the initialization vector. The decryption operation, shown in Figure 3.17, is described by:

$$M_i = E_{i-1} \oplus D_K(E_i)$$

The CBC mode is generally useful for nonreal-time encryption of files, for example, to calculate the signature of a message (or its MAC). In fact, this is the method indicated in the various standards for securing financial and banking transactions: ANSI X9.9 (1986), ANSI X9.19 (1986), ISO 8731-1 (1987), and ISO/IEC 9797-1 (1999) as well as in the ESP protocol of IPSec.

The CFB and OFB modes are more appropriate for the real-time encryption of a character stream, such as in the case of a client connected to a server.

In CFB encryption, the encryption of a block of clear text of m bits is done in units of n bits ($n = 1$, 8, or 64 bits), with $n \leq m$, in n/m cycles. At each cycle, n bits of the clear message, M_i, are combined, with the help of an Exclusive OR, with the left most n bits of the previously encrypted block E_{i-1} to yield the new n bits of the new encrypted block E_i. These same n bits are then concatenated to the feedback bits in a shift register, and then all the bits of this register are shifted n positions to the left. The n left most bits of the register are ignored, while the remainder of the register content is encrypted, and the n left most bits are used in the encryption of the next n bits of the clear text. The decryption operation is identical to the roles of M_i and E_i transposed. Depicted in Figure 3.18 is the encryption, and illustrated in Figure 3.19 is the decryption.

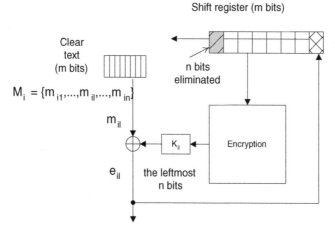

FIGURE 3.18
Encryption in the CFB mode of a block of m bits and n bits of feedback.

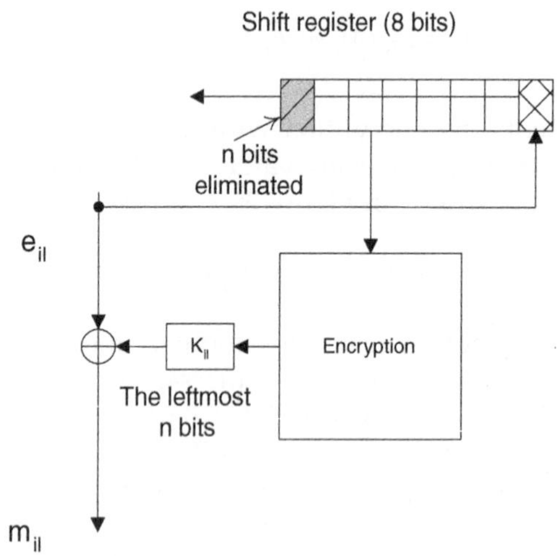

Shift register (8 bits)

FIGURE 3.19
Decryption in the CFB mode of a block of m bits with n bits in the feedback loop.

It is seen that the block encryption algorithm acts on both sides. The decryption operation is sensitive to bit errors, because one bit error in the encrypted text affects the decryption of $(m/n + 1)$ blocks, the present one and the next (m/n). In this mode of operation, the initialization vector needs to be changed after each message to prevent cryptanalysis.

In the case of $n = m$, the shift register can be eliminated, and the encryption is done as illustrated in Figure 3.20. Thus, the encrypted block E_i is given by

$$E_i = M_i \oplus E_K(E_{i-1})$$

where $E_K()$ represents encryption with the secret key K.

The decryption is obtained with another Exclusive OR operation, as follows:

$$M_i = E_i \oplus E_K(E_{i-1})$$

which is shown in Figure 3.21.

The CFB mode can be used to calculate the MAC of a message as the last block encrypted two consecutive times. This method is also indicated in ANSI X9.9 (1986) for the authentication of banking messages, as well as ANSI X9.19 (1986), ISO 8731-1 (1987), and ISO/IEC 9797-2 (2002). In the encryption of a telnet stream with SKIPJACK, $m = 64$ bits, and $n = 32$ or 8 bits, depending on whether integrity is provided. These modes are denoted as CFB-8 without integrity and CFB-32 with integrity.

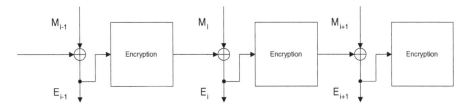

FIGURE 3.20
Encryption in the CFB mode for a block of n bits with a feedback of n bits.

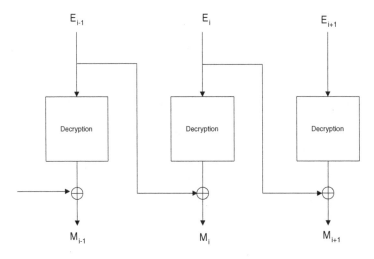

FIGURE 3.21
Decryption in the CFB mode for a block of n bits with a feedback of n bits.

Finally, the OFB mode is similar to the CFB mode, except that the n bits in the feedback loop result from the encryption and are not in the ciphertext transmitted to the destination. This is illustrated in Figures 3.22 and 3.23 for the encryption and decryption, respectively.

The OFB mode is adapted to situations where the transmission systems insert significant errors, because the effects of such errors are confined: a single bit error in the ciphertext affects only one bit in the recovered text. However, to avoid the loss of synchronization, the values in the shift registers should be identical. Thus, any system that incorporates the OFB mode must be able to detect the loss of synchronization and have a mechanism with which to reinitialize the shift registers on both sides with the same value.

In the case where $n = m$, the encryption operation is represented in Figure 3.24 and is described by

$$E_i = M_i \oplus S_i$$
$$S_i = E_K(S_{i-1})$$

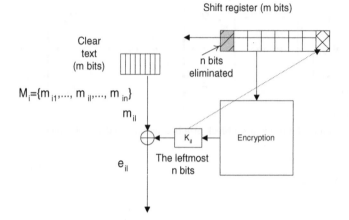

FIGURE 3.22
Encryption in the OFB mode of a block of *m* bits with a feedback of *n* bits.

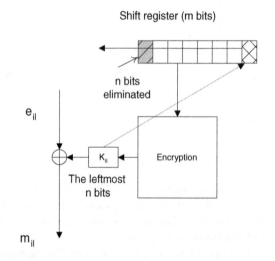

FIGURE 3.23
Decryption in the OFB mode of a block of *m* bits with a feedback of *n* bits.

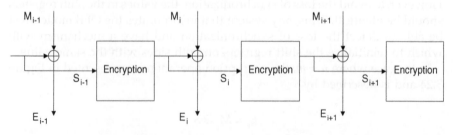

FIGURE 3.24
Encryption in the OFB mode with a block of *n* bits and a feedback of *n* bits.

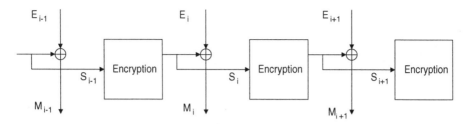

FIGURE 3.25
Decryption in the OFB mode for a block of *n* bits with a feedback of *n* bits.

The algorithm approaches a permutation of *m* bits that, on average, repeats itself every $2^m - 1$ cycles. Therefore, it is recommended to utilize the OFB mode only with $n = m$, i.e., the feedback size equal to the block size, to increase the security of the operation.

The decryption is described by

$$M_i = E_i \oplus S_i$$

$$S_i = E_K(S_{i-1})$$

and it takes place as indicated in Figure 3.25.

3.25.2 Examples of Symmetric Block Encryption Algorithms

3.25.2.1 Advanced Encryption Standard (AES)

The AES is the new symmetric encryption algorithm that will replace DES. It is published by NIST as FIPS 197 and is based on the algorithm Rijndael that was developed by Joan Daemen of Proton World International and Vincent Rijmen from the Catholic University of Leuven (Katholieke Universiteit Leuven). It is a block code with blocks of 128, 192, or 256 bits. The corresponding key lengths are 128, 192, and 256 bits, respectively.

The selection in October 2000 came after two rounds of testing following an NIST invitation for submission to cryptographers from around the world. In the first round, 15 algorithms were retained for evaluation. In the second round of evaluation, five finalists were retained: RC6, MARS, Rijndael, Serpent, and Twofish. All the second-round algorithms showed a good margin of security. The criteria used to separate them related to algorithmic performance: speed of computation in software and hardware implementations (including specialized chips), suitability to smart cards (low memory requirements), etc. Results from the evaluation and the rationale for the selection were documented in a public report by NIST (Nechvatal et al., 2000).

3.25.2.2 Data Encryption Standard (DES)

The DES is one of the most widely used algorithms in the commercial world for applications such as the encryption of financial documents, the management of cryptographic keys, and the authentication of electronic transactions. This algorithm was developed by IBM and then adopted as a U.S. standard in 1977. It was published in FIPS 81 and then adopted by ANSI in ANSI X3.92 (1981) under the name of Data Encryption Algorithm. This algorithm reached the end of its useful life and is expected to be replaced by the AES.

The DES operates by encrypting blocks of 64 bits of clear text to produce blocks of 64 bits of ciphertext. The encryption and decryption are based on the same algorithm, with some minor differences in the generation of sub-keys.

The key length is 64 bits, with 8 bits for parity control, which gives an effective length of 56 bits. The operation of DES consists of 16 rounds of identical operations, each round including a text substitution followed by a bit-by-bit permutation of the text, based on the key. If the number of rounds is fewer than 16, DES can be broken by a clear-text attack, which is easier to conduct than an exhaustive search.

3.25.2.3 Triple DES

The vulnerability of DES to an exhaustive attack encouraged the search of other, surer algorithms until a new standard is available. Given the considerable investment in the software and hardware implementations of DES, triple DES uses DES three successive times with two different keys. Represented in Figure 3.26 are the schema used in triple DES.

The use of three stages doubles the effective length of the key to 112 bits. The operation "encryption-decryption-encryption" aims at preserving compatibility with DES, because if the same key is used in all operations, the first two cancel each other. As there are several ways to attack the algorithm, it is recommended that three independent keys be used (Schneier, 1996a, pp. 359–360).

3.25.2.4 IDEA

The International Data Encryption Algorithm (IDEA) was invented by Xuejia Lai and James Massey circa 1991. The algorithm takes blocks of 64 bits of the clear text, divides them into subblocks of 16 bits each, and encrypts them with a key 128 bits long. The same algorithm is used for encryption and decryption. The IDEA is clearly superior to DES but has not been a commercial success. The patent is held by a Swiss company, Ascom-Tech AG, and is not subject to U.S. export control.

3.25.2.5 SKIPJACK

SKIPJACK is an algorithm developed by the NSA for several single-chip processors such as Clipper, Capstone, and Fortezza. Clipper is a tamper-

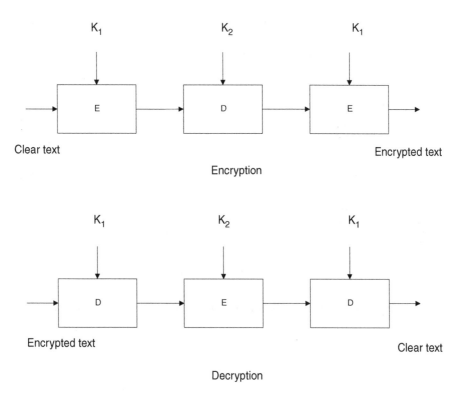

FIGURE 3.26
Operation of triple DES.

resistant, very large scale integration (VLSI) chip used to encrypt voice conversation. Capstone provides the cryptographic functions needed for secure e-commerce and is used in Fortezza applications. SKIPJACK is an iterative block cipher with a block size of 64 bits and a key of 80 bits. It can be used in any of the four modes ECB, CBC, CFB (with a feedback of 8, 16, 32, or 64 bits), and OFB, with a feedback of 64 bits.

3.26 Appendix II: Principles of Public Key Encryption

The most popular algorithms for public cryptography are those of Rivest, Shamir, and Adleman (1978), Rabin (1979), and ElGamal (1985). Nevertheless, the overwhelming majority of proposed systems in commercial systems is based on the RSA algorithm.

It should be noted that RSADSI was founded in 1982 to commercialize the RSA algorithm for public key cryptography. However, its exclusive rights ended with the expiration of the patent on September 20, 2000.

3.26.1 RSA

Consider two odd prime numbers **p** and **q** with a product $N = p \times q$. **N** is the modulus used in the computation, which is public, while the values **p** and **q** are kept secret.

Let $\varphi(n)$ be the Euler totient function of **N**. By definition, $\varphi(n)$ is the number of elements formed by the complete set of residues that are relatively prime to **N**. This set is called the reduced set of residues modulo **N**.

If **N** is a prime, $\varphi(N) = N - 1$. However, because $N = p \times q$ by construction, while **p** and **q** are primes, then

$$\varphi(N) = (p-1)\,(q-1)$$

According to Fermat's little theorem, if **m** is a prime, and **a** is not a multiple of **m**, the

$$a^{m-1} \equiv 1 \;(\mathrm{mod}\;m)$$

Euler generalized this theorem in the following form:

$$a^{\varphi(N)} \equiv 1 \;(\mathrm{mod}\;N)$$

Choose the integers **e, d** both less than $\varphi(N)$ such that the greatest common divisor of $(e, \varphi(N)) = 1$ and $e \times d \equiv 1 \bmod (\varphi(N)) = 1 \bmod ((p-1)(q-1))$.

Let **X, Y** be two numbers less than **N**:

$$Y = X^e \bmod N \text{ with} \qquad\qquad 0 \le X < N$$

$$X = Y^d \bmod N \text{ with} \qquad\qquad 0 \le Y < N$$

because, by applying Fermat's theorem:

$$Y^d \bmod N = (X^e)^d \bmod N = X^{ed} \bmod N = X^{\varphi(N)} \equiv 1 \;(\mathrm{mod}\;N) = 1 \bmod N$$

To start the process, a block of data is interpreted as an integer. To do so, the total block is considered an ordered sequence of bits (of length, say, λ). The integer is considered to be the sum of the bits by giving the first bit the weight of $2^{\lambda-1}$, the second bit the weight of $2^{\lambda-2}$, and so on, until the last bit, which will have the weight of $2^0 = 1$.

The block size must be such that the largest number does not exceed modulo **N**. Incomplete blocks must be completed by padding bits with either 1 or 0 bits. Further padding blocks may also be added.

The public key of the algorithm **Pk** is the number **e**, along with **n**, while the secret key **Sk** is the number **d**. RSA achieves its security from the difficulty of factoring **N**. The number of bits of **N** are considered to be the key size of the RSA algorithm. The selection of the primes **p** and **q** must make this factorization as difficult as possible.

Once the keys are generated, it is preferred that, for reasons of security, the values of **p** and **q** as well as all intermediate values, such as the product (**p**− 1)(**q**− 1) be deleted. Nevertheless, the preservation of the values of **p** and **q** locally can double or even quadruple the speed of decryption.

3.26.1.1 Practical Considerations

To increase the speed of signature verification, suggested values for the exponent **e** of the public key are 3 or $2^{16} + 1$ (65,537) (Menezes et al., 1997, p. 437). Other variants designed to speed decryption and signing are discussed in Boneh and Shacham (2002).

For short-term confidentiality, the modulus **N** should be at least 768 bits. For long-term confidentiality (5 to 10 years), at least 1024 bits should be used. Currently, it is believed that confidentiality with a key of 2048 bits would last about 15 years.

3.26.2 Public Key Cryptography Standards (PKCS)

PKCS are business standards developed by RSA Laboratories in collaboration with many other companies working in the area of cryptography. They are used in many aspects of public key cryptography that are based on the RSA algorithm. At the time of writing this section, their number reached 15.

PKCS #1 (IETF RFC 2437, 1998) defines the mechanisms for data encryption and signature using the RSA algorithm. These procedures are then utilized for constructing the signatures and electronic envelopes described in PKCS #7. In particular, PKCS #1 defines an encryption scheme based on the Optimal Asymmetric Encryption Padding (OAEP) of Bellare and Rogaway. PKCS #2 and #4 were incorporated in PKCS #1.

PKCS #3 defines the key exchange protocol using the Diffie–Hellman algorithm.

PKCS #5 describes a method for encrypting information using a secret key derived from a password. For hashing, the method utilizes either MD2 or MD5 to compute the key, starting with the password and then encrypting the key with DES in the CBC mode.

PKCS #6 is syntax for X.509 certificates.

PKCS #7 (IETF RFC 2315, 1998) defines the syntax of a message encrypted using the Basic Encoding Rules (BER) of ASN.1 (Abstract Syntax Notation 1) (Steedman, 1993) of ITU-T Recommendation X.209 (1988). These messages are formed with the help of six content types:

1. *Data,* for clear data
2. *SignedData,* for signed data
3. *EnvelopedData,* for clear data with numeric envelopes
4. *SignedAndEnvelopedData,* for data that are signed and enveloped
5. *DigestedData,* for digests
6. *EncryptedData,* for encrypted data

The secure messaging protocol, S/MIME (Secure Multipurpose Internet Mail Extensions), as well as the messages of the SET protocol, designed to secure bank card payments over the Internet, utilize the PKSC #7 specifications.

PKCS #8 describes a format for sending information related to private keys.

PKCS #9 defines the optional attributes that could be added to other protocols of the series. The following items are considered: the certificates of PKCS #6, the electronically signed messages of PKCS #7, and the information on private keys as defined in PKCS #8.

PKCS #10 (IETF RFC 2986, 2000) describes the syntax for certification requests to a certification authority. The certification request must contain details on the identity of the candidate for certification, the distinguished name of the candidate, his or her public key, and optionally, a list of supplementary attributes, a signature of the preceding information to verify the public key, and an identifier of the algorithm used for the signature so that the authority could proceed with the necessary verifications. The version adopted by the IETF is called CMS (Cryptographic Message Syntax).

PKCS #11 defines a cryptographic interface called Cryptoki (Cryptographic Token Interface Standard) between portable devices such as smart cards or PCMCIA cards and the security layers.

PKCS #12 describes syntax for the storage and transport of public keys, certificates, and other user secrets. In enterprise networks, VeriSign transmits key pairs to individuals via password-protected PKCS #12 files. Microsoft utilizes this syntax in the new version of NT Server 5.0.

PKCS #13 describes a cryptographic system using elliptic curves.

PKCS #15 describes a format to allow the portability of cryptographic credentials, such as keys, certificates, passwords, and PINs, among applications and among portable devices such as smart cards.

Note that even though the specifications of PKCS #1, #7, and #10 were described in IETF documents, this organization has not accepted them as standards, because they mandate the utilization of algorithms that RSADSI does not offer free of charge.

Also note that in PKCS #11 and #15, the word *token* is used to indicate a portable device capable of storing persistent data.

3.26.3 Pretty Good Privacy (PGP)

PGP is considered to be the commercial system with security closest to the military grade. It is described in one of the IETF documents, namely, RFC 1991 (1996). PGP consists of six functions:

- Public key exchange using RSA with MD5 hashing
- Data compression with ZIP, which reduces file sizes and redundancies before encryption (Reduction of size augments the speeds for processing and transmission, while reduction of redundancies makes cryptanalysis more difficult.)
- Message encryption with IDEA
- Encryption of the user's secret key using the digest of a sentence instead of a password
- ASCII "armor" protects the binary message for any mutilations that might be caused by Internet messaging systems. (This armor is constructed by dividing the bits of three consecutive octets into four groups of 6 bits each and then by coding each group using a 7-bit character according to a given table. A checksum is then added to detect potential errors.)
- Message segmentation

Although the IETF worked on PGP, it has not adopted PGP as a standard yet, because it incorporates protocols that have patent protections, such as IDEA and RSA. Current activities in the IETF attempt to use the framework of PGP but with protocols that circumvent these restrictions.

3.26.4 Elliptic Curve Cryptography (ECC)

Elliptic curves have been studied in algebraic geometry and number theory. They have been applied in factoring integers, in proving primality, in coding theory, and in cryptography (Menezes, 1993). Elliptic curve cryptography (ECC) is a public key cryptosystem where the computations take place on an elliptic curve. These cryptosystems are variants of the Diffie–Hellman and DSA algorithms, thereby giving rise to the Elliptic Curve Diffie–Hellman algorithm (ECDH) and the Elliptic Curve Digital Signal Algorithm (ECDSA), respectively. They can be used to create digital signatures and to establish keys for symmetric cryptography. The ECDSA algorithm is now an ANSI standard (X9.62) (1998).

The elliptic curves are defined over the finite field of the integer numbers modulo, a primary number p [the Gallois field GF(p)] or that of binary polynomials [GF(2^m)]. The key size is the size of the prime number or the binary polynomial in bits. Cryptosystems over GF(2^m) appear to be slower than those over GF(p), but there is no consensus on that point. Their main advantage, however, is that addition over GF(2^m) does not require integer multiplications, which reduces the cost of the integrated circuits implementing the computations.

The ECDSA is used for digital signing, while the ECDH can be used to secure online key exchange. Perfect forward secrecy is achieved with the ephemeral mode of ECDH, i.e., the key is for short-term use. Diffie–Hellman and ECDH are comparable in speed, but RSA is much slower because of the generation of the key pair.

Typical key sizes are in the range of 160 to 200 bits. The advantage of elliptic curve cryptography is that key lengths are shorter than those for existing public key schemes that provide equivalent security. For example, the level of security of 1024-bit RSA can be achieved with elliptic curves with a key size in the range of 171 to 180 bits (Wiener, 1998). This is an important factor in wireless communications and whenever bandwidth is a scarce resource.

Given in Table 3.14 are various computation times for digital signatures with RSA, DSA, and ECDSA on a 200-MHz Pentium Pro (Agnew, 2000).

The results show that RSA is slower for signing and much faster for signature verification than DSA and ECDSA. Thus, from a computational speed viewpoint, RSA is more suitable for certificate verification, while Diffie–Hellman, ECDH, and ECDSA are more suitable for online communication.

Finally, in Table 3.15, the key lengths of RSA and elliptic cryptography are compared for the same amount of security measured in terms of effort to break the system (Menezes, 1993).

TABLE 3.14

Computation Times for Digital Signatures with the RSA, DSA, and ECDSA Algorithms

	Timings in msec (on a 200-MHz Pentium Pro)		
Operation	RSA with N = 1024 and e = 3	DSA with 1024 bits	ECDSA over GF(p) with 168 bits
Sign	43	7	5
Verify	0.6	27	19
Key generation	1100	7	17
Parameter generation	0	6500	High

Source: From Agnew, G.B., in *Electronic Commerce Technology Trends: Challenges and Opportunities*, IBM Press, Toronto, Canada, 2000, 69–85.

TABLE 3.15

Comparison of Public Key Systems in Terms of Key
Length (in bits) for the Same Security Level

RSA	Elliptic Curve	Reduction Factor RSA/ECC
512	106	5:1
1,024	160	7:1
2,048	211	10:1
5,120	320	16:1
21,000	600	35:1

Source: From Menezes, A., *Elliptic Curve Public Key
Cryptosystems*, Kluwer, Dordrecht, 1993. With permission.

3.27 Appendix III: Principles of the Digital Signature Algorithm (DSA)

According to the DSA defined in ANSI X9.30:1 (1997), the signature of a
message M is the pair of numbers r and s computed as follows:

$$r = (g^k \bmod p) \bmod q \quad \text{and}$$

$$s = \{k^{-1}\,[H(M) + x\,r]\} \bmod q$$

where:
p and q are primes such that $2^{511} < p < 2^{1024}$, $2^{159} < q < 2^{160}$, and q is a prime
divisor of $(p - 1)$, i.e., $(p - 1) = mq$ for some integer m,

$g = h^{(p-1)/q} \bmod p$ is a generator polynomial modulo p of order q, with h
any integer $1 < h < (p - 1)$ such that $h^{(p-1)/q} \bmod p > 1$. By Fermat's little theorem, $g^q = h^{(p-1)} \bmod p = 1$ because $g < p$.
Thus, each time the exponent is a multiple of q, the result will
be equal to 1 (mod p).

x and k are randomly generated integers between 0 and q (i.e., $0 < x, k < q$).

x is the private key of the sender, while the public key y is given
by $y = g^x \bmod p$

k^{-1} is the multiplicative inverse of $k \bmod q$, i.e., $(k^{-1} \times k) \bmod q = 1$,
where $0 < k, k^{-1} < q$ and,

$H()$ is the SHA-1 hash function

To verify the signature, the verifier computes the following:

$$w = s^{-1} \bmod q$$

$$u_1 = H(M)\, w \bmod q$$

$$u_2 = r\, w \bmod q$$

$$v = (g^{u_1}\, y^{u_2} \bmod p) \bmod q$$

If $v = r$, the signature is valid.
To show this, we have:

$$v = \{ (g^{[H(M)w \bmod q]}\, y^{rw \bmod q}) \bmod p \} \bmod q$$

$$= (g^{[H(M)w \bmod q]}\, g^{xrw \bmod q)} \bmod p \} \bmod q$$

$$= \{ g^{[H(M)+xr]w \bmod q]} \bmod p \} \bmod q$$

$$= (g^{ksw \bmod q} \bmod p)\, \bmod q$$

$$= (g^{k \bmod q} \bmod p)\, \bmod q$$

$$= (g^{k} \bmod p) \bmod q, \text{ since the generator is of order } q \text{ by construction,}$$

$$= r$$

Note that the random variable k is also transmitted with the signature. This means that if the verifier knows the signer's private key, they will be able to pass additional information through the channel established through the value of k.

3.28 Appendix IV: Comparative Data

This appendix contains data that RSADSI publishes from time to time to illustrate the performances of some encryption algorithms on different computational platforms.

It is important not to take these numbers strictly, because the conditions of measurement were not completely defined, especially because RSADSI does not maintain a permanent site for these comparative data.

3.28.1 Performance Data for JSAFE 1.1

The data in Table 3.16 through Table 3.18 were collected for the cryptographic library JSAFE 1.1 on a Pentium microprocessor running at 166 MHz.

TABLE 3.16

Hashing with JSAFE 1.1

Algorithm	Setup Time (msec)	Speed of Execution (K octets/sec)
MD5	<1	3031.530
SHA-1	<1	1565.576

TABLE 3.17

Symmetric Encryption with JSAFE 1.1

Algorithm	Setup Time (msec)	Speed of Execution (K octets/sec)
DES	3	370
Triple DES	4	250
RC2	3	480
RC4	2	2510.3
RC5	2	1530

TABLE 3.18

Public Key Encryption with JSAFE 1.1

Algorithm	Size of the Key (bits)			
	512	768	1024	2048
	Duration (sec)			
RSA key generation	1.899	3.536	9.370	60.826
RSA encryption	0.002	0.006	0.004	0.0012
RSA decryption	0.030	0.080	0.173	1.256
RSA signature	0.030	0.080	0.173	1.256
RSA signature verification	0.002	0.006	0.004	0.0012
Diffie–Hellman parameter generation	7.195	33.112	—	—
Diffie–Hellman key exchange	0.147	0.466	—	—

3.28.2 Performance for S/WAN

Table 3.19 through Table 3.21 display performance results for S/WAN. S/WAN implements IPSec to ensure interoperability among firewalls and other TCP/IP products. It supports encryption at the IP level.

TABLE 3.19

Computation Speed (in K octets/sec) of Hashing Algorithms with S/WAN

	Platform			
Algorithm	Intel 486 at 33 MHz	Pentium at 90 MHz	Power MAC 8100 at 80 MHz	Sun SPARC Classic at 50 MHz
MD2	28	140	149	47
MD5	320	1100	2700	756

TABLE 3.20

Computation Speed (in K octets/sec) of Symmetric Encryption with S/WAN

	Setup Time (msec)				Computation Speed (K octets/sec)			
Platform Algorithm	Intel 486 at 33 MHz	Pentium at 90 MHz	Power MAC 8100 at 80 MHz	Sun SPARC Classic at 50 MHz	Intel 486 at 33 MHz	Pentium at 90 MHz	Power MAC 8100 at 80 MHz	Sun SPARC Classic at 50 MHz
DES	2.08	0.5	0.36	0.78	32	116	197	73
RC2	0.35	0.08	0.05	0.2	121	384	770	192
RC4	0.28	0.13	0.2	0.27	931	1920	3840	1270
RC4 with MAC	0.41	0.15	0.23	0.45	794	1530	430	200

TABLE 3.21

Computation Speed (in K octets/sec) of Public Key Encryption with S/WAN

	Duration (sec)							
Size of the Public Key	512 bits				1024 bits			
Platform	Intel 486 at 33 MHz	Pentium at 90 MHz	Power MAC 8100 at 80 MHz	Sun SPARC Classic at 50 MHz	Intel 486 at 33 MHz	Pentium at 90 MHz	Power MAC 8100 at 80 MHz	Sun SPARC Classic at 50 MHz
Generation of the RSA key	6.8	1.7	1.3	7.3	61	16	12	68

Size of the Public Key	Speed (octets/sec)							
	512 bits				1024 bits			
Platform	Intel 486 at 33 MHz	Pentium at 90 MHz	Power MAC 8100 at 80 MHz	Sun SPARC Classic at 50 MHz	Intel 486 at 33 MHz	Pentium at 90 MHz	Power MAC 8100 at 80 MHz	Sun SPARC Classic at 50 MHz
RSA encryption	850	3900	5100	890	470	2200	3000	490
RSA decryption	410	2100	2100	450	230	1100	1200	240

Size of the Private Key	Speed (octets/sec)							
	512 bits				1024 bits			
Platform	Intel 486 at 33 MHz	Pentium at 90 MHz	Power MAC 8100 at 80 MHz	Sun SPARC Classic at 50 MHz	Intel 486 at 33 MHz	Pentium at 90 MHz	Power MAC 8100 at 80 MHz	Sun SPARC Classic at 50 MHz
RSA encryption	97	420	580	110	36	170	230	39
RSA decryption	47	210	320	53	17	83	100	19

Key Size	Duration (sec)							
	256 bits				512 bits			
Platform	Intel 486 at 33 MHz	Pentium at 90 MHz	Power MAC 8100 at 80 MHz	Sun SPARC Classic at 50 MHz	Intel 486 at 33 MHz	Pentium at 90 MHz	Power MAC 8100 at 80 MHz	Sun SPARC Classic at 50 MHz
Diffie–Hellman parameter generation	49	7.5	5.3	30	440	65	50	280
Diffie–Hellman key exchange	0.31	0.074	0.053	0.29	1.7	0.37	0.28	1.5

3.28.3 Performance for BSAFE™ 3.0

BSAFE is a general purpose, low-level cryptographic tool kit that contains many industry standard algorithms. Table 3.22 through Table 3.24 give the performance for BSAFE 3.0.

TABLE 3.22

Computational Speed (in K octets/sec) of Hashing
Algorithms with BSAFE™ 3.0

	Platform			
Algorithm	Pentium at 90 MHz	Power MAC at 80 MHz	Sun SPARC Station 4 at 110 MHz	Digital Alpha Station at 255 MHz
MD5	13,156	3,108	5,140	11,714
SHA-1	2,530	1,181	2,000	5,975

TABLE 3.23

Computational Speed (in K octets/sec) of Symmetric Encryption with BSAFE™ 3.0

Platform Algorithm	Setup Time (µsec)				Computation Speed (K octets/sec)			
	Pentium at 90 MHz	Power MAC at 80 MHz	Sun SPARC Station 4 at 110 MHz	Digital Alpha Station at 255 MHz	Pentium at 90 MHz	Power MAC at 80 MHz	Sun SPARC Station 4 at 110 MHz	Digital Alpha Station at 255 MHz
DES	19	31	24	12	963	470	667	1686
Triple DES	60	77	70	28	330	168	247	634
RC2	45	85	56	39	1017	610	806	1606
RC4	65	110	112	54	7800	3116	2000	5390
RC5	50	66	43	19	3000	1165	1400	4192

3.28.4 Performance for BSAFE™ 4.1

Table 3.25 through Table 3.27 display results of tests conducted on the cryptographic library BSAFE™ 4.1 on a Pentium II processor running at 233 MHz.

Questions

1. What are the major security vulnerabilities in a client/server communication?

2. What are the services needed to secure data exchanges in e-commerce?

TABLE 3.24

Computational Speed (in K octets/sec) of Public Key Encryption with BSAFE™ 3.0

Size of the Public Key												
					Duration (sec)							
	512 bits				**768 bits**				**1024 bits**			
Platform Algorithm	Pentium at 90 MHz	Power MAC at 80 MHz	Sun SPARC at 110 MHz	Digital Alpha at 255 MHz	Pentium at 90 MHz	Power MAC at 80 MHz	Sun SPARC at 110 MHz	Digital Alpha at 255 MHz	Pentium at 90 MHz	Power MAC at 80 MHz	Sun SPARC at 110 MHz	Digital Alpha at 255 MHz
Generation of RSA key	0.45	1.1	1.2	0.26	1.5	4.5	4.1	0.59	3.8	13	11	1.28

Key Size								
					Duration (sec)			
	512 bits				**768 bits**			
Platform RSA Operation	Pentium at 90 MHz	Power MAC at 80 MHz	Sun SPARC Station 4 at 110 MHz	Digital Alpha Station at 255 MHz	Pentium at 90 MHz	Power MAC at 80 MHz	Sun SPARC Station 4 at 110 MHz	Digital Alpha Station at 255 MHz
Encryption	0.0027	0.008	0.0076	0.0098	0.0053	0.017	0.015	0.0017
Decryption	0.024	0.076	0.071	0.008	0.066	0.22	0.212	0.024
Signature	0.024	0.076	0.071	0.008	0.066	0.22	0.212	0.024
Verification	0.0027	0.008	0.0076	0.0098	0.0053	0.017	0.015	0.0017

TABLE 3.24 (continued)

Computational Speed (in K octets/sec) of Public Key Encryption with BSAFE™ 3.0

Key Size	1024 bits				2048 bits			
Platform RSA Operation	Pentium at 90 MHz	Power MAC at 80 MHz	Sun SPARC Station 4 at 110 MHz	Digital Alpha Station at 255 MHz	Pentium at 90 MHz	Power MAC at 80 MHz	Sun SPARC Station 4 at 110 MHz	Digital Alpha Station at 255 MHz
Encryption	0.0086	0.029	0.026	0.0026	0.031	0.11	0.096	0.0088
Decryption	0.14	0.534	0.461	0.043	0.93	3.6	3.4	0.29
Signature	0.14	0.534	0.461	0.043	0.93	3.6	3.4	0.29
Verification	0.0086	0.029	0.026	0.0026	0.031	0.11	0.096	0.0088

Duration (sec)

Key Size	512 bits				1024 bits			
Platform Diffie–Hellman Operation	Pentium at 90 MHz	Power MAC at 80 MHz	Sun SPARC Station 4 at 110 MHz	Digital Alpha Station at 255 MHz	Pentium at 90 MHz	Power MAC at 80 MHz	Sun SPARC Station 4 at 110 MHz	Digital Alpha Station at 255 MHz
Parameter generation	11	35	—	5.2	200	733	—	67
Key exchange	0.07	0.242	0.224	0.02	0.47	1.8	1.68	0.139

Duration (sec)

Duration (sec)

Key Size	512 bits				768 bits				1024 bits			
Platform	Pentium	Power MAC	Sun SPARC	Digital Alpha	Pentium	Power MAC	Sun SPARC	Digital Alpha	Pentium	Power MAC	Sun SPARC	Digital Alpha
Clock, MHz	90	80	110	255	90	80	110	255	90	80	110	255
Generation of the DSA key	1.8	5.9	5.6	0.74	7.2	26	25	2.44	13	49	45	4.18

Duration (sec)

Key Size	512 bits			
Platform DSA Operation	Pentium at 90 MHz	Power MAC at 80 MHz	Sun SPARC Station 4 at 110 MHz	Digital Alpha Station at 255 MHz
Signature	0.029	0.087	0.086	0.011
Verification	0.052	0.17	0.16	0.018

Key Size	768 bits			
Platform DSA Operation	Pentium at 90 MHz	Power MAC at 80 MHz	Sun SPARC Station 4 at 110 MHz	Digital Alpha Station at 255 MHz
Signature	0.053	0.179	0.172	0.019
Verification	0.1	0.356	0.338	0.033

Duration (sec)

Key Size	1024 bits			
Platform DSA Operation	Pentium at 90 MHz	Power MAC at 80 MHz	Sun SPARC Station 4 at 110 MHz	Digital Alpha Station at 255 MHz
Signature	0.086	0.306	0.29	0.028
Verification	0.17	0.62	0.58	0.052

Key Size	2048 bits			
Platform DSA Operation	Pentium at 90 MHz	Power MAC at 80 MHz	Sun SPARC Station 4 at 110 MHz	Digital Alpha Station at 255 MHz
Signature	0.3	1.15	1.08	0.94
Verification	0.6	2.3	2.17	0.18

TABLE 3.25

Computational Speed (in K octets/sec) of
Hashing Algorithms with BSAFE™ 4.1

Algorithm	Computation Speed (K octets/sec)
MD5	36,250
SHA-1	20,428

TABLE 3.26

Computational Speed (in K octets/sec) of Symmetric
Encryption Algorithms with BSAFE™ 4.1

Algorithm	Setup Time (μsec)	Computation Speed (K octets/sec)	
		Encryption	Decryption
DES	6.3	4,386	4,557
Triple DES	22	1,596	1,620
RC4	29.8	27,325	28,132
RC5 block of 128 bits, 12 rounds	352	4,576	4,691
RC5 block of 64 bits, 12 rounds	12.2	11,242	12,641

TABLE 3.27

Computational Speed (in K octets/sec) of Public Key Encryption
Algorithms with BSAFE™ 4.1

	Key Size (bits)		
	512	1024	2048
RSA encryption	10.5	4.23	43.6
RSA decryption	5.28	2.87	1.4

3. What factors affect the strength of encryption?

4. What is needed to offer nonrepudiation services?

5. What conditions favor denial of service attacks?

6. Which of the following items is not in a digital public key certificate: (a) the subject's public key; (b) the digital signature of the certification authority; (c) the subject's private key; or (d) the digital certificate serial number?

7. Using the case of AES as a starting point, define a process to select a new encryption algorithm.

8. Compare public key encryption and symmetric encryption in terms of advantages and disadvantages. How can the strong points of each be combined?

9. What are the reasons for the current interest in elliptic curve cryptography (ECC)?

10. In your opinion, are both the tunnel mode and transport mode needed for IPSec?

11. From your experience, give some examples for security protocols at the application layer.

12. Speculate on the reasons that led to the declassification of the SKIP-JACK algorithm.

13. Discuss some potential applications for blind signatures.

14. Discuss some of the vulnerabilities of biometric identification systems.

15. Speculate on the reasons that security directory services in the Telecommunications Management Network (TMN) as defined in ANSI T1.252 are based on X.500 and not on LDAP.

16. What are the problems facing cross-certification? How are financial institutions attempting to solve them?

4

Business-to-Business Commerce

ABSTRACT

Business-to-business commerce is part of a set of measures to dematerialize the exchange of electronic data to improve the integration of different departments within the same enterprise or with commercial partners. This need for coordination became acutely apparent after the introduction of just-in-time production and with the emergence of global enterprises concerned about maintaining, if not enhancing, the speed of response as well as controlling operational cost, despite the breadth of area of geographic coverage. The dematerialization of business-to-business traffic started with tasks related to procurement. With the advancements in information technology (hardware and software) and in telecommunications, the focus expanded gradually to other technical or managerial areas to coordinate the separate elements of data or applications that were assembled as need arose to satisfy specific customer requests. Currently, the efforts aim at having a uniform architecture for the flow of information, end-to-end, along the supply chain.

From a strictly technical viewpoint, the technology that allowed the development of the dematerialized exchange of data relies on the following supports:

- The protocols that manage the exchange of electronic files
- The format of messages related to the supply chain
- The nature of the telecommunication networks
- The information technology platform of the exchanges

To each of the preceding elements corresponds a set of security techniques in the form of products and standards (de jure or de facto). In addition, management aspects cover the following:

- Service offers (cataloging, order taking, payment, billing, logistics, etc.)
- Policies for flow control (purchase policies, traceability of orders, merchandise reception, security)
- Management of electronic documents (archival, retrieval, backup)
- Management of legal responsibilities [In some professions (notaries, bailiffs, etc.), it is essential to ensure integrity and to preserve the archives for a duration defined by law. Data directly or indirectly related to financial results must likewise be retrieved in case of financial or legal audit.]

The security of business-to-business commerce in its immaterial form must take into account the ensemble of these aspects: the specific security techniques for each element taken on its own, and the techniques to reduce the risks affecting management of the overall service.

This chapter focuses primarily on the technical aspects of business-to-business commerce and its security. An overview of the problem to be solved precedes the study of the data structures, exchange protocols, and security techniques. Data structuring can be done in two ways, either in the form of alphanumeric strings or in the form of documents or forms. The alphanumeric technique is historically the first and is known as electronic data interchange (EDI), whereby commercial data are coded in the form of alphanumeric strings based on standardized conventions. The Internet favors the document approach, especially since the development of the metalanguage XML (Extensible Markup Language), from which dialects can be defined to fit specific applications. We describe these two approaches that continue to coexist as well as harmonization efforts. We will describe the various messaging protocols, in particular, X.400 messaging and Internet messaging. We next study the security of the exchanges and some aspects related to the financial EDI (credit transfers, electronic billing). In the end, we review the standardization of business-to-business commerce in terms of main players and perspectives for evolution.

4.1   Overview of Business-to-Business Commerce

An enterprise is the focus of convergent relationships with suppliers, partners, clients, and banks that result in information exchanges. However, a large number of data are reproduced from one form to another inside a given enterprise. When these data reentries are done manually, they can be a source of errors that must later be detected and corrected. In addition, paper documents must be organized and archived for legal and fiscal reasons. As a result, the cost of processing contract documents in paper can reach 7% of the total transaction cost (Breton, 1994; Dupoirier, 1995). The first objective of the dematerialization of business-to-business commerce is to eliminate this additional cost through the exchange of structured and predefined data among the information systems involved in the conduct of businesses, thereby streamlining the tasks of billing, account management, inventory management, etc. (Sandoval, 1990; Kimberley, 1991; Charmot, 1997b; Troulet-Lambert, 1997). A new need emerged as a consequence of restructuring the supply chain as enterprises focused on their core competencies and out-sourced nonessential activities. Just-in-time management, in particular, requires unimpeded and continuous circulation of the information to coordinate production planning with the product delivery and market predictions in real time.

Depicted in Figure 4.1 are the exchanges related to catalog consultation, offer and purchasing transactions, shipment notices, merchandise reception, and financial data flowing within the banking network.

Consider the electronic equivalent of these exchanges by assuming that a given purchaser has identified an item on the basis of the information available online. The purchaser puts the order in an electronic document that is sent directly to the data input system of the supplier. The supplier's software responds to the purchaser with a price quotation that the program of the purchaser receives. If the quote is accepted, the exchanges can continue, and the purchaser's software composes, starting with the data in the offer request and in the price quote, a purchase order that it sends to the information system of the supplier. After its reception at the supplier side, the purchase order is then translated into the internal formats of the supplier and then routed to the various departments, such as accounting, the factory, or the warehouse. An electronic receipt is also sent to the purchaser to confirm the order. Once the order is filled, a shipment notification is constructed and transmitted to the purchaser as well as to the accounting department of the supplier. Reception of the shipment notification or the arrival of the invoice triggers the purchaser's software to create a receiving file.

If the information system of the shipping company is also integrated in the same information system, the documents originating from the transporter (freight letter, delivery notification, etc.) will be composed in an automatic manner on the basis of the data from the supplier. Reception of the delivery notification automatically starts the accounting procedures of the buyer. The receipt notification is reconciled electronically with the initial purchase order and the invoice so as to prepare the payment instructions for the purchaser's bank.

If the banking settlement is done by electronic credit transfer, the banks are responsible for carrying the payment notification to the supplier within the time frame specified by the terms and conditions of the purchase. The information system of the supplier must, in its turn, reconcile the payment with the invoice so as to keep the accounting information up to date. Other back-office services include the preparation of accounting and tax packages and archiving of the data associated with the transaction.

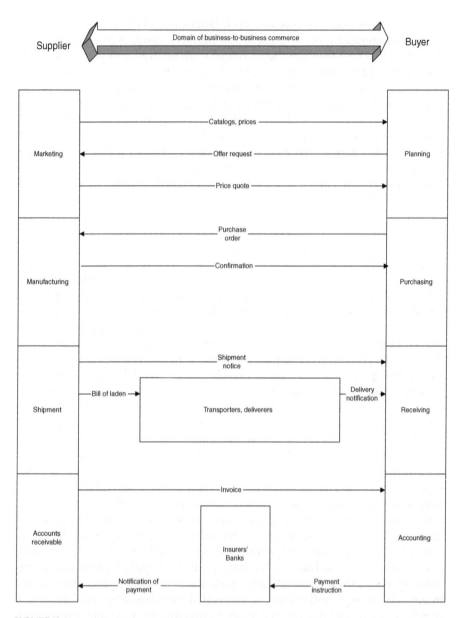

FIGURE 4.1
Message exchanges in business-to-business e-commerce.

Note that the responsibilities of each party should be defined in inter-change agreements. These rules are usually defined in interchange agreements that describe the rules that govern the business transactions and the responsibilities of each party, in addition to the rules for identification and authentication of the various entities. The agreements also specify what is needed to ensure reception of the documents and procedures for conserving electronic documents. Furthermore, in financial EDI, the agreements specify the legal and technical responsibilities of the banks as well as those of the client enterprises concerning the evidence to be presented, the procedures for conserving electronic documents, and the opposability to third parties.

4.2 Examples of Business-to-Business Electronic Commerce

4.2.1 A Short History of Business-to-Business Electronic Commerce

The first attempts at business-to-business electronic commerce (e-commerce) took place in the U.S. in the 1960s, with the aim of improving military supply logistics. Civilian applications soon followed in railroad, truck transportation, civil aviation, international payments (credit transfers, credit cards, and the management of customs). As each industrial group was devising its rules for structuring data without consultation with the others, the U.S. Transportation Data Coordinating Committee (TDCC) was formed to work on the convergence of the various specifications. Its first document, published in 1975, covered transport by air, by road, by railroads, and by maritime or river transport. A bit later, the food and warehouse industries in the U.S. issued their respective standards, UCS (Uniform Communication Standards) and WINS (Warehouse Information Network Standard). Finally, large automobile manufacturers, such as General Motors, and retailers and others with wide distribution networks, such as K-Mart, J.C. Penney, and the National Wholesale Druggists Association (NWDA), imposed their rules to their subcontractors and their billing agents. To avoid the proliferation of sector or proprietary rules, the American National Standards Institute (ANSI) established in 1982 a syntax common to the different business sectors. This syntax is known as the ANSI X12 standard, which is widely followed in North America.

In Canada, large business companies established a standard system for messages and communications, and in 1984, they formed the EDI Canadian Council (EDICC), which united the large distributors (stores, drugstores, warehouses, retailers). In 1986, Telecom Canada proposed a translation service between the internal messaging format of each organization and the X12 format, as well as a secured electronic mailbox.

In the U.K., the Department of Customs and Excise developed the first EDI for customs, known as the London Airport Cargo EDP Scheme (LACES) at Heathrow airport in 1971. The objective was to speed the processing of documents used in the internal trade (Tweddle, 1988; Walker, 1988). This activity, known as the Simplification of International Trade Procedures (SITPRO), produced the Trade Data Interchange (TDI), which was then submitted to the United Nations (U.N.) Economic Commission for Europe to facilitate international trade. It adopted this document as the U.N. Trade Data Interchange (UN-TDI), which evolved into the General Purpose Trade Data Interchange (GTDI) in 1981.

Although similar in form and function, the different syntaxes developed in North America and in Europe diverged in several important aspects, complicating the tasks of information systems developers as well as users (for example, the subcontractors working for different groups on the two sides of the Atlantic). It was, therefore, necessary to investigate the possibility of making a worldwide homogeneous system. The experts from both sides met under the aegis of the U.N. Joint Electronic Data Interchange (UN-JEDI) initiative to reach a consensus and harmonize both standards. This endeavor generated worldwide agreement on the definition of the data elements and use of the EDI for Administration, Commerce and Transport (EDIFACT) language. This agreement was adopted by the U.N. in 1987 and thus by the International Organization for Standardization (ISO) as ISO 9735 (1988). Teletransmission of customs forms using EDIFACT is regularly used within the countries of the European Union since January 1, 1993, the date of opening the borders among member states.

4.2.2 Banking Applications

Without waiting for an agreement across sectors, the Society for Worldwide Interbank Financial Telecommunications (SWIFT) was established in 1987 by 239 banks in 15 countries, with the objective of relaying the interbank messages related to international fund transfers. The aim was to replace paper and telex communications with electronic messaging. However, the installation of the system required considerable efforts, because until the late 1990s, about half of the messages required manual correction. The SWIFT system contains 200 messages that cover all aspects of international finance: cash, retail, large amounts, settlements of real estate transactions, currency operations, treasury, derivatives, international trade, etc. The SWIFT syntax contains codes with which to identify the parties and the processing of the payment instructions in each country (Remacle, 1996).

4.2.3 Aeronautical Applications

The SITA (Société Internationale de Télécommunications Aéronautiques — International Society for Aeronautical Telecommunications) has a network that today connects 350 airline companies and 100 related enterprises. This network allows for the exchange of data concerning reservations, tariffs, boardings, etc., according to the standards of the IATA (International Air Transport Association). These are CARGO-IMP (CARGO Interchange Message Procedures) for freight and AIR-IMP (AIR Interline Message Procedures) for passengers. Another SITA service allows for the selection, purchase, and localization of spare parts used in aviation. Since March 1994, SITA decided to use EDIFACT and ANSI X12 to structure the International Forwarding and Transport Message (IFTM) services and to employ X.400

messaging systems as well as the file transfer protocols OFTP ODETTE[1] File Transfer Protocol and the FTP (file transfer protocol) used over the Internet.

Last, in 1987, Air France, Iberia, and Lufthansa, established a centralized interactive system for reservations of air transport (Amadeus) to link travel agents, airline companies, hotel chains, and car rental companies. The settlement of travel documents among airline companies (changing airline companies after the ticket was issued, trips of several legs on different airlines) is done through the BSP (Bank Settlement Payment) system.

4.2.4   Applications in the Automotive Industry

The worldwide automotive industry is organized around a small number of manufacturers (General Motors, Ford, DaimlerChrysler, Toyota, Renault, etc.) that obtain automotive components from several thousands of suppliers organized in a three-tiered pyramid, as shown in Figure 4.2. The first tier is formed by around 100 entities that are supplied by the second tier of about 5,000 firms. Last, the third tier comprises about 50,000 suppliers, generally small or medium enterprises that work simultaneously with several car manufacturers. Without standardization of the tools, the third-tier suppliers have to invest in training and maintenance of multiple programs for computer-aided design and communication to be able to work with the different automobile manufacturers with whom they partner.

In 1984, the European automobile manufacturers formed ODETTE for the exchange of information between suppliers and car manufacturers. This program supposed standardization of the content and the structure of the documents and use of common transmission protocols. The 20-odd messages in ODETTE were defined according to the syntax rules of EDIFACT by selecting data elements from those in the ISO 7372 (1993) data dictionary. The OFTP is the file transfer protocol used by all the partners participating in the ODETTE organization (de Galzain, 1989).

The GALIA (Groupement pour l'Amélioration des Liens dans l'Industrie Automobile — Group for the Improvement of Ties in the Automobile Industry) is another automobile group with a mission to overcome the operational difficulties that distributors face as a result of multiplicity of languages and legislation, and to improve the exchanges among subcontractors and the general contractor. This group introduced its own exchange procedures.

ANX® (Automotive Network eXchange) is the network of the Automotive Industry Action Group (AAIG), which was formed in 1982 to define rules for exchanging information among partners in the North American car industry. The ANX is based on the Transmission Control Protocol/Internet Protocol (TCP/IP) stack. Experience has underlined the excessive cost of encryption for small and medium enterprises, incompatibility problems among the authentication certificates or the implementations of the protocol IPSec (Internet Protocol Security), as well as difficulties in guaranteeing end-

---

[1] ODETTE stands for Organization for Data Exchange and Teletransmission in Europe.

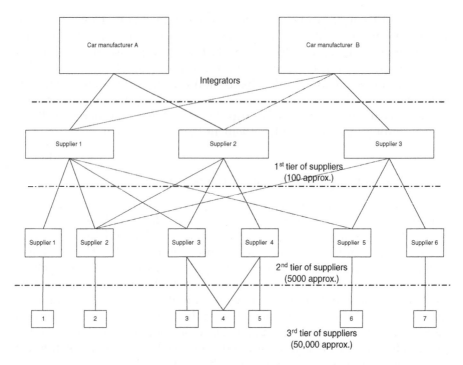

FIGURE 4.2
Pyramidal structure of the automobile industry as integrators and three tiers of suppliers.

to-end quality of service or in localizing and repairing troubles (Borchers and Demski, 2000).

Finally, car manufacturers established Covisint, a virtual marketplace for the procurement of inputs through bids, which can be integrated with the information systems of the enterprises.

4.2.5   Other Examples

The multinationals in the area of industrial chemistry employ the procedures of the CEFIC (Conseil Européen des Fédérations de l'Industrie Chimique — European Council of Industrial Chemistry Federations) in Europe and the CIDX (Chemical Industry Document Exchange) in the U.S. Retail distributors in Europe use the standard EANCOM made by the EAN (European Article Numbering) association for the automatic identification of items, etc.

4.2.6   Effect of the Internet

The Internet stimulated new forms of business-to-business e-commerce, especially for fragmented markets, such as for restaurant businesses. The search for suppliers and the localization of goods or services are particularly facilitated with online catalogs reflecting the terms of commercial agreements among partners (preferential prices, availability intervals, etc.). Furthermore, the integration of these catalogs with enterprise information systems provides better management of data related to purchases (orders, delivery notices, payments, etc.) and tighter coupling with inventory controls and management reporting capabilities.

Harmonization efforts now proceed along two lines. The first is that of convergence of EDIFACT/EDI and XML, which is an area of collaboration of the U.N. CEFACT (Center for Trade Facilitation and Electronic Business) and the OASIS (Organization for the Advancement of Structured Information Standards) consortium as well as the IETF (Internet Engineering Task Force). The second line concerns technologies based on XML and the activities that take place within the World Wide Web Consortium (W3C).

The Internet also contributed to the evolution of the platforms of business-to-business commerce, as discussed in the next section.

4.3 Business-to-Business Electronic Commerce Platforms

The exchanges in business-to-business e-commerce depend on the organization of the supply chain. This chain is called *vertical* if the procured goods intervene directly into the production and *horizontal* if they cover several industries, in which case, the goods are called indirect. In the first case, the purchases are called strategic, while the purchases in the second case are for maintenance, repair, and operations (MRO). Another criterion of distinction is the duration of the contracts among commercial partners: long duration for daily production or temporary for emergencies. Thus, by taking into account the two criteria of urgency of need and the strategic aspects of goods and exchanges services, there are four types of platforms for business-to-business e-commerce: the exchanges for excess inventory, EDI systems, MRO hubs, and generalist catalogs.

The traditional EDI was basically for the purchase of strategic goods directly related to the chain of production or of service creation in a specific sector (automotive, chemistry, steel). However, nonstrategic purchases (equipment and office furniture, travels, etc.), which often represent the large bulk of the volume purchases, remained managed in a traditional way. With the growth of the Internet, many attempts were made to fill this gap through e-procurement with online exchanges, MRO hubs, or yield managers (Kaplan and Sawhney, 2000; Phillips and Meeker, 2000). In these new platforms, all participants (suppliers and consumers) agree to open their information systems and enhance their security infrastructures. Access to the applications is through a Web client. Figure 4.3 depicts the evolution of the traditional EDI systems into these new forms that give different types of service. Thus, exchanges constitute a neutral platform that does not favor

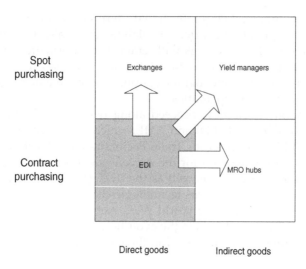

FIGURE 4.3
Platforms for business-to-business e-commerce.

the buyer or the vendor; in contrast, MRO hubs select their suppliers according to criteria that the buyer specifies, while yield managers bring together
buyers and sellers through catalogs or auctions. The latter mechanism constitutes a prized method with which to link offer and demand and to determine price. Intermediation sites had a fleeting popularity within the
deregulated sectors (telecommunications, energy, etc.).

Despite the expected benefits of automated procedures, the rate of penetration of the traditional EDI in industrialized countries varies between 1%
in the U.S. and 5% in some European countries. Within the countries of the
European Union, traditional EDI is mostly used for intranational business,
with applications directed toward intrasector transactions, in which the risk
of litigation is at a minimum. They can be found in the banking sector, in
transportation, retail distribution, aeronautical manufacturing (for example,
the Aérospatiale consortium) and, as stated earlier, the automobile industry
(Landais, 1997; del Pilar Barea Martinez, 1997). The hurdles impeding business-to-business e-commerce are the subject of the next section.

4.4   Obstacles Facing Business-to-Business Electronic Commerce

Three main factors slowed the wide-scale use of traditional EDI systems.
The first is due to the complexity of implementation. The introduction of
EDI does not come easy but requires substantial investments in equipment,
software, and training. The redesign of internal procedures brings about

major changes in the organization of work and in the power relationships within the enterprise. The magnitude and the cost of the task often discouraged small and medium enterprises. The second reason arises from uncertainties regarding the performance of the transport network. The third type of difficulties, which is not the least, relates to the legal status of contracts over electronic media and on the evidentiary value of dematerialized documents. Large corporations use private contracts among commercial partners (or interchange agreements) to define the framework for bilateral electronic transactions: the responsibilities of each party, the rules for identification and authentication of the various entities, and the ways to preserve and archive the electronic documents. This approach may be too burdensome for small- and medium-sized enterprises. In reality, the existing judicial systems were developed in a context where business practice was relying on paper documents (for example, the legal requirement for a handwritten signature). The dematerialization of the support raises the question of the admissibility of electronic documents as evidence, their evidentiary value, and their long-term readability and preservation. Other aspects relate to the identification of the contracting parties, the validity of electronic signatures, the time-stamping of the operations, and the authentication of the origins. This is why the adoption of laws regulating the use of electronic documents and accepting the use of digital signatures in recent years opens evidentiary law and commercial laws to technical advances (Bresse et al. 1997, pp. 162–166).

At the height of the dot.com craze, many predicted that the novel Internet-based platforms would soon supplant the traditional EDI in business-to-business commerce. In fact, while the volume of enterprise purchases over the Internet expanded, use of EDI and private value-added networks is also on the rise. The data in Figure 4.4 show that, instead of being wiped-out IP technologies, EDI and private value-added networks continued to thrive and increased their revenues by about 20% per year, while Internet-based marketplaces floundered, and their numbers dwindled dramatically (Clapaud, 2002).

The arrival of XML and its derivatives raised hopes of putting business-to-business e-commerce within the reach of small- and medium-sized enterprises by democratizing the EDI, for so long reserved to large firms and to government institutions. The expected benefits, however, did not materialize. Innovations do not always meet the expectations of their promoters, particularly if their superiority over the embedded base is not overwhelming. The resistance of the traditional EDI can be explained by the magnitude of the effort needed to use the new technologies and to redefine the rules of dialogue for each commercial sector. It can be explained as well by the significance of the investments already made in the existing technology that is capable of satisfying the expressed needs. Finally, the installation of a new information infrastructure always faces considerable organizational difficulties, irrespective of the technology. It calls for the reorganization of work processes and the modification of the relationships among partners, with concomitant revisions in the enterprise software for billing, accounting,

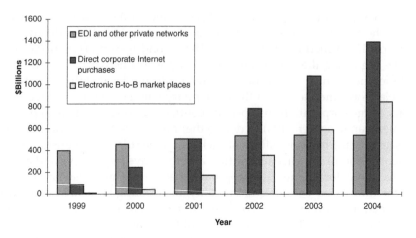

FIGURE 4.4
Evolution of the different approaches of business-to-business e-commerce (1999–2004). (From Yankee Group and the *Wall Street Journal*, May 21, 2001, p. R18.)

inventory management, etc. We can clearly see the origin of the hybrid architecture of the Web EDI, where the messages of the traditional EDI are routed over IP networks. For the user, a simple browser is sufficient to access an information infrastructure that hides the complexities of the traditional EDI. All of this justifies the attention given in this chapter to the traditional EDI, together with the new forms built around XML.

4.5   Business-to-Business Electronic Commerce Systems

The systems of business-to-business e-commerce rest on four foundations (O'Callaghan and Turner, 1995):

1. Structuring of the exchange data according to a common format
2. Security following common procedures
3. Transmission of the data on telecommunications networks
4. Reception and reconversion to the format used internally by the receiving organization

The implementation and the management of such a system includes other aspects, such as the internal organizations of the enterprises, the training of subjects, the follow-up of the operations and maintenance, etc. Sketched in Figure 4.5 is the architecture of an EDI service.

The implementation and the management of such systems include additional aspects, such as the internal reorganization of the enterprise, user training, monitoring, and maintenance (Jackson, 1988). This is why the evo-

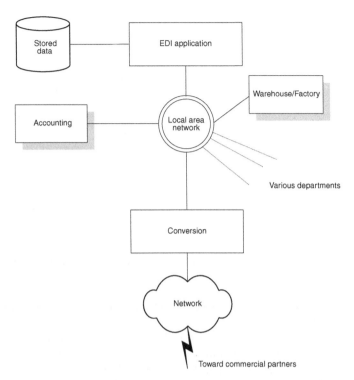

FIGURE 4.5

Components of business-to-business e-commerce systems.

lution of systems of business-to-business e-commerce is not independent of efforts for enterprise application integration (EAI) or for the development of collaborative applications. As shown in Figure 4.6, the inclusion of XML and its variants in the information infrastructure of e-commerce or collaborative applications and the arrival of new mechanisms for communication contributes to the large movement to federate all applications around enterprise portals.

The purpose of this chapter is not to present the information architecture to tie a number of applications around a common core. Nevertheless, it may be useful to keep in mind that as far as the architecture of business-to-business e-commerce is concerned, the infrastructure used depends on what exists and the degree of heterogeneity of the systems encountered.

4.5.1   Generation and Reception of Structured Data

This process concerns the extraction of data related to the transaction (for example, payment instructions) from the appropriate database and the conversion of that data to the common format shared among the different partners using specialized software. The software performs two functions: it converts the representation of the data contained in the internal file to the

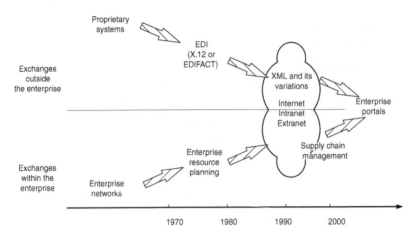

FIGURE 4.6
Portals as the convergence point of collaborative applications within and outside the enterprise.

common format, and it reorganizes the data according to the specifications of the common structure.

The traditional representation of the data exchanges relies on a standard alphanumeric format. Yet, in many sectors, such as automobile or public construction, the exchanges include data of different types (text, images, sound, etc.). The markups of the metalanguage XML assist in determining the various elements of a file so that they can be treated in the most suitable fashions. After restructuring, the data are often secured before transmission. At the destination, the exchanged data are deciphered and reconverted to the "in-house" format of the recipient and then directed toward the specific application that is capable of processing the data.

In the case of EDI, the software that implements the functions of conversion and reorganization of data and of interfacing with internal applications is often called "translator" or "converter" software and is positioned as illustrated in Figure 4.7.

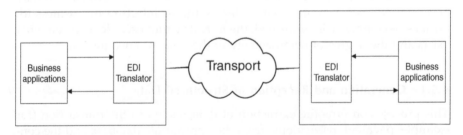

FIGURE 4.7
Positioning of the EDI translators.

Conversion software has two interfaces: one oriented toward the human operator and the other toward the transport network. The requirement to have an interface compatible with the various internal messaging systems in the enterprises — such as cc:Mail from Lotus®, MS-Exchange of Microsoft®, and others — and with the messaging systems in the transport network increased the number of interfaces. Obviously, this increases the cost of the software and enhances the risks of technical failures. The move to the TCP/IP protocol stack, both internal and external to the organization, is expected to reduce the number of interfaces in EDI systems.

4.5.2 Management of the Distribution

The distribution must be managed internally in the enterprise as well as in the external telecommunications network. The received messages must be directed toward the processing application, while the messages toward partners usually contain segments that originate from many applications used in different departments. The communication protocol can vary from X.25, to IP, SNA, or frame relay. The network can also be formed by private lines rented from or offered by a provider of value-added network (VAN) services.

The operator of the distribution network can also offer a protocol conversion service among the various parties, which is of particular interest to those who do not have EDI. Note, however, that use of the Internet as the transport network will clearly change the nature of value-added interventions by directing them toward security services, such as certification of the participants or to notarization of the transactions.

4.5.3 Management of Security

Security is the third aspect of the exchanges of business-to-business commerce. It covers the security of the network using well-established techniques (firewalls, monitoring, passwords, etc.) as well as security of the messages. As explained in the preceding chapter, the services are identification, confidentiality, integrity, authentication, and nonrepudiation. In addition, surveillance of the network and monitoring of the exchanges allow tracing of the trajectory of transactions (with the help of acknowledgments) and collection of data that can be useful in evaluating the performance of the system.

4.6 Structured Alphanumeric Data

The objective here is not to give an exhaustive treatment of all alphanumeric EDI systems, but rather to present the essential notions so as to understand the means of securing the EDI and the problems of integrating with the

Internet. Accordingly, this section focuses on the standard systems, namely, X12 and EDIFACT. As mentioned earlier, X12 is the structuring method most frequently used in North America (the U.S. and Canada), whereas EDIFACT is the norm used in Europe.

Each of these two systems defines a representation of the contents of administrative, commercial, or transport documents as alphanumeric language strings with the help of standardized conventions. The items mentioned in the documents are described with the help of elementary data common to all sectors (purchase order, credit notes, instruction for payment, freight reservation, customs form, etc.). By combining these elementary data, it is possible to form composite data and data segments that can be organized according to precise rules to constitute the canonical messages. The differences between X12 and EDIFACT reside in the definitions of the data elements, the syntax rules, as well as the procedures employed to secure the exchanges.

4.6.1   Definitions

The basic units of an alphanumeric EDI exchange are *data elements* that are defined in a dictionary of elementary data. From a functional viewpoint, the data element is either a *service element* or an *application element*. Service elements contain the information that structures the transmission and are utilized in service segments. In contrast, application elements relate to the heart of the end-to-end transaction, i.e., the data defined and agreed upon by the two parties of the transaction.

A *segment* is a logical set that includes a series of elements, simple or compound, and may include other segments. The order, content, the maximum number of repetitions of the constituents, and the way these repetitions should be organized are defined in the segment dictionary. To express a precise functionality, for example, a purchase order or a payment instruction, the segments are combined and organized in a group of segments.

There are two types of segments: segments denoted as *control segments* in X12 or *service segments* in EDIFACT and *application data segments*. Control (or service) segments are used to structure the content and to distinguish the various parties. The application data segments contain the application data organized by function. It is the entity in charge of managing the application responsible for specifying the coding and the organization of the application data segments.

A *transaction set* (X12) or *message* (EDIFACT) is the set of structured segments in the order defined in the directory of standard messages. These messages represent functions that are common to all activity sectors. For example, billing is based on a universal commercial practice and does not depend on the type of activity. There are two classes of messages:

- *Service messages*, formed with service segments, with the role of correcting syntax errors or application errors
- *Application messages*, formed with application segments

Messages may be mandatory or optional and may be repeated.

In general, messages consist of three distinct zones: header, detail (body), and trailer. The segment groups forming the "detail" concern different transactions that may be included in the same message.

The *functional group* gives the possibility of putting several messages within the same structure. Finally, the *interchange* is the external envelope of all messages originating from the same application, although they may relate to independent transactions.

4.6.2 ANSI X12

ANSI ASC X12 was defined by the ANSI Accredited Standards Committee (ASC) X12. The standardization of ANSI X12 is more advanced than that of EDIFACT; unfortunately, the syntax of X12 is positional, which makes it incompatible with EDIFACT.

Within the X12 terminology, a transaction set corresponds to the useful transmission of the *useful* content of the paper document (purchase order, invoice, etc.) between the computers of two organizations. Each transaction set consists of three parts: a header, a detail (body), and a trailer. The header announces the characteristics of the transmission, while the trailer contains control elements to verify the integrity of the information transmitted.

The body is organized in lines or segments that describe one particular aspect of the total action. In turn, a segment is composed of data elements and codes associated with the function to be performed. The order in which the elements are arranged, their composition, as well as the significance of the codes are defined in the data dictionary. For example, an asterisk (*) separates two consecutive elements, and two asterisks indicate that an optional element was eliminated. If the omitted element (or elements) was at the end of the segment, it would be replaced by the end-of-segment (N/L) to indicate the return to a new line.

As mentioned earlier, there are two types of segments: control segments and data segments. The segment headers and trailers as well as the repetition loops of segments form the control segments. Table 4.1 gives some of the most frequently used transaction sets.

TABLE 4.1

Examples of X12 Transaction Sets

Code	Meaning
810	Invoice
820	Payment order/remittance advice
824	Application advice (results for an attempt to modify a transaction)
827	Financial return notice (impossibility of carrying an 820 transaction)
850	Purchase order
855	Purchase order acknowledgment
856	Ship notice/manifest
997	Functional acknowledgment (to indicate that the received message is syntactically correct)

In general, the software for translating to the EDI format ("the EDI document") does not support all transaction sets and is limited to those used in the target domain of activity.

Several transaction sets can be combined in an X12 interchange in the following manner:

ISA* Interchange header

    GS* Functional group header

        ST* Transaction set header

           Segments (i.e., Purchase Order 1)

        SE* End of transaction set

        ST* Transaction set header

           Segments (i.e., Purchase Order 2)

        SE* End of transaction set

    GE* End of functional group

    GS* Functional group header

        ST* Transaction set header

           Segments (i.e., Purchase Order 3)

        SE* End of transaction set

        ST* Transaction set header

           Segments (i.e., Purchase Order 4)

        SE* End of transaction set

    GE* End of functional group

IEA* Interchange trailer

4.6.3   EDIFACT

EDIFACT is described in the following documents (Charmot, 1997a):

- The vocabulary or the data elements (ISO 7372, 1993)
- Directives for the composition of messages (document Trade/WP 4/R.840/Rev2)
- Syntax rules for structuring the canonical messages (ISO 9735, 1988)
- Directives for using the syntax (document Trade/WP 4/R.530)
- U.N. Rules of Conduct for Interchange of Trade Data by Teletransmission (UNCID)
- Dictionary of data elements, a subset of ISO 7372
- Dictionary of composite data elements
- Dictionary of canonical messages
- U.N. Code List (UNCL), a dictionary of recognized codes

Taking into account the individual practices of each country, it is obvious that the design of standardized messages for a given sector is a long-term effort. Messages are classified according to progress in standardization as follows: status "0" is for messages still being defined; status of "1" indicates that the message is undergoing trials; and stable and standardized messages have the status of "2."

The ISO 9735 (1988) recognizes several character sets: Level A UNOA ensures compatibility with telex terminals, while Level B UBOB has a richer character set (capital and small letters for the basic Latin alphabet, numerals, as well as some special characters); Level C UNOC for Latin alphabet Number 1 (ISO/IEC 8859-1, 1998); Level D UNOD Latin alphabet Number 2 (ISO/IEC 8859-2, 1987); Level E UNOE for Latin/Cyrillic alphabet (ISO/IEC 8859-5, 1999); and Level F UNOF for Latin/Greek alphabet (ISO/IEC 8859-7, 1999). ISO/IEC 8859-6 (1999) and ISO/IEC 8859-8 (1999), in particular, define the representation of Latin/Arabic and Latin/Hebrew characters, respectively, thereby associating right-to-left writing movements with left-to-right movements.

Service segments allow for the structuring of content and the distinguishing of its many parts. They start with a *tag*. The service elements are identified with the letters *UN*, for example, UNH designates the header of a message, while UNT designates its trailer. The tags are defined by the standard EDIFACT in the directory entry of the corresponding segment. Qualifiers can be used together with tags to give specific meaning to the functions of other data elements or segments, which is useful if several functions are represented within the same message. If optional elements are omitted, they will be replaced by the corresponding separator (data element or component of a composite data element) defined for the segment at hand.

In the following sections, the details will be given that are necessary to understand the security mechanisms of EDIFACT that will be presented later in this chapter.

4.6.3.1 UNB/UNZ and UIB/UIZ Segments

The pair of service segments (control segments) UNB/UNZ define, respectively, the beginning and the end of a *batch* exchange, i.e., they bracket the envelope. The segments UIB/UIZ are the header and trailer segments proposed to envelop the interactive exchanges. An optional segment, service string advice UNA, is used if it is needed to redefine the component data element separator or the data element separator or to change the notational sign of the decimal, which is the "," by default. The UNA consists of a sequence of nine characters (UNA followed by six characters according to a specific format that ISO 9735 (1988) specified.)

The UNB defines the transmission characteristics using the following elements:

- Syntax identifier
- Address of the sender
- Address of the recipient
- Date and time of preparation
- Unique interchange control reference
- Recipient reference or password
- Reference to the application used, or, in the case of a single message, the message type (e.g., invoice)
- Processing priority
- Notification if acknowledgment is requested or denied
- Identifier of the communication agreement controlling the interchange
- Test indicator in the case that the interchange is a probe sent to verify the connection between the two ends

The UNZ segment is the interchange trailer that signals the end of the exchange. It includes two elements: the total number of messages of functional groups in the interchange, and the same reference to the interchange as in the UNB.

These elements help check that the delivery of the interchange did not encounter problems.

4.6.3.2 UNH/UNT Segments

The service (control) segments UNH/UNT play the equivalent role of the preceding doubles but for the messages instead of the interchanges. Thus, each message begins with a header segment UNH and ends with a trailer segment UNT. The structure of an EDIFACT message is given in Figure 4.8.

The message header UNH contains control data such as follows:

Header	Data segments	Data segment	....	Message trailer

FIGURE 4.8
Structure of an EDIFACT message.

- The sender's unique message reference number
- The message type, the version number of the standard used to compose the message
- The agency that controls the message specification, maintenance, and publication
- A common access reference, for example, the name of the message file, which will be used to cross-reference subsequent transfers of data

The segment UNT indicates the end of the message and contains the two elements for verification: the number of segments in the message and the message reference number present in the UNH segment.

4.6.3.3 UNS Segment

The service segment UNS is a special segment used to separate the body or detail of a message from the header on one side and from the trailer on the other side.

4.6.3.4 UNG/UNE Segments

The header of a functional group UNG is a service segment that contains the following elements:

- Identifier of the message type in the functional group
- Identifier of the sender's application (for example, the division or department of the enterprise that prepared the functional group)
- Identifier of the recipient's application
- Date and type of preparation
- Functional group reference number
- The controlling agency that specified the message
- The version of the standard that was used to compose the message
- Application password according to the recipient's request (optional)

The UNE service segment closes a functional group. It contains two verification elements: the number of messages in the functional group and the reference number for the functional group that is included in the header.

4.6.3.5 UNO/UNP Segments

Although it is possible to encapsulate a multiformat object, difficulties may arise at the level of handling the various objects and synchronizing them at the destination. To resolve this problem, two new EDIFACT service segments were defined (Charmot, 1997b).

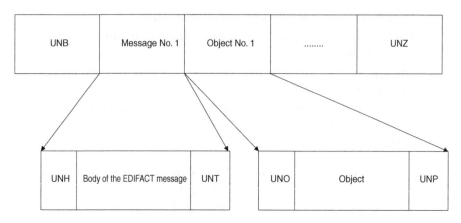

FIGURE 4.9
Encapsulation of a non-EDIFACT object within an interchange.

Depicted in Figure 4.9 is an object surrounded by the two segments, UNO and UNP, that bound the object within an interchange. The UNO identifies the start of the object and can be used to specify its nature. The UNP also gives the sizes of the objects in octets to verify the quality at reception.

4.6.3.6   Structure of an Interchange

An interchange is structured in the following manner:

Segment	Type
Service string advice **UNA**	Optional
Interchange header **UNB**	Mandatory
Functional group header **UNG**	Optional
Message header **UNH**	Mandatory
Message data	
Message trailer **UNT**	Mandatory
Functional group trailer **UNE**	Optional
Interchange trailer **UNZ**	Mandatory

The structure of an interchange without functional groups is given in Figure 4.10. Figure 4.11 depicts the structure of an interchange with functional groups.

4.6.3.7   Partial List of EDIFACT Messages

Table 4.2 contains some EDIFACT messages used in commercial transactions. These examples are given for illustrative purposes and are arranged in alphabetical order.

Interchange Header	Message No. 1	Message No. 2	....	Interchange Trailer

FIGURE 4.10
Structure of an EDIFACT interchange without functional groups.

Interchange Header	Functional Group Header	EDIFACT message	....	Functional Group Header	Functional Group Header	EDIFACT message	....	Functional Group Trailer	Interchange Trailer

FIGURE 4.11
Structure of an EDIFACT interchange with functional groups.

4.6.3.8 *Interactive EDIFACT*

The discussion thus far essentially relates to batch EDI that is encountered in asynchronous and nonreal-time transactions. Yet, real-time transactions, such as reserving plane tickets or querying catalogs, are characterized by shorter response times and a rate of transactions that may be higher than in traditional applications. An EDI that is open or interactive raises technical and judicial questions because the dynamic negotiations must be either notarized or supervised by third parties. The framework for online EDI was drawn for the first time by ISO/IEC 14662, adopted in October 1996 (Troulet-Lambert, 1997). ISO 9735-3 (1998) introduced a syntax for interactive EDI-FACT exchanges. However, most of the current mobilization aims at the development of the language of electronic business XML (ebXML), which mixes traditional EDIFACT with XML.

4.6.4 Structural Comparison between X12 and EDIFACT

Figure 4.12 depicts the correspondence of the control elements of EDIFACT (ISO 9735, 1988) and ANSI X12 (Kimberley, 1991, p. 127).

Shown in Table 4.3 is the correspondence between the terms used in header segments of the EDI interchanges in EDIFACT with those of ANSI X12, using Table K-1/X.435 of ITU-T Recommendation X.435.

4.7 Structured Documents or Forms

Structuring the dialogues as documents or forms allows consideration of all the data exchanged in the context of a commercial transaction, independent of their formats (text, graphics, images, sound, audio, video) and not only

TABLE 4.2

EDIFACT Messages (in alphabetical order)

Message Name	Function
CREADV	Credit advice
CREEXT (CREADV + REMADV)	Extended credit advice
DEBADV	Debit advice
DESADV	Dispatch advice
DOCADV	Documentary credit advice
DOCINF	Documentary credit issuance information
IFTMAN	Arrival notice
INVOIC	Invoice
ORDCHG	Purchase order change request
ORDERS	Purchase order
PAYEXT (PAYORD + REMADV)	Extended payment order
PAYMUL	Multiple payment order
PAYORD	Payment order
REMADV	Remittance advice

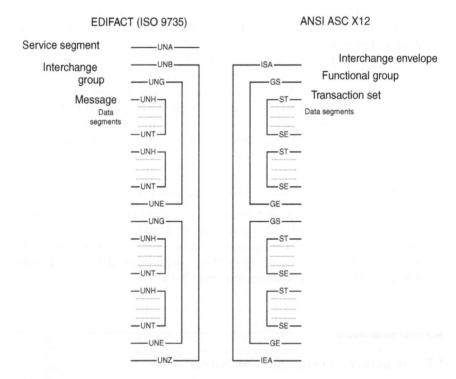

FIGURE 4.12

Comparison of the structures of EDIFACT and ANSI X12.

TABLE 4.3

Comparison of the Terms Utilized in the Headers of the EDI Interchanges

EDIFACT	ANSI X12
Interchange header (UNA and UNB)	Interchange header (ISA)
Functional group header (UNG)	Functional group header (GS)
Message header (UNH)	Transaction set header (ST)
Service string advice	1. Data element separator
	2. Segment terminator
	3. Subelement separator
Syntax identifier	1. Interchange standard identifier
	2. Interchange version ID
Interchange sender	Interchange sender ID
Interchange recipient	Interchange receiver ID
Date/time of preparation	1. Interchange date
	2. Interchange time
Interchange control reference	Interchange control number
Recipient's reference, password	Security information
Application reference	—
Processing priority code	—
Acknowledgment request	Acknowledgment requested
Communications agreement ID	—
Test indicator	Test indicator
—	Authorization information

considering the data that can be represented by alphanumeric characters. Stimulated by the success of the Internet, this approach requires the standardization of various elements forming the contents of documents as well as of the organization of that content. Thus, one can speak of Electronic Form Interchange instead of Electronic Data Interchange to distinguish this approach from the traditional EDI.

There are several ways to structure documents, such as with SGML (Standard Generalized Markup Language), ODA (Open Document Architecture), or CGM (Computer Graphics Metafile) for image documents. However, the most popular approach today is one of the descendants of SGML — XML.

SGML was adopted in 1986 to provide a systematic way to describe the logical structure of document contents with the help of models called Document Type Definition (DTD). Each DTD defines the class of documents that share the same way of organizing the information and contains the rules for interpreting and validating the exchanged electronic documents. The automatic analysis of a DTD can be considered a way to validate SGML.

4.7.1 SGML

SGML is a declarative metalanguage that specifies a way of describing the logical structure of a generic document with markups. Interpretation of these markups follows the indications of the DTD document (Dupoirier, 1995, pp.

107–120). In fact, the DTD defines the hierarchical composition of the document in chapters and paragraphs and serves as a reference and a model to all documents that are members of the same family. The syntactical analysis of the SGML document in light of the associated DTD is independent of the information processing platform, which means that SGML simultaneously defines a methodological framework for the exchange of documents.

SGML directly inspired several efforts to develop automatic documentation, particularly HTML. A key concept of HTML is the utilization of a uniform resource locator (URL) to locate document sources accessible on the Internet, which was one of the factors responsible for the enormous success of the Web.

HyTime (Hypermedia/Time-Based Document Structuring Language) is another ISO standard (ISO/IEC 10744, 1992) that extends the application of SGML to hypermedia exchanges with the help of a meta-DTD that ties the elementary DTDs together.

SGML is associated with another ISO activity, namely, DSSSL (Document Style Semantics and Specification Language), that became ISO 10179 (1996). DSSSL aims to prepare SGML documents that can be used directly for printing: pagination, typography, and imposition. DSSSL has the advantage of accommodating non-Latin alphabets (for example, Arabic, which requires mechanisms for right-to-left composition).

4.7.2 XML

In 1994, a contribution was presented to ISO/IEC JTC1/SC18/WG8 to replace alphanumeric EDI with SGML (ISO, 1994), but this proposal was judged to be too radical. It was up to the W3C to define XML on the basis of SGML to facilitate the definition of multitype files that can be shared among applications using the broadcast possibilities of the Internet. The XML, as a subset of SGML, can be the starting point of the definition of specialized markup languages for any types of documents (Bryan, 1998; Michard, 1999). In Table 4.4, some of these XML derivatives for financial services are listed.

Clearly, the development of XML stimulated new forms of business-to-business e-commerce by enabling the exchange of commercial data coded according to commonly admitted conventions.

4.7.3 Integration of XML with Alphanumeric EDI

The XML can contribute to enhance EDI solutions in several ways. It allows the attachment of multimedia objects to the exchanged messages and adds interactivity to EDI. Furthermore, it gives recipients the possibility to interpret and process nonstandardized segments or messages by attaching their DTDs. It also increases the possibility of automatic translation of the data into the recipient language. Finally, it provides a mechanism with which to

TABLE 4.4

Some Derivatives of XML for Financial Services

Acronym	Title	Remarks
FinXML	Fixed Income Markup Language	—
FIXML	Financial Information Exchange Markup Language	—
FpML	Financial Products Markup Language	Over-the-counter (OTC) derivatives
FXML	Financial Exchange Markup Language	—
IFX	Interactive Financial Exchange	Synthesis of the two specifications for online financial exchanges OFX and Gold
IRML	Investment Research Markup Language	—
MDDL	Market Data Definition Language	Description of marketing research
NewsML	Electronic News Markup Language	Multimedia financial news
NTM	Network Trade Model	Stock and risk evaluation
OFX	Open Financial Exchange	Online exchange of payment messages among financial institutions; used for electronic billing
RIXML	Research Information Exchange Markup Language	Document indexing
STPML	Straight-Through Processing Extensible Markup Language	—
SwiftML	Society for Worldwide Interbank Financial Telecommunications Markup Language	Language used by SWIFT
xBRL	Extensible Business Reporting Language	Financial reporting
XFRML	Extensible Financial Reporting Markup Language	—

recall documents using their URLs, thereby avoiding their attachments in the exchanges.

The integration of XML and EDI consists of standardizing aggregate elements (or "classes") that are combined to form messages, segments, or forms sent to the destination with the necessary instructions for interpreting them. A global repository or library of objects contains the core components of the messages sent (data structures or objects). The sender queries the repository to define the partner's profile and the scenarios of the exchanges. The requests are routed to specialized libraries that contain EDIFACT/X12 standards, the appropriate industry agreements or standards, or the elementary data and the collaborative agreements between the two organizations involved. A glossary defines the structures of the documents, their logical components, and the business processes, by referring to the dictionaries of each data category. The APIs have the task of managing the different conversions. Represented in Figure 4.13 is the architecture of the XML/EDI convergence. Notice, however, that the use of XML increases the sizes of the files by at least 20%.

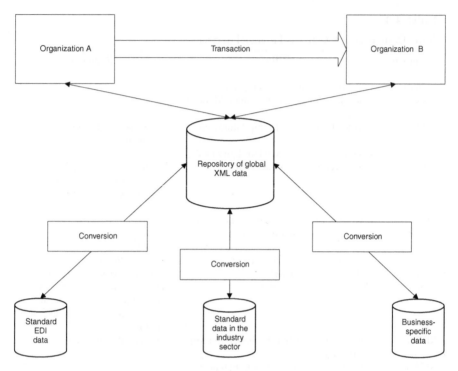

FIGURE 4.13
Exchanges involved in the XML/EDI integration.

The dialogue between an XML client and an EDI application takes place through an EDI converter and an XML server (see Figure 4.14). This arrangement takes into consideration prior investment in alphanumeric EDI and progress achieved in the Universal Data Element Framework (UDEF), financed by the U.S. federal government within the CALS (Continuous Acquisition and Life-Cycle Support) initiative, as well as specification of data elements in ISOIEC 11179 (1999).

Let us now review the main efforts to achieve the convergence of XML and EDI, in alphabetical order.

4.7.3.1   BizTalk®

Shepherded by Microsoft®, BizTalk defines a framework called BizTalk Framework to use XML in e-commerce. The rules for BizTalk concern the definition and the publications of XML schemas and the use of XML messages for existing applications to communicate.

4.7.3.2   Commerce XML (cXML)

Commerce XML (cXML) is Ariba's initiative for standardizing access to catalogs, including personalized catalogs.

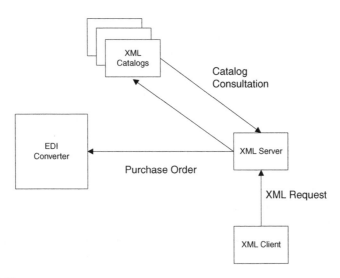

FIGURE 4.14
EDI/XML integration.

4.7.3.3   Electronic Business XML (ebXML)

With ebXML, a complete and secured framework for commercial exchanges that can substitue for alphanumeric EDI is defined. It was started jointly in September 1999 by the OASIS consortium and the CEFACT. The ebXML is a variant of XML for business-to-business e-commerce. In this framework, the central repository of the traditional EDI is replaced by a distributed database on a network. The framework of reference gives scenarios for typical exchanges (selection-purchase-payment-delivery) for each sector considered (automotive, chemistry, etc.) based on which are defined the subjects of the transactions, the messages exchanged, their order in the exchange, and the roles of each of the participants.

One of the first applications of ebXML was in the areas of retail and logistics, such as the project e-TEXML in the French lingerie sector.

4.7.3.4   SAML (Security Assertion Markup Language)

The SAML (Security Assertion Markup Language) is defined by OASIS to describe, using XML, user profiles and requests for authentication before an individual or an object is authorized access to a given service (OASIS, 2002). When a user identifies itself to an SAML server, the server attaches to the request a description of the user's access rights in SAML within a Simple Object Access Protocol (SOAP) envelope. This envelope will be sent to the server of any Web service that the user would later invoke. The SAML structures are transported using the POST method of HTTP or within SOAP messages coded in XML.

With SAML, the user access profile can be propagated from one application to the other during the same session. By ensuring the interoperability of proprietary solutions of rights management and by establishing equivalencies among administrative domains, a single authentication of the requester per session is possible; this is called Single Sign-On (SSO). As a consequence, after being authenticated in one administrative domain, the user can access the resources in other domains associated with the first without requiring additional authentication. The need for a single authentication is tied to the development of enterprise portals, and it requires that security policies (identification, authentication of identity and of attributes, etc.) be strengthened.

4.7.3.5 SOAP (Simple Object Access Protocol)

The SOAP is a universal protocol for the exchange of messages and requests that relies on HTTP and XML to structure the exchange of data among Web services (Box et al., 2000). SOAP defines a mechanism for a structured and typed exchange of information in a decentralized and distributed environment by generalizing the Remote Procedure Call (RPC). This technique allows a program to start the execution of a procedure on a remote server. Version 1.1 of 2000 has interfaces according to J2EE and Net. SOAP is independent of the information technology platform, but suffers from a lack of service guarantees concerning the transport of exchanges. Accordingly, it is better adapted to the internal applications of the enterprise.

The SOAP messages are XML documents formed of three elements: an envelope, coding rules, and conventions with which to make remote procedure calls and interpret their responses. The envelope shows the type of message and includes declarations on the name space or additional attributes. It consists of a header and a body. The presence of the body is mandatory, while the header is optional — it supplies elements with which to interpret the exchanges without prior agreements between the two parties of a communication. In contrast, the body is a container for application data intended for the ultimate recipient of the message.

4.7.3.6 UDDI (Universal Description, Discovery, and Integration)

The UDDI is a public or private directory service used to localize remote components or services available online. UDDI aims to overcome difficulties due to the heterogeneity of directories by supplying standard elements to answer requests regarding the availability of products or services, information on the purchases (prices, delivery delays, etc.), as well as the existence of substitute products. With UDDI, commercial offers can be organized by vendors in "white pages" or by sector and geographic zone in "yellow pages."

4.7.3.7   WSDL (Web Services Description Language)

WSDL is a language for describing the mechanisms for invoking services available on the Web. This way, WSDL offers a uniform method for calling services (Christensen et al., 2001).

4.8   EDI Messaging

Enterprise networks use many proprietary messaging systems: Microsoft Exchange, Lotus cc:Mail, etc. Many messaging systems were standardized, for example, X.400 of the ITU-T (International Telecommunication Union — Telecommunication Standardization Sector) and its continuation in F.435/ X.435, and for the Internet side, SMTP (Simple Mail Transfer Protocol) of the IETF RFC 821 (1982), and its MIME (Multipurpose Internet Mail Extensions) described in the IETF RFC 2045 (1996). Both messaging systems follow the same model of a user agent and a message transfer agent, even though the connection protocols between the components are not identical.

In the following, the messaging systems will be studied from the point of view of the service security that they offer. For further details on each messaging system, the reader can consult the literature on electronic messaging, for example, Palme (1995) and Bouillant (1998), or the text of the various standards.

4.8.1   X.400

The X.400 messaging system was one of the first alternatives to proprietary messaging systems and to third-party networks. The X.400 is often used as a generic name to include a whole series of ITU-T recommendations. Among these recommendations, X.420 specifies the P2 protocol to envelop messages. Therefore, P2 can be used to encapsulate EDI messages before their transmission. Starting in 1988, messaging systems conforming to the X.400 series of recommendations could, at least in theory, offer all the necessary security services: identification, authentication, integrity, confidentiality, and non-repudiation of the origin. The architecture of these security services is available in Recommendation X.402.

An X.400 implementation is required to offer some basic security services, the most important of which are as follows (see Table 4/X.400):

- Access management
- Time-stamping of messages
- Message sequencing to correlate later notifications with each originating message

- Message content type indication and content type of attached objects
- Nondelivery report

The following are among the optional security functions:

- Secure access to the messaging system and mutual authentication of peer entities
- Message labeling according to the security policy
- Probes to verify end-to-end connectivity
- Proof of submission, which is the confirmation that the message transfer agents (MTAs) located in the various network nodes received the service for delivery to the originally specified recipient; the corresponding security service is nonrepudiation of submission
- Proof of delivery, i.e., confirmation that the message was delivered to the destination

These security services apply uniformly to all content of the EDI interchange. Application to a part of the content (a message or a segment) requires some of the EDIFACT security techniques that will be discussed later.

Message labeling permits each message to be treated according to the required degree of security. Furthermore, X.400 messaging systems can announce contractual levels of service quality, according to the degree of urgency of the traffic. These objectives are defined in Recommendation F.410 and are reproduced in Table 4.5.

In 2000, a decree generalized the combination of EDIFACT/X.400 in the relationship between enterprises and the government of France.

Adaptations of X.400 messaging to EDI were defined in 1991 through Recommendation F.435/X.435. Recommendation X.435 specifies a messaging system called "EDI messaging" capable of transporting EDI contents. New messaging procedures, denoted as PEDI (Protocol EDI), were introduced to replace the P2 protocol in EDI applications and to allow the formation of an EDI message from several objects. The wide use of the Internet made this approach obsolete.

4.8.2 Internet (SMTP/MIME)

An enterprise that already uses the Internet for messaging and internal distribution of documents is obviously interested in reusing it for its EDI applications. Unless there is a total change, which is not likely, even with the introduction of XML, this approach will be taken in at least two steps: a retrofit of IP to accommodate the requirements of EDI and a long-term activity to merge the two approaches, possibly after the XML/EDI integration is completed.

TABLE 4.5

Transfer Delay Objectives in X.400 Messaging Systems

Grade of Delivery	Time for Delivery of 95% of the Messages	Time for Forced Nondelivery
Urgent	45 minutes	4 hours
Normal	4 hours	24 hours
Nonurgent	24 hours	36 hours

Thus, the various MIME implementations will have to conform to the complete specifications defined in IETF RFCs 2045 through 2049 (1996). The encapsulating protocol of RFC 2046 is compatible with the basic messaging protocol SMTP that is on top of the TCP/IP layers. This encapsulation allows the inclusion of different object types as well as the transmission of non-ASCII text ASCII, which makes it suitable for EDI. MIME allows separation of the body of a message into distinct parts, separated by delimiters. The delimiter is a demarcation line that can be defined as a sequence of characters that does not appear anywhere else in the message. The lines that follow the delimiter define the properties of the object for the recipient applications.

An EDI message that will use MIME will include, first, the SMTP message header, as defined in IETF RFC 822 (1982) (the sequence of fields From, To, Date, Subject, etc.). The header precedes a series of declarations that indicates the content type, the initial representation of the characters, the coding used to protect the text from being mutilated in the Internet, and finally, the succession of body parts, each separated from the other with delimiters.

The coding type for EDI files has to be exclusively base64 to oblige the sender's user agent to convert the text to the 7-bit ASCII code and append a "carriage return" and a "new line" <CR><LF> at the end of the file. The presence of the field "Content-Transfer-Encoding: base64" in the header allows the destination user agent to perform the inverse processing and to recover the initial content.

To illustrate with an example, and without attempting to specify all the details of MIME, a skeleton message can take the following form:

```
To: < recipient's electonic address>
Subject:
From: <sender's electronic address>
Date:
Mime-version: 1.0
Content-Type: multipart/mixed; boundary = "abxyxms0n"

 — abxyxms0n
Content-Type: text/plain;charset = "ISO-8859-1"
```

```
<This is preamble to an EDI message that can contain
accents>
 — abxyxms0n
Content-Type: Application/<EDI standard>
Content-Transfer-Encoding: base64

<This is the EDI interchange coded according to the
standard indicated in the Content-Type>
 — abxyxms0n
Content-Type: Application/DRAWING; id = 260; name =
"Cost"
Content-Transfer-Encoding: base64
<Here is the graphical file Cost>
 — abxyxms0n —
```

The content types must be registered by IANA (Internet Assigned Numbers Authority), the Internet registration authority. Three content types are recognized for EDI applications: EDI-X12, EDIFACT, and EDI-CONSENT. This last category is used for proprietary EDI applications.

It should be noted that, contrary to X.400 messaging, the messages included with this basic format are in the clear, and thus are easily readable. Also, there is no guarantee that the message will reach its destination.

4.9   Security of EDI

Security of EDI includes technical and managerial aspects. The technical part has many aspects related to the security of the exchanges as well as of network elements and network element management systems (Kwon et al., 1997). In the following, we will restrict ourselves to the security of the exchange, the other aspects being outside the scope of this book.

Many security procedures for EDI messages, whether coded according to X12 or to EDIFACT, were standardized. These procedures cover services for authentication, message integrity, confidentiality, and nonrepudiation of the origin.

The IETF proposals are distinct from those adopted by the UN/ECE (United Nations Economic Commission for Europe) or by ANSI. The ITEF proposed three different security mechanisms to accommodate the various messaging protocols: PGP (Pretty Good Privacy)/MIME, S/MIME (Secure Multipurpose Internet Mail Extensions), and S-HTTP (Secure HyperText Transfer Protocol).

4.9.1 X12 Security

Secure X12 transmissions use the security structures defined in X12.58 issued in December 1997. Before then, security consisted of using a password before each transmission and in each direction. Should a recipient want to respond, a new connection would have to be established in the inverse direction. Value-added networks provided security services by establishing the two unidirectional circuits instead of the two parties transparently.

Authentication uses the DES (Data Encryption Standard) algorithm to calculate a Message Authentication Code (MAC). For protection against replay attacks, the standard recommends inserting in the message to be authenticated a combination of a date of composition and a unique identifier (such as a purchase order number). Nonrepudiation employs a public key encryption along with time-stamping.

The standard generalizes the concept of digital signature in the form of assurances that express the business intent. An assurance is contained in newly defined segments: the S3A or S4A segments and the SVA segments. These segments are added before calculating the MAC or signing the message. The combination S4A/SVA frames the unsecured transaction before encryption or authentication and offers a first level of protection for the functional group. The combination S3A/SVA offers a second level. Each level has its own keys. Furthermore, at each security level, the keys utilized for a service (for example, authentication or encryption) should be different.

The standard allows an optional compression of the message before encryption in addition to an optional filtering of the encrypted or compressed data. Filtering prevents the occurrence of binary sequences that may incorrectly activate the control functions of the transmission systems. Three types of filters are recognized: the conversion of each binary into two hexadecimal characters; the conversion of the binary data into a string of printable ASCII characters, which produces an expansion of the required bandwidth by about 23%; and, finally, an ASCII/Baudot filter to convert the binary data to a string of characters that belong to ASCII and Baudot character sets, a procedure that results in an expansion of about 86%.

A transmission secured by these segments is illustrated as follows:

ISA* Interchange header

    GS* Functional group header

    S3S* Security header — Level 1

    S3A* Assurance header — Level 1

        ST* Transaction set header

        S4S* Security header — Level 2

        S4A* Assurance header — Level 2

            Segment details (e.g., purchase order)

        SVA* Security value — Level 2

        S4E* Security trailer — Level 2

　　　　　SE* Transaction set trailer

　　　SVA* Security value — Level 1

　　　S3E* Security trailer — Level 1

　　　GE* Functional group trailer

　IEA* Interchange trailer

Note, however, that the security mechanism gives the same protection to all parts of a transaction. This is a difference with EDIFACT services that offer finer resolution and permit the possibility of different types of protection according to the different fields of a transaction.

X12 can directly utilize the X.509 certificates that can be delivered by a certification authority. Level 3 is considered to be sufficient for EDI applications (CommerceNet, 1997).

4.9.2　EDIFACT Security

Security of EDIFACT follows ISO 7498-2 (1989). This standard is the result of the European research program TEDIS (Trade Electronic Data Interchange System) that lasted from 1988 to 1994. The services offered are message integrity, authentication of the origin, and nonrepudiation (at the origin and at the destination). Confidentiality is not offered explicitly but may be constructed with the other services.

EDIFACT security services can be offered in two ways: by sending security segments "in band" or "out-of-band." In the in-band approach, the security segments flow jointly with the messages to be protected, whereas in the out-of-band approach, separate security messages are used.

ISO 9735-5 (2002) considers the first case and defines the segments that must be inserted as well as the means to distinguish them from user traffic.

The out-of-band security measures rely on the AUTACK message, defined in ISO 9735-6 (2002). This message contains the hash of the EDIFACT structure that is mentioned or the signature of this same structure. The AUTACK message can also be used as an acknowledgment when it is sent by the recipient or by an entity that has the authority to act on behalf of the recipient. Used in this manner, the message confirms reception of the EDIFACT structure that was sent, confirms the integrity of its content at destination, and helps establish nonrepudiation at the destination.

Finally, the management of EDIFACT certificates (inscription, renewal, replacement, revocation, delivery) as well as the generation, distribution, and management of keys are performed with the KEYMAN message defined in ISO 9735-9 (2002). The KEYMAN messages can also refer to the certification path and the revocation list defined according to X.509. These references are transported in a binary format between the segments UNO and UNP.

It should be noted that the EDIFACT certificates are different from X.509 certificates both in format and in method of management. To resolve this issue, the European Commission sponsored the DEDICA (Directory-Based

EDI Certificate Access and Management) project to allow access to secure EDI with X.509 certificates. A DEDICA gateway performs the necessary conversions, thereby saving the users from having to obtain and maintain two sets of certificates.

4.9.2.1 Security of EDIFACT Documents Using In-Band Segments

Figure 4.15 shows how security segments are inserted in the initial structure. Each USH/UST corresponds to a given security service. This permits the possibility of varying the offered service for distinct parts of the interchange; thus, a given interchange can include data from several transactions that do not require the same degree of protection. The price of this flexibility is a further complication of the protocol, a complication that may seem excessive, because the same effect could have been obtained by sending consecutive messages instead of a single complex message.

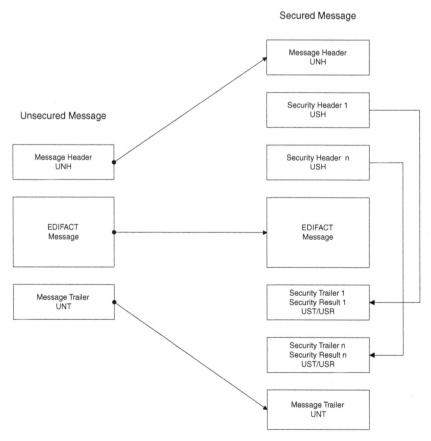

FIGURE 4.15
Security of EDIFACT messages with in-band segments.

Each security structure begins with a header USH and ends with a trailer UST. It contains the following segments: USA, USC, USH, and USR, according to the structure shown in Table 4.6. These segments inform the recipient of the invoked level of security, of the security mechanisms utilized and their parameters, as well as the extent of the protected domain. These various segments will be recognized and processed by applications that demand security functions and will be safely ignored by other applications. They will be inserted in the interchange between the UNH and UNT segments for a message or between the UNO and UNP segments in the case of a non-EDIFACT object.

The advantage of this selective protection is that the consequences of a security breach would be limited to the concerned party without compromising the whole interchange. Another advantage is to allow selective access to the different data included within the interchange. For example, in a payment instruction, the intermediary banks do not need to know the identities of the creditor and the debtor.

Protection against loss or duplication of messages relies on optional fields that contain a sequential number counter and a time stamp.

The following subsections give a more precise definition of the content of these various segments groups.

4.9.2.1.1 Segment Group Number 1

This is a group of segments (called the security header) that defines the security services, the security mechanisms, and the elements necessary to carry out the validation calculations. The group can be repeated for each

TABLE 4.6

Security Structure for "In-Band" Segments

Label	Segment Name	Type	Maximum Number of Repetitions
Segment Group 1		O	99
USH	Security header	M	1
USA	Security algorithm	O	3
Segment Group 2		O	2
USC	Certificate	M	1
USA	Security algorithm	O	3
USR	Result to be validated	O	1
....			
Segment Group *n*		O	99
UST	Security trailer	M	1
USR	Result to be validated	O	1

Note: M = mandatory; O = optional.

service offered (e.g., integrity, authentication of the origin, nonrepudiation of the origin) to the same message or if the same security service is benefiting several messages.

The security header segment USH contains the following:

- A mandatory security reference number
- The scope of the security application
- An indication whether an acknowledgment is requested
- The role of the intermediary providing the security services (e.g., document issuer, notary, witness, contracting party)
- Details concerning the sender and the receiver: name of the symmetrical encryption key or of the certificate in case of a public key encryption
- A time stamp (date and time when the security services were applied)

To each security service invoked, there is a corresponding USA segment that defines the algorithm utilized and its parameters.

4.9.2.1.2 Segment Group Number 2

This group contains the elements pertinent to public key encryption, in particular, the certificates. Two repetitions are allowed to communicate sender and recipient certificates. The latter case occurs if the sender utilizes the public key of the recipient instead of his or her private key.

If the recipient does not have the public key of the certification authority, the group will contain references to the sender's public key through the certificate of authenticity; the hashing function that the certification authority employs for producing the certificate; and the public key algorithm that the certification authority uses to sign, as well as its public key. These references will be contained in three USA segments. The segment USR will contain the signature of the public authority that will be used to verify the transmission integrity.

If the recipient knows the public key of the certification authority, it will be communicated by reference (for example, the name of the file where it resides) so that the recipient could extract it from a secure database.

4.9.2.1.3 Segment Group n

The segment UST is the security trailer that separates the message from the security structure. It is followed by the segment USR that contains the outcome that the validation results should match when the computations are done according to the specifications of the header segment USH.

4.9.2.1.4   Examples

The following two examples illustrate how the mechanisms indicated above can be used to provide security services for the protection of EDIFACT messages.

Example 1: Nonrepudiation of the Origin and Message Integrity

There are several ways to ensure nonrepudiation of the origin. Assume that the mechanism used is to condense the message with a one-way hashing function, and then encrypt the digest with a public key algorithm. Assume also that the certification authority utilizes a different hash function to sign the certificate of authenticity accorded to the sender. Finally, assume that the hash function used by the sender was already the subject of a bilateral agreement with the recipient.

There are now four algorithms: the hash function and the public key algorithm that the sender utilizes to condense the messages and encrypt the digest; the hash function and the public key algorithm that the certification authority utilizes to calculate the digest of the certificate and to sign it, respectively. Because the sender's hash algorithm is the subject of a prior bilateral agreement, only three USA segments will be necessary. The sender's public key is obviously contained in the certificate that the public authority signed.

Two USR segments contain the data needed to verify the integrity of the transmission, namely, the encrypted digest of the EDIFACT message and the signature of the certification authority.

Example 2: Nonrepudiation of the Origin and Confidentiality

For confidentiality, the EDIFACT message can be encrypted using a symmetrical key that will then be encrypted with the public key of the recipient. This is, in fact, the procedure specified by the protocol ETEBAC5. This is the protocol that French banks use to secure file exchanges with their client businesses. This method was also adopted in the SET (Secure Electronic Transaction) protocol for securing bankcard transactions over open networks, which will be discussed in Chapter 7.

Two groups of Number 2 segments are needed: the first relates to the sender's public key certificate and the second to the recipient's public key that the sender will use to encrypt the secret symmetrical key. To each group corresponds a USR segment that contains the result to be validated to verify the integrity of the received message.

Four USA segments are involved. The first segment, which is within Segment Group Number 1 , contains the parameters of the symmetrical algorithm. The other three segments, respectively, identify the public key algorithm of the sender, the hash function utilized by the certification authority to condense the sender's certificate before signing it, and, finally, the public key algorithm and the public key that the authority uses to sign the sender's certificate.

4.9.2.2 Security of EDIFACT Documents with Out-of-Band Segments: The AUTACK Message

The AUTACK message is used to provide the same levels of security as with in-band segments, including acknowledgment of reception. Authentication is ensured by symmetrical encryption of the digest of the structure to be protected or by exchange of certificates. For nonrepudiation of the origin, the digest is encrypted with a public key. Depicted in Table 4.7 is the security structure of the AUTACK message.

TABLE 4.7

Security Structure of the AUTACK Message

Tag	Segment Name	Type	Maximum Number of Repetitions
UNH	Message header	M	1
Segment Group 1		O	99
USH	Security header	M	1
USA	Security algorithm	O	3
Segment Group 2		O	2
USC	Certificate	M	1
USA	Security algorithm	O	3
USR	Security result to confirm the validity of security as certified by the certification authority	O	1
USB	Identification of AUTACK (type, function, time stamp, sender, recipient)	M	1
Segment Group 3		M	9999
USX	Security references to what is being secured	M	1
USY	Security references (result to be verified)	M	9
Segment Group 4		M	99
UST	Security trailer	M	1
USR	Security result that verifies the security of the AUTACK message	O	1
UNT	Message trailer	M	1

Note: M = mandatory; O = optional.

Segment USX points to the EDIFACT structure to be secured and records the date and time of its creation. Each USX segment corresponds to one or several USY segments that contain the results to be matched to verify the validity of the message to which USX refers.

Segment USR of Segment Group Number 4 is optional; it is only necessary when the AUTACK message is to be verified. For example, if USY contains the encrypted digest, USR may be omitted. However, if the AUTACK message is used as a secure acknowledgment, USR will contain the encrypted digest of the message. The whole structure will be enclosed in a pair of segments USH/UST (see Example 5 below).

Segment USB identifies the sender and the recipient of the interchange to be secured. It also contains a time stamp and indicates if the sender requested an acknowledgment.

Example 3: Authentication of the Origin of the EDIFACT Message, of the Origin of the AUTACK Message, and of the Integrity of the EDIFACT Message

In this example, a symmetrical algorithm is used with a secret key that both parties already have.

Segment USH contains, among other details, references to the symmetrical key; Segment USY contains the digest of the EDIFACT message; and Segment USR in Group 4 contains the encrypted digest of the AUTACK message. When the recipient verifies that the value of the computed encrypted digest is identical to the value in the message, this will be a verification of the authenticity of the origin of both EDIFACT and AUTACK messages. The integrity of the EDIFACT message is verified if the value of its digest computed after reception is the same as the value indicated in the USY segment.

Example 4: Nonrepudiation of the Origin

Segment USC holds the sender certificate and the public key of the certification authority. Three USA segments are needed to identify, respectively, the public key algorithm that the sender uses for signature, the hashing function, and the public key algorithm that the certification authority employs to compute the digest of the sender certificate and to sign the certificate.

Segment USR of Group 2 contains the signature of the certification authority that should be verified at reception. The USB has the security parameters and indicates if an acknowledgment is required, the security date and time, as well as the identity of the sender and the recipient.

Segment USX points to the secured EDIFACT message, and the USY segment has the sender signatures that the recipient will have to verify. The segment USR of Group 4 has the signature of the sender as applied to the AUTACK message.

Nonrepudiation of the origin is the consequence of authentication of the origin of the EDIFACT and AUTACK messages, as verified with the sender's signature on both. Note that the sender's signature is used twice; accordingly, the USR segment of Group 4 is redundant and can be eliminated at the origin.

Example 5: An Interchange Comprises Two Messages — Nonrepudiation of the Origin of the First and Authentication of the Origin of the Second

The transmission will contain two sets of Segment Group Number 1, one for each required service. The first service is offered in the same manner as for Example 4. Authentication of the origin of the second message is the other service. A USA segment will define the symmetrical algorithm used, its parameters, and its mode of operation. Because the sender's signature is verified for each message, securing the AUTACK message is not necessary.

Example 6: Nonrepudiation of the Origin with a Request of Acknowledgment

The transmission will contain two sets of Segment Group Number 1, one for each required service. The first will relate to nonrepudiation of the origin and the second to the request of a secure acknowledgment.

Segment USC of Segment Group Number 2 contains the sender certificate and the public key of the certification authority. Three consecutive USA segments identify, respectively, the public key algorithm that the sender uses, the hash function, and the public key algorithm, both used by the certification authority to generate the digest and sign the sender's certificate.

Segment USB contains the time stamp of the AUTACK message, the credentials of both parties, and the references to the message.

Segment USX includes references to the message to be secured. Segment USY has the digest of the message encrypted with the sender's private key.

Two sets of Segment Group Number 4 are used, one for the EDIFACT message and the other for the AUTACK message. The USR segment of the last Segment Group Number 4 will contain the signature of the AUTACK message to verify that it arrived safely at the destination.

4.9.3   IETF Proposals

By adding security services, Internet messaging systems are catching up with most of the X.400 functions, with a few exceptions regarding authentication of the route and guarantees on delivery times. Similar to X.400, proposals in the IETF for secure messaging apply to all the messages of an interchange; they do not permit a finer resolution at the level of individual messages or segments. Another limitation of this approach is that the structure of MIME does not include referrals to different segments, which precludes cross-referencing of segments.

The outlines of the various IETF proposals are as follows (IETF RFC 3335, 2002):

- The sending organization sends the encrypted data and the signature of the message (message digest encrypted with the sender's private key) in an envelope constructed according to PGP/MIME, S/MIME, or S-HTTP, and requests an acknowledgment.

- The recipient organization recovers the symmetrical encryption key and its parameters (for example, the initialization vector) with its private key, decrypts the message, and checks the signature with the public key of the sender, thereby simultaneously verifying the integrity of the data and the authenticity of the sender.

- The recipient sends back a signed acknowledgment by encrypting its digest with the recipient's private key; the acknowledgment, in turn, contains the digest and the identifier of the received message.

In IETF jargon, the digest is called *message integrity check* (MIC) and the acknowledgment is the *message disposition notification*. If the acknowledgment is signed, it will be in the multipart format of MIME to accommodate the acknowledgment and the signature. The sender can then verify the integrity of the acknowledgment, thereby proving the following:

- The recipient authenticated the sender.

- The recipient recognizes having received the message that corresponds to the mentioned identifier.

- The message was received, with its integrity intact.

- The recipient cannot deny having sent the acknowledgment.

For encryption, PGP/MIME uses PGP or OpenPGP, the first being described in IETF RFC 1991 (1996), while the second is in IETF RFC 2440 (1998). PGP is considered to be the closest commercial algorithm to military-grade performance (Garfinkel, 1995); its adoption as an IETF standard floundered because of its use of patented techniques, particularly the RSA key exchange and IDEA encryption, something that the IETF avoids by principle. OpenPGP got around this difficulty.

TABLE 4.8

Reference Documents for S/MIME

Document Number	Title
RFC 2633	S/MIME Version 3 message specification
RFC 2632	S/MIME Version 3 certificate handling
RFC 2634	Enhanced security services for S/MIME
RFC 2437	PKCS #1: RSA Encryption Version 2.0
RFC 2315	PKCS #7: Cryptographic Message Syntax Version 1.5

S/MIME is now in its third version, and its use is described by several RFCs; the more important are given in Table 4.8. Its second version was not standardized by the IETF, because it requires use of a patented key exchange mechanism that is not freely available.

S/MIME allows a choice to be made between two symmetrical encryption algorithms DEC in the CBC mode or triple DES. The use of RC2 and RC5 is optional, but usage of RC2 with a key of 40 bits is not recommended. Hashing is done with MD5 or SHA-1. The digest, called MIC, and the symmetrical session keys are encrypted using the RSA algorithm with keys of 512 or 1024 bits. Note that in a multipart MIME message, computation of the digest takes into account the content of all parts, including the headers (IETF RFC 2045, 1996).

The following two examples show how secured messaging is used on the Internet.

4.9.3.1   *PGP/MIME Encrypted and Signed*

In this case, the EDI data are processed as a multipart content that is encrypted and signed.

The protocol stack is defined in IETF RFC 2015 (1996) for PGP and IETF RFC 3156 (2001) for OpenPGP as follows:

RFC 822/2045
> RFC 1847 (multipart/encrypted)
>> RFC 2015/RFC 3156 (application/pgp-encrypted)
>>> « Version 1 »
>>> RFC 1767 (application/<EDI type used>) (encrypted)
>>> RFC 2015 (application/pgp-signature) (encrypted)

Thus, the message will take the following form:

```
To: <recipient's electronic address>
Subject:
From: <sender's electronic address>
Date:
```

```
Mime-version: 1.0
Content-Type: multipart/encrypted;
 protocol = "application/pgp-encrypted";boundary =
"delimiter";
 — delimiter
Content-Type: application/pgp-encrypted
Version: 1
 — delimiter
Content-Type: application/octet-stream
 — — -BEGIN PGP MESSAGE — — -
Version 2.6.2
*Content-Type: multipart/signed; micalg = pgp-<hashing
symbol>;
protocol = "application/pgp-signature"; boundary =
"delimiter for the signature";
*
* — delimiter for the signature
*&Content-Type: Application/<EDI standard utilized>
*&Content-Transfer-Encoding: <coding>
*&...
*&<EDI object>
*
* — delimiter for the signature
*Content-Type: Application/pgp-signature
*
* — — -BEGIN PGP MESSAGE — — -
*Version 2.6.2
*
*<signature>
* — — -END PGP MESSAGE — — -
*
* — delimiter for the signature
 — — -END PGP MESSAGE — — -
 — delimiter —
```

In this example, the signature is calculated using all the lines that begin with the symbols *&, while all the lines that begin with the symbol * are encrypted

4.9.3.2   S/MIME Message Encrypted and Signed

The messages are put in the format that PKCS #7 specified. The PKCS #7 is a general syntax for signing or encrypting data.

The content of an S/MIME utilized for EDI can be one of two types: *SignedData* and *EnvelopedData*. The first content type is used to send the signatures, while the second is reserved for encrypted messages. The signature has several attributes, among which are a time stamp and identifiers of the algorithms used.

A secured message can be sent in several formats, for example, *multipart/signed, application/pkcs7-mime,* or *signed-data.* To encrypt a message, an "envelope" has to be used with a content type of *enveloped-data* using the format *application/pkcs7-mime.*

Signing and encrypting a message is done in two steps: signing the content and enveloping the data. The structure uses a pile of RFC specifications, as follows:

RFC 822/2045

    RFC 2633 (application/x-pkcs7-mime)

       RFC 1847 (multipart/signed) (encrypted)

          RFC 1767 (application/<EDI type used>) (encrypted)

          RFC 2633 (application/x-pkcs7-signature) (encrypted)

Accordingly, the message will take the following form:

```
To: < recipient's electronic address>
Subject:
From: <sender's electronic address>
Date:
Mime-version: 1.0
Content-Type: application/pkcs7-mime; smime-type =
enveloped-data;
 name = smime.p7m
Content-Transfer-Encoding: base64
Content-Disposition: attachment; filename = smime.p7m
*Content-Type: multipart/signed; boundary = "encryption
delimiter";
* micalg = sha1;
* protocol = "application/x- pkcs7-signature"
*
* — encryption delimiter
*&Content-Type:Application/<EDI standard>
```

```
*&Content-Transfer-Encoding: <encoding type>
*&
*&<Here message of EDI object>
*
* - encryption delimiter
*Content-Type: multipart/signed;
* protocol = "application/x-pkcs7-signature";boundary =
"signature delimiter"
*
* - signature delimiter
*Content-Type:application/pkcs7-signature;name =
smime.p7s
*Content-Transfer-Encoding: base64
*Content-Disposition:attachment;filename = smime.p7s
*
*<signature>
*
* - signature delimiter -
 - encryption delimiter -
```

As explained earlier in the presentation of MIME, specifying a *base64* coding in the field Content-Transfer-Encoding protects the transport of binary files.

The hashing algorithm is indicated by the variable *micalg*. It can be either SHA-1 or MD5. The digest is computed for all lines that start with *& above. Note that SHA-1 is recommended to avoid the weaknesses in MD5.

All the lines that start with the symbol * are encrypted, and accordingly, all that lies between the segments ISA and IEA for X12 and the segments UNA/UNB and UNZ for EDIFACT is encrypted. However, the invisibility of the headers that ensue may stifle the efforts of the operators of value-added networks to follow the routing of messages to guarantee their arrival at destination.

The optional parameter *name* assures compatibility with previous versions. The parameter *filename* is also optional, but it plays two roles: transmitting information on the content beyond gateways that do not recognize S/MIME and facilitating file management in the recipient's computer. Naming conventions for the files help in the realization of these two objectives.

4.9.4   Protocol Stacks for EDI Messaging

Figure 4.16 summarizes the different protocol stacks for EDI messaging without security, while Figure 4.17 depicts a synthetic view of the protocol

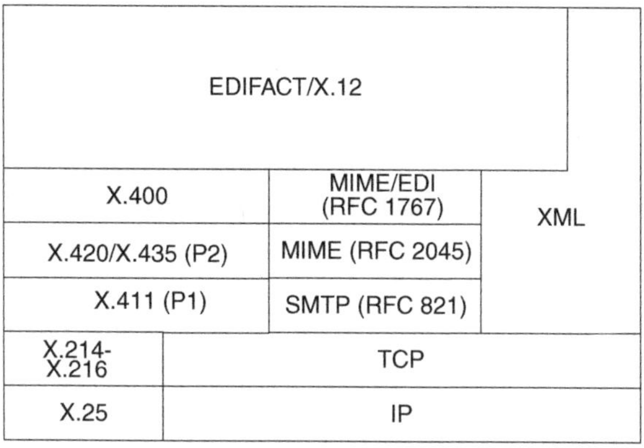

EDIFACT/X.12			
X.400	MIME/EDI (RFC 1767)	XML	
X.420/X.435 (P2)	MIME (RFC 2045)		
X.411 (P1)	SMTP (RFC 821)		
X.214-X.216	TCP		
X.25	IP		

FIGURE 4.16

Protocol stacks for EDI messaging (without security).

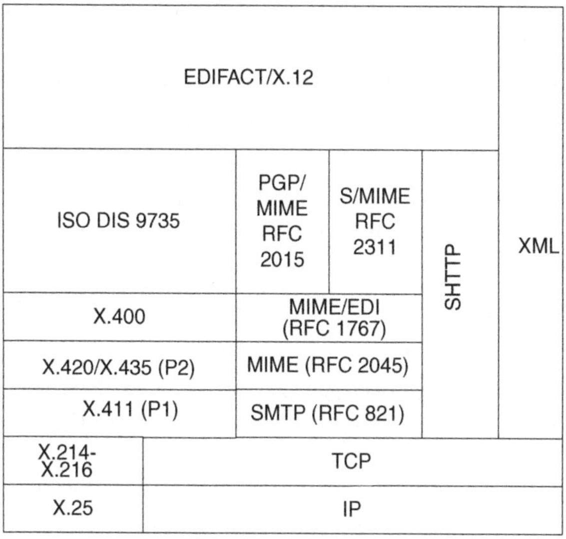

EDIFACT/X.12					
ISO DIS 9735	PGP/ MIME RFC 2015	S/MIME RFC 2311	SHTTP	XML	
X.400	MIME/EDI (RFC 1767)				
X.420/X.435 (P2)	MIME (RFC 2045)				
X.411 (P1)	SMTP (RFC 821)				
X.214-X.216	TCP				
X.25	IP				

FIGURE 4.17

Synthetic view of the protocol stack for secure EDI.

stack for a secure EDI. A more rigorous comparison of the various protocols can be found in specialized references on electronic messaging, such as Palme (1995) or Bouillant (1998).

4.9.5 Interoperability of Secured EDI and S/MIME

The EDIINT (Electronic Data Interchange — Internet Integration) group of the IETF pondered the interoperability of EDI translation software with

S/MIME message systems. The CommerceNet consortium sponsored interoperability tests that were completed in August 1997. These tests covered the following scenarios:

- Sending and receiving a MIME message containing an unsecured EDI interchange with a signed or unsigned acknowledgment
- Generating and exchanging a public key certificate of the following types:
  - PKCS #7
  - X.509 Versions 1 and 3
  - Self-certification
  - VeriSign or Entrust
- Sending and receiving EDI messages encrypted with the RC2 symmetrical algorithm and with a signed or unsigned acknowledgment
- Sending and receiving a MIME message containing an unencrypted but signed EDI interchange
- Sending and receiving signed and encrypted EDI messages
- Sending and receiving signed and unsigned acknowledgments

The EDI data exchanged were formed from multipart EDIFACT and X12 purchase orders with different types of segment separators, whose dimensions varied from 1.166 to 1206.193 K octets.

Six companies participated in the tests. These companies were, in alphabetical order, Actra Business Systems, Atlas Products International, Digital Equipment Corporation (DEC), Harbinger Corporation, Premenos, and Sterling Commerce. Premenos is the producer of TrustedLink Templar, one of the software packages that is best positioned in the market of secure EDI.

Not surprisingly, a few problems were uncovered. One problem involves a signed message that comprises several parts, when the message path includes message transfer agents that are different from SMTP (for example, X.400 or Microsoft Exchange). In this instance, the validity of the signature may be compromised because of the successive conversions of the message. By signing the multiple-part message before encryption, MIME headers are part of the signed message, which would be protected from any tampering that the messaging gateways might cause.

Since these tests were conducted, there have been many mergers and acquisitions involving these companies. For example, Compaq acquired DEC and then merged with HP; Harbinger Corporation absorbed Premenos; and Sterling Commerce acquired a fellow French company, Comfirst, a producer of messaging systems and servers for the Internet and Minitel as well as EDI middleware, before being bought by Computer Associates, etc.

4.9.6 Security of XML Exchanges

The security of XML exchanges is the subject of several recommendations and technical reports from the W3C. Thus, XML Encryption allows selective encryption by identifying the sections to be encrypted with specialized tags, while the processing of signatures is specified in XML Digital Signature (XML-DSIG). Treatment of encryption keys is available in a technical report entitled XML Key Management Specification (XKMS). This document has two parts: XML Key Information Service Specification (X-KISS) and XML Key Registration Service Specification (X-KRSS). X-KISS provides a client the possibility of avoiding the computational load of the security tasks, such as encryption, signature, authentication, etc., by delegating them to specialized servers with large computational power. X-KRSS specifies the protocol to register data about public keys so they can be made available to all secured services on the Web. Finally, the elements used for authentication are defined with SAML within SOAP messages.

4.10 Relation of EDI with Electronic Funds Transfer

An enterprise conducting e-commerce must track its cash flow among its various departments as well as its accounts with its bank. The activities correspond to the following:

1. Internal exchanges within the enterprise from the accounting department to the finance department, concerning, for example, notification of the due date for payment to suppliers, notification of salaries payment due date, and files for customers' invoices

2. Exchanges between the enterprise finance department and its bank concerning the various payment instruments, such as bills of exchange, check remittances, and credit transfers

3. Exchanges between the bank and the finance department concerning the account statement, unpaid invoices, and statements of drafts (bills of exchange) and promissory notes

4. Internal exchanges within the enterprise from the finance department to the accounting department about notification of payment to suppliers, notification of salary settlement, and reconciliation of internal accounting ledgers and with the bank statement

Shown in Figure 4.18 are the above exchanges, with the numbers referring to the numbered list above.

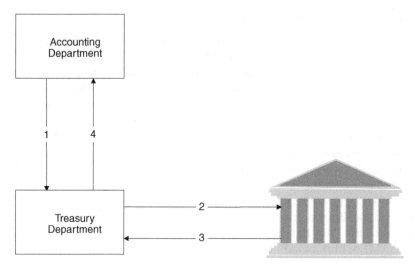

FIGURE 4.18
Exchanges within an enterprise as well as between the enterprise and its bank.

An enterprise carries out three types of accounts reconciliation to harmonize the circuits of production and distribution with accounting (Dragon et al., 1997, pp. 378–380):

- Business reconciliation of the settlements with the invoices
- Financial reconciliation to adjust the actual cash flow with the forecast
- Account reconciliation between the bank statement and the enterprise records

The migration for paper-based to electronic support offers the possibility of joining these loops to facilitate tracking and to reduce delays in settlements. Figure 4.19 shows the trajectories of the financial EDI messages as well as those of the EDI control messages are shown.

Each enterprise assembles its transaction records and presents them, in a common format, to the settlement computer through the telecommunication network (in early systems, messengers would carry magnetic tapes). The computer groups all valid requests, sorts them according to the drawer bank, and then sends them to their respective banks to debit the drawer's account. Because all parties now share the same reference numbers, all types of reconciliation will be easier, particularly if they are automated. These exchanges are sketched in Figure 4.20.

Financial EDI opened the way for automatic processing of fund transfers between an enterprise and its bank. Today, many enterprises utilize credit transfers for paying salaries, compensations, pensions, and benefits, as well as for direct-debit notification, because the integration of the bank settlement

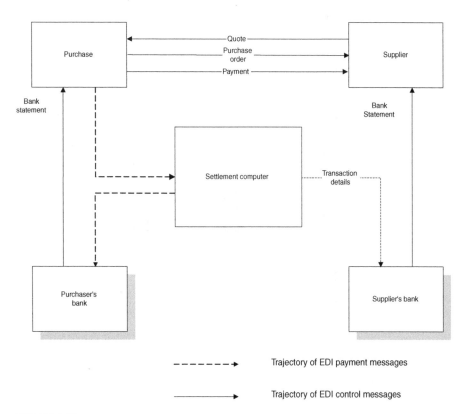

FIGURE 4.19
Trajectories of the EDI financial and control messages.

system with the business circuits allows for better management of the settlement speed.

Manual settlements introduce three types of delays. The first type is due to the postal service; the second is the result of the manual treatment of the check by the beneficiary before depositing it in a bank account. Finally, the third delay comes from the bank settlement. If D is the day that the beneficiary deposits the payment instrument into a bank account, the settlement takes place on day $D + 1$, but the beneficiary's account is only credited on day $D + n$, where n varies according to the banking system and to whether the banks are in the same state, but is usually between 2 to 4 days. During this time, the drawee's bank has the amount working for its own account for $(n - 1)$ days.

Although the manual treatment allowed the payer to keep the funds for a longer period, the advantages of the electronic funds transfer would be shared between both the payer and the beneficiary. Each could take the reduction of the float as well as the faster notifications into account in the management of cash flow, either by modifying the payment date or by negotiating the terms for payment.

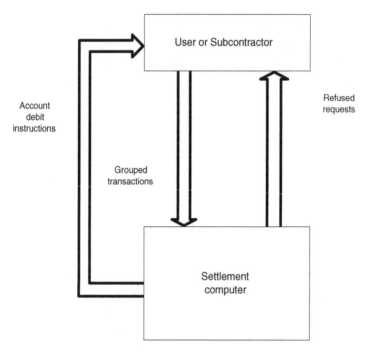

FIGURE 4.20
Exchanges involved in bank settlements.

4.10.1 Funds Transfer with EDIFACT

Many financial EDIFACT messages were standardized. These messages exclude interbank exchanges and focus on the exchanges between banks and their clients as well as the exchanges among businesses. This is because settlement architectures differ from one country to another, and there is no justification to reconsider them just to conform to EDIFACT specifications. However, normalized messages can encapsulate proprietary messages to give a common interface among the systems of the various units.

There are three categories of messages for financial EDIFACT:

- Simple messages, such as PAYORD and REMADV, that describe a single operation with abundant comments
- Detailed or "extended" messages, such as PAYEXT, that allow the juxtaposition of the details to the basic operation, for example, identifications of the invoices, of the nature of settlements, of the reasons for nonpayment, etc.
- Multiple messages, for example, PAYMUL, that include several financial transactions of the same kind

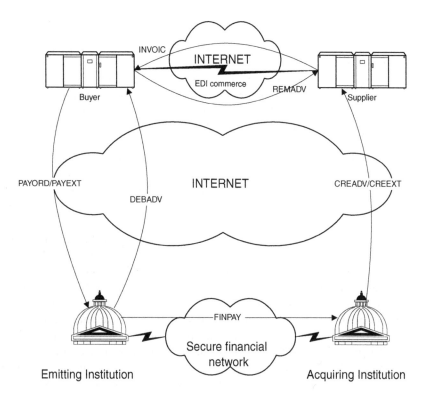

FIGURE 4.21
Credit transfer using the messages of financial EDIFACT.

The diagram in Figure 4.21 represents the way that these messages can be utilized to effect a credit transfer in response to an invoice. The two Internet "clouds" could be either the same public network (if the exchanges are secured) or one or several private networks or value-added networks (Cafiero, 1991; Hendry, 1993, pp. 125–131).

In this transaction, the information that was initially in PAYORD (or PAYEXT) is copied in CREADV (or CREEXT). The client can also send REMADV directly to the supplier with the content of the CREEXT message.

The FINPAY message is for interbanking credit transfer through the SWIFT network and is used for international payments.

The REMADV message is only needed when the remittance and the payment instructions are separated in time. In such a case, the remittance advice informs the supplier that a payment will be made at the date defined in the supply contract. The figure sketches a bare-bones illustration of the operation and does not take into account all potential problems, such as the lack of funds, identification errors, etc., that will trigger the exchange of the appropriate messages.

4.10.2 Funds Transfer with X12

Fund transfers with X12 follow the preceding outline with the Transaction Set 820 replacing the messages PAYORD/PAYEXT and DEBADV. Following each exchange, the recipient may use the Transaction Set 997 to inform the sender that the received message is syntactically correct, thereby ensuring the regular progress of each stage of the transaction. If the credit transfer cannot be done, the Transaction Set 827 is sent. Of course, all of these messages can be secured with the procedures defined in X12.58 or those that are based on S/MIME. In a pilot experiment, the time needed for the transmission and the processing of these exchanges, assuming error-free transmissions, varied between 12 minutes per 100 instructions and 58 minutes for 1000 instructions, instead of the average 2 to 4 days for manual settlement (Segev et al., 1996).

The delay between the order of a stock trade and its settlement increases the transaction costs and the financial risks. This is why the Security Industry Association (SIA) in the U.S. set June 2005 as the date when all securities trades must be completed by the end of the day following the trade, down from the current 3 days (the so-called T + 1 settlement). However, between 20 and 30% of all trades fail, mostly because of data entry errors (Manchester, 2002). Thus, the ambitious target is only possible by full automation of the end-to-end processing of the transaction. This concept of Straight -Through Processing (STP) covers activities from the front office, including gathering intelligence, through to the final settlement passing by order taking and execution.

4.11 Electronic Billing

Electronic billing is called Electronic Bill Payment and Presentment (EBPP) or Electronic Invoice Payment and Presentation (EIPP). This is a logical step after the dematerialization of commercial exchanges and has been around since the Minitel. While this feature appeals to large bill producers (utilities and telecommunication companies), it may also interest the public at large, because it gives subscribers the capability of viewing their bills on the screen and of paying them online. From a business-to-business perspective, the electronic links among buyers, suppliers, and their respective bankers provide firms with easier tools for financial management and inventory management.

The establishment of electronic billing on the Internet requires a larger opening of the information systems of the parties and uniformity in the message formats used. This is to group bills from different suppliers and consolidate them in a single bill that will be presented to the firm in a form that is compatible with its information system. These steps are shown in Figure 4.22.

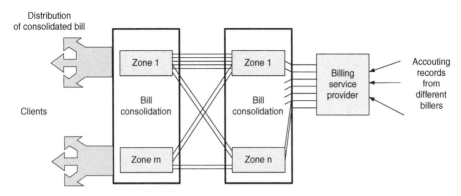

FIGURE 4.22
Grouping and consolidation of electronic bills.

The payment phase depicted in Figure 4.23 relies on collectors associated with credit institutions to collect the payments (with credit transfers, electronic checks, etc.) and deposit them in acquirer banks.

CheckFree, one of the foremost payment intermediaries in the U.S., captured 90% of the market for online billing in this country. Other actors are consortia of banks as well as individual entities, such as Billserv and Online Resources. In the consumer space, MasterCard offers a special service for banks called Remote Payment and Presentment Service (RPPS) on the basis of OFX. Finally, the operators of the settlement networks ACH and CHIPS are pushing the idea of using their networks to deliver electronic bills and payments.

4.12 EDI Integration with Business Processes

Depicted in Figure 4.24 is a simplified synthesis of all the EDI elements in a commercial transaction. The basis is the real-life efforts of a telephone operator (Bell Atlantic, today Verizon) to streamline its internal processes and reduce the processing cost of commercial invoices by at least 20% (Sivori, 1996). Of course, in this simplified view, many elements of a real transaction were discarded for the sake of clarity.

The integration of EDI in business processes clearly poses logistic and legal challenges, namely, its integration with the internal processes of the corporation (the "back-office") and the legal status of the EDI. Standards help to identify and track parcels or crates in automated warehouses, particularly the Multi Industries Transport Label (MITL) and the Serial Shipment Container Code (SSCC).

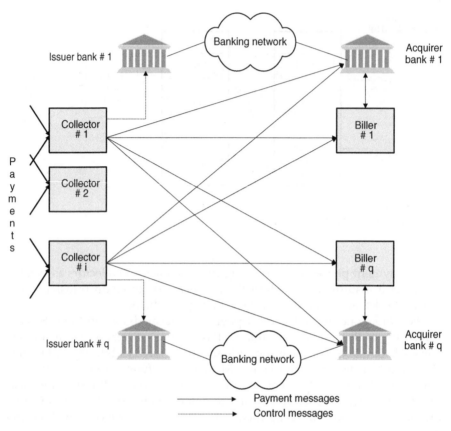

FIGURE 4.23
Online payment of electronic bills.

The problem is how to integrate these offers with existing processes in an economical and efficient manner. Probably, solutions envisaged for large enterprises will not be identical to those for small and medium enterprises.

4.13 Standardization of the Exchanges of Business-to-Business Electronic Commerce

4.13.1 EDI/EDIFACT

In the U.S., DISA (Data Interchange Standards Association) has acted as the secretariat for the ANSI accredited committee (ASC) X12 since 1987 and is responsible for the publication and distribution of the standards. As in all ANSI committees, the technical activities for development and maintenance of X12 take place in subcommittees that represent the various sectors inter-

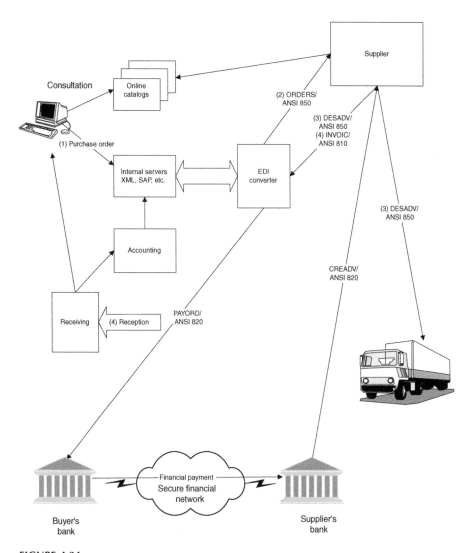

FIGURE 4.24
Synthesis for EDIFACT and X12. The numbers in parentheses show the sequence of the messages.

ested in standardization (manufacturing, purchasing, finance, transportation, public administration, health care, etc.). For example, in the health-care field, ANSI turns to the Health Industry Business Communications Council (HIBCC). This organization groups the various institutions interested in the use of EDI for health care, in particular, the Healthcare EDI Coalition (HEDIC) and the National Wholesale Druggists Association (NWDA).

The work on EDIFACT is within the scope of the Commerce Division of the UN/ECE. Starting in March 1997, the activities were moved to the Center for Facilitation of procedures and practices in Administration, Commerce and Transport (CEFACT), now called the U.N. Centre for Trade Facilitation and Electronic Business, which was attached to the Committee for Development of Trade (CDT). To take into account the rapid development of e-commerce, a new structure was formed in 1998 and given the name UN/EDIFACT Work Group (EWG). It is this structure that supports the work on ebXML to perpetuate the legacy of traditional EDI. Figure 4.25 depicts the current organization (Charmot, 1997b, pp. 40–42).

Although the world is divided into six regions to encourage regional activities, participation in the standardization of EDIFACT comes primarily from Western Europe and North America. Groups of experts associated with industry or professional organizations initiate the propositions that

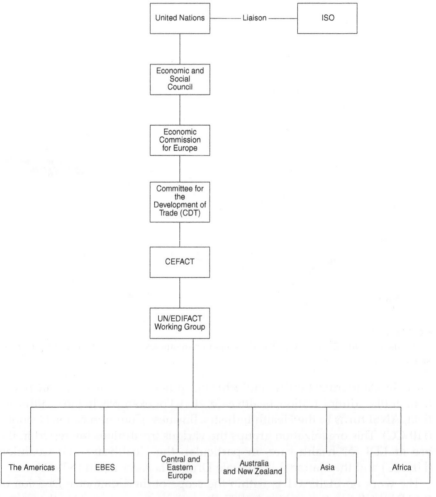

FIGURE 4.25
Standardization of EDIFACT within the United Nations system.

will be approved at the national level before being submitted to such international organizations such as UN/ECE and ISO to harmonize the various proposals.

Among the main national organizations, arranged in alphabetical order by country, are the following:

1. In France, the committees EDIFRANCE and SIMPROFRANCE (*Comité Français pour la Simplification des Procédures du Commerce International* — French Committee for the Simplification of Procedures for International Commerce) as well as the AFNOR (*Association Française de Normalisation* — French Association for Standardization)

2. In Germany, DIN (*Deutsches Institüt für Normung e.V.* — German Institute for Standardization)

3. In the U.K., the committee SITPRO that represents British interests in international bodies and promotes EDI on the national level

4. In the U.S., ANSI X12, which manages the North American EDIFACT Board

Standard development is animated by rapporteurs supported by regional "EDIFACT Boards." The rapporteurs meet twice a year by organizing Joint Rapporteurs Team (JRT) meetings. Beginning in June 1998, EWG replaced the JRTs.

The West European EDIFACT Board (WEEB) was formed in 1988 to coordinate the activities at the West European level. It included members of the European Commission, the European Free Trade Association (EFTA), the European Committee for Standardization [*Comité Européen de Normalisation* (CEN)], European industry associations, and international organizations such as SWIFT. Simultaneously, the European Commission started the TEDIS program to promote the utilization of EDI in Europe. Within the framework of these activities, the European Commission played the role of the WEEB secretariat from 1988 to 1994, when TEDIS ended.

In September 1995, CEN hosted the European Board for EDI Standardization (EBES) (*http://www.cenorm.be/ebes*), where development of EDIFACT messages continued in various fields such as commerce, transportation, customs, banking, statistics, construction, insurance, tourism, health care, government purchases, social administration, and employment. The work from the experts group carries the label EBES Expert Group (EEG), although many documents in circulation still carry the older name of Message Development Group (MDG).

The ISO committee responsible for the standardization of EDIFACT is Technical Committee (TC) 154.

4.13.2 XML/EDI Integration

The main organizations involved in the integration of XML/EDI are mentioned below in alphabetical order.

4.13.2.1 CEFACT

CEFACT of the U.N. promotes a generic definition for the exchanges encountered in business-to-business e-commerce through the ebXML. This generic framework must be further defined for each sector individually. Shown in Figure 4.25 is the position of the CEFACT within the U.N. system.

4.13.2.2 CommerceNet

CommerceNet (*http://www.commerce.net*) is a nonprofit consortium of U.S. companies formed to promote global solutions for e-commerce (Tenenbaum et al., 1995). Action of this consortium takes place along three axes:

1. Development of XML catalogs and the associated DTDs
2. Demonstration of test beds of practical XML solutions
3. Interworking of XML with alphanumeric EDI

DISA collaborates with CommerceNet on the subject of the convergence of X12 and XML.

4.13.2.3 IETF (Internet Engineering Task Force)

The EDIINT group of the IETF defined ways to secure the transmission of structured data related to business-to-business commerce using EDI, EDIFACT, or XML with SMTP messaging as content types defined by the MIME extensions.

4.13.2.4 Open Buying on the Internet (OBI)

Open Buying on the Internet (OBI) is an initiative of the OBI Consortium, started in 1997, with a major push by American Express, to encourage business purchases over the Internet. The messages are based on transaction set 850 (Purchase Order), enveloped to form an OBI object and then sent using HTTP. Security is achieved with the Secure Sockets Layer (SSL) protocol and with X.509 certificates that implement Version 3 of the specifications.

4.13.2.5 Open Trading Protocol (OTP) Consortium

The OTP (Open Trading Protocol) Consortium, which started in 1997 as well, advocates the use of the OTP, which uses XML, thereby competing head-to-head with the OBI Consortium.

4.13.2.6 Organization for the Advancement of Structured Information Standards (OASIS)

OASIS is a consortium grouping the main software developers and system integrators (Sun Microsystems, IBM, Microsoft). It is developing the ebXML language in collaboration with the CEFACT to define XML messages for commerce starting from the existing EDIFACT. OASIS is also responsible for the development of the Universal Business Language (UBL) to describe the content of messages stored in model libraries, thereby facilitating the creation of XML schemas for groups of transactions (orders, bills, etc.).

It was mentioned earlier that OASIS defined SAML to describe with XML the data elements to be verified before authorizing access, by an individual or a group, to a given service. The SAML structures are typically included in HTTP forms sent with the POST method or in SOAP messages coded in XML.

4.13.2.7 RosettaNet

The RosettaNet (*http://www.rosettanet.org*) consortium primarily focuses on the supply chain for the electronics industry, where the legacy EDI is almost negligible, which allowed a greenfield start. The application is to tie together the manufacturers of electronic components (e.g., Intel), the equipment manufacturers (e.g., Cisco Systems, HP, Siemens, Toshiba), the system integrators, the wholesale dealers (e.g., Ingram Micro), and the retailers (e.g., CompUSA). By linking the information systems of all the participants of the supply chain, it becomes much easier to know the state of the supply at any moment and to coordinate the capacities of each sector to avoid buildups or shortages. To align the processes used along the chain on the same reference, the description of the business processes uses Partner Interface Processes (PIPs). These are, in effect, DTDs, where the objects and the transaction data models are described in XML. There are tools to verify the behavior of an XML script with the different sectorial PIPs already available.

In 2002, the Uniform Code Council (UCC) absorbed RosettaNet. The UCC covers some 23 industries associated with public warehousing and grocery industries.

4.13.3 XML

Standardization of XML is the work of the W3C, which started with the joint support of the Massachusetts Institute of Technology (MIT) in the U.S., the *Institut National de Recherche en Informatique et en Automatique* (INRIA) in France, and Keio, Shonan-Fujisawa University, in Japan.

In 2002, the European Research Consortium for Informatics and Mathematics (ERCIM) replaced INRIA as the European representative in W3C. The ERCIM was founded in 1989 and unites 15 European research institutes specializing in information technology (*http://www.ercim.org*).

4.14 Summary

Automation and dematerialization of exchanges are means to improve productivity. However, they do not depend solely on telecommunication and information processing technologies. By modifying the course of the information flows within the enterprise and by establishing collaborative relations among separate organizations, business-to-business e-commerce is causing a major reorientation in the ways information is managed.

The magnitude of the investments that large enterprises have already committed to EDI, in one form or another, will continue to delay the development and large-scale adoption of XML-based solutions, as long as the legacy systems satisfy the needs. No enterprise can afford to rethink its mode of operation every few months to follow the latest fashion. Without stability, it is highly unlikely that a commercial network would be able to survive. Yet, systems using the Internet exhibited excessive ambitions or offered incomplete solutions. Integration in a network of the internal processes of several enterprises requires substantial prior thinking, particularly because the question of security over open networks like the Internet is not completely resolved. Furthermore, the cornerstone of the Internet is a network without intelligence, mere pipes connecting the end points; yet, commercial links include many intermediaries, such as banks, that intervene directly in the conduct of the exchanges. Time is needed for a sufficient number of participants to accept a new networked organization and for the turnover to be significant. It should be expected, therefore, that solutions that use XML will gradually replace existing solutions as the infrastructure is renewed and as new requirements arise that the legacy systems cannot meet. Some of these needs include consultation of multimedia catalogs, multimedia exchanges, and quasi-real-time queries of inventories or of market status. Standardization of the information flow is essential to avoid ending up with a complicated architecture.

Questions

1. What are the possible exception conditions that can affect order fulfillment in business-to-business e-commerce?
2. What are the main benefits and main costs related to business-to-business e-commerce?
3. Compare the limitations and advantages of the platforms for traditional EDI as compared to Web-based platforms.

4. Compare and contrast business-to-business and business-to-consumer e-commerce in terms of procurement processes, order sizes, market sizes, and work flow.

5

SSL (Secure Sockets Layer)

ABSTRACT

SSL (Secure Sockets Layer) protocol is a generalist protocol for securing exchanges; however, it is widely used in electronic commerce (e-commerce) applications (Freier et al., 1996). SSL was integrated in 1994 in Netscape Navigator to secure communication between a client and a server over an open network such as the Internet. The first public version of SSL was 2.0, because Version 1.0 was a test version used internally within Netscape. Version 3.0, which is currently used, corrected some of the weaknesses that were discovered in the preceding version. Version 3.0 served as the starting point for the protocol TLS (Transport Layer Security) of the IETF (Internet Engineering Task Force), defined in IETF RFC 2246 (1999). TLS was, in turn, adapted to wireless communication in the WTLS (Wireless TLS) protocol. Both TLS and WTLS will be described in the next chapter.

This chapter presents the SSL protocol in some detail, including descriptions of the exchanges and an estimation of the duration of each message during session establishment. The chapter contains an appendix that gives the structure of SSL messages. For the interested reader, the book by Rescorla contains abundant information concerning the implementation of secure systems with SSL and TLS (Rescorla, 2001).

5.1 General Presentation of the SSL Protocol

SSL is a general protocol that can be applied to secure any exchanges between two points. This explains why it surpassed the S-HTTP (Secure HyperText Transfer Protocol) of IETF RFC 2660 (1999), which can only protect HTTP exchanges. In contrast, the SET (Secure Electronic Transaction) protocol, which was developed by Visa and MasterCard and which will be the subject of the Chapter 7, is exclusive to bankcard transactions.

5.1.1 Functional Architecture

SSL secures the exchanges between a client and a server in a transparent manner by being between the application and the transport layers of the OSI (Open Systems Interconnection) reference model. It may be possible to qualify SSL as a "presentation protocol." Figure 5.1 shows the location of SSL in the TCP/IP protocol stack.

Figure 5.2 shows the correspondence between SSL and other Internet protocols. SSL does not function on top of the UDP (User Datagram Protocol), because the latter does not offer a reliable transport, which may lead to IP packet losses. The flow interruptions that may result would be interpreted as security breaks that would force disconnection of the communication.

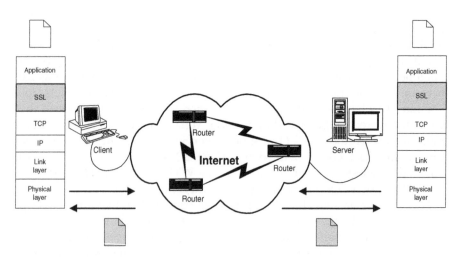

FIGURE 5.1
Functional model of SSL.

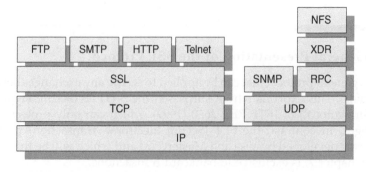

FIGURE 5.2
Position of SSL protocol in the TCP/IP stack. (RPC = Remote Procedure Call, XDR = External Reference Data.)

Accordingly, SSL cannot protect the following protocols: SNMP (Simple Network Management Protocol), NFS (Network File System), DNS (Domain Name Service), and the voice on IP protocol ITU-T Recommendation H.323.

IANA (Internet Assigned Numbers Authority) allotted specific IP ports to some applications for communicating with SSL. These ports are given in Table 5.1. The applications listed in Table 5.2 use the shown port by widespread convention, although IANA has yet to bless that usage officially.

5.1.2 SSL Security Services

SSL offers three security services: authentication, integrity, and confidentiality. With the help of a digital signature, it is possible to provide the elements necessary for a nonrepudiation service. SSL defines a framework to use the encryption and hashing algorithms that were negotiated between the two parties. This flexible structure is open to the integration of new algorithms as the users adopt them.

TABLE 5.1

IP Ports Assigned to Secure Applications with SSL

Secure Protocol	Port	Nonsecure Protocol	Application
S-HTTP	443	HTTP	Secure request-response transactions
SSMTP	465	SMTP	E-mail
SNNTP	563	NNTP	Network news
SSL-LDAP	636	LDAP	Light version of X.500
SPOP3	995	POP3	Remote access of mailbox with message download

Note: S-HTTP = Secure HyperText Transfer Protocol; SSMTP = Secure Simple Mail Transfer Protocol; SNNTP = Secure Network News Transfer Protocol; LDAP = Lightweight Directory Access Protocol; and POP = Post Office Protocol.

TABLE 5.2

IP Ports Used without Formal Attribution to Secure Applications with SSL

Secure Protocol	Port	Nonsecure Protocol	Application
FTP-DATA	889	FTP	File transfer
FTPS	990	FTP	Control of file transfer
IMAPS	991	IMAP4	Remote access to mailbox with or without downloading of messages
TELNETS	992	Telnet	Remote access to a computer
IRCS	993	IRC	Internet chat; text conferencing

Note: IMAP = Internet Message Access Protocol; IRC = Internet Relay Chat.

5.1.2.1   Authentication

Authentication uses a certificate that conforms to ITU-T Recommendation X.509 Version 3. It takes place only at the session establishment and before the first set of data is transmitted. This service was optional in Version 2.0 of SSL and became mandatory for the server in Version 3.0. This choice stimulated the usage of SSL without a complex infrastructure to manage the keys for the general public. However, the server can require the client to authenticate itself and may refuse to establish the session because the certificate is lacking.

Authentication can be static or dynamic. In a static authentication, the client extracts the public encryption key from the certificate that the server sends. In dynamic authentication, in contrast, the server certificate contains a signature key with which to sign a temporary encryption key. Through this *ephemeral* key exchange, a pair of temporary keys can be used to protect the data exchanged to construct the symmetric encryption key for the session. Thus, the static key contained in the server certification allows the client to authenticate and verify the integrity of the public ephemeral key that the server generates.

Dynamic authentication was first used to satisfy the U.S. export regulations that limit RSA encryption key lengths to 512 bits. Accordingly, a longer (1024-bit) key extracted from the server certificate is used to sign a temporary 512-bit key with the strong key. Furthermore, it can be used to protect against any compromise of the server's static key, in which case every SSL session that was ever established with that key would be exposed.

In addition to RSA, other algorithms for key exchange are the Diffie–Hellman algorithm, and the previously secret algorithms that the National Security Agency (NSA) developed for cryptographic applications on the PCMCIA card called "Fortezza" (NSA, 1994). The use of ephemeral Diffie–Hellman algorithm helps establish *perfect forward secrecy*: they are the ephemeral keys that cannot be recovered once destroyed. The Fortezza card uses a variant of the Diffie–Hellman algorithm called the Key Exchange Algorithm (KEA) for key agreement and the block cipher algorithm called SKIPJACK for encryption. Signatures are computed with the DSA, and message digests are derived with the SHA-1 hash function. It should be noted, however, that the integration of SSL with Fortezza is not without problems (Rescorla, 2001, pp. 126–128).

5.1.2.2   Confidentiality

Message confidentiality is based on the utilization of the symmetric encryption algorithms with key size of 40 or 128 bits. The same algorithm is used on both sides, but each side uses its own key that it shares with the other party. These keys are called the *client_write_key* on the client side and the *server_write_key* on the server side. The algorithms that can be used are DES (Data Encryption Standard), triple DES (3DES), DES40 (which is the same as DES but with a key size of 40 bits), RC2, RC4 with a key of 128 bits (RC4-128) or of 40 bits (RC4-40) for export outside the U.S., IDEA (International Data Encryption Algorithm), and the algorithm SKIPJACK of Fortezza.

The algorithms DES40 and RC4-40 were deliberately weakened using keys of 40 bits to satisfy the restrictions that the U.S. government imposed on the export of cryptographic software as defined in International Traffic in Arms Regulation (ITAR). We have seen that according to the same regulations, the usage of the RSA for key exchange was limited to 512 bits and not 1024 bits. Since then, the legislation was somewhat relaxed in 1999.

5.1.2.3 *Integrity*

Integrity of the data is assured with the application of hash functions; they essential employ the HMAC (Hashed Message Authentication Code) procedure to ensure better protection against attacks (Bellare et al., 1996). The hash functions utilized can be either SHA (Secure Hash Algorithm) or MD5 (but the strength of the latter has been questioned). The digest is treated by a series of operations that depend on a secret key, and the result is called MAC (Message Authentication Code).[1] This operation serves as well for authentication, because the secrets utilized in the encryption of the digest are known only to the two parties.

5.2 SSL Subprotocols

The SSL protocol is composed of four subprotocols: Handshake, Record, ChangeCipherSpec (CCS), and Alert. Figure 5.3 depicts the arrangement of the various components. It shows that the protocol Record is on top of the transport layer, while the other three protocols are between the application and the Record layer.

The Handshake protocol is responsible for the authentication of the communicating parties, for the negotiation of the encryption and hash algorithms, and for the exchange of a secret, the PreMasterSecret. The function of the CCS protocol is to signal to the Record layer any changes in the security parameters. Finally, the Alert protocol indicates errors encountered during message verification as well as any incompatibility that may arise during the Handshake. The Record protocol applies all the negotiated security parameters to protect the application data as well as the messages originating from the Handshake, the CCS, or Alert protocols.

---

[1] Note that SSL uses the word MAC for two different computations: a digest and the HMAC. This book will continue to use the terminology of the SSL documents to avoid any confusion.

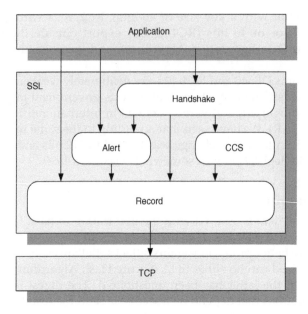

FIGURE 5.3
Protocol stack for the SSL subprotocols.

5.2.1   SSL Exchanges

The exchanges described in SSL happen in two phases:

1. During the preliminary phase, identification of the parties, negotiation of the cryptographic attributes, and the generation and sharing of keys take place.
2. During the exchange of data, the security depends on the algorithms and parameters negotiated in the preliminary phase.

At any moment, it is possible to signal an intrusion or an error.

SSL builds and adapts to the new context of security the notion of a TCP/IP application session. Whenever a client connects to a server, an SSL session is started. If the client connects to another process on the server, a new session is started without interrupting the initial session. If the client returns later to the first process and wishes to conserve the cryptographic choices already made, the client will ask the first server to resume the old session instead of starting a new one. Thus, in SSL, a session is an association between two entities that have a common set of parameters and of cryptographic attributes. To limit the risks of attacks through message interception, the text that defines SSL suggests limiting the age of the session identifier to a maximum of 24 hours; however, the exact duration is at the discretion of the server. In addition, the interrupted session can be resumed only if the proper suspension procedures are used.

The concept of a connection in SSL was introduced to allow an application to refresh (i.e., modify) certain security attributes (e.g., the encryption key) without affecting all the other attributes that were negotiated at the start of a session. A session can contain several connections under the control of the applications. The concepts of SSL sessions and connections can be illustrated with the help of their state variables and the associated security parameters.

5.2.1.1 State Variables of an SSL Session

An SSL session is uniquely identified with the following six state variables:

- A *session ID*, which is an arbitrary sequence of 32 octets that the server selects to identify an active session or a session that can be reactivated.
- The *peer certificate*, which is that of the correspondent, and it conforms to Version 3 of X.509. The value of the certificate can be null if the correspondent is not certified or if the Fortezza algorithm was used.
- The *compression method* — SSL allows the possibility of negotiating a data compression method. For the time being, no method was selected and, as a consequence, this field remains blank.
- The *cipher spec*, which defines the encryption and hash algorithms used out of a preestablished list.
- The *MasterSecret*, with a size of 48 octets, is shared between the client and the server. This parameter is used to generate all other secrets; therefore, it remains valid for the whole session.
- A flag denoted *is resumable* to describe whether the session can be used to open new connections.

Five elements define a cipher suite:

- The type of encryption, whether it is a stream code or a block code.
- The algorithm of encryption — the choice can be made among RC2, RC4, DES with a key of 40 bits or of 64 bits, or the Fortezza algorithm (at one time, a classified algorithm in the U.S.). Note that it is also possible not to encrypt the exchanges.
- The hashing algorithm that can be either MD5 or SHA — it is also allowed to choose no hashing at all.
- The size of the digest.
- A binary value indicating the permissibility to export the encryption algorithm according to U.S. law on export of cryptography

Negotiation of a cipher suite is done in the open during session establishment. Table 5.3 gives the ensemble of algorithms supported by SSL. The recognized cipher suites are presented in Table 5.4.

TABLE 5.3

Algorithms Negotiated by the Handshake Protocol

Function	Algorithm
Key exchange	RSA, Fortezza, Diffie–Hellman
Stream symmetric encryption	RC4 with keys of 40 or 120 bits
Block symmetric encryption	DES, DES40, 3DES RC2, IDEA, Fortezza
Hashing	MD5, SHA

TABLE 5.4

Recognized Cipher Suites in SSL

Key Exchange	Symmetric Encryption	Hashing	Signature
RSA	Without encryption	MD5 or SHA	—
	RC4-40	MD5	—
	RC4-128	MD5 or SHA	—
	RC2 CBC 40	MD5	—
	IDEA CBC	SHA	—
	DES40 CBC	SHA	—
	DES CBC	SHA	—
	3DES EDE CBC	SHA	—
Diffie–Hellman	DES40 CBC	SHA	DSS or RSA
	DES CBC	SHA	DSS or RSA
	3DES EDE CBC	SHA	DSS or RSA
Diffie–Hellman ephemeral	DES40 CBC	SHA	DSS or RSA
	DES CBC	SHA	DSS or RSA
	3DES EDE CBC	SHA	DSS or RSA

Note: SSL allows the exchange of keys with Diffie–Hellman without authentication of the parties ("Anonymous Diffie–Hellman"). This mode is vulnerable to man-in-the-middle attacks, which makes it unsuitable for commercial applications.

5.2.1.2   State Variables for an SSL Connection

The parameters that define the state of a connection during an SSL session are those that will be "refreshed" when a new connection is established. These parameters are as follows:

- Two random numbers (server_random and client_random) of 32 octets each: These numbers are generated by the server and the client, respectively, at the establishment of a session and for each new connection. The secret key will be derived using these random numbers, which means that these numbers are exchanged in the clear at the opening of a session. In contrast, during the establishment of an additional connection when the session is active, the process of encryption is fully functioning, and the numbers are transmitted encrypted. The use of these numbers protects against replay attacks using ancient messages.

- Two secret keys, server_MAC_write_secret and client_MAC_write_secret: These keys will be employed within the hash functions to calculate the message authentication code, or MAC. The size of the MAC depends on the hash algorithm used, which will be 16 octets for SHA or 20 octets for MD5.

- Two keys for the symmetric encryption of data, one for the server side and the other for the client side: While the same algorithm is used by both parties, each can use its own key, server_write_key or client_ write_key, respectively, provided that they share it with the other side. The key size depends on the encryption algorithm selected and the legislation on cryptography.

- Two initialization vectors for symmetric encryption in the Cipher Block Chaining (CBC) mode: One vector is for the server side and the other for the client side. Their sizes depend on the selected algorithms.

- Two sequence numbers, one for the server and the other for the client, each coded over 8 octets: These sequence numbers are now maintained separately for each connection and are incremented whenever a message is sent on this connection. This mechanism offers some protection against replay attacks because it prevents the reuse of already emitted messages.

Each connection has its own cryptographic parameters (keys and initialization vectors), but all connections of the same session share the MasterSecret. In addition, the confidentiality keys for each direction remain independent of each other.

5.2.2 Synopsis of Parameters Computation

Illustrated in Figure 5.4 is the computation of MasterSecret, starting from the PreMasterSecret and the parameters client_random and server_random. The value of the MasterSecret will remain constant throughout the session. The parameters client_random and server_random are exchanged in the clear, while the PreMasterSecret is exchanged confidentially with the help of the key exchange algorithm. The computation of the key from encrypting the hash key and the intialization vectors begins from the variables MasterSecret, client_random, and server_random, in the manner depicted in Figure 5.5.

The opening of a new connection will lead to the recalculation of the variables client_random and server_random, although the value of the MasterSecret remains unchanged. As a consequence, the variables client_write_key, respectively, server_write_key will be recomputed at the opening of a connection.

Note that for each new connection, each of the communicating entities uses a symmetric encryption key different from its partner. Thus, a flow in one direction is encrypted with a different key from corresponding flow in the opposite direction.

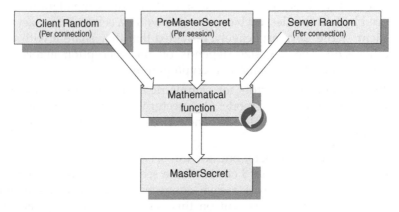

FIGURE 5.4
Construction of the MasterSecret at the start-up of a session.

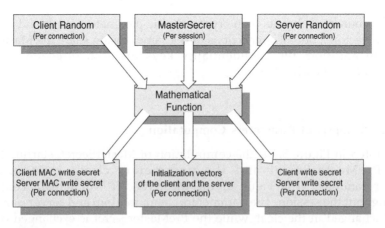

FIGURE 5.5
Generation of the secrets and the initialization vectors at the start-up of a session or a connection.

5.2.3   The Handshake Protocol

5.2.3.1   General Operation

The Handshake protocol begins with the mandatory authentication of the server, authentication of the client being optional. Once the authentication is established, both parties move to the negotiation phase to select the cipher suite that will be used throughout the session. Thus, the Handshake protocol conditions the whole process of secure data transfer, which makes it a prime target for potential aggressors.

Shown in Table 5.5 is a chronological list of the messages of the Handshake protocol and their significance. Figure 5.6 illustrates the exchanges during the establishment of a session. The structure of the protocol messages is presented in Appendix 5.1.

5.2.3.2   Opening of a New Session

A session that the client initiates begins with the transmission of the Client-Hello message to the server. The server can also take the initiative by sending the HelloRequest. This message does not contain any information and is used only to alert the client that the server is ready. The subsequent exchanges take place in the same manner, irrespective of the way the session started.

According to SSL Version 3.0 procedures, the exchanges that take place during the opening of a new session include the following four stages:

*   Identification of the cipher suites available at each site
*   Authentication of the server
*   Exchange of secrets
*   Verification and confirmation of the exchanged messages

5.2.3.3   Identification of the Cipher Suites

The first messages exchanged between the client and the server, ClientHello and ServerHello, allow the negotiation of the encryption algorithms and the characteristic secrets of the session. Both the client and the server must begin by choosing the version of the SSL protocol that is common to both parties (Version 3.0 or the default, Version 2.0). The ClientHello message, even if it specifies Version 3.0, is encapsulated in a Record message in the format defined by Version 2.0. This feature gives servers the possibility of responding even if they have not been upgrades from Version 2.0 to 3.0.

After sending the ClientHello message, the client waits for the arrival of the ServerHello message. This message must indicate the version of the protocol and the unique session identifier that the server selected. The session identifier will be used to resume an already established session, as will be shown later. The message also contains the random number server_random. By exchanging the random numbers, client_random and

TABLE 5.5

The Messages of the Handshake Protocol in Chronological Order

Message	Message Type	Direction of Transmission	Meaning
HelloRequest	O	Server → Client	Notice to the client to begin the Handshake
ClientHello	M	Client → Server	This message contains: The version of the SSL protocol The random number Client_random The session identifier: session_ID The list of cipher suites that the client selects The list of compression methods that the client selects
ServerHello	M	Server → Client	This message contains: The version of the SSL protocol The random number: Server_random The session identifier: session_ID A cipher suite A compression method
Certificate	O	Server → Client Client → Server	This message contains the server's certificate or the client's certificate if the client has one and the server has requested it
ServerKeyExchange	O	Server → Client	The server sends this message if it does not have a certificate or owns only a signature certificate
CertificateRequest	O	Server → Client	The server sends this message to request the client's certificate
ServerHelloDone	M	Server → Client	This message informs the client that the transmission of the ServerHello and subsequent messages have ended
ClientKeyExchange	M	Client → Server	Message containing the PreMasterSecret encrypted with the server's public key
CertificateVerify	O	Client → Server	Message allowing the explicit verification of the client's certificate
Finished	M	Server → Client Client → Server	This message indicates the end of the Handshake and the beginning of data transmission, protected with the newly negotiated parameters

Note: M = mandatory; O = optional.

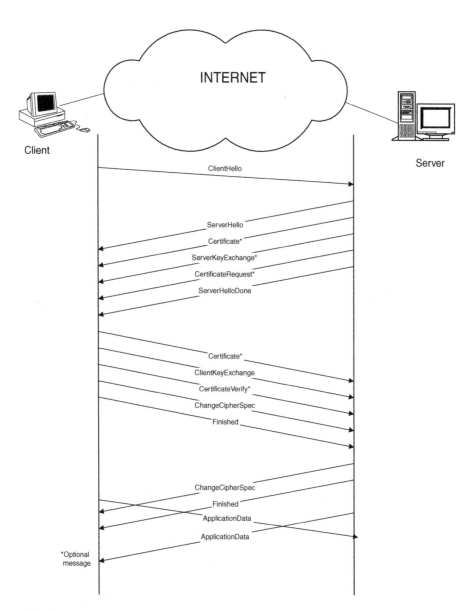

FIGURE 5.6
Messages exchanged during the establishment of a new session. (From Sherif et al., *Proc. ISCC'98*, pp. 353–358, ©1998 IEEE. With permission.)

server_random, each party will be able to reproduce the secrets of its correspondent, thereby sharing the secrets.

The message must include the cipher suite that the server retained from among the choices the client suggested. This suite will be used to ensure the confidentiality and the integrity of the data. If no common suite is available, the server (or the client, according to the case) will generate the error message close_notify as specified by the Alert protocol, and the session will be dropped. Note, however, that some servers do not send the message close_notify as required but close the TCP session without warning.

It is important to remember that this negotiation takes place in the clear. An intruder may attempt to intercept the ClientHello message and replace it by a fake message with less-robust algorithms. Protection against this attack can be achieved with the Finished message that is exchanged at the end of the Handshake, as will be shown later.

5.2.3.4   *Authentication of the Server*

In this second phase, the server authenticates itself by sending the Certificate message with the following:

- The Version 3 X.509 certificate of the server, which includes the public key of the cipher suite previously selected (The protocol Alert will indicate an error if the certificate is not included.)
- The certification path and the certificate of the certification authority

As a reminder, an X.509 certificate has the following form:

$$\text{Certificate} = \text{Name}_{\text{Server}} + \text{Name}_{\text{CA}} + \text{Pk}_{\text{Server}} + \text{SK}_{\text{CA}}\{H[\text{Name}_{\text{Server}} + \text{Name}_{\text{CA}} + \text{Pk}_{\text{Server}}]\}$$

where: $\text{Name}_{\text{Server}}$ = the server name;
$\quad\quad\quad\text{Name}_{\text{CA}}$ = the name of the certification authority;
$\quad\quad\quad\text{PK}_{\text{Server}}$ = the server public key;
$\quad\quad\quad\text{SK}_{\text{CA}}\{x\}$ = the signature of x with the private key of the certification authority; and
$\quad\quad\quad H(y)$ = the digest of y.

If the server does not have a certificate, then instead of the Certificate message, it transmits the ServerKeyExchange message. However, the servers used in e-commerce have an interest in being authenticated.

Even if a server presents a certificate, it may be obliged to transmit the ServerKeyExchange message under the following conditions:

- The Fortezza key exchange algorithm is used: In this case, the Server-KeyExchange message is sent without a signed key, because it is provided in the certificate. The message contains a random value that is used as part of the KEA key agreement process.
- The RSA key exchanged is ephemeral: The X.509 certificate is for signature and the ServerKeyExchange message contains the ephemeral public key of the server and its signature to verify the integrity of the exchanged key.
- The ephemeral Diffie–Hellman method is used for key exchange: In this case as well, a signed key (usually with the DSA algorithm) is

included in the ServerKeyExchange message to verify its integrity before establishing the joint key.

Thus, for dynamic authentication (i.e., a temporary key), the server must have a signature certificate, and the server transmits two consecutive messages: the Certificate message, then the ServerKeyExchange message. The Certificate message includes the public key for signature of the server. The client verifies the certificate with the public key of the certification authority. The ServerKeyExchange message contains the public parameters of the algorithm used to exchange the secret key for symmetric encryption. The server signs these parameters using the private key that corresponds to the public key already recovered in the first step. Upon receiving this second message, the client ensures that the public parameters of the key exchange algorithm are those of the server, by verifying the signature using the public key already received. This message contains the digest of the concatenation of the variables client_random and server_random, as well as the other server parameters. Hashing is done by the chosen algorithm, either MD5 or SHA.

After authenticating itself, the server can ask the client to do the same. The server next sends the CertificateRequest message that contains a list of the types of certificates requested, arranged in the order of the server's preference for the certification authorities.

Following these messages, the server sends the ServerHelloDone message that signifies to the client that it is finished and that it is waiting for a response.

5.2.3.5   Exchange of Secrets

If the server asks the client to authenticate itself using the CertificateRequest message, the client must respond by including its certificate, if it has one, in the Certificate message. If the client is not able to give a certificate, the answer will be no_certificate. This is merely a warning and not a fatal error, unless the server requires client authentication, in which case, the connection will be interrupted. Otherwise, user authentication will take place at the application level with a login and a password.

The client sends next the ClientKeyExchange message with the content depending on the algorithm used for key exchange and the type of certificate, according to the following:

- If the RSA algorithm is used, the PreMasterSecret is encrypted, either with the public key of the server or with a temporary key contained in the ServerKeyExchange message, as explained earlier.

- If the key exchange is according to the Diffie–Hellman, the message ClientKeyExchange contains the public key that the client sends to the server. Each site can perform separately the necessary computations that will yield a shared secret, with a value of PreMasterSecret.

- If the key exchange is based on the Fortezza algorithm, a Token Encryption Key (TEK) is calculated from the parameters that the server sent in the ServerKeyExhange message. This key will then be used to encrypt the client key, client_write_key, the secret PreMasterSecret, and the initialization vectors..

If the client has a certificate for digital signature, it will then send an explicit confirmation with CertificateVerify message. This message contains the digest of all previous messages starting from ClientHello, with the exception of the container message. This digest is encrypted with the client's private key. The purpose of this message is to give the server the ability to authenticate the client by verifying its signature. It is not sent in Fortezza's Handshake.

Accordingly, the content of the CertificateVerify message will be as follows:

hash{MasterSecret + pad_2 + hash(messages_already_sent_but_this_one + MasterSecret + pad_1)}

where hash is either MD5 or SHA, depending on the case. The fields pad_1 and pad_2 contain, respectively, 48 repetitions of the octets 0x36 and 0x5C in the case of MD5 and 40 repetitions for SHA.

The client then invokes the protocol ChangeCipherSpec to start the encryption of the exchanges with the choices made in the previous two phases.

The client immediately sends the Finished message. This message is a hash of all the Handshake messages that the client sent, starting with ClientHello; the hash uses the cryptographic attributes just negotiated. The purpose is to foil all man-in-the-middle attacks by verifying the integrity of the ensemble of exchanges.

The hash is calculated with the following formula:

hash{MasterSecret + pad_2 + hash(handshake_messages + Sender + MasterSecret + pad_1)}

where hash is either MD5 or SHA. The field handshake_messages contains the exchanged messages with the exception of the ChangeCipherSpec and the current message. Note the presence of the field Sender, which contains the identity of the sender.

Having transmitted the Finished message, the client begins sending the encrypted application messages without waiting for an acknowledgment.

Note that the message ChangeCipherSpec is not included in the computation, because it is not part of the Handshake protocol. For security reasons, it is preferable not to process the received Finished message unless it follows the ChangeCipherSpec message (Wagner and Schneier, 1996). Also, the MasterSecret is included in the MAC of CertificateVerify, ChangeCipherSpec, and Finished.

5.2.3.6 Verification and Confirmation by the Server

Upon receipt of the Finished message, the server attempts to reproduce the same hash with the message that it previously received. It compares the result with the content of the Finished message that just arrived from the client. This step will allow detection of any intruder that would have intercepted and modified the messages.

The server sends, in turn, the message ChangeCipherSpec and Finished. Here, again, the Finished is generated from all the messages that were sent and transmitted encrypted. The server starts to send the application data.

The transmission of the data just after sending the Finished message can open the following security hole. An intruder can modify either the ClientHello message or the ServerHello message to force the use of a weak cipher suite, recover the PreMasterSecret, and divert the Finished message. In this way, the intruder can substitute for one or both parties. To defend against this attack, it is sufficient to wait and not begin data transmission until the Finished messages are exchanged from both sides (Mitchell et al., 1998).

5.2.3.7 Summary: Session Establishment

Figures 5.7 and 5.8 show, respectively, the global state machines of the client and the server during the establishment of a session. The notation used in these figures is the one commonly used in the area of formal description techniques of communications protocol. Thus, "processA!message1" means that processA is sending message1 and "processB?message2" shows that processB received message2. The process SSL has two interfaces, one with the applications and the other with TCP transmission protocol; the latter is the one represented in the figures.

5.2.3.8 *Connection Establishment*

Establishment of the first SSL connection is the same as the establishment of a session as explained above. If an SSL was already established, TCP flows can transit in both directions. Thus, establishment of a new connection consists of refreshing the parameters client_random and server_random with the ClientHello and ServerHello messages, while preserving the encryption and hashing algorithms already selected. A new authentication is avoided, and, contrary to what happens for session establishment, the ClientHello and ServerHello messages are encrypted. Figure 5.9 depicts the exchanges.

The ClientHello message contains the session identifier of the session that will carry the connection. Should this identifier be absent from the server's tables, either because it is incorrect or because the session that it refers to has expired, the client is not rejected, and the server starts a new complete Handshake to establish a new session.

The client and the server confirm their agreement by sending the ChangeCipherSpec message from each side and end the abbreviated Handshake with the Finished message as before.

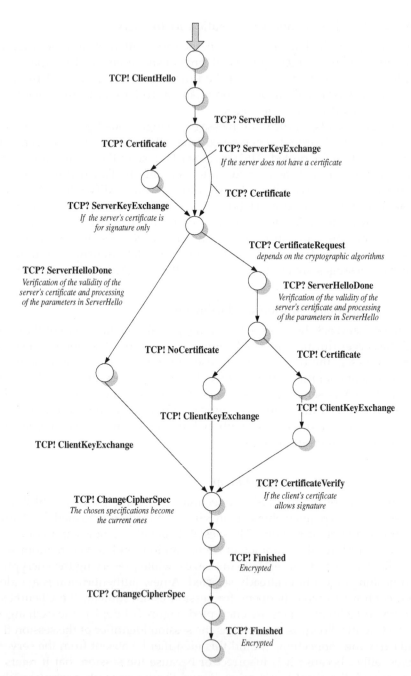

FIGURE 5.7
Session establishment as seen by the client's state machine.

The specifications are not clear concerning the appropriate action if the new exchanges introduce a new cipher suite. Similarly, it is not clear if in a

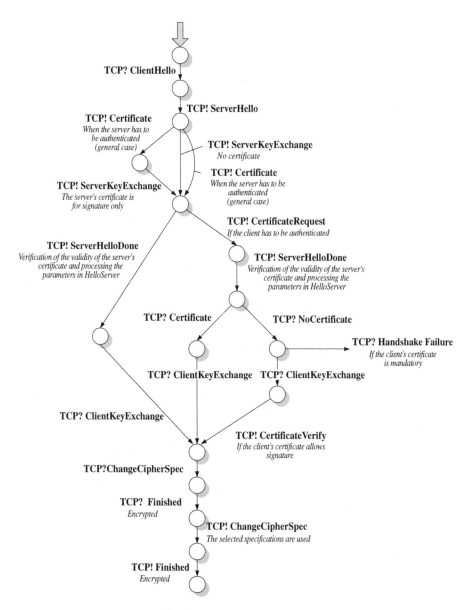

FIGURE 5.8
Session establishment as seen by the server's state machine.

session previously established with Version 3.0, the new connection can be of Version 2.0. However, downgrading the version can degrade the security due to the weaknesses of Version 2.0. Therefore, it is not recommended. In fact, some attacks can be initiated by forcing such a downgrade (Rescorla, 2001, pp. 137, 308).

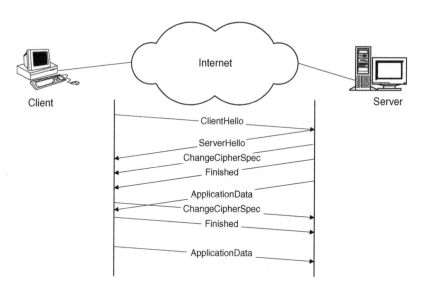

FIGURE 5.9
Exchanges during a connection establishment.

5.2.4 The ChangeCipherSpec Protocol

The ChangeCipherSpec (CCS) protocol consists of a single message that has the same name as the protocol and whose size is 1 octet. It indicates to the protocol Record that the encryption can start with the cryptographic algorithms that were negotiated. Before this message, the encryption was the task of the Handshake protocol. Following receipt of this message, the protocol Record layer on the transmit side will have to modify its write attributes, i.e., the method of encrypting the sent messages. On the receive side of the far-end entity, the protocol Record layer will modify its read attributes to be able to decipher the received messages.

5.2.5 The Record Protocol

As seen above, the Record protocol intervenes only after transmission of the ChangeCipherSpec message. During the session establishment, the role of the protocol Record layer is to encapsulate the Handshake data and to transmit them without modification toward the TCP layer. During the data encryption phase, the Record protocol receives data from the upper layers (Handshake, Alert, CCS, HTTP, FTP, etc.) and transmits them to the TCP after performing, in order, the following functions:

1. Segmentation of the data in blocks of maximum size of 214 octets
2. Data compression, a function considered but not supported in the current version of the specifications

3. Generation of the digest to ensure the integrity service

4. Data encryption to ensure the confidentiality service

The Record protocol adds a header of 5 octets to each message received from the upper layers. This header indicates the message type according to its origin: subprotocols Handshake, Alert, CCS, or application data, such as HTTP and FTP. It signals, as well, the version of the SSL protocol used and the length of the encapsulated data blocks. Computations for the MAC associated with the integrity service code include the sequence number for each fragment sent to detect potential replay or reordering attack. It is not possible, however, to correct the effects of these attacks because the Record protocol receives a field dedicated to transport these numbers.

The inverse tasks are performed on the receive side: decryption, integrity verification, decompression, and reassembly. If the computed digest is not identical to the one that was received, the Record protocol invokes the Alert protocol to relay the error message to the transmit unit.

The operation of the Record layer is depicted in Figure 5.10.

5.2.6 The Alert Protocol

The role of the Alert protocol is essentially to generate alarm messages after any errors and to signal the change of states, such as the closing of a con-

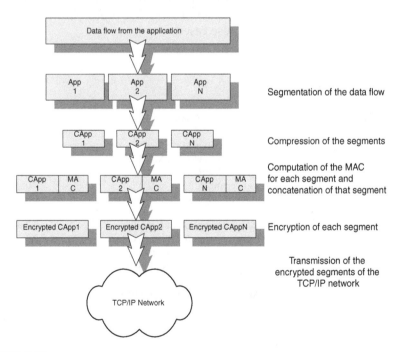

FIGURE 5.10
Functional representation of the Record layer.

nection. Like all the messages coming from the upper layers, the Record layer encrypts the messages using the encryption attributes in place. Depending on the seriousness of the threat, the alarm can be a simple warning or can cause disconnection of the session. A warning message is for caution and does not require a specific action. In contrast, a fatal message forces the transmit side to close the connection immediately, without waiting for an acknowledgment from the other party. From its side, the receiver will close the connection as soon as the alarm message arrives. Because "fatal" messages force session disconnects, SSL is vulnerable to denial-of-service attacks should an intruder succeed in substituting the canonical messages with nonconforming messages that would provoke session disconnects.

The Alert protocol can be invoked in one of the following ways:

- By the application, for example, to indicate the end of a connection
- By the Handshake protocol if it encounters a problem
- By the Record protocol directly, for example, if the integrity of a message is in question

Table 5.6 lists the messages of the protocol Alert arranged in alphabetical order.

Notice that there are no messages related to differences in the choice of cipher suites in a session and a connection that it carries or between a resumed session and the initial session. In the TLS protocol, however, new messages are defined to cover these cases.

TABLE 5.6

Messages of the Alert Protocol

Message	Context	Type
bad_certificate	Failure of a certificate verification	Fatal
bad_record_mac	Reception of an incorrect MAC	Fatal
certificate_expired	Expired certificate	Fatal
certificate_revoked	Revoked certificate	Fatal
certificate_unknown	Invalid certificate for other reasons than the above	Fatal
close_notify	Voluntary interruption of a session	Fatal
decompression_failure	The decompression function received improper data (data too long)	Fatal
handshake_failure	Inability to negotiate common parameters	Fatal
illegal_parameter	A parameter in the Handshake is out of range or is inconsistent with other parameters	Fatal
no_certificate	Negative response to a certificate request	Warning/Fatal
unexpected_message	Inopportune arrival of a message	Fatal
unsupported_certificate	The received certificate is not supported	Warning/Fatal

5.2.7 Summary

The tasks accomplished by the various SSL subprotocols are illustrated in Figure 5.11.

5.3 Example of SSL Processing

This section presents a detailed analysis of the processing of an SSL transaction to illustrate the preceding discussion. This example will allow the

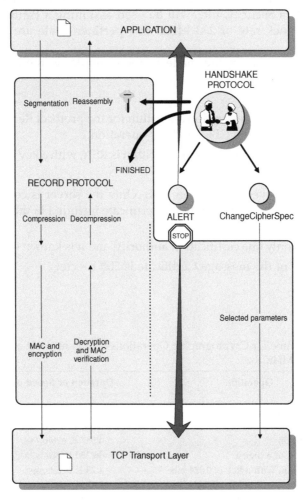

FIGURE 5.11
Functions of the SSL subprotocols.

evaluation of the cryptographic computation load of SLL for the following cases:

- Session establishment
- Connection establishment
- Application processing

5.3.1 Assumptions

The performance of SSL depends on several factors, particularly the performance of the processor used, the type of cipher suite selected, as well as the length of the certification path. Results of the benchmarks BSAFE4.0 from RSA Data Security, Inc., will be used assuming a Pentium II™ from Intel at the clock rate of 233 MHz. The pertinent data are presented in Table 5.7.

The following choices are made in the example on the analysis of the cryptographic load SSL:

- The symmetric encryption algorithm for the protocol Record is DES, with 64 bits in the CBC mode of operation.
- The asymmetric encryption algorithm is RSA, with a key of 1024 bits.
- The hash function is MD5.
- There is a level certification path. Only the server is certified, and the exchanges contain only the certificate included in the Certificate message.
- There is only one certification authority, and it is known to the client.
- The size of the message Certificate is 500 K octets.

TABLE 5.7

Processing Time for Cryptographic Operations on a Pentium II at the Clock Rate of 233 MHz

Operation	Duration or Speed of Execution
DES: Key generation	6.3 µsec
DES: Encryption	4,386 K octets/sec
DES: Decryption	4,557 K octets/sec
MD5: Creation of a digest	36,250 K octets/sec
RSA: Encryption with a key of 1024 bits	4.23 K octets/sec
RSA: Decryption with a key of 1024 bits	2.87 K octets/sec
SHA: Creation of a digest	20,428 K octets/sec

5.3.2 Establishment of a New Session

5.3.2.1 Message Size

All the Handshake messages have a header of 4 octets that describes the message type coded over 1 octet and the length coded over 3 octets. The ClientHello and ServerHello messages are sent in the clear during a request for session opening and are not encrypted. These messages contain the following fields:

- The protocol version coded over 2 octets, the first of which indicates the version and the second the subversion.
- The random numbers client_random (server_random), each with length of 32 octets. They consist of two fields: the first has 4 octets and contains the time as indicated by the client's internal clock and expressed in universal time. The size of the second field is 28 octets, and it contains a random number by a random number generator.
- The session identifier, which has a variable length between 0 and 32 octets and is preceded by a field that defines its length. During session establishment, the total length is 1 octet because the session identifier is null, while for connection establishment, the total length is 33 octets.
- The cipher suite list that contains the list of suites the client can accept or that the server selected. Every suite is coded over 2 octets. Given that SSL accepts up to $2^{15} - 1$ different cipher suites, the list length will not exceed $2^{16} - 2$ octets, and 2 octets are sufficient to code the list length. Thus, the list is preceded by a field of 2 octets that defines the size of the cipher list in octets. A client can propose several cipher suites at the beginning of a session. To open a connection within an established session, the client is confined to the cipher suite used in the session at hand.
- The list of the compression methods that the client can accept. Each of the methods listed is coded over 1 octet. SSL can accept up to 255 distinct methods; however, no method has yet been standardized.

Summarized in Table 5.8 are the previous sizes.

5.3.2.2 ClientHello Message

The length of the ClientHello message for the first opening of a session is

$$L = 2 + 32 + 1 + 4 + 1 = 40 \text{ octets}$$

to which the Handshake protocol overhead is added, to give a total size of

TABLE 5.8

Field Sizes in the ClientHello and ServerHello Messages

Field Name	Maximum Field Size (octets)	Payload Length in the Example (octets)	Number of Octets where the Length is Coded	Total Length (octets)
Session ID	32	32	1	33
Cipher suite	65,534	2	2	4
List of compression methods	255	0	1	1

$$L' = 40 + 4 = 44 \text{ octets}$$

5.3.2.3 ServerHello Message

The size of the ServerHello message is

$$L = 2 + 32 + 33 + 4 + 1 = 72 \text{ octets}$$

With the Handshake overhead, the total length is

$$L' = 70 + 4 = 76 \text{ octets}$$

Because both messages are sent in the clear, their processing times are negligible.

5.3.2.4 Certificate Message

In this example, the certification path has only one level. The client, therefore, merely has to verify the signatures of the server and the certification authority. The X.509 certificate takes the following form:

$$\text{Certificate} = \text{Name}_{\text{Server}} + \text{Name}_{\text{CA}} + \text{Pk}_{\text{Server}} + \text{RSA}_{\text{Sk(CA)}} \{\text{MD5}\{\text{Name}_{\text{Server}} + \text{Name}_{\text{CA}} + \text{Pk}_{\text{Server}}\}\}$$

where $\text{RSA}_{\text{Sk(CA)}}\{ \}$ represents RSA encryption with the private key of the certification authority, $S_{K(CA)}$. The client has to do the following:

1. Decrypt with the public RSA key $\text{Pk}_{(AC)}$ the signature of the certification authority. The digest of the MD5 algorithm is 16 octets; therefore, this operation will take

$$t_1 = 16 \text{ octets}/4.23 \text{ K octets/sec} = 3.78 \text{ msec}$$

2. Calculate the hash of $(Name_{Server} + Name_{CA} + Pk_{Server})$. In the conditions of the example, this computation will take

$$t_2 = 500 \text{ octets}/20{,}428 \text{ K octets/sec} = 0.025 \text{ msec}$$

Upon reception of the certificate from the server, the client has to verify the two signatures, and this operation will take

$$t_3 = 2 \times (3.78 + 0.025) = 7.61 \text{ msec}$$

5.3.2.5   ClientKeyExchange Message

The size of the parameter PreMasterSecret is 48 octets. Its encryption by the client with the RSA algorithm requires

$$t_4 = 48 \text{ octets}/4.23 \text{ K octets/sec} = 11.34 \text{ msec}$$

In contrast, the decryption by the server takes

$$t_5 = 48 \text{ octets}/2.87 \text{ K octets/sec} = 16.72 \text{ msec}$$

The total duration is

$$t_6 = 28.06 \text{ msec}$$

5.3.2.6   *Calculation of the Cipher Suite*

Once the PreMasterSecret is exchanged, the client and the server will compute the MasterSecret:

MasterSecret =

MD5 (PreMasterSecret + SHA ('A' + PreMasterSecret + client_random + server_random)) +

MD5 (PreMasterSecret + SHA ('BB' + PreMasterSecret + client_random + server_random)) +

MD5 (PreMasterSecret + SHA ('CCC' + PreMasterSecret + client_random + server_random))

The size of the field

'A' + PreMasterSecret + client_random + server_random

('A' represents the character *A*) is 113 octets. As a consequence, the time necessary to generate the digest with the SHA algorithm is

$$t_7 = 113 \text{ octets}/20{,}428 \text{ K octets/sec} = 0.006 \text{ msec}$$

Next, MD5 hashing is applied to the field of length 68 octets defined by

PreMasterSecret + SHA('A' + PreMasterSecret + client_random + server_random)

which requires

$$t_8 = 68 \text{ octets}/36{,}250 \text{ K octets/sec} = 0.002 \text{ msec}$$

Being repeated three times, the total time necessary to calculate the MasterSecret (by either the client or the server) is

$$t_9 = 3 \times (0.006 + 0.002) = 0.024 \text{ msec}$$

Once the MasterSecret is obtained, the encryption key is deduced from the following encryption block:

KeyBlock =

MD5 (MasterSecret + SHA('A' + MasterSecret + server_random + client_random)) +

MD5 (MasterSecret + SHA ('BB' + MasterSecret + server_random + client_random)) +

MD5 (MasterSecret + SHA ('CCC' + MasterSecret + server_random + client_random)) + [...]

The computation is repeated as many times as necessary for the encryption block to be large enough to extract the following elements:

```
client_MAC_write_secret[hash_size]
server_MAC_write_secret[hash_size]
client_write_key[key_length]
server_write_key[key_length]
client_write_IV[size_of_ the initialization vector]/*
For Fortezza only. It cannot be exported from the U.S.*/
server_write_IV[size_of_ the initialization vector]]/*
For Fortezza only. It cannot be exported from the U.S. */
```

As for the cipher suite, it includes the DES algorithm with a key of 64 bits and the MD5 algorithm (nonexportable version). Thus, it is necessary to have

$$2 \times (16 + 8 + 8) = 64 \text{ octets}$$

Four operations of the following type are needed:

MD5 (MasterSecret + SHA (A_ + MasterSecret + server_random + client_random))

From the results calculated during the construction of the *MasterSecret*, the time to create the decryption key is, therefore,

$$t_{10} = 4 \times 0.008 \text{ msec} = 0.032 \text{ msec}$$

The total time for computation of a cipher block is, therefore,

$$t_{11} = t_9 + t_{10} = 0.024 + 0.032 = 0.056 \text{ msec}$$

5.3.2.7   ServerHelloDone Message

The server sends this message to the client to indicate that it is expecting a response. This message has no content.

5.3.2.8   Finished Message

The role of this message is to verify that the key exchange and the authentication were performed correctly. Each party verifies the integrity of the exchanges by applying a hash function to all the transmitted and received messages in addition to the MasterSecret parameter.

The length of the Finished message is 36 octets, and it contains two digests: the first results from the MD5 function (16 octets) and the second from the SHA function (20 octets). Each digest is calculated with the help of the following formula:

hash{MasterSecret + pad_2 + hash(handshake_messages + Sender + MasterSecret + pad_1)}

where hash is either MD5 or SHA, depending on the case. The field handshake_messages comprises all the messages exchanged during the Handshake. These messages are shown in Table 5.9 for the example studied. Note the absence of the ChangeCipherSpec message, because the latter is not a message of the Handshake protocol.

Note that the total lengths of the exchanged Handshake messages are included in the calculation of the Finished message and are different on each side. This is because the generation of the second Finished message has to

TABLE 5.9

Length of the Handshake Messages (for the Example
under Consideration)

Message	Length (octets)
ClientHello	44
ServerHello	76
ServerHelloDone	4
Server Certificate	Variable, on the average 500
ClientKeyExchange	52
Total	676

take into consideration the first. For session establishment, it is the client
that sends the first Finished message (see Figure 5.6); for connection establishment, it is the server (see Figure 5.9). This aspect was neglected in the
calculations of this example because its effect on the results is minimal.

The sizes of the Sender and MasterSecret fields are 4 and 48 octets, respectively. The fields pad_1 and pad_2 contain, respectively, 48 repetitions of the
octets $0x36$ and $0x5c$ for MD5, and 40 repetitions for SHA.

For MD5 hashing, the first hashing operation takes

$$t_{12} = (676 + 4 + 48 + 48)/36{,}250 = 0.021 \text{ msec}$$

while the second corresponds to

$$t_{13} = (48 + 48 + 16)/36{,}250 = 0.003 \text{ msec}$$

Similarly, for SHA, the first hash operation corresponds to

$$t_{-12} = (676 + 4 + 48 + 40)/20{,}428 = 0.038 \text{ msec}$$

while the second takes

$$t_{-13} = (48 + 40 + 20)/20{,}428 = 0.005 \text{ msec}$$

Thus, the time for forming the Finished message is

$$t_{14} = (t_{12} + t_{-12} + t_{13} + t_{-13}) = 0.067 \text{ msec}$$

5.3.2.9   Processing at the Record Layer

The Finished message is transmitted to the Record layer, which calculates
the MAC and then encrypts the whole with the symmetric DES algorithm
with a key of 64 bits.

The MAC computation for the Finished message is done with the following
equation (see §7.2.3.1 of SSL specifications in Freier et al., 1996):

MAC = hash (MAC_write_secret + pad _2 + hash (MAC_write_secret + pad_1 + seq_num + length + content))

Here, hash represents the MD5 hash function. The other elements are as follows:

- MAC_write_secret is an element extracted from the KeyBlock, with a size of 16 octets for MD5.
- pad_1 and pad_2 are two fixed elements of 48 octets of length for MD5.
- seq_num is the sequence number of the exchanges of a connection coded over 8 octets.
- length is the length of the fragment coded on a maximum of 2 octets (in this example, 1 octet is enough).
- content is the transmitted data, of maximum length of 16,384 octets in the clear and of 17,408 octets compressed $(2^{14} + 2^{10})$ (for the message Finished, the size is 36 octets).

By applying the previous formula, the computation time for the MAC of the Finished message is found to be

t_{15} = {(16 + 48 + 8 + 1 + 36)/36,250 K octets/sec + (16 + 48 + 16)/36,250 K octets/sec} = 0.005 msec

The Finished and its MAC are encrypted with DES in

t_{16} = (36 + 16)/3241 K octets/sec + initialization time of the algorithm = 0.016 msec + 0.0063 msec= 0.022 msec

The total time for generating Finished is

$$t_{17} = 0.067 + 0.022 = 0.089 \text{ msec}$$

In the conditions of the example, the total time necessary for the cryptographic treatment of a session establishment is as follows:

$$\tau\text{session} = t_3 + t_6 + 2 \times (t_{11} + t_{17})$$

$$= 7.42 + 28.06 + 2 \times (0.056 + 0.089) = 35.96 \text{ msec} \approx 36 \text{ msec}$$

This time is divided between the client and the server in the ratio 19.068 ($\approx$19.1) msec for the client to 16.865 ($\approx$16.9) msec for the server.

5.3.3   Processing of Application Data

Once the Handshake is terminated, the application data are encrypted with the new security parameters as negotiated. These data will be subjected to four operations at the Record layer: fragmentation in blocks of maximum size of 16 K octets, compression, hashing, and finally, encryption with a symmetric algorithm.

5.3.3.1   MAC Computation and Encryption

Assume a block of size 16 K octets. Just as in the previous cases, the compression operation will not be considered because the algorithmsecaccepted by SSL were not determined. Assume that the hashing algorithm is MD5 and that the encryption algorithm is DES.

The computation of the MAC is done with the same formula:

MAC = hash (MAC_write_secret + pad _2 + hash (MAC_write_secret + pad_1 + seq_num + length + content))

The time taken to generate the MAC is, therefore,

$t_{15} = \{(16 + 48 + 8 + 2 + 16{,}384)/36{,}250$ K octets/sec + (16 + 48 + 16)/36,250 K octets/sec} = 0.452 msec

The encryption time for the fragment and its MAC is

$t_{16} = (16{,}384 + 16)$ octets/4386 K octets/sec = 3.74 msec

The time for the protocol Record to process the application data is, accordingly, 0.452 + 3.74 = 4.192 msec.

5.3.3.2   Decryption and Verification of the Data

The time to decrypt the encrypted data fragment (with the MAC included) is

(16,384 + 16) octets/4557 K octets/sec = 3.74 msec

The time to verify that the MAC is identical to the time for its construction, i.e., 0.45 msec.

The time for decryption and verification of a message by the Record is, thus, 4.19 msec.

The time to process a message of 16 K octets is

$$T = 4.192 \text{ msec} + 4.19 \text{ msec} = 8.382 \text{ msec}$$

5.3.4 Connection Establishment

Consider two cases:

1. The connection is established in an existing session.
2. The connection establishment requires resumption of a suspended session.

5.3.4.1 *Connection Establishment in an Existing Session*

The lengths of the ClientHello and ServerHello messages are 44 and 76 octets, respectively. These messages will be considered as normal data and treated as such by the Record layer using the parameters defined for the existing session.

5.3.4.1.1 *MAC Computation and Encryption*

By substituting with the appropriate values in the MAC formula:

MAC = hash (MAC_write_secret + pad _2 + hash (MAC_write_secret + pad_1 + seq_num + length + content))

The computation time of the MAC for ClientHello is

$t_{15} = \{(16 + 48 + 8 + 1 + 44)/36{,}250$ K octets/sec $+ (16 + 48 + 16)/36{,}250$ K octets/sec$\} = 0.005$ msec

The time for encrypting the message and its MAC is

$$t_{16} = (44 + 16) \text{ octets}/4386 \text{ K octets/sec} = 0.014 \text{ msec}$$

The processing time of ClientHello as application data by the Record protocol is thus 0.019 msec. The corresponding time for the ServerHello message, with a length of 76 octets, is 0.027 msec (0.006 + 0.021).

5.3.4.1.2 *Data Decryption and Verification*

The time for message decryption and verification (MAC included) is

$$(44 + 16) \text{ octets}/4557 \text{ K octets/sec} = 0.013 \text{ msec for ClientHello}$$

and

$$(76 + 16) \text{ octets}/4557 \text{ K octets/sec} = 0.02 \text{ msec for ServerHello}$$

The verification of the MAC takes the same time as its construction: 0.019 msec for ClientHello and 0.027 msec for ServerHello. This gives the time for decryption and verification as 0.032 msec for ClientHello and 0.047 msec for ServerHello.

The time for cryptographic computation for a new connection establishment is

$$tconnection = 0.024 + 0.034 + 0.037 + 0.055 = 0.125 \text{ msec}$$

which is distributed between the client and the server as follows: 0.066 msec for the client and 0.059 msec for the server.

5.3.4.2 Session Refresh

Refresh of a current session is done with a simplified Handshake, as shown in Figure 5.12

The ClientHello and ServerHello messages are sent in the clear. However, in contrast to the case of a session establishment, there is no need for authentication; as a consequence, there is no certificate verification.

There is a need to recalculate the KeyBlock starting from the current MasterSecret and to transmit a Finished message. This message contains all the exchanged HandShake messages. Its size will be much smaller than during the session establishment, because it does not contain a certificate.

To calculate the time for cryptographic processing of the Finished message, note that the size of the message becomes 76 + 76 = 152 octets. This is because

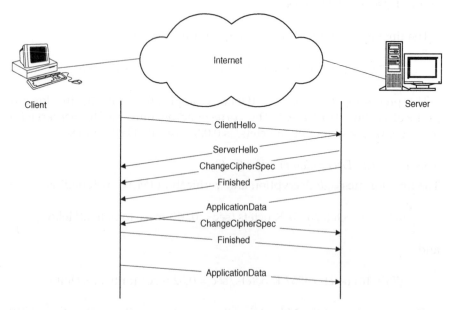

FIGURE 5.12
Simplified Handshake.

the ClientHello message must now contain the Session_ID, which brings the size of its content to

$$L = 2 + 32 + 33 + 4 + 1 = 72 \text{ octets}$$

With the Handshake header, the message length is

$$L' = 70 + 4 = 76 \text{ octets}$$

The length of the ServerHello message is also 76 octets. In this case, the values of t_{12} and t_{-12} become

$$t_{12} = (152 + 4 + 48 + 48) \text{ octets}/36{,}250 = 0.007 \text{ msec}$$

and

$$t_{-12} = (152 + 4 + 48 + 48) \text{ octets}/20{,}428 = 0.0123 \text{ msec}$$

The other times remain unchanged, i.e.,

$$t_{13} = (48 + 48 + 16) \text{ octets}/36{,}250 = 0.003 \text{ msec}$$

$$t_{-13} = (48 + 40 + 20) \text{ octets}/12{,}962 = 0.005 \text{ msec}$$

The cryptographic processing of the Finished message takes

$$t_{14} = (t_{12} + t_{-12} + t_{13} + t_{-13}) = 0.027 \text{ msec}$$

which is an order of magnitude less than in the case of a session establishment. In particular, this duration is less than the time needed to construct the KeyBlock, which is always $t_{11} = 0.056$ msec.

The total time to establish a connection is

$$\text{tconnection} = 2 \times (0.056 \text{ msec} + 0.027 \text{ msec}) \approx 2 \times 0.08 = 0.16 \text{ msec}$$

Shown in Figures 5.13 and 5.14 are the state machines of SSL for a session refresh from the client side and server side, respectively.

5.3.4.3 Summary

The results in Table 5.10 show that, for the example at hand, the cryptographic load for the client side is slightly higher than that on the server side. These computations were done for a fairly typical case, where only the server has a certificate that the client verifies. Even in this case, the overall load on the server will be considerable because it processes many simultaneous requests for session establishment. In general, the load on the server can be

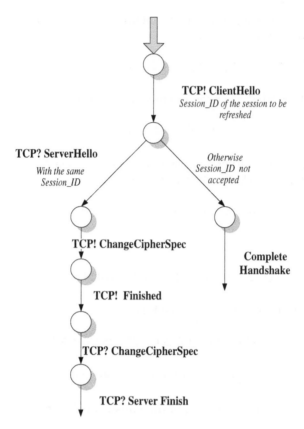

FIGURE 5.13
State machine for session resumption as seen from the client side.

twice or three times the load on the client; when the client has an authentication certificate, the client load will be quadrupled, while that of the server increases by about 20% (Rescorla, 2001, pp. 187–188). In particular, usage of the DSA signature algorithm with the key exchange through ephemeral Diffie–Hellman multiplies by a factor between five and seven the load on the server (Rescorla, 2001, p. 192).

5.4 Performance Acceleration

The performance of an e-commerce server is measured by three criteria: the number of simultaneous encrypted sessions (the number of transactions per second), the response time to requests, and the throughput available for each of these sessions. However, the SSL protocol is avid for computing resources due to the use of public key algorithms, particularly the ephemeral Dif-

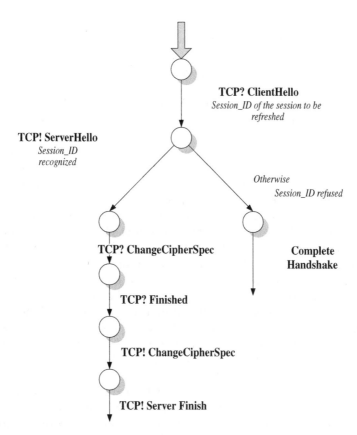

FIGURE 5.14
State machine for service resumption from the server side.

TABLE 5.10

Cryptographic Processing Times for SSL

Operation	Computation Time for the Client (msec)	Computation Time for the Server (msec)	Total (msec)
Establishment of a new session (Handshake)	19.1	16.9	36
Session refresh (simplified Handshake)	0.08	0.08	0.16
Establishment of a new connection	0.066	0.059	0.125
Computation time for 16 K octets of data (encryption, decryption, construction, and verification of the MAC)	4.2	4.2	8.4

fie–Hellman algorithm; this load increases with the number of simultaneous sessions saturating the server and leading to a substantial drop in the performance.

To protect the service quality, operators of e-commerce sites add to their servers *SSL accelerators* that relieve them of the cryptographic computations. These accelerators come either in the form of separate machines or as boards to be inserted in the server. Separate machines act as proxies between the server and the clients and can be organized in series or in parallel to increase the computing power by distributing the load over several machines. A cluster arrangement has an additional advantage: it can ensure recovery in case of failure by automatic replacement of the failed machine to another one in reserve. To succeed in this recovery, each machine must know the instantaneous states of all other connections involving machines of the same cluster (Rescorla, 2001, pp. 204–209; Rescorla et al., 2002). However, inter-machine communications add another computational surcharge that can mask the gain in speed if the number of machines per cluster exceeds three to four machines. In addition, because SSL is a point-to-point protocol, the proxy has to terminate the SSL link, i.e., decrypt the messages from the client to extract the information to be distributed to the other machines (or to a mirror machine) so as to ensure recovery in case of failure, before forwarding them to the server. Clearly, it becomes imperative to protect all these machines against potential physical or logical intrusions; otherwise, the whole security edifice will be affected.

The choice of an accelerator board is tricky: expected performance gains vary with the board, the server configuration, and the traffic profile. While the average response time improves, significant delays may remain in some situations. The occupancy rates of CPUs and, as a consequence, their responses to unexpected traffic peaks vary. The same can be said about their speeds, with which their performances saturate and their tolerance to failures increases. Their use requires a good study of the service environment before their insertions; otherwise, they may be useless if not damaging (Bontoux, 2002).

5.5   Implementations

There are many commercial offers for SSL, most of which use Version 3.0 of SSL (SSLv3), with RSA for key exchange and traffic encryption. In general, they comprise two modules: the first for the cryptographic functions and the second for the SSL protocol library. This software architecture allows modification of the properties of the composite SSL server, according to legal restrictions or technical choices.

The OpenSSL project is a collaborative effort to develop a commercial-quality implementation of the SSL and TLS protocols starting from the library that Eric A. Young and Tim J. Hudson created under the name of SSLeay. Participants are all European volunteers, mostly from Danemark, Germany, Sweden, and the U.K., which avoid U.S. restrictions on the export of encryption software. The code is available free of charge and can be downloaded from *http://www.openssl.org*.

The site *http://www2.psy.uq.edu.au/~ftp/Crypto/* provides information on SSLeay and SSL, such as a list of bugs discovered in some commercial implementations, which can be found at *http://www2.psy.uq.edu.au/~ftp/Crypto/ssleay/vendor-bugs.html*.

5.6 Summary

SSL supplies a relatively simple mechanism to protect exchanges between two points over TPC. The modular architecture of SSL allows the evolution of some parts of the protocol without affecting the whole structure. Therefore, it is possible to introduce new improved algorithms and to take into account the specific requirements of national legislations. The technical evolution of SSL ended finally in TLS, a standard from the IETF that will be presented in the next chapter.

During the Handshake, the client and the server negotiate the cipher suite and establish a shared secret (MasterSecret) and cryptology parameters. Data are segmented using the Record protocol; each fragment is encrypted individually with protection of its integrity. To protect against replay attacks or reordering attacks, computation of the MAC includes the sequence numbers of the fragments to be transmitted.

The main computational load in SSL comes from the cryptography, particularly during session establishment. The ephemeral Diffie–Hellman algorithm for key establishment is exceptionally costly and should be avoided if the perfect forward security is not absolutely necessary. Among the measures used for load reduction are session resumption to avoid a new Handshake and use of accelerators. The latter solution, however, creates some difficulties, because SSL is not adapted to multipoint relations. Thus, the presence of additional machines between the client and the server to carry out the cryptographic operations adds new security risks. The same can be said for applications that involve several actors simultaneously, such as a client, a merchant, and a gateway with banking networks. To take into account various intermediaries in business relations, several SSL associations must be established in parallel which would substantially increase the complexity of the operation. Another weakness, which is structural, is due to the fact that the encryption parameters are not necessarily updated during a session, so that the longer the session, the higher is the chance that the keys can be broken.

SSL, originally proposed by Netscape, is now available in many commercial or free offers. OpenSSL is such a free software developed in Europe to avoid the restrictions that the U.S. imposes on encryption software produced in their territories. Note that some countries limit the protection offered by SSL by limiting the key length or restricting the choice of encryption algorithms.

Questions

1. What are the main limitations of SSL?
2. What are the main differences between an SSL session and an SSL connection?
3. What are the advantages and disadvantages of the ephemeral mode of operation in SSL?
4. What makes the design of SSL accelerators so difficult?

Appendix 5.1: Structures of the Handshake Messages

A5.1 Messages of the Handshake

The SSL specifications concerning the messages of the Handshake. are provided here for reference. Each message contains fields that are either simple fields of the type uint8, uint16, uint24, or opaque, or are compound fields made up of simple fields. The element types uint8, uint16, and uint24 correspond to elements of 1, 2, and 3 octets, respectively. An element of the opaque type contains 1 octet of encrypted data.

According to the notations used in the SSL specifications, an array of constant length of element type Type_Simple will be written as follows:

$$\text{Type\_Simple } T[n]$$

where n is the number of octets of the array (and not the number of elements of type Type_Simple that are present). Thus, if Type_Simple is of type uint16 (i.e., formed of 2 octets), an array T containing two elements of type Type_Simple, will have a value n of 4.

Variable-length arrays are denoted as

$$\text{Type\_Simple } T <n_1...n_2>$$

where n_1 and n_2 are, respectively, the minimum and maximum number of octets in the array T.

All the messages exchanged throughout the Handshake protocol have a header with two fields:

- A message identifier coded over 1 octet
- A message length coded over 3 octets

A5.1.1 Header

```
struct {
  HandshakeType msg_type;
  uint24 length;
  select (HandshakeType){
```

```
        case hello_request: HelloRequest;
        case client_hello: ClientHello;
        case server_hello: ServerHello;
        case certificate: Certificate;
        case server_key_exchange: ServerKeyExchange;
        case certificate_request: CertificateRequest;
        case server_hello_done: ServerHelloDone;
        case certificate_request: CertificateRequest;
        case client_key_exchange: ClientKeyExchange;
        case finished: Finished;
    } body;
} Handshake;
enum {
    hello_request(0), client_hello(1), server_hello(2),
    certificate(11), server_key_exchange(12),
    certificate_request(13),
    server_hello_done(14), certificate_verify(15),
    client_key_exchange(16),
    finished(20), (255)
} HandshakeType;
```

A5.1.2 *HelloRequest*

```
    struct {} HelloRequest;
```

A5.1.3 *ClientHello*

```
    struct {
        uint32 gmt_unix_time;
        opaque random_bytes[28];
    } Random
    opaque SessionID<0..32>;
    uint8 CipherSuite[2];
    enum {null(0), (255)} CompressionMethod;
    struct {
        ProtocolVersion client_version;/* The version */
        Random random;/* Random number generated by the
                client */
```

```
      SessionID session_id;/* Session identifier
   CipherSuite cipher_suites <2..2^16-1>;/*Cipher suite
            that the client proposes */
   CompressionMethod compression_methods<1..2^8-1>;/
            *Compression method */
} ClientHello;
```

A5.1.4 *ServerHello*

```
struct {
    ProtocolVersion server_version;
    Random random;
    SessionID session_id;
    CipherSuite cipher_suite;
    CompressionMethod compression_method;
} ServerHello;
```

A5.1.5 Certificate

```
opaque ASN.1Cert<2^24-1>
struct {
    ASN.1Cert certificate_list<1..2^24-1>;
} Certificate;
```

A5.1.6 *ServerKeyExchange*

```
enum {rsa, diffie_hellman, fortezza_dms}
            KeyExchangeAlgorithm
struct {
    opaque RSA_modulus<1..2^16-1>;
    opaque RSA_exponent<1..2^16-1>;
} ServerRSAParams;
struct {
 public-key-encrypted PreMasterSecret pre_master_secret;
}EncryptedPreMasterSecret;
struct {
    opaque DH_p<1..2^16-1>;
```

```
        opaque DH_g<1..2^16-1>;
        opaque DH_Ys<1..2^16-1>;
} ServerDHParams;

struct {
    opaque r_s [128];
} ServerFortezzaParams

struct {
select (KeyExchangeAlgorithm) {
    case rsa:
        ServerRSAParams params;
        Signature signed_params;
    case diffie_hellman:
        ServerDHParams params;
        Signature signed_params;
    case fortezza_dms:
        ServerFortezzaParams params;
} ServerKeyExchange;

enum {anonymous, rsa, dsa) SignatureAlgorithm;
digitally-signed struct{
select (SignatureAlgorithm) {
    case anonymous:;
    case rsa:
        opaque md5_hash[16];
        opaque sha_hash[20];
    case dsa:
        opaque sha_hash[20];
};
} Signature;
```

A5.1.7 *CertificateRequest*

```
enum {
RSA_sign(1), DSS_sign(2) RSA_fixed_DH(3),DSS_fixed_DH(4),
```

```
RSA_ephemeral_DH(5), DSS_ephemeral_DH(6),Fortezza_dms(20),
        (255)
} CertificateType;
opaque DistinguishedName<3..2¹⁶-1>;
struct {
    CertificateType certificate_types<1..2⁸-1>;
    DistinguishedName certification_authorities<3..2¹⁶-1>;
}CertificateRequest;
```

A5.1.8 *ServerHelloDone*

```
struct {} ServerHelloDone;
```

A5.1.9 *ClientKeyExchange*

```
struct {
    ProtocolVersion client_version;
    opaque random[46];
} PreMasterSecret;
struct {
public-key-encrypted PreMasterSecret
pre_master_secret;
} EncryptedPreMasterSecret;
struct {
    opaque y_c<0..128>;
    opaque r_c[128];
    opaque y_signature[20];
    opaque wrapped_client_write_key[12];
    opaque wrapped_server_write_key[12];
    opaque client_write_iv[24];
    opaque server_write_iv[24];
    opaque master_secret_iv[24];
    opaque encrypted_preMasterSecret[48];
}FortezzaKeys;
enum (implicit, explicit) PublicValueEncoding;
struct {
    select (PublicValueEncoding) {
        case implicit: struct {};
```

```
            case explicit: opaque DH_Yc<1..2^16-1>;
    } dh_public;
    } ClientDiffieHellman Public;
    struct {
        select (KeyExchangeAlgorithm) {
            case rsa: EncryptedPreMasterSecret;
            case diffie_hellman:ClientDiffieHellmanPublic;
            case fortezza_dms: FortezzaKeys;
        } exchange_keys;
    } ClientKeyExchange;
```

A5.1.10 *CertificateVerify*

```
    struct {
    Signature signature;
    } CertificateVerify;
```

A5.1.11 *Finished*

```
    struct{
        opaque md5_hash[16];
        opaque sha_hash[20];
    } Finished;
```

6

TLS (Transport Layer Security) and WTLS (Wireless Transport Layer Security)

ABSTRACT

Presented in this chapter are two protocols: TLS (Transport Layer Security) and WTLS (Wireless Transport Layer Security), both derived from SSL. TLS arose from the need to standardize SSL, a task that the IETF accomplished by producing RFC 2246 (1999). WTLS is the Wireless Application Protocol (WAP) Forum's approach to secure transactions in mobile networks (Wireless Application Protocol Forum, 1999). Nevertheless, numerous incompatibilities persist among implementations of SSL and TLS because of the divergence between Netscape's implementation of SSL, which many consider the reference implementation, and the protocol specifications that Netscape wrote (Rescorla, 2001, pp. 50, 79, 89). On the other side, because of constraints on wireless communications, WTLS is, from the outset, incompatible with TLS or SSL, which necessarily reflects itself in network planning and in the administration of end-to-end security.

6.1   From SSL to TLS

Although TLS exchanges follow the same schemes as SSL, the two protocols differ on several points. The main differences concern the following items:

- The instant at which the encryption of the sent data starts
- The available cipher suites
- The method of computation of the MasterSecret and derivation of the keys
- The number of Alert messages
- The reaction to Record blocks of unknown types

Let us have a closer look at these differences.

6.1.1　Start of the Encryption of Transmitted Data

In contrast to SSL, which allows one side to start transmitting encrypted data just after having sent the message Finished, TLS requires that both Finished messages be exchanged before authorizing that transmission. This delay allows plugging of a hole through which an intruder could modifiy either the ClientHello message or the ServerHello message to force the use of a weaker cipher suite, recover the PreMasterSecret, divert the Finished message, and undermine the security edifice (Mitchell et al., 1998).

6.1.2　The Available Cipher Suite

The IETF RFC 2246 (1999) that defines TLS adopted all the cipher suites in the SSL specifications with the exception of those associated with Fortezza. Nevertheless, the coded points used by the latter were reserved to avoid any possible collision between the two protocols.

Just after the approval of IETF RFC 2246 in January 1999, U.S. policy with respect to the export of encryption software was partially liberalized. It is now legal to use a key of 56 bits for symmetrical encryption or a key of 1024 bits to sign key exchanges with the help of the RSA or ephemeral Diffie–Hellman algorithms. To take advantage of this liberalization, it was proposed to add the cipher suites in Table 6.1 to the exportable versions and, for internal usage in the U.S., to include RC4 at 128 bits (RC4-128) for symmetrical encryption, SHA-1 for hashing, and ephemeral Diffie–Hellman for the exchange of keys (Banes and Harrington, 2001). As of September 2003, the IETF had not yet adopted these proposals, while the Internet draft in question expired in January 2000.

6.1.3　Computation of MasterSecret and the Derivation of Keys

The construction of the secrets in TLS is similar to that of SSL Version 3, but the organizations of the computations are different. More precisely, the computation of PreMasterSecret uses the following formula:

$$\text{MasterSecret} = \text{PRF}(\text{PreMasterSecret}, \text{"master secret," client\_random} \parallel \text{server\_random})$$

where
　　PRF is a pseudorandom function of the form PRF (secret, label, seed)
　　$\parallel$ is an operator to indicate concatenation
　　The label is a series of characters, in this case, the string "master secret"
　　The seed is a random number formed by concatenation of client_random and of de server_random.

TABLE 6.1

Cipher Suites Added to TLS after the Liberalization of the Export Regulations in the U.S.

Key Exchange	Symmetric Encryption	Hash	Signature
RSA (1024 bits)	DES CBC	SHA	—
	RC4-56	SHA	—
Ephemeral Diffie–Hellman	DES CBC	SHA	DSS
(1024 bits)	RC4-56	SHA	DSS

Thus, by a simple change of label, TLS allows the generation of new keys, without recourse, to new random numbers. From the description of SSL, we remember that the values client_random and server_random are produced by the client and the server, respectively, and are exchanged with the messages ClientHello and ServerHello.

The function PRF is constructed as follows:

PRF (secret, label, seed) =
$$P\_MD5 (S1, label \parallel seed) \oplus P\_SHA\text{-}1(S2, label \parallel seed)$$
where

P_hash (secret, random) = HMAC_hash [secret, A(1) $\parallel$ random] $\parallel$
    HMAC_hash [secret, A(2) $\parallel$ random] $\parallel$
    HMAC_hash [secret, A(3) $\parallel$ random] $\parallel$...
    $\oplus$ is the Exclusive OR random = label $\parallel$ seed
    P_hash represents P_MD5 or P_SHA-1,
        depending on whether the MD5 or SHA-1
        algorithms are used
    HMAC_hash indicates either HMAC_MD5 or
        HMAC_SHA-1 according to the algorithm
        used

Note that four iterations of the function PRF are sufficient to get 64 octets. In this case, all the octets of the output from P_MD5 will be used, but only 64 out of the 80 octets that P_SHA-1 produces will be used. In contrast, to get 80 octets, there is a need to have five iterations from P_MD5 and four iterations from P_SHA-1.

The values A(1), A(2), A(3),..., are computed recursively from the value of the seed and the label, as follows:

$$A(0) = label \parallel seed$$

$$A(i) = HMAC\_hash [A(i-1)]$$

The secrets S1 and S2 are formed from PreMasterSecret by dividing it into two equal parts. Thus, we have

$$LS1 = LS2 = ceiling\ (LS/2)$$

The function ceiling (x) gives the smallest integer equal to or greater than x. LS, LS1, and LS2 are, respectively, the lengths of the initial secret and of S1 and of S2. If LS is odd, the equality of the sizes is achieved by adding an octet at the beginning of the second half, where the last octet of the first half is copied.

For a secret k and a message m, the hashed message authentication code HMAC_hash is computed with a hash function H() according to the following formula:

$$\text{HMAC\_hash}\ (k,m) = H\left[\bar{k} \oplus \text{opad} \,\|\, H(\bar{k} \oplus \text{ipad}, m)\right]$$

As explained in Chapter 3, the secret $\bar{k}$ is derived from the initial key k of L bits by padding with a series of "0" bits to form the block size required by the hash algorithm. The variables opad and ipad are the constants for outer padding and inner padding, respectively, 0x36 and 0x5C. These octets are repeated 64 times to form a block for either the MD5 or SHA-1 algorithms.

Having obtained the MasterSecret, the PRF is reapplied to form the block from which will be extracted the symmetric encryption keys, the hash keys, and the initialization vectors of the client and the server.

$$\text{KeyBlock} = \text{PRF (MasterSecret, "key expansion," server\_random} \,\| \\ \text{client\_random)}$$

Here, the label is the string "key expansion," while the seed is formed of the concatenation of server_random and of client_random. Note that their order is the inverse of their order in the computation of MasterSecret. The computation is repeated sufficient times to produce all the parameters of the cipher suite chosen, i.e.,

```
client_write_MAC_secret(hash_size)
server_write_MAC_secret(hash_size)
client_write_key(key_length)
server_write_key(key_length
/* For nonexportable algorithms only/
client_write_IV(size_of the_initialization_vector)
client_write_IV(size_of the_initialization_vector)
```

Consider, for example, the nonexportable cipher suite {RSA, DES in CBC mode, MD5}. The size of the secret to compute HMAC is 16 octets and that of each symmetric key or each initialization vector is 8 octets, bringing the total to 64 octets. It will be obtained after four iterations.

6.1.4   Alert Messages

The range of messages of the Alert protocol in TLS was expanded with 12 new messages, as listed in Table 6.2 in alphabetical order.

TABLE 6.2

Additional Messages in the Alert Protocol of TLS

Message	Context	Type
access_denied	Valid certificate received, but access was refused due to unsatisfactory checks	Fatal
decode_error	Message not decoded for incorrect size or for out-of-range parameter	Fatal
decrypt_error	Failure of one of the cryptographic operations in the Handshake, for example, the decryption of ClientKeyExchange, signature verification, or verification of the Finished message	Fatal or warning
decryption_failed	Failure in decryption of a block	Fatal
export_restriction	Negotiation violating a U.S. export restriction	Fatal
internal_error	Internal error independent of the protocol or the peer	Fatal
insufficient_security	Cipher suite that the client proposes is less than what the server requires	Fatal
no_renegotiation	Renegotiation refused following the initial contact	Warning
protocol_version	Protocol version that the client requests is recognized but not supported	Fatal
record_overflow	The size of the block used exceeds the specifications	Fatal
unknown_ca	Unknown certification authority	Fatal
user_canceled	Handshake is being canceled for some reason unrelated to a protocol failure	Warning

Just like for SSL, the messages of the Alert protocol are not authenticated, which exposes them to a truncation attack, whereby an attacker sends false Alert messages with the same sequence numbers as packets of encrypted data, thereby causing their removal of the legitimate data from the stream (Saarinen, 2000).

6.1.5   Responses to Record Blocks of Unknown Type

The computational load on a server during the establishment of a SSL/TLS session can be substantial, particularly after receiving the ClientKeyExchange message. By submerging the server with requests, an attacker can overwhelm the processing resources of the server. However, contrary to SSL, a TLS implementation can ignore received Record blocks if their types are unknown (§6 of IETF RFC 2246, 1999). This property allows resistance against a denial-of-service attack during the establishment of a TLS session (Dean and Stubblefield, 2001).

This defense consists of sending a *puzzle* to the client when the rate of server occupancy exceeds a certain threshold. As a consequence, the client is obliged to space its transmissions to resolve the puzzle, allowing the server some respite for recuperation.

The puzzle is a simple cryptographic problem. The server selects a random number x, computes its hash $h(x)$, and then forms a new number x' by zeroing the n lower-order bits of x. The triplet $\{n, x', h(x)\}$ forms the puzzle that is sent following the Certificate message. The client has to discover the value of x by trying all the values for the missing n bits and then comparing hash for each combination with the received hash $h(x)$. The average number of hashes that the client computes is 2^{n-1}, while the server computes only two values: the first is part of the puzzle and the second verifies that the hash of the value that the client discovers corresponds to the hash that was sent. These exchanges are shown in Figure 6.1.

Thus, as long as the server congestion lasts, only the clients that will be able to react to the puzzle will have a chance to connect; the others will ignore the puzzle. After a certain time, the server will disconnect the connections without traffic, thereby freeing the resources allocated to these aborted sessions. Once the congestion is cleared, the server will have all the resources needed to respond to all connection attempts following the normal TLS procedures, i.e., without imposing the puzzles on the clients.

6.2 WTLS

WTLS is the result of a complete revision of TLS to meet the constraints of the wireless environment. Thus, it must do the following:

- Sustain flow interruptions due to packet loss and function over transport protocols that are not reliable, particularly UDP (User Datagram Protocol)
- Operate even when the round-trip transmission delays are important
- Take into account the reduced computation power or storage capacity of mobile terminals

The modifications associated with these requirements made WTLS totally incompatible with SSL and TLS, which imposes a burden on the end-to-end security of transactions. Given the complexity raised by the potential solutions to end-to-end security, the WAP Forum changed its course and adopted the mobile profile of TCP that the IETF is developing for wireless links, while preserving WTLS to maintain backwards compatibility.

6.2.1 Architecture

Depicted in Figure 6.2 is the position of WTLS in the protocol stack of WAP 1.0 and its correspondence with the TCP/IP stack. Access pages are described

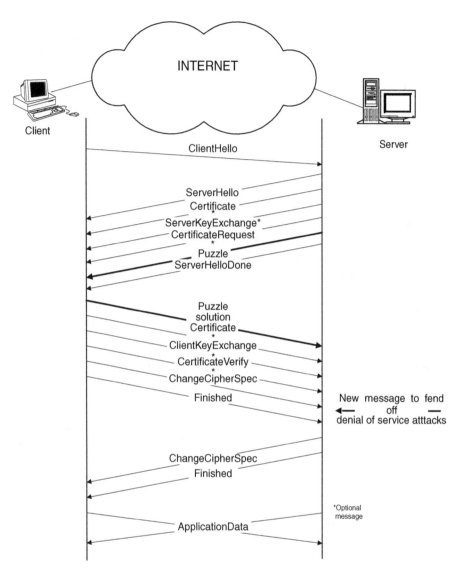

FIGURE 6.1
The TLS session establishment with protection against denial-of-service attacks with puzzles.

using the Wireless Markup Language (WML), which is more adapted than HTML to the constraints of wireless communication, such as limited bandwidth and terminals with reduced resources (in terms of memory, computation power, battery life, narrow screen, keyboard not suitable for text input, etc.). The WAP Forum works to consider all types of access technologies to the air interface. For the GSM (Global System for Mobile Communication) networks, this means the Short Message System (SMS) for bit rates of 9.6 kbit/sec and the General Packet Radio Service (GPRS) for bit rates between

28 and 56 kbit/sec. Cellular technologies of the third generation (3G) or UMTS (Universal Mobile Telecommunication System) are considered in WAP 2.0, which supports all the TCP/IP stack and switches for navigation on micro-browsers, and to the XHTML (eXtensible HyperText Markup Language) of the W3C using a mobile profile called XHTMLMP (XHTML Mobile Profile).

Use of WTLS in banking transactions requires a public-key infrastructure under the control of financial institutions. A wireless identification module (WIM) in the smart cards of the terminal will hold the necessary keys and the public-key identification certificate.

6.2.2   From TLS to WTLS

Just like SSL, WTLS is composed of four protocols: Handshake, Record, ChangeCipherSpec (CCS), and Alert. Figure 6.3 depicts its position in the protocol stack between the WAP application called Wireless Transaction

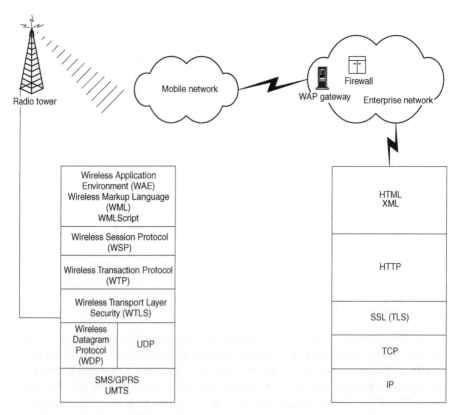

FIGURE 6.2
Position of WTLS in the protocol stack of WAP 1.0, and its correspondence with the TCP/IP stack.

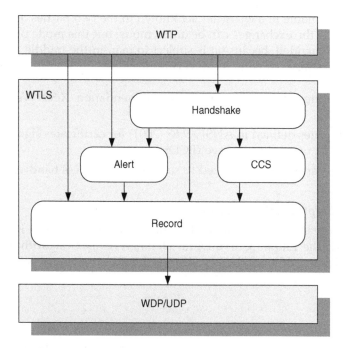

FIGURE 6.3
Protocol stack for WTLS subprotocols.

Protocol (WTP) and the transport layers WDP or UDP. In contrast to SSL/ TLS, WDP, and not Record, is responsible for data fragmentation. From the point of view of efficient use of available bandwidth, the combination WAP/ WTP/UTP reduces by more than half the number of packets that HTTP/ TCP/IP use to perform the same functions.

The main modifications that WTLS introduces to SSL/TSL relate to the following elements:

1. The format of the identifiers and the certificates
2. The cryptographic algorithms
3. The content of some Handshake messages
4. The exchange protocol during the Handshake
5. The calculation of secrets
6. The size of the parameters
7. The Alert messages
8. The role of the Record protocol

6.2.2.1 The Formats of Identifiers and Certificates

The WTLS identifiers are of several types: X.509 distinguished names, SHA- 1 hash of the public key, a secret binary key known only to the two parties, and a textual name in a character set known to the two parties. Just like for SSL and TLS, the exchanges can be anonymous, but this mode of operation is not recommended, because it is subject to man-in-the-middle attacks.

WTLS recognizes three types of certificates:

1. Certificates defined in ITU-T Recommendation X.509 Version 3 (1996)
2. Certificates defined in ANSI X9.68 (2001) for certificates signed with the DSA over elliptic curve (ECDSA)
3. WTLS certificates optimized to save on the wireless bandwidth

6.2.2.2 Cryptographic Algorithms

Table 6.3 contains key-exchange algorithms used by WTLS. It is seen that WTLS adds the Elliptic Curve Diffie–Hellman (ECDH) algorithm to those used in TLS. In addition, the key size for the RSA exchanges is 512, 768, or 1024 bits.

TABLE 6.3

Key Exchange Algorithms Negotiated by the Handshake Protocol of WTLS

Code for the Key Exchange Suite	Description	Key Size Limit (bits)
NULL	No exchange; the length of the PreMasterSecret is zero	—
SHARED_SECRET	The secret is exchanged by other means than WTLS, for example, it may be cast in smart card of the mobile terminal	None
DH_anon	Diffie–Hellman key exchange without authentication; each side sends to the other its temporary Diffie–Hellman public key; each party calculates the PreMasterSecret using its private key and the public key of the counterpart	None
DH_anon_512	Same as DH_anon	512
DH_anon_768	Same as DH_anon	768
RSA_anon	RSA without authentication; the PreMasterSecret is the secret value that the client generates, encrypted with the public key of the server, appended with the server's public key	None
RSA_anon_512	Same as RSA_anon	512
RSA_anon_768	Same as RSA_anon	768
RSA	RSA with authentication with certificates	None
RSA_512	RSA with authentication with certificates	512
RSA_768	RSA with authentication with certificates	768
ECDH_anon	ECDH without authentication	None
ECDH_anon_113	ECDH without authentication	113
ECDH_anon_131	ECDH without authentication	131
ECDH_ECDSA	ECDH with certificates signed by ECDSA	None

TABLE 6.4

Encryption Algorithms Used in WTLS

Representation of the Cipher	Exportable	Effective Length of the Key (bits)	Use in SSL/ TLS
NULL	Yes	—	Yes
RC5_CBC_40	Yes	40	—
RC5_CBC_56	Yes	56	—
RC5_CBC	No	128	—
DES_CBC_40	Yes	40[a]	Yes
DES_CBC	No	56	Yes
3DES_CBC_EDE	No	168	Yes
IDEA_CBC_40	Yes	40	—
IDEA_CBC_56	Yes	56	—
IDEA_CBC	No	128	Yes

[a] The encryption uses only 35 bits of the key, and the remaining form parity bits (Saarinen, 2000).

Table 6.4 contains the encryption algorithms used in WTLS. They all use blocks of 64 bits with an initialization vector of the same size. Notice the presence of several variations of the RC5 protocols as well as of IDEA. Some algorithms were deliberately weakened to satisfy the restrictions of the U.S. authorities. Similar to SSL and TLS, the same algorithm is used by both sides, but each part generates its own key and then shares it with its partner.

Finally, the hash functions SHA-1 or MD5 are used to verify the integrity of the transport. However, Table 6.5 presents much weaker procedures than those used in SSL/TLS.

6.2.2.3   The Content of Some Handshake Messages

The ClientHello message adds the following fields to those already present in SSL/TLS:

- A list of the key-exchange methods that the client accepts, in order of decreasing preference (In SSL/TLS, the whole cipher suite is negotiated at once.)
- The list of identifiers that the client accepts, in order of decreasing preference
- The list of certificates
- The list of compression methods
- The sequence number mode for packets
- The number of exchanged bits before the cryptographic parameters are refreshed (key_refresh)

TABLE 6.5

Hashing Algorithms Used in WTLS

Representation of the Hash Function	Description	Key Size (bits)	Size of the Hash (bits)	Use in SSL/ TLS
SHA_0	No hashed MAC	—	—	—
SHA_40	Only the first 40 bits of the output are used	160	40	—
SHA_80	Only the first 80 bits of the output are used	160	80	—
SHA	All output bits are used	160	160	Yes
SHA_XOR_40	The input data are divided into blocks of 5 octets; an Exclusive OR operation is successfully performed on different blocks; this function is not recommended and does not protect stream ciphers (Saarinen, 2000); it is kept for terminals with little computing power	0	40	—
MD5_40	Only the first 40 bits of the output are used	128	40	—
MD5_80	Only the first 80 bits of the output are used	128	80	—
MD5	All output bits are used	128	128	Yes

The additional fields in the ServerHello message contain the following information:

- The sequence number mode for packets
- The interval in exchanged bits between two consecutive updates of the cryptographic parameters (key_refresh)

The renewal of the cryptographic parameters (encryption key, key for hashed MAC calculations, and initialization vectors) occurs after the exchange of $2 \times$ key_refresh bits. This cadence is negotiated at the beginning of a session. Modification of the encryption parameters during a session reduces the risk due to breakage of short keys. The variables client_random and server_random remain constant throughout the duration of the associated connection.

6.2.2.4　The Exchange Protocol during the Handshake

The main modifications are as follows:

1. Messages going in the same direction can be consolidated in one transmission. This consolidation is mandatory when the transport is not reliable, and it is used for text exchanges in SMS.

2. Retransmission of messages is allowed under some conditions.

3. The ClientKeyExchange message is no longer mandatory. In fact, it is redundant when the key exchange uses ECDH, and the client is certified.

4. The packet sequence number mode can be selected. When the transport lay is TCP, each side maintains the sequence numbers separately for each connection and increments them with each message transmission for that connection. This is the same for SSL/TLS, and the mode is called implicit. In contrast, when the transport layer is UDP or any other unreliable protocol, the explicit mode is mandatory. In this mode, the sequence numbers are sent in the clear in Record messages so that they can be used to compute the hash values. A third choice is not to have any numbers, but this mode is not recommended, because it is exposed to replay attacks, in particular when the higher layers do not include necessary defenses.

6.2.2.5   Calculation of Secrets

6.2.2.5.1   Computation of the PreMasterSecret

In the case of a shared secret, the PreMasterSecret is stored in the terminal smart cards; otherwise, it is exchanged using one of hte algorithms of Table 6.3.

The size of the PreMasterSecret varies according to the method used for the key exchange.

6.2.2.5.2   Computations of MasterSecret

The PRF is constructed with one hash function only. Thus, we have

$$\text{PRF (secret, label, seed)} = \text{P\_hash (secret, label} \parallel \text{seed)}$$

where

P_hash (secret, rand) = HMAC_hash [secret, A(1) ‖ rand] ‖
                HMAC_hash [secret, A(2) ‖ rand] ‖
                HMAC_hash [secret, A(3) ‖ rand] ‖ …
                rand = label ‖ seed
                P_hash represents P_MD5 or P_SHA-1, depend-
                    ing on whether MD5 or SHA-1 is used
                HMAC_hash indicates HMAC_MD5 or
                    HMAC_SHA-1, depending on the case

In this case, the seed is a random number, and the label is a character string. In this particular case, the label is the text "master secret." Once that MasterSecret is obtained, the encryption key can be extracted for the encryption block:

KeyBlock = PRF {MasterSecret, expansion label, sequence number ‖

server_random ‖ client_random}

The block is recomputed at regular intervals according to the period for key refresh. On the server side, the expansion label is *server expansion*, while on the client side, it is *client expansion*. The operations are repeated enough times so that the output can be divided into the required fields. From the client side, the partitions are the following:

```
client_write_MAC_secret[size of the hash key]
client_write_encryption_key[size of the encryption key]
client_write_IV[size of the intialization vector]
```

On the server side, the partitions are as follows:

```
server_write_MAC_secret[size of the hash key]
server_write_encryption_key[size of the encryption key]
server_write_IV[size of the intialization vector]
```

For the exportable encryption algorithms, the encryption key and the initialization vector are capped according to the following formulae:

```
final_client_write_key = PRF(client_write_key, "client
    write key", client_random || server_random)
client_write_iv = PRF("","client write IV", client
    sequence number || client_random || server_random)
final_server_write_key = PRF(server_write_key, "server
    write key", client_random || server_random)
server_write_iv = PRF("","server write IV", server
    sequence number || client_random || server_random)
```

Thus, for these algorithms, the pseudorandom function used in the last step of the computation of the initial vectors does not contain a secret. As a consequence, their value depends on variables that were sent in the clear, i.e., the sequence numbers of the client or the server and the random numbers client_random and server_random.

Finally, the initialization vector for the CBC encryption mode is formed from the vector calculated above using an Exclusive OR operation applied on a new vector S. Vector S is constructed by concatenating the two octets of the sequence numbers enough times to form the number of octets needed in the initialization vector. For example, if we start with an initialization vector IV_0 of 64 bits, the initialization vector used IV will be calculated with the corresponding vector S according to the following formula:

$$IV = IV_0 \oplus (S \parallel S \parallel S \parallel S)$$

TABLE 6.6

Comparison of the Sizes of Some Variables in SSL/TLS and WTLS

	Size (octets)	
Variable	SSL/TLS	WTLS
symmetric encryption key	5–16	5–21
client_random	32	16
session identifier	3	2
MasterSecret	48	20
sequence number	8	2
PreMasterSecret	48	Variable (20 for RSA)
server_random	32	16

6.2.2.6 Parameter Sizes

As shown in Table 6.6, WTLS parameters were shortened compared to those of TLS/SSL.

As a result, the size of the Finished message is reduced to 12 instead of 36 octets.

6.2.2.7 Alert Messages

Table 6.7 contains an alphabetical list of the Alert messages that WTLS added to those of the TLS. The message close_notify of SSL/TLS was separated into connection_close_notify and session_close_notifiy to distinguish between the terminations of connections from the closure of a session. WTLS introduced another type of message (critical message) whose significance is left for the recipient to define. As for SSL/TLS, the Alert messages are susceptible to truncation attacks because they are not authenticated.

6.2.2.8 Record

As was indicated earlier, in WTLS, data fragmentation is the responsibility of the WDP and not the Record protocol, as is the case for SSL/TLS. The role of the Record is to add an unencrypted header of 1 to 5 octets to each message arriving from the higher layers, in order to indicate the message type and to signal the presence of optional fields. These include 2 octets for the sequence number in the case of explicit numbering, and 2 octets for coding the length of the block of encapsulated data.

6.2.3 Service Constraints

WAP was designed so that mobile workers could maintain contact with the information systems of their enterprises, banks, etc. However, we see that TLS and WTLS are incompatible. This has two consequences. From the server side, applications equally accessible to the mobile and sedentary workers must be written and maintained in two formats: HTML or XML on one side

TABLE 6.7

Additional Alert Messages in WTLS

Message	Context	Type
connection_close_notify	Voluntary termination of the connection	Fatal
decompression_failure	Input data for the decompression function are invalid (for example, too long)	Fatal
disabled_key_id	Keys that the client supplied were administratively disabled	Critical alert
duplicate_finished_received	The server received a second Finished in an abbreviated Handshake	Warning
key_exchange_disabled	Key exchange was disabled to protect an anonymous key exchange in progress	— [a]
session_close_notify	Voluntary termination of the session	Fatal
session_not_ready	Secure session not available for administrative reasons	Critical alert
unknown_key_id	None of the client keys are recognized	Fatal
unknown_parameter_index	Server does not know the indicator that the client supplied for the available key exchange suite	Critical alert

[a] Left to the appreciation of the sender.

and WML on the other. From the network side, a gateway must intervene to ensure interoperability between TLS and WTLS. For example, the message transmitted from the mobile terminal and secured by WTLS must be decrypted by the gateway and then encrypted again according to TLS. This conversion is not risk-free, because for a short interval, the message is in the clear, so end-to-end security cannot be guaranteed without additional efforts to protect the information in the clear during the conversion.

6.2.3.1 Possible Location of the WAP/Web Gateway

The conversion gateway can be located with respect to the WAP platform and the enterprise firewalls according to several possibilities:

1. The enterprise manages the access by deploying the gateway and the WAP platform behind its firewall. In this case, the security problem related to the WTLS/TLS conversion is resolved. This is shown as Case I in Figure 6.4. Mobile users connect to the telephone number that their employer supplied, and the WTLS protocol is used up to the gateway. Either the radio link extends to the enterprise network or the employer houses its equipment in a secure location that the mobile operator provides. In either case, the employer takes care of all the links in the security chain. The advantage for the enterprise is that it is completely independent of the mobile network whose only role is to supply the radio channel. However, the enterprise must be able to master all the necessary skills to manage and maintain the system and evolve it with technology.

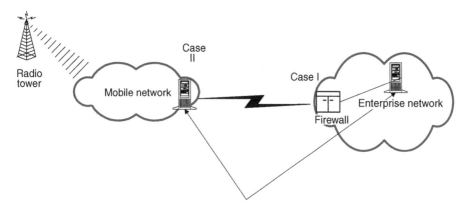

Possible location of the WAP/web gateway

FIGURE 6.4
Possible locations of the WAP/Web gateway between the mobile network and the enterprise network.

2. The mobile operator is responsible for the radio communication, including the conversion from WTLS to TLS, as shown in Case II of Figure 6.4. The telephone operator must be a trusted party and be capable of carrying out the responsibilities that befall it.

3. The mobile operator is only responsible for the radio link. The gateway WAP/Web is hosted at the Internet Service Provider (ISP) of the enterprise before the firewall on the enterprise's premises. This corresponds to Case III of Figure 6.5, in which the provider controls end-to-end security.

4. Finally, as shown in Case IV of Figure 6.5, it is possible for the enterprise to rely completely on its ISP for the whole security: the WAP/Web conversion, application hosting, and firewall.

None of these configurations is totally satisfactory. This is why the problem was also addressed from a protocol point of view. We present two approaches in the following. The first is called Integrated Transport Layer Security (ITLS) and the second Network-Assisted End-To-End Authentication (NAETEA).

6.2.3.2 ITLS

The solution ITLS consists of encrypting the message twice using two different keys. The first key is a secret shared between the mobile terminal and the WAP/Web gateway; the second is the key between the mobile terminal and the server (Kwon et al., 2001). This solution taxes the terminal in terms of computation power management. It requires some modifications in the WTLS protocol at the client level to exchange a secret with the gateway. More

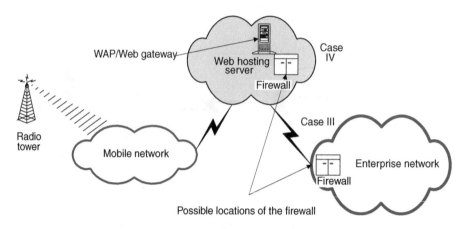

FIGURE 6.5
Possible arrangements of the firewall and the WAP/Web gateway between the enterprise and its Internet access provider.

precisely, the WAP gateway must transmit to the client a new message IntCertificate following the Certificate message coming from the server, while the client responds with another new message IntClientKeyExchange. A third message Hash_Handshake is used to verify the integrity of the session establishment between the client and the gateway. Finally, the gateway intervenes to modify the content of the ClientHello, ServerHello, and Finished messages, as well as the Record exchanges with user's data. These modifications are depicted in Figure 6.6.

6.2.3.3 NAETEA

The NAETEA solution has the advantage of avoiding any changes to WTLS and of saving the terminals from additional crytopgraphic computations through an intervention of the mobile network, once the connection encrypted with WTLS is established. In this protocol, the terminal shares a secret session key with the network (Ks) (established with WTLS) and has a pair of public/private keys for digital signature PSK_{SIGT} and SSK_{SIGT}. The Web server has a pair of public/private encryption keys PK_W and SK_W and a pair for signature PL_{SIGW} and SSK_{SIGW}. The corresponding network keys are PK_N/SK_N and PSK_{SIGN}/SSK_{SIGN}, respectively. In the following, K{X} represents the encryption of the content X with the key K, and ‖ indicates concatenation:

1. The terminal produces a random number $Rand_T$ and encrypts it with the public encryption key of the Web server PK_W. This operation gives $PK_W \{Rand_T\}$.
2. The terminal adds to the cryptogram its certificate and encrypts the whole with the secret key shared with the network:

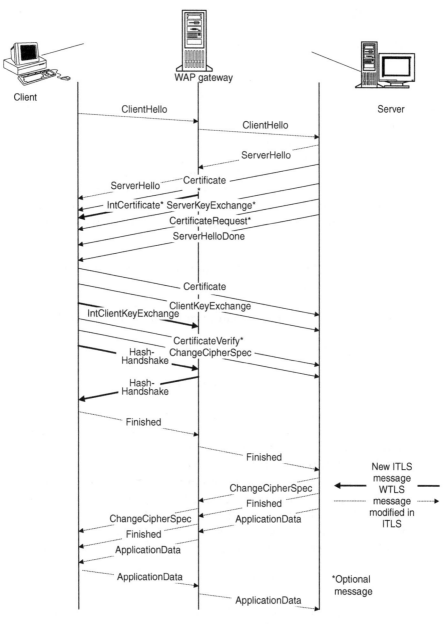

FIGURE 6.6
Exchanges of the ITLS protocol.

$$m1 = Ks\{PK_W \{Rand_T\} \parallel \text{signature certificate of the terminal}\}$$

3. The terminal signs the message with its private signature key SK_{SIGT} and a hash function H:

$$\text{sigT} = \text{SK}_{\text{SIGT}}\{\text{H(m11 } \| \text{ identifier of the Web server)}\}$$

4. The message M1 sent to the network is formed as follows:

$$\text{M1} = (\text{m1 } \| \text{ sigT } \| \text{ signature certificate of the terminal } \| \text{ identifier of the Web server})$$

5. After receiving the message M1, the network verifies the integrity of the signature sigT with the help of the public signature key extracted from the received certificate. Next, it decrypts m1 with the key Ks to verify that the certificate is identical to the one that was encrypted. Once this verification is accomplished, then the network can forward the data received in full confidence to the Web server. It makes a new message by signing the data with its private signature key KS_{SIGR}:

$$\text{m21} = (\text{KP}_{\text{W}}\{\text{Rand}_{\text{T}}\} \| \text{ sigT } \| \text{ terminal certificate } \| \text{ identifier of the Web server})$$

$$\text{sigN} = \text{KS}_{\text{SIGN}}\{\text{H(m21, signature certificate of the network)}\}$$

6. The message M2 is constructed as follows:

$$\text{M2} = \text{m21 } \| \text{ sigN } \| \text{ network certificate}$$

7. Having received the message M2, the Web server examines the network signature to ensure that the message M2 really comes from the network. Once this verification is done, it can extract the random number Rand_{T} with its private encryption key SK_{W}.

8. The Web server generates another random number Rand_{W} and constructs the session key $K_{\text{WT}} = \text{H(Rand}_{\text{T}} \| \text{ Rand}_{\text{W}})$ and a hash H $(\text{Rand}_{\text{T}} \| K_{\text{WT}})$. This session key will serve to encrypt the exchanges between the server and the mobile terminal. The concatenation of the two random numbers and the server encryption certificate gives m31:

$$\text{m31} = \text{Rand}_{\text{W}} \| \text{ H(Rand}_{\text{T}} \| K_{\text{WT}}) \| \text{ signature certificate of the Web server}$$

9. It computes sigW $= \text{SK}_{\text{SIGW}}\{\text{H(m3)}\}$ with its private signature key SK_{SIGW}. The message M3 returned to the network is formed by concatenation as follows:

$$\text{m31 } \| \text{ sigW } \| \text{ signature certificate of the Web server}$$

and the whole is encrypted with the public encryption key of the network PK_{N}. Thus, we have the following:

$$M3 = PK_N \{m31 \parallel sigW \parallel \text{signature certificate of the Web server} \}$$

10. The network decrypts the message M3 with its private key SK_N, verifies the signature, and prepares accordingly the message to the terminal. This message is encrypted with the key Ks established with the terminal.

$$M4 = Ks \{Rand_W \parallel H(Rand_T \parallel K_{WT}) \parallel \text{encryption certificate}$$
$$\text{of the network}\} \parallel \text{encryption certificate of the network}$$

11. The terminal proceeds to authenticate the message by comparing the certificate received in the clear with the one encrypted with the key Ks. Once this is verified, the terminal extracts the random number $Rand_W$, then calculates the encryption key K_{WT} with $H(Rand_T \parallel Rand_W)$. Finally, it compares the value of the hash $H(Rand_T \parallel K_{WT})$ with the one enclosed in the message M4. If both values are identical, the terminal can be confident that it is the server that received the random number $Rand_T$ that it originally sent and, therefore, that the key K_{WT} can guarantee the confidentiality of the exchanges with the server.

The exchanges associated with this protocol are depicted in Figure 6.7.

6.3 Summary

The success of SSL led to its normalization with TLS and its metamorphosis as WTLS for mobile communications. Nevertheless, there are some differences between TLS and SSL, and WTLS is totally incompatible with SSL/TLS, because WTLS is optimized to save bandwidth. Unfortunately, the

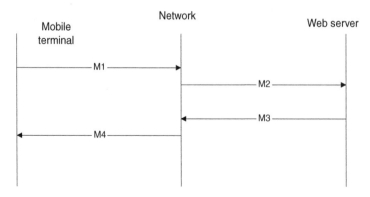

FIGURE 6.7
End-to-end authentication across WTLS and TLS with the NAETEA protocol.

service architecture becomes complex when WTLS and SSL/TLS are associated with the same connection. In addition, the operation of WTLS raised some concerns with respect to the weaknesses of certain algorithms retained. Thus, the new version of the WAP specifications returned to TLS by adopting a profile that is particular to wireless transactions.

Questions

1. What is the advantage of delaying data transmission until both sides exchanged and verified their Finished messages?
2. How can a denial-of-service attack be mounted against SSL/TLS/WTLS? How can the use of cryptographic puzzles offer some protection?
3. List some sources of incompatibilities between WTLS and TLS.
4. Why are some weak encryption algorithms used in WTLS?
5. Compare and contrast solutions to the problem of the unencrypted message in WTLS/TLS conversion.
6. In the NAETEA protocol:
   a. What is the function of the shared key between the terminal and the mobile network?
   b. How is the terminal authenticated by the Web server?
   c. Why is the session key between the mobile terminal and the Web server based on two random numbers?
   d. How is the session key protected from the mobile network?

7

The SET Protocol

ABSTRACT

This chapter discusses the SET (Secure Electronic Transaction) protocol designed to secure bankcard transactions executed over open networks such as the Internet. This protocol was sponsored jointly by Visa and MasterCard in collaboration with the main players in the informatics world, such as IBM, GTE, Microsoft, SAIC (Science Applications International Corporation), Terisa Systems, and VeriSign (SET Specification, 1997). The common objective was to encourage the use of credit cards for online payments and to avoid market fragmentation into a multitude of incompatible protocols.[1]

SET operates at the application layer, independently of the transport layer, a property that distinguishes it from SSL. In practice, however, SET is usually considered for securing transports that conform to the TCP protocol. SET focuses only on the payment and excludes the search and selection of goods. In a SET transaction, cardholders make payments without inserting their cards in any reader; rather they present a certificate previously awarded by a certification authority. This certificate is stored in the hard disk of a computer or on a diskette and allows the authentication of the cardholder with public key cryptography.

Because SET is a U.S. invention, its initial focus was on security through software implementation. The adaptations of SET to integrated circuit cards (chip or smart cards) originated from Europe. One of these projects, Cyber-COMM, took into account the specific needs of French bankcards. In addition, SETCo published two documents to adapt SET to smart cards (SETCo, 1999a, 1999b).

---

[1] A fully owned subsidiary of SAIC, Tenth Mountain Systems, Inc., was the first SET Compliance Certification Authority (SET SCCA). The SAIC also owns the InterNIC (Internet Network Information Center), which, until 1999, was the sole administrator of Internet domain names .com, .org, and .net. It is also the owner of Telcordia (formerly BellCore).

7.1 SET Architecture

The fundamental principle that guided SET architects was to secure bankcard transactions over the Internet without modifying the existing banking circuits for authorization and remote collection.

As previously discussed, bankcard networks have authorization servers that filter abusive transactions according to precise criteria, such as if an expense ceiling is reached or if an excessive number of transactions are conducted in a given interval. Thus, before authorizing a bankcard transaction, the merchant has to query the authorization servers of the card scheme. Later in the settlement phase, the merchant is paid the amount that corresponds to the goods or services. However, various bankcard systems require that financial settlement be made only after shipment of the acquired goods. It is only then that the merchant sends the bank a clearance request to recover the outstanding debts; this request is then forwarded through the banking networks to the issuer bank. Note that in some bankcard systems, the authorization and settlement requests can be combined in the same operation for each transaction. Other systems allow the grouping of transactions, so that several authorization and settlement requests can be sent at the same time, for example, at the end of the workday. If the merchant is to reimburse the buyer, either because the product was returned or was defective, the merchant gives its bank instructions to credit the client's account.

To allow the scheme to work on a worldwide level, SET introduced two new entities: the certification authority to certify the actors and the payment gateway. The latter manages the border between the Internet and the network of bankcards.

Thus, there are six participants in SET:

1. The cardholder, whose card will have to conform to the SET specifications and be issued by an issuer institution, typically a bank affiliated with Visa or MasterCard
2. The merchant's server
3. The payment gateway
4. The certification authority
5. The issuer institution of the cardholder's bankcard
6. The acquirer institution, which is the merchant's bank

Figure 7.1 depicts the functional architecture of SET. The cardholder, the merchant, the certification authority, and the payment gateway are connected through the Internet. The client does not establish a direct connection to the payment gateway but uses a tunnel that goes through the merchant's server, using the tunneling technique. Each of the participants first must obtain a certificate from a certification authority that conforms to the SET specifica-

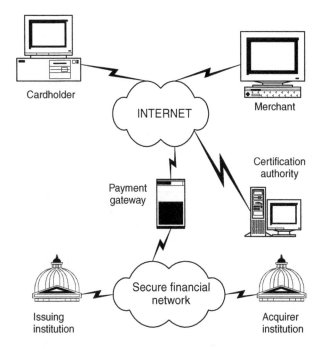

FIGURE 7.1
The various actors in a SET transaction.

tions. These certificates are enclosed in each of the messages exchanged among the cardholder, the merchant, and the payment gateway.

The issuer and acquirer institutions are linked with a closed and secure bank network. The payment gateway is the bridge between the open and the closed networks, which protects access to the banking network. The gateway has two interfaces, one conforming to the SET specifications, on the Internet side, and the other to the proprietary protocol on the secure financial network side.

SET secures the exchanges between the client and the merchant and, at the same time, the exchanges between the merchant and the payment gateway. The payment gateway, as in the case of remote payments with the Minitel, manages the payments on behalf of the banks that issue the bankcards and the acquirer banks. As a consequence, the gateway must be approved by the banking authorities; if not, the financial institution is responsible for carrying out these functions.

Figure 7.2 depicts the position of SET in the protocol stack TCP/IP (Sherif et al., 1997).

SET is a transaction-oriented protocol, and it functions in the request/response mode; i.e., messages are in pairs. The message structure follows the DER (Distinguished Encoding Rules) of the ASN.1 (Abstract Syntax Notation 1) (Steedman, 1993). The ASN.1 version used dates from 1995; as described

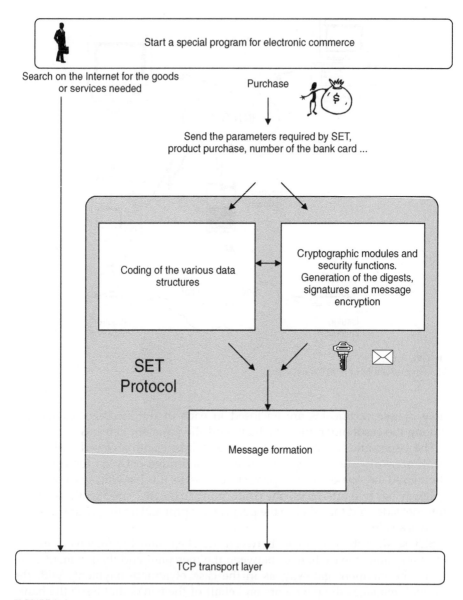

FIGURE 7.2
Position of SET in the TCP/IP protocol stack.

in ISO/IEC 8824-1 (1998), 8824-2 (1998), 8824-3 (1998), and 8224-4 (1998); whereas the DER rules are in ISO/IEC 8825-1 (1998). The messages are encapsulated with MIME as described in PKCS #7 (IETF RFC 2315, 1998). In particular, SET uses the following PKSC #7 structures:

- *SignedData*, for data that are signed
- *EnvelopedData*, for clear data that are in a digital envelope
- *DigestedData*, for digests
- *EncryptedData*, for encrypted data

7.2 Security Services of SET

SET transactions provide the following services:

- Registration of the cardholders and merchants with the certification authority
- Delivery of certificates to the cardholders and merchants
- Authentication, confidentiality, and integrity of the purchase transactions
- Payment authorization
- Payment capture to initiate the request for financial clearance on behalf of the merchants

SET employs the techniques of public key cryptography to simultaneously guarantee the following:

- Confidentiality of the exchanges, i.e., that they cannot be read online by an entity external to the transactions
- Integrity of the data exchanged among the client, the merchant, and the acquirer bank
- Identification of the participants
- Authentication of the participants

A necessary but not sufficient condition for nonrepudiation of the transactions is that the cardholder be certified. Other conditions include a trusted time-stamping mechanism and an irreproachable certification authority. Finally, because it is the payment gateway that verifies the exactness of the payment instructions and not the merchant, the gateway will be called upon to arbitrate disputes. Note that, however, to facilitate deployment of the SET protocol, buyer certification is optional in Version 1.0 of the specifications.

Message confidentiality is achieved with symmetric encryption algorithms (also called secret key encryption algorithms). The secret key is distributed with public-key cryptography algorithms. For instance, when the payment gateway wants to send confidential information to the merchant, it generates a symmetric encryption key with which it encrypts the data. This same key is encrypted with the public key of the merchant, who, being the only entity with the corresponding private key, is the only party capable of retrieving the symmetric key and decrypting the data.

Message integrity aims at guaranteeing that the received data are exactly what the sender transmitted and that they were not corrupted by mischief or error during their transit on the network. SET uses the digital signature of the sender to ensure message integrity, i.e., the digest of the message encrypted with the private key of the sender. Any entity having access to the corresponding public key is able to verify the message integrity by comparing the calculated digest of the message with the one obtained by decrypting the signature. In this way, if the public/private key pair is unique and attempts to steal the sender's identity were not successful, the digital signature simultaneously guarantees the sender's identity and the data integrity.

The identification of the participants in a SET transaction corresponds to a preestablished relation between an encryption key and an entity. Each entity attaches to its message, whether encrypted or not, a digital signature that only that entity could have generated but that can be verified by the peer entities. As seen above, in SET, the sending entity constructs this signature by encryption with its private key using a public-key algorithm; this signature is verified with the corresponding public key.

In SET, the certificate signed by the certification authority gives credence to the association of a public key with its owner, thus providing authentication. The authentication procedures of SET are based on Version 3 of ITU-T Recommendation X.509 (1996). Each certificate contains the identity of its owner, a public key related to the public-key encryption algorithm used, and the signature of the authority that issued the certificate. For mutual authentication, two parties have to go backward along the certification path until they encounter a common authority.

7.2.1   Cryptographic Algorithms

As shown in Table 7.1, SET employs existing cryptographic algorithms. It is seen that SET uses SHA-1 for ordinary hashing as well as HMAC-SHA-1 for keyed hashing.

TABLE 7.1

Cryptographic Algorithms Used in SET

Algorithm	Services
DES	Confidentiality
RSA	Authentication, identification, and integrity
SHA-1	Hashing
HMAC-SHA-1	Keyed hashing

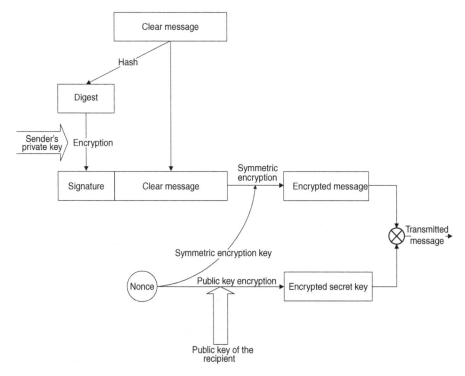

FIGURE 7.3
Processing of a typical message in SET.

In a SET transaction, some participating entities, such as the merchants, have two pairs of public encryption keys: one for signing documents and the other for exchanging keys during the identification phase. The first pair is called the signature key, and the second is called the key-encryption key (or key for key exchange). Figure 7.3 summarizes the main steps for processing a typical message to implement the services of authentication, confidentiality, and integrity. The private key used for signature is from the pair of signature keys, whereas the public key for encrypting the symmetric key comes from the pair of encryption keys.

Once created, the digest is coded according to the DER rule and then arranged to conform to the content type DigestedData and to a digest with the various fields, as shown in Table 7.2. As stated earlier, this syntax is based on the PKCS #7 (IETF RFC 2315, 1998) syntax as redefined in Book 2 of the SET specifications. The digest length then jumps from 20 to 46 octets, which corresponds to an increase in the transmission overhead of more than 100%.

The ensemble formed with the digest and the clear message is encrypted with the Data Encryption Standard (DES) algorithm using as a symmetric key a random number of 8 octets (64 bits) generated anew for each message. This key is then sent to the recipient within a digital "envelope" with the format

TABLE 7.2

Digest for the PKCS #7 Format in SET

Field	Length (octets)
Header of DigestedData	2
Version	3
Header of DigestAlgorithm	2
Algorithm	7
Parameters of the algorithm	2
ContentInfo	2
ContentType	6
Header of hash	2
The hash proper	20
Total	46

specified in PKSC #7. This envelope is, in turn, encrypted with the public RSA key of the message's recipient. The length of the RSA key is 1024 bits.

To give additional protection to the symmetric DES key, before encrypting the DES key with the public key of the recipient, the numerical envelope is processed with the OAEP (Optimal Asymmetric Encryption Padding) technique that Bellare and Rogaway (1995) proposed. The OAEP processing makes all bits of the cryptogram equally resistant to attack. The cryptogram is formed of 128 octets that include the symmetric key, the buyer's PAN (Primary Account Number), and other variables that are combined with padding, random numbers, and the like, to offer enhanced resistance to brute-force attacks (SET, 1997, Book 3, pp. 15–23). Thus, the SET protocol uses the private key of the sender for the digital signature and the public key of the recipient to encrypt the digital envelope that contains the DES key used to encrypt the clear text. If the cardholder is certified for signature, the certificate is attached to the message encrypted with the symmetric key, as shown in Figure 7.4. In this case, each message is authenticated.

7.2.2 The Method of the Dual Signature

One innovation that SET introduced is the procedure for dual signature. This procedure links the elements of two messages, each sent to a different recipient with the same signature to avoid unnecessary exchanges. With this procedure, each recipient can read the message that is addressed to him or her and verify the integrity of the other message without knowing its content. For example, suppose that a buyer simultaneously sends the purchase order to the merchant and the payment instructions to the bank (through the payment gateway). Of course, the buyer would like the payment instructions to be executed only after the merchant accepted the purchase order.

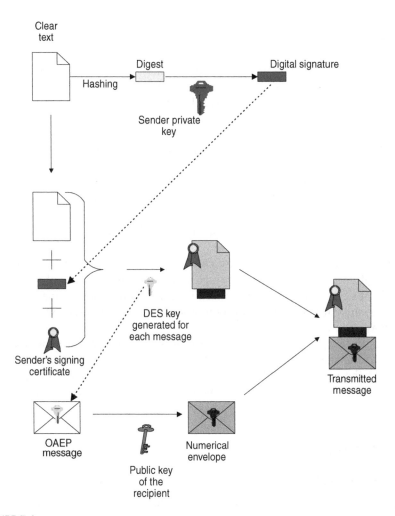

FIGURE 7.4
Cryptographic processing of SET messages sent by the buyer.

Let m1 be the message destined to the merchant and m2 be the message for the payment gateway (by way of the merchant), respectively. The buyer concatenates the digests of both messages to construct a new m3:

$$m3 = H(m1) \parallel H(m2)$$

The buyer then applies the hashing function H() to the new message m3. Accordingly, the message sent to the merchant is composed from the following elements:

$$m1, H(m1), \{m2, H(m2)\} \, PK_G, \{H(m3)\} \, SK_C, H(m1) \oplus H(m2)$$

where PK_G is the public key of the payment gateway, SK_C is the private key of the buyer, and $\oplus$ is the Exclusive OR operation.

The merchant reads the message m1 that is addressed to him and verifies its integrity by comparing the received H(m1) with the recalculated value. Following verification, the merchant decrypts H(m3) with the client's public key, PK_C, and extracts H(m2) by the Exclusive OR operation $H(m1) \oplus \{H(m1) \oplus H(m2)\}$. This allows it to concatenate the computed digest H(m1) with the extracted digest H(m2) to reconstitute m3 = H(m1) $\|$ H(m2). If H(m3) gives the result obtained by decrypting $\{H(m3)\}$ SK_C with the client's public key, then the integrity of the message m2 to the payment gateway was established without accessing the content of m2. If the merchant accepts the integrity of the message, it sends to the gateway

$$\{m2, H(m2)\}\ PK_{G,}\ \{H(m3)\}\ SK_{C,}\ H(m1)$$

As a consequence, the gateway can extract H(m2) and, by concatenating H(m1), it verifies that the quantity H(m3) is the same as that obtained with the client's public key, which is an indication that the merchant accepted the offer, because it is the merchant who relays the client's offer to the payment gateway.

Thus, the payment is not executed unless the merchant accepts the offer, whereas the sale is effective only if the bank approves the payment instruction. At the same time, the payment gateway cannot access the purchase order, and the payment instructions remain opaque to the merchant.

7.3 Certification

In this section, SET procedures for the certification of the cardholder and of the merchant are presented. It is important to note that this "certification" has a different meaning than that of the "certification" of SET components. The objective of the latter is to verify the conformance of a SET implementation to the specifications.

7.3.1 Certificate Management

The architecture for certificate management is shown in Figure 7.5. This figure shows the nine elements that intervene in the certification path, starting from the root certification authority and ending with the participants in a SET transaction.

The root certification authority is at the top of the SET certification structure. It is the ultimate verifier of the authenticity of the participants and controls the delivery of the electronic certificates. CertCo & Digital Trust

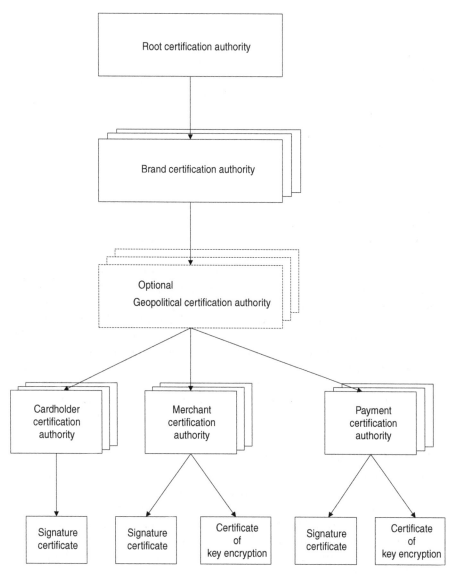

FIGURE 7.5
Hierarchy of certificate management.

Company was selected to fulfill this role. It is responsible for issuing the root key for public key cryptography of length 2048 bits that serves to generate the necessary certificates to authenticate all the other actors. CertCo was founded in December 1996 as a subsidiary of Bankers Trust, which merged in 1998 with the Deutsche Bank. However, CertCo was not supposed to be the sole guardian of the key, as the key was intended to be divided into fragments, with each fragment probably being owned by one of the large card issuers.

The brand certificate authorities are in the next level of the certification hierarchy. This is the level of the various organizations for management of bankcards, principally Visa and MasterCard, or credit cards, such as American Express. These authorities can delegate their responsibilities in a given zone to geopolitical certification authorities whose role is to accredit local authorities. It should be noted that, for all of these authorities, the size of the keys for public key cryptography is 1024 bits.

Figure 7.5 shows that end entities in SET have two types of certificates: signature certificates and key encryption certificates. Signature certificates are used to verify the signatures, while key encryption certificates are used when the secret key used for the symmetric encryption is encrypted. A pair of keys corresponds to each type of certificate, so that there are signature keys and key encryption keys; for legal reasons, the signature keys are larger. Because the cardholder is not allowed to have a certificate other than a signature certificate, cardholders have to use the keys that the other parties send them to encrypt and exchange symmetric keys.

The merchant and the payment gateway, in contrast, can have the two types of certificates. Local certification authorities own a third type of certificate to show that they are allowed to certify end entities. The authority that certifies the payment gateway needs to have the additional certificate to sign the certificate revocation list. In general, the acquirer bank certifies the payment gateway. Authorities of the three higher hierarchical levels have only two certificates: the first authorizes them to certify the lower hierarchical levels, while the second gives them the right to sign the revocation lists. Although the protocol specifies the procedures for certification, the procedures for certificate revocation are left to the individual authorities.

7.3.1.1   Cardholder Certificate

By registering with a certification authority, the cardholder can obtain a signature certificate with the authority's signature. This cardholder certificate guarantees correspondence between a pair of signature keys and the attributes of the cardholder [name, account number, expiration date of the card, personal identification number (PIN), etc.]. According to Version 1.0 of SET specifications, the requirements on cardholder certification depend on the policy of the issuer bank.

The certificate does not contain the name of the cardholder but instead contains an encrypted value that acts as a pseudonym. Similarly, the certificate does not contain the financial references of the payment but contains their digest calculated with a one-way hash function starting with the account number, the expiration date, and the PIN:

$$\text{Hash} = \text{H(account number} \parallel \text{expiration date} \parallel \text{PIN)}$$

where $\parallel$ represents the concatenation operation.

Because the bank knows the financial references, it is able to ascertain whether the certificate really belongs to the cardholder.

7.3.1.2 Merchant Certificates

The merchant needs two pairs of keys: one for message encryption and the other for signature. To reduce the processing needed to deduce the certification path, the merchant may stock on its computer the certificates of the payment gateway. The time savings are important, particularly if the merchant is in touch with several acquirer institutions, because there is a need to reconstruct the certification path for each one of them at every exchange. This is the case, for example, for international commerce. As a consequence, the total number of certificates that a merchant needs may depend on the number of gateways with which it is dealing.

7.3.1.3 Certificate of Financial Agents

The payment gateway is usually certified by the acquirer bank. The certificate for key encryption contains a public encryption key that is used to encrypt the DES key used for encrypting the messages, such as the payment instructions.

If the acquirer bank expects to play the role of a certification authority, it has to own four certificates: a signature certificate, a key encryption certificate, a certificate for signing certificates, and a certificate to sign the certificate revocation list. The acquirer bank, in turn, is certified, either by a geopolitical certification authority, should there be one, or by the brand certification authority that is situated in the next level.

The financial institution that issues the cardholders' cards must have a certificate from the brand certification authority that allows it to certify its clients.

7.3.1.4 Certificates of the Root Authority

The certificates of the root authority are self-signed and are accessible by all end entities of SET. The signature certificate contains the digest of the master key as well as a replacement key. This replacement key assures service continuity, because each time the master key is renewed, new certificates are issued with the replacement key as the master key. The validity of a new certificate is verified by comparing the digest of the new key in the new certificate with its digest in the old certificate.

An end entity can, at any time, verify the validity of the certificate of the root authority by querying its own certification authority. The latter is supposed to know the digest of the certificate of the root authority and, therefore, can verify if the two digests are identical. If the comparison shows that the root certificate is not valid, the user can request a new certificate.

TABLE 7.3

Schedule for Certificate Renewal

Entity	Signature Certificate	Key Encryption Certificate	Certificate for Signature Certificate	Certificate for Signature CRL[a]
Cardholder	3 years			
Merchant	1 year	1 year		
Payment gateway	1 year	1 year		
Cardholder certification authority	1 year	1 year	4 years	
Merchant certification authority	1 year	1 year	2 years	
Payment gateway certification authority	1 year	1 year	2 years	2 years
Geopolitical certification authority			5 years	2 years
Brand certification authority			6 years	2 years
Root authority			7 years	2 years

[a] CRL = Certificate Revocation List.

7.3.1.5 Certificate Durations

SET specifications (Book 2, Appendix T, pp. 566–569) suggest that various certificates be renewed according to the schedule shown in Table 7.3. The private keys of all entities are renewed annually, except that of the cardholder, which lasts 3 years.

7.3.2 Registration of the Participants

7.3.2.1 Cardholder Registration

Registration can be by e-mail or on the Web by clicking on a special button. Registration of the cardholder comprises the three phases shown in Figure 7.6: initialization, request for the registration form, and presentation of the filled-in form. Listed in Table 7.4 are the SET messages exchanged between the certification authority and the cardholder during registration.

TABLE 7.4

Messages for Registering the Cardholder

Message	Direction of Transmission	Significance
CardCInitReq	Cardholder → Certification authority	Request to initiate a registration
CardCInitRes	Certification authority → Cardholder	Response to the request to initiate a registration transaction
RegFormReq	Cardholder → Certification authority	Request for a registration form
RegFormRes	Certification authority → Cardholder	Response to the request for a registration form
CertReq	Cardholder → Certification authority	Certificate request
CertRes	Certification authority → Cardholder	Response to a certificate request

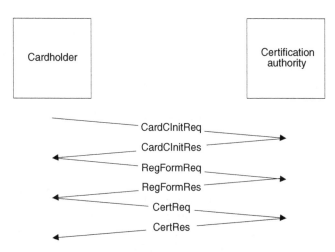

FIGURE 7.6
Messages exchanged during cardholder registration.

Details of cardholder registration are presented as follows.

7.3.2.1.1 Initialization

The registration starts when the cardholder software sends a CardCInitReq to the certification authority to begin the certification procedure. The certification authority responds with the message CardCInitRes, which is signed with the authority's private signature key and which contains the certification authority's certificates for key exchange and signature. Figure 7.7 depicts the composition of the CardCInitRes message.

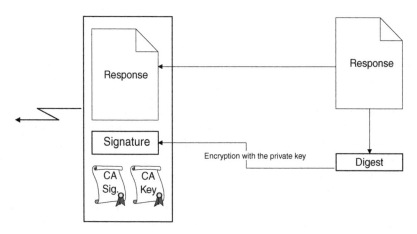

FIGURE 7.7
Formation of the CardCInitRes message.

Upon reception of this message, the cardholder software verifies the authority's signature, either by using the public key that was obtained in a safe manner or by following the certification path to the root authority.

After verification of the certificate, the cardholder is able to protect the confidentiality of the subsequent exchanges. Each message is encrypted with a secret encryption key using the DES algorithm. This secret key will be encrypted with the public key of the authority, so that only the latter is able to decrypt. This processing is identical to that depicted in Figure 7.2.

7.3.2.1.2 Request for the Registration Form

The cardholder must complete a registration form to which the issuer of the bankcard has agreed. To obtain this form, two sets of exchanges between the cardholder software and the certification authority are needed. The first message is RegFormReq and is used to request the registration form. It contains the identity of the financial institution or the issuer bank and the cardholder's payment data, particularly the PAN. The message is encrypted with a random secret key by applying the symmetric encryption algorithm DES. Next, the secret key and the card number are encrypted with the public key of the authority. Figure 7.8 depicts the formation of the RegFormReq message. Note that the messages originating from the cardholder are not signed as yet, because the requested signature was not yet awarded. [Also, note that the first four numbers of the PAN form the BIN (Bank Identification Number). They identify the issuer bank and the card brand (Visa, Master-Card, etc).]

After receiving and processing the RegFormReq message, the certification authority extracts the pertinent registration form from its archives, if it already has it, or retrieves it from competent sources. It then signs it with its private signature key and encloses its response in the RegFormRes message.

7.3.2.1.3 Completing and Sending the Form

The cardholder's software verifies the authority's signature, following the certification path all the way back to the root authority, if necessary. Once the authenticity of the response is verified, the cardholder's software extracts the registration form and verifies its integrity by decrypting the signature with the certification authority's public key and comparing the decrypted digest with the digest recalculated on the basis of the received form. If everything is correct, the cardholder enters the required data in the form to request the certificate. The cardholder's software generates the pair of private and public keys and three random numbers. The first random number is the secret key that will serve to encrypt the message before its transmission with the symmetric algorithm DES. The second random number is the symmetric key that the certification authority will use to encrypt the certificate. The third random number is attached to the registration form and will be used by the certification authority as explained below.

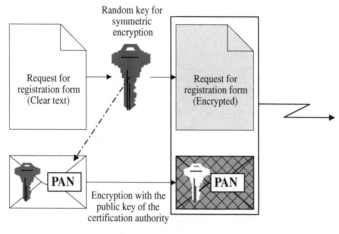

Cardholder

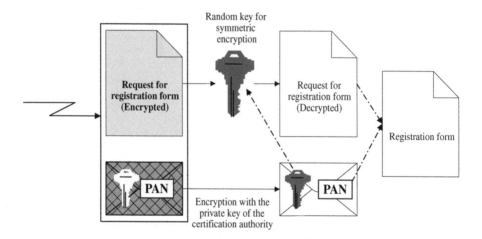

Certification authority

FIGURE 7.8
Processing of the RegFormReq message by the cardholder's software and the certification authority.

Thus, the CertReq that the cardholder transmits contains the registration form, the cardholder's public key for signature that was just generated, and the second secret key. The whole message is hashed and the digest encrypted with the cardholder's private signature key to produce the message signature. Next, the message is encrypted with the first secret key, which, in turn, is encrypted with the public key of the certification authority. Figure 7.9 summarizes the processing steps used to form the CertReq message.

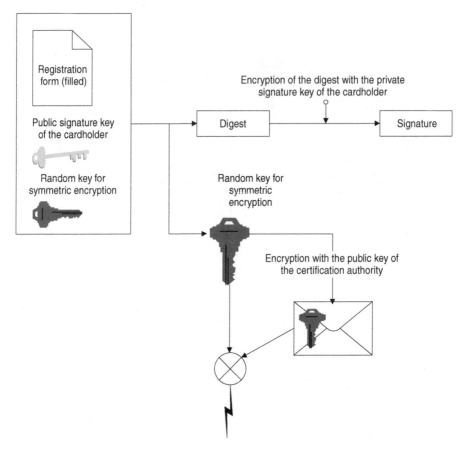

FIGURE 7.9
Formation of the CertReq message.

At this stage, the authority has all the information necessary to check the certification request according to the regulations of the card issuer institution; these rules are independent of SET. Once the verification is done, the authority proceeds to issue the certificate. The authority uses the third random number it received from the cardholder to construct another secret that it keeps to itself. It then computes the hash of this new secret, the PAN, and the card expiration date, i.e., H(PAN, expiration date, random number). This hash is included in the certificate in order to establish a protected link between the certificate and the user's data.

Note that the use of the random number, or *salting*, defends against dictionary attacks on the account number. In fact, the possible values for the expiration date are limited and could be guessed on the basis of the certificate expiration dates. Similarly, the number of issuing banks is limited. Without the random number (or *salt*), it would be relatively easy to construct a dictionary of all possible numbers formed of the concatenation of account numbers and expiration dates and then systematically compare the calculated hash values with the value in the certificate. When the two numbers match, the PAN would be obtained.

The authority signs the certificate with its private signature key and assigns it an expiration date, according to the operating rules. The third random number that the cardholder generated will be resent to the cardholder (to protect against replay attacks). The CertRes is encrypted with the symmetric DES that the cardholder included in the CertReq message. After receiving and processing the CertRes message, the cardholder stores the certificate in order to be able to participate in SET transactions.

7.3.2.2 Merchant Registration

Merchants must be registered by a certification authority before they can exchange SET messages with buyers and payment gateways. In a manner similar to the client's registration, the merchant engages in a series of exchanges of SET messages with its certification authority to obtain a signature certificate and a key encryption certificate. Table 7.5 contains the set of messages that the merchant's server exchanges with the certification authority during the merchant's registration and certification. These exchanges are depicted in Figure 7.10.

The merchant's software starts the conversation by sending the Me-AqCInitReq message. This command is simultaneously a request for registration and for the registration form. The authority responds by attaching the form to the Me-AqCInitRes message that also contains the authority's signature certificate so that the merchant can verify the signature by following the certification path all the way back to the root authority.

According to SET, the merchant is obliged to have two certificates: the first is for signatures, whereas the second is for encryption of the symmetric keys. As a consequence, the merchant's software has to produce two pairs of public/private keys: the first for signature and the second for encryption. Thus, the merchant's software transmits the filled-out registration form with the two public keys attached, as well as a signature calculated over the whole

TABLE 7.5

Messages for Merchant's Registration

Message	Direction of Transmission	Significance
Me-AqCInitReq	Merchant → Certification authority	Initialization request for certification
Me-AqCInitRes	Certification authority → Merchant	Response to the initialization request
CertReq	Merchant → Certification authority	Certificate request
CertRes	Certification authority → Merchant	Answer to the certificate request

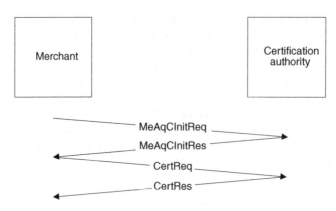

FIGURE 7.10
Merchant's registration and certification.

transmission with the private key for signature. The whole message is encrypted with the secret key according to the DES algorithm. This secret key and the merchant's account number next are put in a digital envelope and then encrypted with the public key for key encryption of the certification authority. The whole set constitutes the CertReq message. However, this is different from the case of the client, because the message contains only one secret key for encryption, namely, the one that was used to encrypt the message during the processing indicated in Figure 7.11.

Following reception of the CertReq message, the authority decrypts the envelope to extract the secret key and then to decipher the certification request. Once the signature is successfully checked, the request is submitted for approval by the acquirer bank according to the rules established by the financial institution; these rules are outside the scope of SET. Once the authorization is obtained, the authority constructs the certificates that the merchant requested, signs them with its private signature key, and then transmits them in the CertRes message, joining its signature certificate. Figure 7.12 shows how this message is prepared and then transmitted to the merchant.

Upon receipt of the CertRes, the merchant's software verifies the signature with the public key for signature of the authority to prevent fraud. After this check, the software extracts the certificates and saves them for later use.

7.4 Purchasing Transaction

In a purchasing transaction, SET intervenes after the client chooses the desired item and selects the means of payment. SET is responsible first for securing the transport of the payment authorization over the network. Next, SET confirms the transaction in a protected manner. Finally, SET is responsible for securing the reimbursement of the merchant through the financial

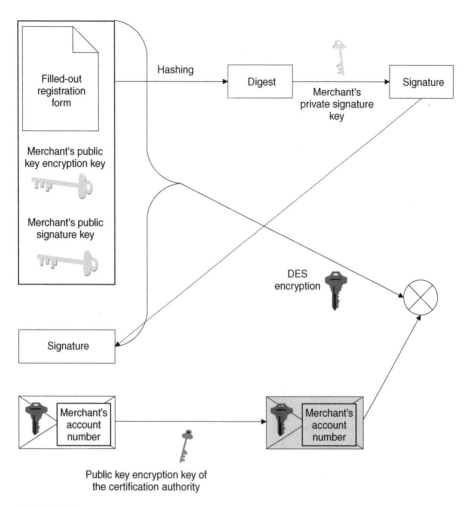

FIGURE 7.11
Processing of the CertReq message at the merchant's site.

clearance process. SET can also secure other operations, for example, payment adjustment, but the principles are the same as for the straightforward case of purchasing and clearance. Therefore, these additional operations will not be discussed, and the reader is invited to refer to SET specifications or to books on SET (Loeb, 1998).

7.4.1 SET Payment Messages

There are two types of payment messages: mandatory and optional. Mandatory messages occur in every purchasing and payment transaction, whereas the optional messages are for complementary services that are not

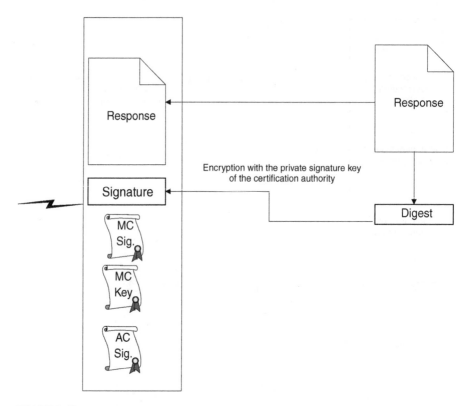

FIGURE 7.12
Composition of the CertRes message by the certification authority. (MC = merchant's certificate; AC = authority's certificate.)

TABLE 7.6

Mandatory Payment Messages in SET

Messages	Direction of Transmission	Meaning
PReq	Cardholder → Merchant	Purchase request
PRes	Merchant → Cardholder	Response to the purchase request
AuthReq	Merchant → Gateway	Authorization request
AuthRes	Gateway → Merchant	Response to the authorization request
CapReq	Merchant → Gateway	Capture request
CapRes	Gateway → Merchant	Response to the capture request

needed in each implementation. Table 7.6 lists the obligatory payment messages, and Table 7.7 lists the optional messages.

Mandatory messages can be grouped into three categories of request/response pairs related to purchase orders, payment authorizations, and financial settlements. The basic exchange for a payment includes the following messages: PReq, AuthReq, AuthRes, PRes, CapReq, and CapRes (Figure

TABLE 7.7

Optional Payment Messages in SET

Messages	Direction of Transmission	Meaning
PInitReq	Cardholder → Merchant	Initialization message to allow the cardholder to obtain the merchant and the payment gateway certificates
PInitRes	Merchant → Cardholder	Response to the PInitReq message
AuthRevReq	Merchant → Gateway	Message used by the merchant to cancel an authorization and to reduce the amount of a transaction already authorized
AuthRevRes	Gateway → Merchant	Response to the AuthRevReq message
InqReq	Cardholder → Merchant	Inquiry on the status of the transaction in progress
InqRes	Merchant → Cardholder	Response to the inquiry in InqReq
CapRevReq	Merchant → Gateway	Message used to cancel a capture
CapRevRes	Gateway → Merchant	Response to the CapRevReq message
CredReq	Merchant → Gateway	Message used to claim credit on a transaction already captured
CredRes	Gateway → Merchant	Response to the CredReq message
CredRevReq	Merchant → Gateway	Request to cancel a CredReq message
CredRevRes	Gateway → Merchant	Response to the CredRevReq message
PCertReq	Merchant → Gateway	Request by the merchant for the payment gateway certificate
PCertRes	Gateway → Merchant	Response to the PCertReq message
BatchAdminReq	Merchant → Gateway	Message for administering capture request
BatchAdminRes	Gateway → Merchant	Response to the BatchAdminReq message
Error	All	Error message

Note: The Error message is sent when the received message is of a type that is not recognized or if the message content is ambiguous and requires clarification.

7.13). The PReq and PRes messages are exchanged between the merchant and the cardholder, whereas the other messages are exchanged between the merchant and the payment gateway.

The messages from the cardholder toward the payment gateway pass by the merchant's server. To preserve the confidentiality of the client's banking data from the merchant, the client has to have the gateway certificate for key encryption to establish an encrypted link with it.

7.4.2 Transaction Progress

7.4.2.1 Initialization

With the optional PInitReq message, the cardholder requests the certificates from the merchant and the payment gateway. This message is transmitted in the clear on the network. The response is the PInitRes message that

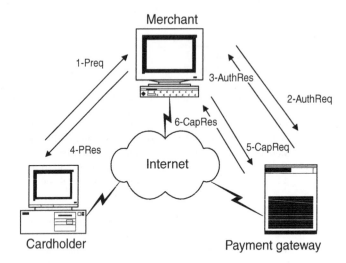

FIGURE 7.13
Mandatory messages in a SET purchase transaction. (From Sherif, M.H. et al., *Proc. ISCC'98*, 1998, pp. 353–358, ©1998 IEEE. With permission.)

contains the merchant's signature certificate, the gateway's key-encryption certificate, and a transaction identifier, TransID. This message is signed with the merchant's private signature key.

Upon reception of the PInitRes message, the cardholder verifies the merchant's signature and the merchant's signature certificate, as well as the key encryption certificate of the gateway by following the trust chain backward to the first trusted authority or even back to the root authority. Once authenticity is proven, the merchant may store the certificates to avoid going back on the itinerary for any new transaction.

7.4.2.2   Order Information and Payment Instruction

The order information (OI) and the payment instructions (PI) are transmitted in the PReq message. The order information identifies the goods to be purchased, whereas the payment instructions contain the PAN, the price of the items, and the transaction identifier TransID, if it was supplied by the merchant. Otherwise, the cardholder software creates the transaction identifier.

The digests of the order information and the payment instructions are concatenated, then encrypted with the private signature key of the cardholder to construct the dual signature of the PReq message. Next, this message is signed with the private signature key of the cardholder. Illustrated in Figure 7.14 is the construction of the dual signature of the PReq message.

The payment instructions, together with the digest and the dual signature, are transmitted after their encryption with a secret key generated with a random number generator. This key, in turn, is encrypted with the public-key encryption key of the payment gateway before transmission, which puts it beyond the merchant's reach. When the cardholder has a signature certif-

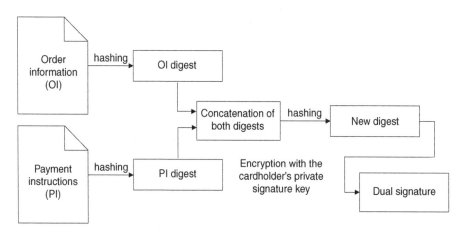

FIGURE 7.14
Dual signature of the PReq message.

icate, this certificate is transmitted as well. Figure 7.15 depicts the components of the PReq message, and it is seen that this is one of the most complex messages of the protocol.

More details on the PReq message are given below to convey an idea of the complexity of the SET protocol and the redundancies that it introduces as a way to guarantee security. Keep in mind that the content of the PReq message depends on whether the cardholder has a signature certificate. If the cardholder is certified, the PReq will be comprised of two parts as indicated in Figure 7.16.

The symbols used in Figure 7.16 are those used in the SET specification. Thus:

- The operator DD(A) designates the digest of A reorganized according to the DigestData content type of PKCS #7 and denoted as digest of A.

- The operator L(A,B) is a linkage operator that represents a link or reference pointer to B attached to A, produced by concatenation of A to the digest of B.

$$L(A,B) = \{A,DD(B)\}$$

- The operator SO(s,t) represents the signature of participant s on message A of type PKCS #7 SignedData.

- The operator EX(r,t,p) represents the following sequence of operations: encryption of message t with a DES key, K; insertion of this key and the variable p in the PKCS #7 envelope with OAEP to form OAEP{(k,p)}; and the encryption of this envelope with r, the public RSA key of the recipient.

- Finally, fields between the square brackets [ ] are optional.

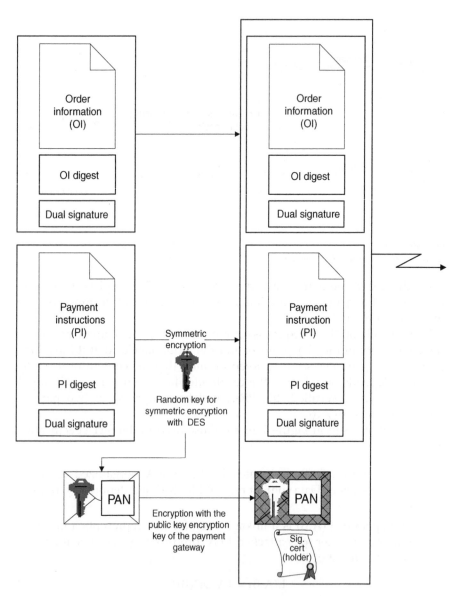

FIGURE 7.15
Construction of the PReq message.

The first component of the PReq message, PIDualSigned, contains the payment instructions and is directed to the acquirer bank with all the necessary indications to identify the cardholder. The second part of the message, OIDualSigned, contains the order information, signed and encrypted, as indicated earlier. Listed in Table 7.8 are the mandatory fields shown in Figure 7.16 and their contents. For more details, the reader is invited to refer to Book 3 of the SET specifications.

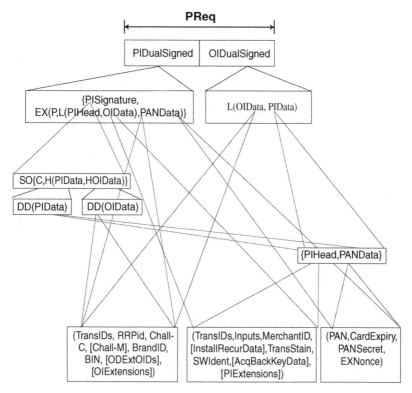

FIGURE 7.16
Composition of the PReq message.

TABLE 7.8

Content of the Mandatory Fields of PReq

Field Name	Contents
Inputs	Digest of the data included in the order information and the purchase amount
OIData	Digest of the order information, the challenge to the merchant to ensure the freshness of the cardholder signature, the card brand, and the issuer bank identifier
OIDualSigned	OIData and digest of PIData
PANData	Card number, expiration date, secret code, nonce to protect against dictionary attacks
PIDualSigned	PISignature and linkage with the order information
PIHead	Merchant identifier, hashed digest HMAC[a] of the global transaction identifier XID and the cardholder secret code, identifier of the software provided, data on the algorithm of the acquirer bank indicated in the gateway certificate
PISignature	Digest of the cardholder signature
TransID	Transaction identifiers

[a] HMAC = Hashed Message Authentication Code.

7.4.2.3   Authorization Request

The authorization request initiates a dialogue between the merchant server and the payment gateway on one side, and between the gateway and the issuer bank on the other side. The merchant server treats the PReq differently depending on whether it contains a signature certificate of the cardholder. If a certificate is available, the server verifies it by going back to the certification path and checking the signature on the order information using the public key of the cardholder mentioned in the certificate. Then, the server verifies the dual signature, and when the security checks are met, it sends back an acknowledgment to the cardholder with the PRes message. This message is transmitted as a clear text and is signed to verify the integrity of transmission upon receipt. The message must contain the merchant's signature certificate.

The merchant's server next relays the payment instructions to the gateway in the AuthReq message. This request is signed with the private signature key of the merchant, and the whole message is encrypted with a symmetric DES key. Next, this key is enveloped electronically, together with the merchant's signature and encryption certificates and, if available, the signature certificate of the cardholder. The envelope then is encrypted with the public encryption key of the payment gateway. Figure 7.17 depicts the different elements that constitute the AuthReq message.

7.4.2.4   Granting Authorization

Upon receiving the AuthReq message, the gateway checks the received certificates by going back along the certification path. It decrypts the envelope of the request for payment authorization with the help of the private key of the pair of key-encryption keys to extract the symmetric encryption key. This symmetric key is then used to decrypt the request. The integrity of the request is verified by the public signature key of the merchant.

The payment gateway performs similar verifications for the payment instructions, extracts the symmetric key used by the cardholder and the cardholder's PAN, and then verifies the dual signature with the public signature key of the cardholder. Furthermore, the gateway must check the consistency of the request for payment authorization with the payment instructions, for example, that the transaction identifier is the same in both parts.

Once this stage is successfully terminated, the payment gateway prepares an authorization request and sends it to the issuer bank through the bankcard network. The steps are not part of SET and will not be explained here. Suffice it to say that bankcards have a spending cap, so if the expense ceiling of a cardholder is reached, a referral procedure kicks in with an attendant intervention by phone before authorizing the payment (Dragon et al., 1997, pp. 171–179). The merchant software is supposed to take care of these eventualities.

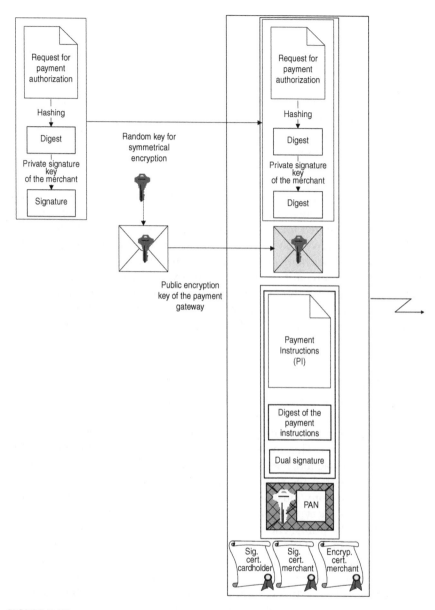

FIGURE 7.17
AuthReq message transmitted to the payment gateway by the merchant.

Once the authorization arrives, the gateway sends the response message
AuthRes. This message is encrypted with the DES algorithm with a secret
random number, which is put in an envelope that is encrypted with the
public key of the key-encryption pair of the merchant. The AuthRes message
comprises two distinct parts:

- One mandatory part is the response of the issuer bank authorizing the payment.

- One optional part contains a capture token signed with the private signature key of the payment gateway. This token must accompany any capture request associated with this transaction to facilitate daily remote collections. The token is encrypted with a second secret key that the gateway generates. As before, the secret key will be inserted in an envelope that is encrypted with the public encryption key of the payment gateway. Thus, this envelope can only be opened by the payment gateway.

When it receives the AuthRes message, the merchant server extracts from this message the response of the issuer bank to the request for authorization as well as the encrypted token. The server decrypts, with its private encryption key, the first envelope to recover the secret key and to decipher the response of the issuer banks to the authorization request.

In contrast, the merchant server cannot recover the capture token, because the DES encryption key remains in the envelope encrypted with the public encryption key of the payment gateway. The envelope containing the DES key and the encrypted token is stored as is and then used to accompany capture requests.

7.4.2.5   Capture

This is the procedure whereby the merchant claims the credit that was implicitly accorded during transaction authorization. The financial settlement is initiated by sending the CapReq message to the payment gateway to request capture or collection. This message includes, among other information, the amount of the transaction, the transaction identifier, the reference to the authorization, the encrypted and signed capture token, and the envelope containing the encryption key previously saved. Similar to all other SET messages, the capture request is encrypted with a secret key that is inserted in an envelope encrypted with the public encryption key of the payment gateway.

This request can take place directly after the purchase or at regular intervals, by grouping several purchases for efficiency reasons. One capture message can thus include several capture tokens from several transactions. The requests are given a sequence number to distinguish among them and to associate a response with its associated request.

Upon receipt of the message, the payment gateway extracts the secret key from the envelope to decrypt the capture request. It then proceeds to check the signature, as indicated previously. The gateway then extracts the envelope that contains the symmetric key used to encrypt the capture token and decrypts it with its private-key encryption key. Thus, it will be the only entity capable of extracting this symmetric key to decrypt the capture token. The payment gateway sends a clearance request using all the collected information to the acquirer bank.

To conclude, the gateway responds to the merchant with the CapRes message. This message contains the response of the acquirer bank with the amount finally credited to the merchant account. This message is signed with the private signature key of the payment gateway. The CapRes message is encrypted with a secret key that is included in the envelope encrypted with the public-key encryption key of the merchant. Verification of the message by the merchant's server proceeds according to the procedures explained several times earlier.

7.5 Optional Procedures in SET

The merchant can initiate several procedures by sending optional requests to the payment gateway. These procedures are as follows:

1. Modification or cancellation of a previous authorization with the help of the exchange of AuthRevReq and AuthRevRes messages. The merchant can place a request at any moment after the authorization is obtained but before the capture request.

2. Modification or cancellation of a capture with the messages CapRevReq and CapRevRes.

3. Refund of the cardholder with the messages CreditReq and CreditRes (This exchange occurs after reconciling the accounts of the merchant with the issuer bank. The response message confirms or rejects the request for refund.)

4. Cancellation of a refund. The optional message CredRevReq gives the merchant the option to request cancellation of a refund already accepted. The message from the gateway CredRevRes contains the decision of the issuer bank regarding the cancellation.

5. Request for the gateway's key encryption certificate. The merchant request is in the PCertReq, whereas the response of the payment gateway is in the PCertRes message.

6. Grouped settlement of a batch of capture tokens. The merchant sends the AdminReq intended to the acquirer bank through the payment gateway by joining the accumulated capture tokens, usually every 24 hours. The message can indicate the start or the end of the processing of a batch of tokens, or an inquiry on the status of a batch. The response from the bank is relayed through the gateway within the AdminRes message.

7.6 SET Implementations

SET is modular in the sense that each phase has its own set of messages, which facilitates segmentation of implementations according to design requirements. For example, with a JAVA platform, the deployment of SET can be done through downloadable JAVA applets.

SET promoters, particularly Visa and MasterCard, funded development of a reference implementation, SETREF (Secure Electronic Transaction Reference Implementation). The SETREF is written in C and operates on UNIX, the various Windows releases, and Windows NT. SETREF is also distributed in a CD with the book by Loeb (1998). SET Secure Electronic Transaction LLC (SETCo) (www.setco.org) maintains SET specifications, manages SET compliance and interoperability testing, and issues licensing agreements.

SETREF utilizes the cryptographic toolkit BSAFE, which is produced by RSADSI, but is subject to U.S. export restrictions . Accordingly, the tests cover symmetric encryption with DES with keys of 40 and 56 bits, and asymmetric encryption with RSA with keys up to 2048 bits. Furthermore, the conformance tests cannot be exported or reexported to Cuba, Iran, Iraq, Libya, North Korea, Sudan, or Syria. Even though it is possible to use SETREF without using the cryptographic modules, this gap is a major inconvenience. To overcome this drawback, the worldwide academic community developed an interface compatible with BSAFE with the use of the CAPI of SSLeay, already mentioned in Chapter 5.

SETREF comprises seven applications: three payment applications and four registration applications. The software architecture is shown in Figure 7.18.

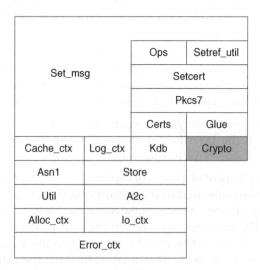

FIGURE 7.18
Software architecture of SETREF.

The main modules of SETREF are as follows:

- *A2c*, a module of conversion from ASN.1 to the DER format
- *Alloc_ctx*, for memory allocation
- *Asn1*, a module for definition of the ASN.1 types
- *Cache_ctx*, which creates a cache to replicate the data from one message to the other and to know the state of a transaction
- *Certs*, a module handling the operations of certification as per X.509
- *Crypto*, which is the cryptographic module that cannot be exported from the U.S.
- *Error_ctx*, a module that gives explicit error messages
- *Io_ctx*, which handles the input/output interfaces
- *Log_ctx*, which logs the operations
- *Set_msg*, the module that acts as the interface between the application program and SET
- *Util*, which provides generic tools for applications programming

SETREF has internal operating modules that the application program does not deal with directly:

- *Glue*, for internal operations
- *Kdb*, a database
- *Ops*, a module of cryptographic operators
- *Pkcs7*, to implement the PKCS #7 functions
- *Setcert*, to process X.509 certificates
- *Setref_util*, a set of generic utilities for SET
- *Store*, a cache database

RSADSI produced S/PAY, a toolbox with demonstration software to show how to integrate SET with payment software.

7.7 Evaluation

To ensure the security of all bankcard payments on the Internet, SET designers attempted to overcome the fundamental weakness of IP Networks (where management and control information are mixed with users' traffic) by requiring strong authentication of each message exchange. This decision increased the complexity of the protocol and the computational overload (Inza, 2000; Schneier, 1998b). Thus, a certification infrastructure must be established, and

TABLE 7.9

Comparison between SET and SSL

SSL	SET
Simple and easy to use	Complex and requires a certification infrastructure
Generalist protocol	Banking payment protocol
Distributed with browsers	A wallet has to be installed on the client side
Authentication infrastructure is not mandatory	Infrastructure for authentication is mandatory
Authentication is at the beginning of the session	Each exchange is authenticated
Point-to-point protocol	Several parties participate in a transaction
The merchant receives all the details of the order and the payment	With the dual signature, access to information is restricted to those who need it

users have to install software (or wallet) of 4 to 6 M octets on their machines. In contrast, SSL is much simpler, and its distribution accompanies a browser. Summarized in Table 7.9 are the main points of contrast between SET and SSL.

The comparative information in Table 7.9 will help us to understand the resistance to SET. In addition, other factors contributed to slowing its utilization, such as the following:

- Legal aspects, particularly when the legislation limits the use of encryption
- Secrets on the cardholder side stored on the hard disk of a computer, which may not give sufficient security (This is one of the factors that stimulated the development of C-SET as an extension of SET, adapted to smart cards, and which is treated in the next chapter.)

Some attempts to simplify SET (from the client's point of view) by using SSL will be presented in the next chapter. Inversely, SET *fácil* (Easy SET) — a solution that the Spanish bank Banesto from the Santander Central Hispano group as well as IBM advocate. It consists of including the Payment intermediary in the bank's domain. As a consequence, the server takes care of all operations that need intense computations, such as processing of a client's certificate. Because of its reduction in computational loads, the software installed on the client's post is called a *thin wallet* (Inza, 2000).

7.8 Summary

The principal characteristics of SET are as follows:

- The merchant keeps the order information that is signed with the private signature key of the client. The merchant also retains the response of the gateway (in the AuthRes message) signed with the private signature key of the gateway. If the client is certified, the merchant has a copy of the cardholder certificate and the public key that it cites. Nevertheless, the merchant does not have the details of the client's bankcard.

- The cardholder receives a response to the purchase order, which is signed by the private signature key of the merchant. The cardholder receives a copy of the merchant signature certificate but not the encryption certificate, which is used for financial settlement.

- The payment gateway knows the financial details of the transaction between the merchant and the cardholder without being aware of the subject of the transaction.

- Each of these transactions has a unique transaction number that is encrypted, which prevents replay attacks.

Nevertheless, the security offered uses complicated means, which translates into a computation overload, and, consequently, the response time may not be adequate.

Questions

1. What is the purpose of the dual signature in SET?
2. Compare the client-based wallet of SET with server-based wallets.
3. Go to the site http://epso.jrc.es/purses.html and summarize the most recent developments in SET.

8

Composite Solutions

ABSTRACT

From the previous chapter, it is clear that Secure Electronic Transaction (SET) imposes a larger computational load than Secure Sockets Layer (SSL) because of the authentication of each message exchange. This makes it too expensive for small payments. In addition, SET is not readily adapted to payments with integrated-circuit cards.

This chapter consists of three parts. First, we discuss attempts to extend SET to smart cards, particularly the Chip-Secured Electronic Transaction (C-SET) protocol. Then we see how the joint operation of SSL and SET allows a reduction of the cryptographic load that SET imposes on client systems and merchant servers. The combination of the two protocols simplifies the certification procedure without sacrificing the protection of the client's financial data. These two extensions retain the fundamental characteristics of SET:

- The cardholder secrets, such as the bankcard number, are hidden from the merchant during a transaction.
- The merchants and, optionally, the clients are certified.

Finally, we discuss Three-Domain Secure (3-D Secure), the most recent Visa initiative to secure remote bankcard payments using only SSL/TLS.

8.1   C-SET and Cyber-COMM

The use of a personal identification number (PIN) associated with an integrated circuit card to authorize banking transactions reduced fraud dramatically; in France, for example, the rate dropped to 0.023% in 1996 and even to 0.018% in 1998. Clearly, banks have an interest in investigating the use of smart cards to secure payments on the Internet. This is the reason SETCo investigated ways to ensure the compatibility of SET with the new generation of chip cards based on the EMV (EuroPay, MasterCard, Visa) specifications. One document discusses the authorization and financial settlement of SET transactions (SETCo, 1999a). Another defines online identification of cardholders using PINs entered from the keyboard or via secure PIN-pad readers (SETCo, 1999b).

Two French consortia were in competition to adapt SET to smart cards. A joint project of the Gie (Groupement d'Intérêt Économique), Cartes Bancaires, and Europay aimed at establishing a new standard — C-SET (Groupement, 1997). The second project, E-Comm, was a commercial offer supported by BNP, Société Générale, Crédit Lyonnais, Visa, France Télécom, and Gemplus. These two efforts converged into Cyber-COMM, which is technically based on C-SET and which includes the BO' cards used in France, as well as the EMV cards.

Let us now consider the technical details of C-SET.

8.1.1   General Architecture of C-SET

According to the framework of C-SET, the cardholder must enter a confidential code in a secure card reader to authorize a payment. This secret code is considered to be the equivalent of a handwritten signature. As a consequence, the software in the card as well as in the bank network will execute the necessary procedures. In contrast, the SET protocol requires the user to click on a mouse to select a course of action. The software that resides on the user's computer is responsible for carrying out the command and retrieving the credentials stored on the user's computer. Clearly, the security of the overall operational environment in SET is more critical for payment security than in the case of C-SET. Finally, storage of secrets in the integrated circuit card protects them against viruses and computer failures.

Figure 8.1 shows how C-SET proposes to amend the SET architecture to accommodate a smart-card reader that reads the data securely.

The C-SET architecture adds the following parties to the actors that intervene in the SET transactions:

- The secure card reader or PIN-pad reader, which is the peripheral device used for the cardholder to enter the PIN securely. It takes the form of a box resembling a calculator with a keyboard and a screen. It includes software and a cryptographic module. The reader must be tamper resistant and must not store any secrets. Thus, the PIN should travel directly from the cardholder to the reader and avoid the computer, which is the weak link in security. Note that secured readers can be installed in specialized locations, for example, in an automated teller machine (ATM).

- The C-SET registration server, which awards the C-SET certificates to the participants.

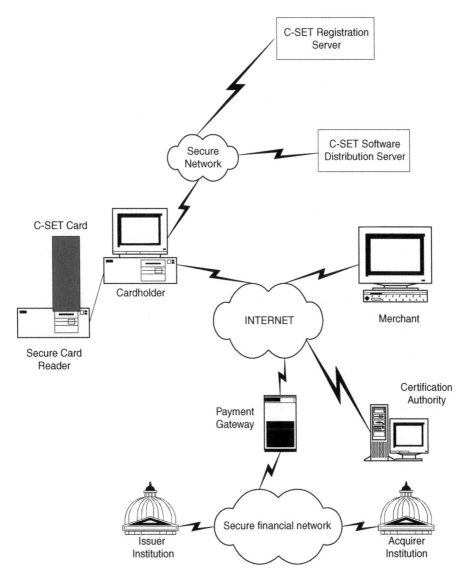

FIGURE 8.1
General architecture of C-SET.

- The C-SET software distribution server, which is an agent of the issuer bank that assures the download or the distribution of the secure software and its installation. This approach allows the automatic integration of other applications as they become available, or as EMV multiapplication cards become available.

Several components can coexist in the same physical machine. In the long run, the readers may become integrated with microcomputers; however, they have to be approved by the Gie Cartes Bancaires.

The C-SET payment gateway functions as the agent of the acquirer bank, as is the case with SET, and as a C-SET/SET converter to give C-SET-certified merchants the opportunity to work with partners who are SET certified.

Figure 8.2 identifies those secured components of C-SET that are shared with SET. It is noted that the scope of C-SET includes production of the card (a subject that will be further explained in Chapter 13) as well as secure distribution of the software.

8.1.2   Cardholder Registration

Registration of the cardholder at the registration server is the means for certification. The steps of registration are illustrated in Figure 8.3 and are outlined below:

1. The cardholder connects the PIN pad to the computer, activates the browser, connects to the issuer C-SET registration server, and sends a registration request.

2. The registration server sends a form to the cardholder.

3. The cardholder fills out the form and sends it back to the registration server.

4. The registration server checks all the necessary information and then downloads the PIN-pad reader verification software to the cardholder's computer and activates it.

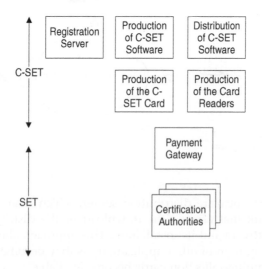

FIGURE 8.2
Secured components of SET and C-SET.

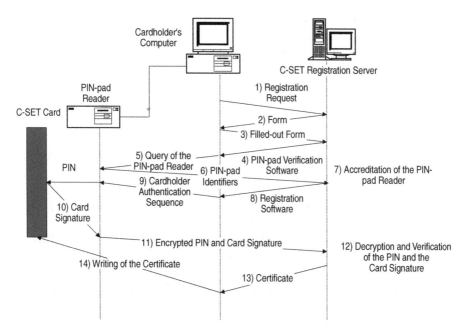

FIGURE 8.3
Cardholder's registration in C-SET.

5. The PIN-pad reader verification software scans the cardholder's computer ports and reads its brand and model.

6. The verification software sends the PIN-pad reader identifiers to the registration server.

7. The registration server checks that this PIN-pad reader is on the approved list. Then, it downloads to the cardholder's computer the registration software that corresponds to the type of PIN-pad reader and activates it.

8. The registration software invites the cardholder to insert the card and then sends to the PIN-pad reader a sequence of instructions (cardholder authentication sequence).

9. The PIN-pad reader receives the authentication sequence and, after performing the necessary checks, executes the authentication sequence. It then asks the cardholder to key in the PIN. If the PIN verification is successful, it sends the card a command to compute a digital signature.

10. The PIN-pad reader receives the response of the card, which includes the digital signature of the card.

11. The PIN-pad reader encrypts the response of the card with public-key encryption and sends the encrypted message to the registration software on the cardholder's computer. The registration software relays the message to the registration server.

12. The registration server receives the encrypted message, verifies the card's signature, and then generates a certificate.

13. The registration software on the cardholder's computer receives the certificate and verifies that it is using the public key to ensure its authenticity.

14. The verification software writes the certificate on the card and then deletes itself.

The following differences between the certification procedures of C-SET and SET are noteworthy:

- A PIN-pad reader of an accredited brand and model is necessary to complete the cardholder's certification.
- The cardholder is asked to insert his or her card and enter the PIN code during registration, which verifies the end-to-end authentication procedure.
- The certificate is written on the smart card and not on the hard disk of the user's computer, as in SET.

8.1.3 Distribution of the Payment Software

Distribution of the C-SET software is under the control of the issuer bank. However, the bank can delegate this responsibility to a qualified agent. The software distribution server can work online or offline (using diskettes, for example). In Figure 8.4, the exchanges during online distribution are presented.

The steps are similar to those for the cardholder's registration. The differences are in Steps 8 and 13:

- In Step 8, the software downloaded is the payment software and the activation software. The activation software takes control of the cardholder's computer.
- In Step 13, the download server sends a signature of the software with a signature certificate. The activation software receives the certified signature and verifies it using the server's public key. If the verification is successful, the software writes the certified signature in the cardholder's computer and then deletes itself.

8.1.4 Purchase and Payment

A purchase and payment transaction involves the exchange of messages among the C-SET card, the PIN-pad card reader, the software installed on the cardholder's computer, the merchant's server, and the C-SET payment

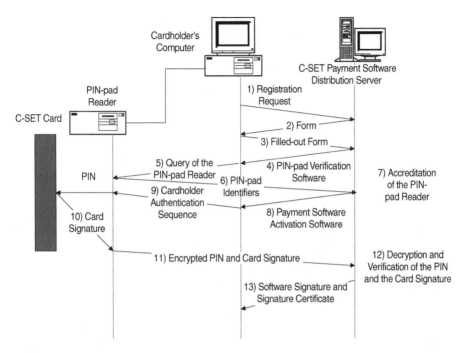

FIGURE 8.4
Online distribution of the C-SET payment software.

gateway that acts as the interface with banking networks. Figure 8.5 presents the exchanges among the various parties during a purchase transaction with payment.

The transaction proceeds as follows:

1. The cardholder's software asks the PIN-pad reader for its identification in order to check that it is connected.

2. The card reader returns its identifiers.

3. The cardholder's software requests the card for the cardholder's certificate.

4. The card returns the certificate.

5. The cardholder sends the purchase order with the certificate to the merchant's server.

6. The merchant server verifies the cardholder's certificate using the public key of the C-SET certification authority. If the verification is successful, it sends the merchant's certificate to the cardholder.

7. The cardholder's software verifies the merchant's certificate. It then displays the merchant's identity (name, address, Internet address, etc.) and the order for the user. Then, it sends the payment instructions and the digest of the order to the PIN-pad reader.

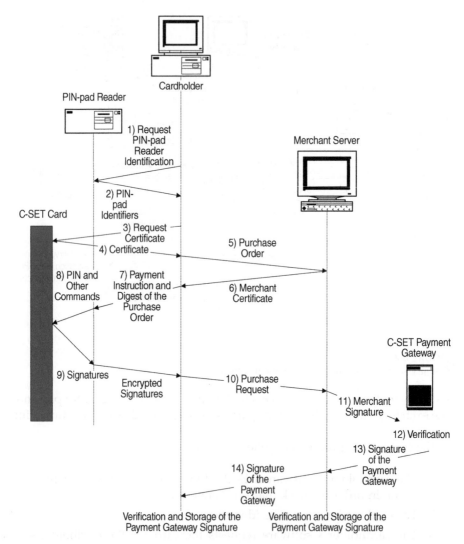

FIGURE 8.5
Message exchanges during a C-SET purchase transaction.

8. The PIN-pad reader displays the amount for the cardholder's approval. This approval is indicated by keying in the PIN, which is sent to the card with a request for the card signatures on the payment instructions and the digest of the purchase order. The PIN and other card coordinates are encrypted with a symmetric DES key that is shared with the gateway. In this manner, the merchant does not have access to the financial data of the cardholder.

9. When the PIN-pad reader receives the messages that the card signed (the payment instructions and the digest of the purchase order), it

encrypts them with the public key of the payment gateway. This assures their confidentiality with respect to the merchant.

10. The cardholder's software sends to the merchant's server a signed purchase request that contains the cardholder's certificate and the encrypted message.

11. The merchant server receives the purchase requests, verifies the order, computes its digest, signs it with its private key, and forwards all the messages to the payment gateway, requesting an authorization from the issuer bank.

12. The payment gateway receives the authorization request and performs the necessary identity verifications. Thus, after decrypting the digests of the purchase order computed by the cardholder and the merchant, respectively, each with its own private encryption key, the payment gateway will confirm that both are identical and send a signed response message to the merchant.

13. The merchant server receives the payment gateway's authorization, authenticates it with the gateway's certificate, and verifies that the digest of the purchase order received from the payment gateway is identical to the one it sent. It stores the signature received and forwards the response to the cardholder.

14. Similarly, the cardholder's software authenticates the gateway's response and stores the response.

The main differences in a payment transaction by C-SET and SET are the following:

- C-SET does not use the dual-signature technique. Protection of the financial data from access by the merchant is done through symmetric encryption between the card and the gateway.

- The cardholder sends the purchase order twice (Messages 5 and 10). The second time, it is accompanied by the encrypted messages to the payment gateway from the card as well as from the PIN-pad readers.

8.1.5 Encryption Algorithms

C-SET utilizes symmetric and public-key encryption to offer the required security services. The algorithms used are listed in Table 8.1. Nonrepudiation is achieved because cardholders must key in their PINs, which triggers the card to sign the payment instructions under the control of the PIN.

To protect the card number PAN (Primary Account Number), a pseudonym is used in the exchange. This pseudonym is calculated from a secret key shared among all banks. From this key, the gateway can reconstruct the account number. The details of this algorithm have not been published. Note

TABLE 8.1

Encryption Algorithms Used in C-SET

Algorithm	Characteristics	Services
DES	Key length = 56 bits	Confidentiality of the card number
RSA	Key length = 1024 bits	Authentication of the cardholders, the merchants, the payment gateway, the software of the cardholder, and the software of the PIN-pad reader; confidentiality of the payment instructions
SHA-1	—	Generation of the digests

that in conformance with French legislation, only the confidentiality of the messages to the payment gateway is protected.

In C-SET, the registration server plays the role of the certification authority and delivers the certificates signed with its private key. The C-SET specifications indicate that the integrity of the distributed software is verified after download using signature keys. Similarly, the integrity of the PIN-pad reader software is verified with the signature keys.

8.1.6 Interoperability of SET and C-SET

C-SET exchanges were defined in such a way that they can be integrated with SET exchanges. This means that a transaction can take place even if one of the parties conforms to SET and the other to C-SET. The security level achieved is the same as in C-SET.

8.1.6.1 Case 1: Cardholder Is C-SET Certified and Merchant Is SET Certified

This case assumes that SET is the merchant's protocol of choice for handling bankcard payments on a global level. A C-SET-certified cardholder chooses to pay with a security solution that conforms to C-SET. In this case, the C-SET/SET converter acts as an agent of the cardholder to translate the C-SET payment instructions into a SET message, such as PReq, that is relayed to the merchant server. In the reverse direction, the merchant's response, for example PRes, is converted to the C-SET format. This role is depicted in Figure 8.6. The converter must be SET-certified as a buyer and C-SET certified as a merchant. If the roles of converter and payment gateway are combined, the converter may be able to use the SET signature certificate of the payment gateway for signing the various SET messages.

8.1.6.2 Case 2: SET-Certified Cardholder and C-SET-Certified Merchant

This case is the reverse of the previous one. A cardholder somewhere in the world with a SET-conforming bankcard makes a purchase from a merchant

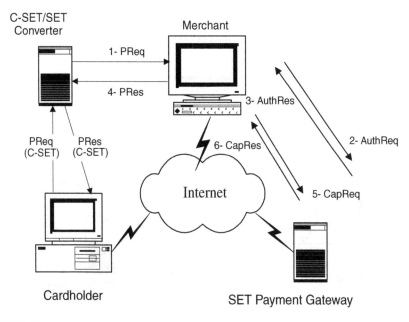

FIGURE 8.6
Role of the C-SET/SET gateway as an agent of the cardholder.

that is C-SET certified. The SET/C-SET converter translates the SET messages into their C-SET equivalents, and vice versa. Figure 8.7 illustrates this role. The converter function is shown separate from the payment gateway function, although both functions are often implemented within the same platform. In this case, the SET/C-SET converter must be SET-certified as a merchant.

8.2 Hybrid SSL/SET Architecture

This section presents another hybrid payment architecture that combines the security advantages of SET with the simplicity of SSL in card transactions. From its inception, the SET protocol was intended to be a solution dedicated to electronic commerce (e-commerce) that would preserve the existing interfaces with bank networks. Accordingly, the protocol gives the buyer the ability to verify the authenticity of the merchant site with the help of certificates. In addition, the method of the dual signature in SET links the purchase order that the client sends to the merchant as well as the payment instructions directed to the issuer bank via the SET payment gateway with the same signature. Thus, the payment instructions are sent encrypted to the merchant, which merely forwards them to the payment gateway. Furthermore, only

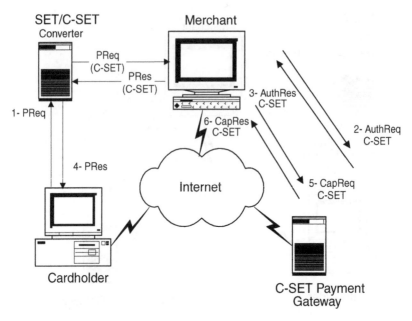

FIGURE 8.7
Role of the SET/C-SET gateway as an agent of the merchant.

the merchant can read the purchase order, which remains opaque to the gateway, and only the payment gateway is allowed to extract the purchaser's bankcard number to submit an authorization request to the bank network.

Nevertheless, establishing the necessary hierarchy of SET authorities introduces a heavy certification structure that causes computational delay for each exchange. Even though the certification of the cardholder is optional in the first version of SET, it is mandatory to verify the certificates at each request/response pair. Because, in general, the certification authorities of the client, the merchant, and the payment gateway are not necessarily the same, it is necessary to track backwards on the certification path until a common higher authority, if not the root authority, is reached, so as to verify the authenticity of the certificates. This heavy operational burden restricts the protocol to transactions beyond a certain minimum value and makes it unsuitable for small amounts.

In contrast, the SSL protocol is session oriented; that is, authentication checks are performed only at the beginning of a session. As already indicated, SSL intervenes at the level of the transport layer to secure the whole connection with the same security parameters (same keys and same hashing and encryption algorithms). Among the other advantages of SSL is the ease of implementation, because it is already available in all browsers. Finally, the computational load of SSL is much lighter than that of SET. All of these factors spurred the use of the SSL protocol to secure consumer-oriented commerce on the Internet, even though SSL remains limited to point-to-point connections.

It is reasonable, then, to look into another architecture relying on converters — this time to bridge SET and SSL. This architecture would retain the benefits of SET while enjoying the simplicity of SSL. Clients and merchants could continue to use their SSL applications, while SET services would be assured on the bank side. This solution necessarily relies on a payment intermediary to manage the conversions between SET and SSL and to act as the proxy and the guarantor of its clients (consumers and merchants) to SET-certification authorities. For example, the payment intermediary could manage the SET certificates for customers and save them from the concomitant administrative loads. At the same time, the intermediary would guarantee the authenticity of those it represents, with respect to each other as well as with respect to the SET authorities. Similar to the Minitel model, the exchanges between a client and a merchant would go through the intermediary. The intermediary could also play the role of a gateway SET/SSL. In this case, the client and the merchant would talk to the payment intermediary on a channel secured by SSL. However, the intermediary would use the protocol SET to communicate with the payment gateway. In this solution, CGI (Common Gateway Interface) scripts present the merchant server to the user as an online electronic payment terminal.

This approach is similar in many aspects to the classical model of the Minitel for the PSTN (Public Switched Telephone Network). For the Internet, it resembles the solution called PayLine that INTRINsec and Experian (formerly SG2) developed. PayLine, in effect, uses a payment intermediary and operates as follows (Barbaux, 1998):

1. After receiving the purchase order from the buyer, the merchant server sends the information related to the purchase to the PayLine intermediary in the form of a "sales slip" on a link secured with SSL and then directs the client to the intermediary.

2. The buyer sends the payment information to the payment intermediary on the SSL-secured link. The payment intermediary verifies the information and, if needed, requests payment authorization from the issuer bank.

3. The intermediary forwards the confirmation to the merchant and then redirects the client to the latter.

To accomplish a transaction, the client has to address two interlocutors over the Internet, namely, the merchant to place the order and the PayLine payment intermediary to send the payment instructions, which the intermediary sends over the secure bankcard network. PayLine certifies the merchants and, if needed, the clients, while the client's credit verification is the responsibility of the banking circles.

A possible variation of this architecture is to associate the merchant's bank with the payment intermediary. This saves the merchant the costs and delays of credit transfers. The intermediary becomes the technical agent of the bank to relay and secure the message exchanges. This solution is implemented in the system SIPS (Service Internet de Paiement Sécurité — Secure Internet Payment Service), from Atos, used by several French Banks. This architecture is also the basis of the Three-Domain SET (3-D SET).

The solutions presented above, and many other similar solutions, can be integrated in the SET/SSL model (Sherif et al., 1997, 1998). Clearly, an additional advantage of this hybrid mode of operation is the ability to change the protocol on the server side without bringing about parallel changes on the client side.

8.2.1 Hybrid Operation SET/SSL

As shown in Figure 8.8, in the hybrid SET/SSL operation, the payment intermediary acts at the intersection of three domains: that of the client, that of the merchant, and that of SET. In this configuration, the payment intermediary plays the following roles:

- SSL server with respect to the client and the merchant
- Web host for the merchant
- SSL certification authority for both (Potential users have to register, and the merchants will verify the privileges granted to the holder of a given card.)

From the SET viewpoint, the payment intermediary acts as the merchant's agent and will be the merchant as far as the SET payment gateway is concerned. The intermediary will have two pairs of private/public keys: one pair for the exchange of data encryption keys and the other pair for signing the transmitted data. Because it has the role of SSL certification authority for the merchant and the client, it will own a third pair of private/public keys and an SSL certificate.

If the intermediary is simultaneously a SET certification authority and an SSL certification authority, it can generate the SET certificates for the client and for the merchants and keep the certificates for use in the SET transactions the parties want to execute. Thus, its role will be similar to the role of C-SET/SET converter, seen earlier in a transaction between a C-SET cardholder and a SET-certified merchant. This solution increases overall security, because users' credentials will be more protected than if they were stored on the hard disks of the client or the merchant. However, the credential owners would not be able to use their SET certificates without passing by the payment intermediary. If the client is SET-certified, which will probably be the case in future versions of SET, the SSL/SET payment intermediary could take charge of client registration with SET authorities and carry out,

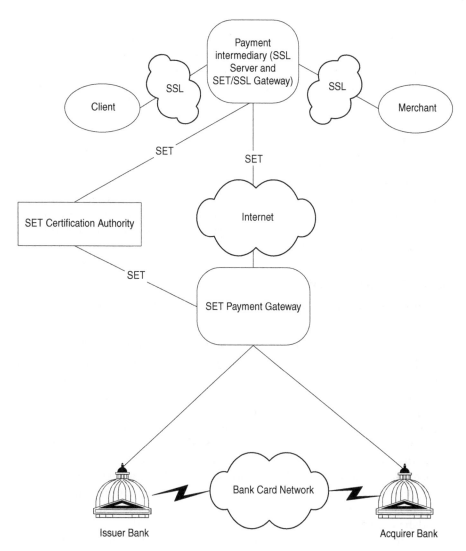

FIGURE 8.8
Hybrid SET/SSL (3-D SET).

in its place, the authentication for each exchange, thereby reducing the computational load on the client machine.

Similarly, the intermediary could take charge of merchant registration with SET authorities and store the merchant's certificates and pairs of private/public keys. In this way, the merchant would avoid downloading the SET software and going through the associated certification procedures; it would need only a browser to present the SSL certificate. For small businesses, this would ease the problem of access to consumer-oriented e-commerce.

Technically, there are three possible solutions, depending on if the merchant site is hosted on the payment intermediary, if the payment intermediary is associated with the SET payment gateway, or if the servers of the merchant, the payment intermediary, and the SET payment gateway are distinct. In the solution denoted as MOSET (Merchant-Originated SET) or MIA (Merchant-Initiated Authorization), the merchant server initiates all SET exchanges, irrespective of the channel on which the order arrives (mail, e-mail, fax, etc.). However, the merchant will have access to all the details of the transaction, including the payment information. When the three servers are independent, a three-party dialogue will be established that involves the consumer, the payment intermediary, and the merchant server. Although the exchanges become more complicated in this case, the basic principle remains unchanged, and this case will not be considered below.

Thus, whatever the approach, each time a consumer chooses to pay with a bankcard, he or she clicks on the button displayed by the browser that leads to the merchant site. In fact, this opens a second page dedicated to the merchant on the server of the payment intermediary.

8.2.2 Transaction Flows

A transaction includes four parts, as discussed in the following.

8.2.2.1 SSL Session between the Client and the Intermediary

The client visits the merchant's site and chooses the desired article. He or she then clicks on the link to the merchant's Web page and ends up on the server of the payment intermediary. After establishing an SSL session with the intermediary server, the client fills out an order form that includes the client name, order details, name of the merchant, the transaction amount, etc., and particularly, the number of the bankcard to be charged. The merchant does not have access to the client's card number.

As explained in Chapter 5, the Handshake protocol is responsible for the authentication of the communicating parties, the negotiation of the hashing and encryption algorithms, and the exchange of secrets. In this case, all Handshake messages shown in Figure 8.9 will be used.

The certificate from the payment intermediary to the client contains the SSL certificate of the intermediary. The return Certificate message from the client includes the client's SSL certificate that the intermediate issued. Verification of the certificate in both directions is much simpler than in the case of SET because of the minimal hierarchy in the certification path.

As explained in Chapter 5, the exchange of RSA keys uses a signature certificate that is included in the ServerKeyExchange message. Accordingly, when a signature certificate is employed, the server transmits two messages in sequence:

1. The Certificate message contains the public key for signature that the server will use, signed with the private signature key of the certification authority. The client verifies the integrity of the certificate with the public signature key of the certification authority.

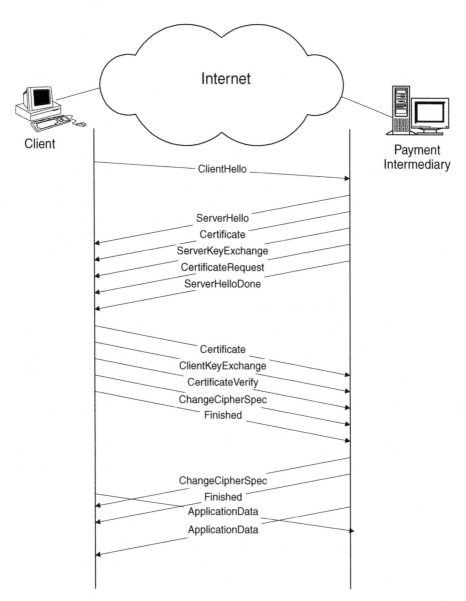

FIGURE 8.9
Handshake between the client and the payment intermediary.

2. The ServerKeyExchange message contains the public parameters of the algorithm for the exchange of the secret key. These parameters are signed with the private signature key of the server that corresponds to the public key obtained in the previous step. Upon receiving this second message, the client verifies the integrity of the parameters.

Note that the number of messages exchanged can be reduced if the two sides reutilize the parameters of a previously established session. This is particularly useful if the client makes all purchases from the same intermediary.

The client and the intermediary can now converse in a secure manner using the SSL secret session key. A message can now contain both the purchase order and the payment instructions. For example, it can contain, among other things, the buyer and merchant identities, the brand of bankcard used, the client account number, and the purchase amount.

The intermediary next establishes another SSL session with the merchant. Once the communication is secured, the intermediary transfers to the merchant a payment order without the financial details or, possibly, without the purchaser identity. The merchant analyzes this request. If it is accepted, the merchant will inform the intermediary of its acceptance, according to the application protocol established between the intermediary and the merchant. Once the intermediary is satisfied with the validity of the response, it initiates a SET communication with the SET payment gateway.

8.2.2.2   Payment Authorization

In this phase of the transaction, the intermediary behaves exactly as a merchant in the SET protocol, exchanging the messages shown in Figure 8.10. Using the data collected from the client and the merchant, the payment intermediary forms the authorization request AuthReq and signs it with the private key of the merchant. The dual signature of SET is constructed with the client private key, if the client is SET-certified; otherwise, the intermediary could use its own signature key.

If the response is contained in AuthRes, the intermediary alerts the merchant that the payment authorization was obtained to deliver the goods to the client. Simultaneously, it signals to the client that the financial transaction was authorized.

8.2.2.3   Notification of the Merchant and the Client

In this phase, the payment intermediary informs the client and the merchant of the authorization results using the SSL session already established. However, if the responses from the SET payment gateway were delayed, the sessions could have expired. In this case, the Handshake protocol would have to be invoked again.

8.2.2.4   Financial Settlement

Once the goods are shipped, the merchant sends the intermediary a request for settlement (clearance). The exchanges follow the application protocol established between the intermediary and the merchant and are transported over a session secured by SSL. In general, because of the delay between Phase 3 and Phase 4, a new SSL session would have to be established.

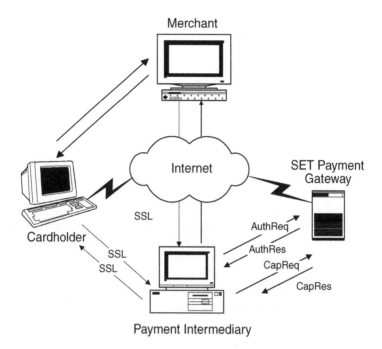

FIGURE 8.10
Role of the payment intermediary in SSL/SET conversions.

Having received the merchant's request, the payment intermediary reconfigures it into the SET CapReq message, which is directed toward the SET payment gateway. The response in the CapRes is converted back to the format common with the merchant and is sent in a session secured by SSL.

8.2.3 Evaluation of the Hybrid Mode SET/SSL

The use of the payment intermediary reduces the cryptographic load on both the client and the merchant by an order of magnitude. This is because SSL is less computationally demanding than SET, and because the intermediary offloads the clients and merchants from the SET procedures. SET requires authentication of each message, which adds delays and computational costs to each transaction. By offering this service, the intermediary can justify investment in more powerful machines. Also, because the SET credentials (certificates, keys, etc.) reside at the site of the payment intermediary, they can be protected from failures in the hardware of the individual users.

Such a solution could be useful for some users. Its low cost and ease of registration would particularly attract small businesses or those that deal with small amounts. Furthermore, because the merchants and the purchasers are obliged to go through the intermediary for their transactions, the intermediary can intervene to arbitrate disputes, provided that all traces of the communications are kept with time stamps. Finally, this solution circumvents the difficulties that restrictive laws in the area of cryptography could pose. If the payment intermediary resides in a country that allows the use of the encryption parameters that SET retained, its connection to people in other countries (merchants and clients) could be done with some of the weak parameters that SSL allows.

The main drawback of this method is that the client would not control the certificate issued in his or her name, because it is stored by the intermediary. Furthermore, if the merchant and the client always go through the same intermediary, in the long run, the intermediary may be able to reconstruct their marketing profiles. In the absence of legislation on the subject, this may be objectionable to individual parties.

One possible solution is for the bank to host the payment intermediary, a solution championed by the Spanish bank Banesto of the Santander Central Hispano group in cooperation with IBM SET *fácil* (Easy SET). On the contrary, in MOSET (also called 3-D SET), the merchant initiates the SET exchanges irrespective of the channel by which the command came (mail, e-mail, fax, etc.). This architecture is called the *thin wallet architecture* because of the reduction in the size of the required client software.

8.3   3-D Secure

While the large-scale deployment of SET, either in its full-blown or light version, was stymied by the complexity of the appropriate cryptographic infrastructure, card fraud is increasing in parallel with increased use of the Internet and mobile telephony. Visa estimates that 0.11% of transactions on the Internet are fraudulent, compared to 0.05% overall (Berman, 2000). In 2000, bank losses in the European Union due to card swindles increased by 48% to reach 0.07% of the turnover (Austin, 2001). In France, the rate of fraud doubled from 1999 to 2000, essentially due to the Internet and the purchase of air time for some contracts for mobile telephony (Le Monde, 2000). In the U.K., card fraud increased by 55% in 2000 over 1999 losses and by 30% in 2001 over the previous year (Brayshaw, 2001; Schneider, 2002). In the U.S.,half of the contested transactions took place online, even though they amounted to merely 1 to 2% of the turnover.

Most of these frauds are attributed to magnetic strip cards, especially identification frauds because the card number and the expiration data can be gleaned from the carbon receipts. This spurred an accelerated conversion to the EMV chip cards, with the target date for conversion set for January 2005. To provide incentive to banks and merchants to upgrade their equipment to conform to EMV, Visa shifted the liability to the non-EMV-compliant party in the case of a fraudulent transaction. Currently, online merchants are fully responsible for "card-not-present" transactions through mail, phone, or the Internet; merchants pay the costs of stolen merchandise and extra penalty fees. Upon conversion, the liability will shift to the party that does not offer the security. A merchant ready for EMV can still accept payments with a magnetic strip card, but the buyer's bank will be responsible in contested cases. Conversely, if the buyer has a chip card but the merchant does not have the means with which to accept it, the merchant's banks will be responsible for any transaction that the cardholder repudiates.

Until then, the banking industry is confronting the fraud problem with several initiatives. One of these initiatives is to authenticate the card in mail-order or telephone-order transactions using a nonembossed value on the card. For Visa and MasterCard, this is a three-digit value on the signature panel in the back of the card, called a Customer Verification Value (CVV2), while for American Express, it is a four-digit code or Confidential Identity (CID) on the face of the card. The cardholder needs to give this value to complete the transaction.

For Internet transactions, Visa introduced in 2001 a solution that is simpler — though less elegant — than SET called 3-D Secure, which is operational since April 1, 2003, within the Verified by Visa program.

The architecture considers three domains: that of the issuer bank, that of the acquirer bank, and the domain of interoperability, where the Visa intermediary for 3-D Secure verifies the parties. Figure 8.11 shows that 3-D Secure substitutes four point-to-point SSL (or TLS) connections for a multipoint connection established among the buyer, the merchant, and the payment gateway. This is why the solution may be called 3-D SSL.

The goal of 3-D Secure is to facilitate remote bankcard payments, irrespective of the access channel — Internet, broadband digital TV, SMS (Short Message Service), WAP, etc. — and to prevent the three main sources of fraud: use of unauthenticated cards, theft of credit card numbers, and fraudulent claims by unethical merchants. At the same time, the use of SSL/TLS adds the services of confidentiality, integrity, and authentication. This model of operation preserves existing banking channels, and the circuits for verification and for financial settlement continue to go through the VisaNet network. This solution avoids the installation of any additional software on users' terminals, while the merchants need only to add a plug-in to their payment server. Visa remains the fulcrum of the transaction, replacing the payment gateway and the certification authority of SET. Visa provides the following functions:

- A directory service that determines whether there is a range of card numbers that includes the primary account number (PAN) being verified

- A certification function to generate the various X.509 certificates to be used

- A repository for authentication history, recording each attempted payment authentication, irrespective of whether it was successful

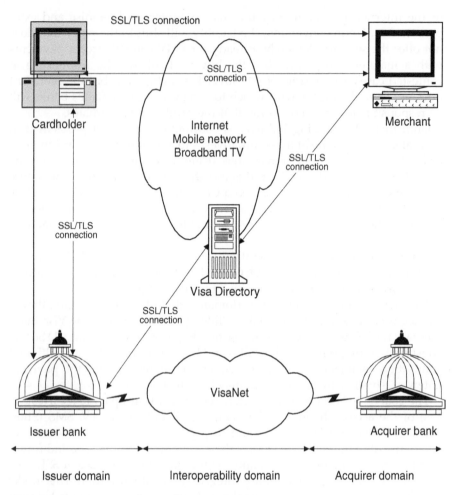

FIGURE 8.11
The domains of 3-D Secure and the SSL/TLS point-to-point connections.

8.3.1   Enrollment

To participate in the Visa Authenticated Payment Program, the issuer and acquirer banks provide their bank identification numbers (BINs) and the URL addresses of their respective servers.

Client enrollment is by whatever channels the issuer bank selected. Similarly, the cardholder authentication mechanism is left to the issuer. It can be a PIN, a smart card, an identity certificate, a biometric measure, etc. However, a password that the user chooses at enrollment is the default mechanism. If needed, the issuer bank will supply all that is necessary (equipment or software) for authentication. Visa will keep all the data related to this card (holder identity, banking coordinates, version, etc.) in its secure directory.

The acquirer bank is responsible for supplying the Merchant Server Plug-in (MPI) and for helping the merchant in its activation.

Finally, the client's browser must be able to offer a conduit for the communication between the merchant and the issuer bank without asking for the buyer's intervention. This feature will be incorporated in Microsoft's Internet Explorer and Netscape's Navigator.

8.3.2 Purchase and Payment Protocol

The negotiation between the buyer and the merchant is outside the scope of 3-D Secure. The message flow for payment authentication is shown in Figure 8.12. The messages are coded in XML (3-D Secure, 2002).

Payment authentication proceeds as follows:

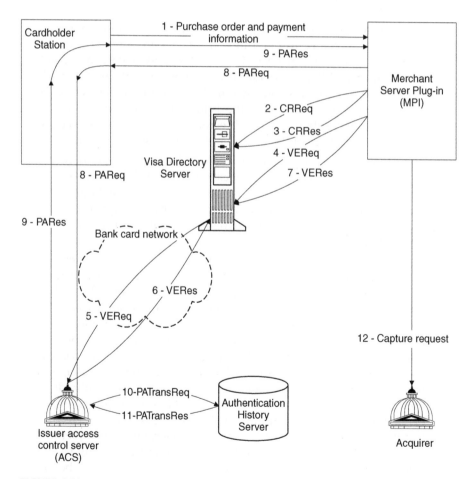

FIGURE 8.12
Flow for payments with 3-D Secure.

1. 3-D Secure is started when the buyer indicates intention to buy by clicking on the corresponding button.

2. After receiving the purchase order with the payment instructions, the merchant plug-in (MPI) requests the list of participating card ranges from Visa's 3-D Secure directory, with the CardRangeRequest (CRReq) message. This is to verify that the PAN is within the range of participating cards.

3. The directory response is in the message CardRangeResponse (CRRes). Depending on the parameters of the initial request, CRRes can contain the entire list of participating card ranges or the changes since the last exchange. The returned information can be used to update the MPI's internal cache. It is worth noting that this exchange may be skipped if the MPI has the ability to store the content of the Visa directory, provided that the local cache is refreshed every 24 hours at least.

4. Next, the MPI sends the VerifyEnrollmentRequest (VEReq) message to the Visa directory to determine if 3-D Secure authentication is available for a particular PAN. The directory checks if the merchant, the acquirer, and the cardholder PAN can be authenticated using 3-D Secure. If the response indicated in the VerifyEnrollmentResponse (VERes) message is negative, the transaction can proceed in the traditional way. Otherwise, the directory will interrogate the access control server (ACS) of the issuer bank to request its authorization.

5. The VERes message from the access control server of the issuer bank indicates whether the holder is registered in the 3-D Secure program. In such a case, it contains the URL address toward which the cardholder browser will post the data provided by the merchant plug-in. If the response is negative, the transaction can continue along the classical lines.

6. Upon receiving the response from the directory, the MPI sends a PayerAuthenticationRequest (PAReq) to the issuer access control server using the URL obtained in the preceding step. The PAReq contains details of the purchase that needs to be approved. The merchant defines the frame for the purchase authentication page (either in an existing window or by opening a secondary pop-up window), while the contents of the page are under the control of the issuer. The transfer of control to the ACS is through the user, and relies on a JavaScript that resides in the page sent to the cardholder's browser. However, a more secure way would be to ask for the user's explicit intervention to continue the 3-D Secure transaction. To protect the data from being inadvertently changed by the browser, the data are coded using Base64 coding. Because in that coding each 3 octets of data are expanded into 4, the data are compressed before coding. Furthermore, for mobile Internet devices, the Condensed

Payer Authentication Request (CPRQ) is used to reduce bandwidth utilization.

7. The issuer ACS establishes an SSL/TLS session with the cardholder work-station and displays the details of the transaction, as obtained from the MPI, and asks the holder for an identification (by a PIN, through biometric means, by inserting a chip card in a card reader, etc.) and for approval of the purchase. The ACS links the PAReq with the VERes messages.

8. After authentication of the holder, the ACS builds a Cardholder Authentication Verification Value (CAVV) over 20 octets. The CAVV includes a cryptogram of 2 octets computed with a Visa algorithm and using, among other operations, symmetric encryption with DES. It gives VisaNet a fast way to validate the integrity of data proving that the authentication took place. It is sent in the message Payer-AuthenticationResponse (PARes) to the MPI signed with the signature key of the ACS. The message uses Base64 coding, and the data are compressed before encoding. For mobile Internet devices, Condensed Payer Authentication Request Response (CPRS) is also used.

9. The response is sent to the Authentication History Server for archiving within the PayerAuthenticationTransactionRequest (PATransReq) message. The confirmation of the archiving is in the PayerAuthenticationTransactionResponse (PATransRes) message.

10. The merchant's server verifies the signature before sending it for settlement.

To summarize, the above exchanges comprise four phases:

- Update of the merchant's cache with the ranges of valid accounts
- Verification that a given card number is enrolled in 3-D Secure and whether authentication is available to it
- Authentication of the cardholder and his or her authorization of a specific transaction
- Archiving of the authentication attempt

Summarized in Table 8.2 are the 3-D Secure messages and their significance.

8.3.3 Clearance and Settlement

Having received the authentication response, the merchant server verifies the signature of the issuer bank and extracts the necessary information to make a capture request. It sends this request to its bank with an indication

TABLE 8.2

3-D Secure Messages

Message	Direction of Transmission	Significance
CRReq	Merchant → Visa directory	Request full or partial list of participating card ranges
CRRes	Visa directory → Merchant	Return full or partial list of participating card ranges or changes from last update
VEReq	Merchant → Visa directory → Issuer	Check if 3-D Secure authentication is available for a particular PAN
VERes	Issuer → Visa directory → Merchant	Indicate if 3-D Secure authentication is available for a particular PAN
PAReq	Merchant → Issuer (via cardholder's browser)	Base64-encoded request to authenticate the payer protected against inadvertent change by the browser
CPRQ	Merchant → Issuer (via cardholder's browser)	Same function as *PAReq* for mobile communications
PARes	Issuer → Merchant (via cardholder's browser)	Base64-encoded response to *PAReq* signed by the issuer access control server
CPRS	Issuer → Merchant (via cardholder's browser)	Same funtion as *PARes* for mobile communications
PATransReq	Issuer → Authentication History Server	Request archiving of *PARes*
PARransRes	Authentication History Server → Issuer	Response to archival request

that the transaction is secured according to 3-D Secure. The acquirer bank requests the approval of a Visa Directory Server. The server compares the data received from the issuer and acquirer banks before according its approval. Once that approval has been given, the settlement follows the usual procedures.

8.3.4   Security

3-D Secure is an attempt to strengthen remote payments with bankcards by relying on an infrastructure under the control of Visa and its associated banks.

The various entities are given X.509 Version 3 certificates, as follows:

1. The Visa directory has a server certificate with respect to the merchant and a client certificate with respect to the issuer.
2. The MPI has a server certificate with respect to the cardholder and a client certificate with respect to the Visa directory and the issuer.
3. The ACS has a server certificate and a signature certificate with which to sign the PARes message; the ACS initiates two sessions using its server certificates: one to the cardholder and the other to the merchant.

The certification authority can be Visa or one of the recognized certification authorities.

The use of SSL/TLS gives the services of confidentiality, integrity, authentication, and, with TLS, nonrepudiation. The mandatory cipher suite is (TLS_RSA_WITH_3DES_EDE_CBC_SHA): SHA-1 for hashing, triple DES for symmetric encryption, and RSA for static signature. Typically, the symmetric key used for encryption of traffic involving the cardholder is 40-bit for DES (80-bit security for triple DES) due to the proliferation of U.S.-exportable browsers. The minimum size of the RSA key for signature is 768 bits, and the recommended value is 1024 bits. An optional cipher suite is (TLS_DH_RSA_WITH_3DES_EDE_CBC_SHA), where the Diffie–Hellman algorithm is used to exchange keys and RSA is used to sign the messages in the key exchange to ensure their integrity. Visa recommends using the 128-bit SSL/TLS cipher suites whenever possible. In particular, the connections among the MPI, the cardholder terminal, and the issuer ACS use HTTPS. It should be noted that each transaction requests at least five point-to-point links secured with SSL/TLS.

In principle, the operation should not impose hardships on the cardholders, while merchants would probably outsource the maintenance of the 3-D Secure security plug-in. However, management of five point-to-point links for each transaction may pose scalability problems. Finally, browsers may be the weak point of the whole construct, because they serve as conduits for exchanges between the merchants and issuer banks, even though they reside on the clients' terminals. Browsers on mobile Internet devices usually do not support JavaScript, or this capability is regularly turned off in corporate networks for security reasons. So, other approaches need to be considered and tested for their security.

8.4   Payments with CD-ROM

A new method by which to secure payments on the Internet consists of exploiting CD-ROM readers to compensate for the lack of smartcard readers in PCs. At the moment of accomplishing the payment transaction, a control associated with the HTML page asks the user to insert the CD-ROM card reader and then enter a PIN. This card is supplied by the bank that holds the user's account and contains the data usually stored in the magnetic strip of the bankcard as well as a secret code for the card. The exchanges with the authorization server are encrypted with the triple DES algorithm on a network for electronic funds transfer. Only the processor for the holder's account is capable of decrypting the messages to extract the card number and verify the PIN.

In this way, execution of the transaction requires both the presence of the CD-ROM and the input of the PIN, so that buyers cannot repudiate their transactions unless they have reported the loss of their CD-ROM. At the same time, the merchant has no access to card data. Finally, the amount of the transaction is debited instantaneously from the holder's account.

This solution, called SafeDebit, was adopted in the U.S. by several operators of electronic funds transfer networks or payment processors: NYCE, Star, Concord, etc.

8.5　Summary

The acceptance of SET for online transactions was limited in the face of the simplicity of SSL and its widespread diffusion in browsers. Cyber-COMM launched a system that uses C-SET to secure online payments with BO' smart cards and ultimately EMV cards. In addition, many ingenious solutions were tried to reduce the load that SET imposes on the client by combining SET with SSL. The architectures of these solutions vary according to the degree of association of the payment intermediary with the merchant server or the SET payment gateway. Finally, to confront increase in fraud, Visa introduced a new architecture, 3-D Secure, that avoids the need to install large software products at cardholder terminals and reduces the complexity of merchant tasks.

Questions

1. Define the protocol stack in the payment intermediary that terminates the SSL connection on the client side and that handles the interaction with the SET payment gateway on the SET side.
2. Define the message exchanges among the different parties in a hybrid SET/SSL operation. Assume that only the mandatory SET messages are used. The parties are the cardholder, the merchant, the payment intermediary, and the SET payment gateway.
3. From the results of Question 2, estimate the duration of N transactions for both SET and the hybrid operation SET/SSL as a function of the length of the messages.
4. Compare and contrast SET *fácil* (Easy SET) and MIA (Merchant-Initiated Authorization).
5. Identify sources of security risks in 3-D Secure.

9

Micropayments and Face-to-Face Commerce

ABSTRACT

This chapter describes micropayment systems that can be used for face-to-face commerce, in addition to the Internet. One encounters this type of commerce at bakeries, pastry shops, newsstands, coffeehouses, tobacco emporiums, grocery stores, public transportation systems, parking meters, and vending machines. In the overwhelming majority of countries, cash is the preferred way to conduct payment for face-to-face commerce, with the proportion of people using fiduciary money varying between 90 and 100%. Checks and bank cards are used rarely for reasons of costs and exposure to risk.

The means of payment presented here have integrated-circuit cards (or microprocessor cards) as physical support. These systems aim to replace cash, while facilitating the personalization of the services offered. In general, they constitute electronic purses on integrated-circuit cards (smart cards or memory cards) that can be used for neighborhood commerce. They contrast with virtual purses, which depend on programs installed in the client's personal computer. Whatever the purse type, the electronic value expresses the quantity of fiduciary money that resides in a rechargeable memory. Recharging the electronic purse with monetary value requires the intervention of a financial institution.

The terminology for these systems is still in flux. Some reserve the term micropayment for transactions with values that lie between 10 cents and U.S.\$10, and use picopayment for values less than 10 cents U.S. This distinction, however, is not universally recognized, and in this book, all payments less than U.S.\$10 are considered micropayments.

In this chapter, we present the common properties of micropayment systems and then illustrate them with the main commercial offers (in alphabetical order): Chipper®, GeldKarte, Mondex, and Proton. Further information on the position of electronic purses with respect to existing or planned payment systems for ten Western European countries is available in a synthetic report from the European Science and Technology Observatory (ETSO) network (Böhle et al., 1999).

9.1   Characteristics of Micropayment Systems

As mentioned in Chapter 2, solutions for replacing cash differ with respect to three fundamental aspects:

- The representation of the purchasing power, which can be indicated by legal monetary units or by consumption units, i.e., jetons
- The means to make the financial value available; thus, what is available to the user may be a smart card or software on a PC
- The level of security available, which differs according to the amounts at risk and the purpose of security; for example, are proofs of payment needed, or is it merely sufficient to prevent utilization by unauthorized persons?

Micropayment systems differ from other means of payment mostly in their operating costs, which have to be commensurate with the small values handled. To reduce operating costs so the payment system is profitable, various approaches are used, such as service prepayment, reduction of computational load, offline authorization, and grouping of micropayments before financial clearance.

Prepayment often takes the form of a card charged with value or a subscription to the service to be consumed, all paid with legal monies. This operation allows the system operator to invest the collected sums and to take advantage of the cash gains due to the time lapse between when the card or the subscription is bought and when the associated service is consumed. The disadvantage for cardholders or subscribers is that they have to mobilize part of their holdings by paying before consumption.

Prepaid card use is expanding in many countries. Thus far, however, the value stored is mostly expressed in terms of jetons (service units, telephone impulses, etc.) and not in legal tender. The intermediaries or brokers, for example, a telephone or pubic transportation company, take the responsibility of issuing and distributing the cards and of collecting the funds that are then transmitted to a credit institution. The new generation of jeton holders will likely be coupled to other monetary and nonmonetary functions in the same multifunctional card.

Service prepayment does not exclude that the internal economy of the system is a credit, i.e., postpayment of services with jetons already prepaid with legal tender. In the PayWord remote payment system, which will be examined in Chapter 10, the purchase is done by credit in jetons that the seller later exchanges against legal tender at the intermediary (the broker). The broker, in contrast, can choose to be prepaid or postpaid, by using the coordinates of the subscribers' bank cards to the micropayment network.

Diminishing the computational intensity or complexity can alleviate the requirements on digital processors, which leads to a reduction in the transaction cost. One way of achieving this alleviation is by preferring symmetric encryption algorithms over public-key algorithms whenever possible. Other reductions are possible by not checking with the verification center before authorizing each individual transaction and by grouping the billing and financial clearance of the transactions. Local verification is done by checking lists that are updated on a daily basis. If public key cryptography is used, algorithms for local verification can use the following elements (Even and Goldreich, 1983):

- The amount available to the subscriber
- The public key, PK_B, of the credit establishment that stored the monetary value in the electronic purse and the private key of the subscriber, SK_P
- The public-key certificate of the subscriber signed by a certification authority or by the subscriber's bank
- The PIN (Personal Identification Number) of the subscriber

Finally, it is possible to carry out periodic monetary clearances through a global bill by grouping the amounts that accumulated in the virtual cash register of a vendor through successive microtransactions. Grouping offers two specific advantages: to prevent the collection cost from exceeding the value of the amounts collected and to give the intermediary the possibility of investing all the sums that collected before giving the payments to the creditors. Grouping offers an advantage to the client in that it can provide partial anonymity for individual transactions.

9.2 Potential Applications

Through a series of pilot experiments, potential applications of electronic purses or jeton holders were studied in a realistic environment (Van Hove, 2001; Westland et al., 1998). The lessons learned suggest that successful applications of electronic purses are those that introduce incremental changes to an existing distribution and collection infrastructure. For example, the purse will be loaded with value from special bank terminals, from Automated Teller Machines (ATMs) modified to be able to recognize the new cards, or with portable secure terminals. Furthermore, the new generation of multiapplication chip cards would be able to assist in mass acceptance by coupling micropayments with other monetary and nonmonetary functions. Among these nonmonetary functions are the maintenance of URL addresses, passwords, and data related to the transaction (such as delivery address and billing address).

Applications in mass transit seem to meet these conditions, particularly the use of contactless cards for ticketing, which explains why they are getting increasing attention. In 1995, the European Commission financed the electronic purse Minipay in Turin, Italy, to promote the use of multiapplication smart-cards in public transportation systems. The electronic purse was associated with other applications, such as the payment of telephone calls from public phones, loyalty programs, some administrative files, and personal data. Today, such schemes for mass transit systems are moving into commercial life in several countries, with the first large-scale ticketing applications already in use in Hong Kong (Octopus system) and in Daejong, South Korea. Several systems are being implemented in main European and Japanese cities (Adams, 2002).

Mass applications have strict requirements on the duration of transactions. The absolute maximum of transaction time is 300 ms, with most transactions taking less than 150 ms. This underlines the fact that RSA encryption is too slow. Current efforts focus on speeding the necessary computations, on using symmetric encryption, or finally on exploring new avenues, such as elliptic curve cryptography (ECC).

Let us now look at the most commonly used electronic purses.

9.3 Chipper®

Chipper® is the electronic purse issued in the Netherlands by the telephone operator KPN with the support of the Postbank as a credit lending institution. The commercial offer is called CyberChipper®, and it allows payment over the Internet as well. Although Chipper technology interested many backers, the commercial prospects remain in doubt.

The architecture of CyberChipper is shown in Figure 9.1.

The payment intermediary, Chipper Central, identifies and authenticates the purse and the user before loading the electronic purse with value. It carries out the same functions with respect to the merchant terminal, as the transactions are grouped in preparation for clearance. This intermediary is connected to the payment gateway Chipper Netherlands. Security Application Modules (SAMs) control the exchanges between the merchant server and the payment intermediary on one side and between the same server and the client terminal on the other side. These modules integrate the security functions with the merchant's cash register to authorize transactions offline by performing the following functions:

- Identification of the card and the cardholder
- Authentication of the card and the cardholder
- Control of a shortened revocation list downloaded from the main authorization center

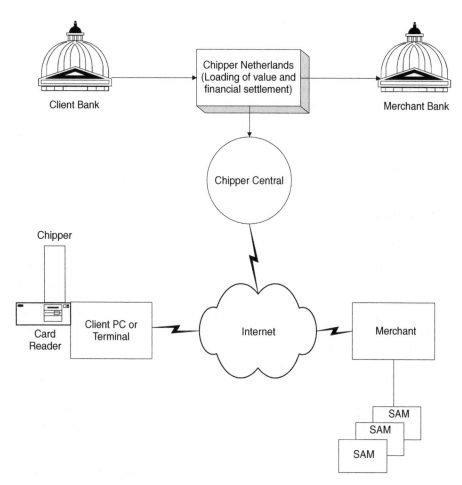

FIGURE 9.1
Architecture of CyberChipper®.

- Automatic initiation of an authorization request, when needed, as defined by the banking security policy
- Generation of the signed receipts for the transaction
- Storage of the transaction records
- Automatic connection, usually at night, with the banking servers to update and synchronize the various banking records

The details of the security protocol were not made public at the time this section was written. It is known, however, that confidentiality is protected with symmetric encryption with the triple Data Encryption Standard (DES), and that public-key encryption with RSA will be introduced for authentication.

The Chipper electronic purse uses the IBM Multifunction Card (MFC) with the ISO/IEC 7816-4 (1995) and ETSI TE9 (1993) specifications. The protocol used is the IBM Smartcard Identification (ISI) protocol, which is proprietary but was used in many university projects in the Netherlands (Cuijpers, 1997). It includes the ST 16SF48 chip produced by SGS Thomson with the following characteristics: 19 K octets of ROM, 288 octets of RAM, and 8 K octets of Electrically Erasable Programmable Read-Only Memory (EEPROM) (Kirschner, 1998). The card reader includes a keyboard, a screen, and applications specific to the purse that allow viewing of the value stored in the card.

Swiss Telecom adopted the Chipper specifications in its purse Smart Scope.

9.4　GeldKarte

The GeldKarte started as the chip version of the Eurocheque card that was first introduced in 1968 as a check guarantee card across banks (GeldKarte, 1995). Originally, the Eurocheque card provided a check with a guarantee of payment by the issuer, if the beneficiary wrote the card references on the check. In the 1980s, magnetic strips were added to the back of the card, and it became a means for cash withdrawal recognized in an international network of ATMs. Nowadays, this card is widely used in Austria, Germany, the Netherlands, and Switzerland (Dragon et al., 1997, pp. 156–157). GeldKarte is also used in the Navigo project of BMS (Billettique Monétique Service) in the Parisian region for micropayments (less than 30 euros) in shops, in the mass transit networks, in parking meters, and for parking. [Note that the purse used in Navigo results from the combination of two preceding purses — Modeus and Moneo.]

The GeldKarte can be used for face-to-face commerce as well as for remote payments on the Internet (which requires a terminal or a PC with a card reader and software to control access). The software displays the amount left on the card, the value of the transaction, and the state of the connection; it should also keep a record of the transactions that were carried out.

Many products are based on the GeldKarte. In 1996, Deutsche Telekom, together with the German train operator, Deutsche Bundesbahn, and the Association of Municipal Transports (VDV), introduced the PayCard, which was later called the T-Card, based on the GeldKarte. GeldKarte is also the building block for the purse Modeus/Moneo that will replace paper tickets with contactless payments in the public transportation systems of the Parisian region around 2004. This is done by integrating the card readers or validators in the gates to allow reading of contactless cards at a distance of 10 cm. The antenna is integrated in the surface of the cards and transmits at a carrier frequency of 13.56 MHz.

Several smart-card manufacturers, such as Gemplus and Giesecke & Devrient, support the specifications of the GeldKarte in their product line. The chip is manufactured by Infineon (ex-Siemens) or Motorola with the following capacities: 12K octets of ROM, a RAM of 256 octets, and 8 K octets of EEPROM (Kirschner, 1998). Access to the data stored in the GeldKarte, the structure of the messages between the card reader and the card, and the logical architecture conform to ISO/IEC 7816-4 (1995) (Althen et al., 1996).

The credit institution financially backing the GeldKarte is the Central Credit Commission ZKA (Zentraler Kreditausschuß).

9.4.1 Registration and Loading of Value

Merchants and cardholders register by signing a contract with the system operator. The merchant's terminal has a SAM in the form of a card (the merchant's card). The identification data are stored in a central file that the system operator maintains.

The card is loaded with value at a special loading terminal using a secured line to the loading center. Loading comprises two distinct phases: a debit operation of a bank account and loading of the purse. The parties in the loading phase are the cardholder, the loading operator (which may be distinct from the issuer bank), and the authorization center.

The debit operation consists of five main steps (Sabatier, 1997, p. 76):

- Card authentication
- Identification and authentication of the cardholder by the operator using the confidential code (PIN)
- Input of the required amount
- Request for authorization from the authorization center
- Update and synchronization of the records of all parties

Figure 9.2 depicts the various operations numbered in a chronological order during the loading of value (Sabatier 1997, p. 77). In this figure, the functions of the loading operator, the authorization server, and the data store were separated from those of the issuer bank, although the same entity may perform all of them.

The value is transferred from the client's account to a clearing account associated with the card, which will be used for compensating the client.

9.4.2 Payment

The merchant terminal controls the exchange of value between the client's card and the merchant card with the help of the SAM that it contains. The payment transaction is considered offline, because it does not depend on the intervention of an authorization server or of the issuer bank. This transaction consists of the following steps: reciprocal authentication of the client's purse and the merchant's SAM, transfer of the debit amount, and production of the encrypted electronic receipts.

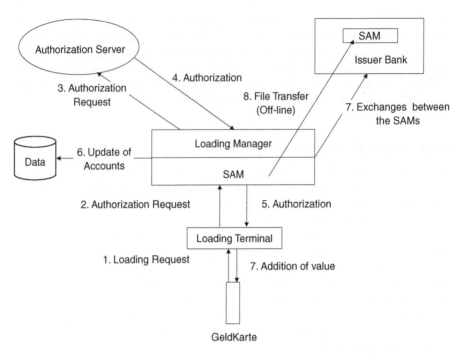

FIGURE 9.2
Charging of the GeldKarte purse.

Reciprocal authentication of the purse application of the cardholder and the merchant card takes the following steps:

1. Following insertion of the client's card in the terminal slot, the merchant's payment manager requests the card for its identification (CID).

2. The SAM of the merchant terminal (the merchant card) obtains a random number RAND and inserts it in the command Start debit process that the merchant terminal sends to the client's card. The use of the random number protects against replay attacks.

3. The card responds to the command by giving its identifier CID, concatenated with its current sequence number SNo, the random number RAND, and the hash H(CID, SNo, RAND), where H() is the hash function of ISO/IEC 10118-2 (1994). The hash is encrypted using DES with the card's secret key K_C, to calculate the Message Authentication Code (MAC).

4. The merchant card derives the card key K_C starting with the master key K_{GM}, using a "diversification" algorithm, and then decrypts the MAC. The integrity of the received message can be verified by comparing the value obtained from the decryption of the MAC with the hash value H(CID, SNo, RAND) from the received message.

The exchange of value takes place in the following steps:

1. The terminal sends to the merchant's SAM the command Start payment.

2. The merchant's SAM responds with the message Debiting. This message contains the merchant identifier MID, the transaction number TNo as defined by the merchant's card, the sequence number SNo of the client's card, and the serial number of the client's Geld-Karte as retrieved from a store within the SAM. The integrity of this information is protected with a MAC using the client's key K_{RD} and is then encrypted with the same key.

3. The terminal transmits the message to the card after adding the value to be transferred, M.

4. The client's card verifies that the MAC of the message corresponds to the received data, thereby proving the authenticity of the merchant. The purse ascertains that the debit is correctly attributed by verifying that the identification of the card corresponds to its own. The presence of the sequence number SNo enables it to detect a replay attack, when the sequence number in the message is not the same as the number that it sent to the terminal.

5. The purse verifies that the amount to be withdrawn is less than the total value stored. If this is the case, it reduces the actual amount by the value of the payment, increments SNo by one unit, records the transaction in a register, and computes a proof of payment. The proof is constructed by concatenating the amount M, the merchant identifier MID, the sequence numbers SNo and TNo, and the card clearing account number ANo. The MAC of the result is calculated with K_{RD}, and the whole set is sent to the merchant card in the message Check payment.

6. The merchant SAM recomputes the amount M', starting from the proof of payment, to verify that the amount paid is the same as the amount requested (M). It will verify that the merchant MID is its own and that the sequence number TNo is the same number that it produced.

7. If all the checks are positive, the merchant's SAM increments the transaction number TNo by one unit, increases the value that the merchant card stores by M, informs the terminal of the success of the transfer, and records the transaction data with a new key, K_{ZD}.

These exchanges are illustrated in Figure 9.3.

To start the clearance process, the merchant presents, offline, to the issuer bank, the encrypted proof of payment to credit its account through the financial settlement circuits. Because the proof includes the transaction

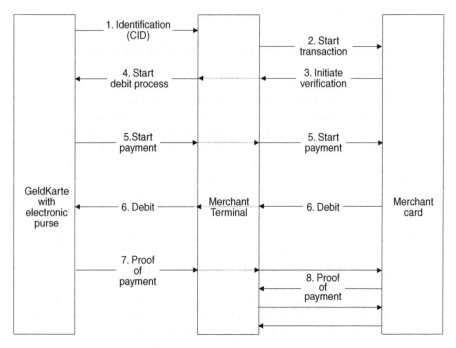

FIGURE 9.3
Message flow during a payment with GeldKarte.

number TNo, the receipt cannot be reused in another fraudulent clearance request.

9.4.3   Security

The GeldKarte protocol uses cryptography to guarantee the following:

- The cardholders and the merchants are authentic.
- The request for financial clearance is used only after the debit of an authenticated GeldKarte and only to the benefit of the card of the mentioned merchant.
- The messages are not reused for fraudulent claims.

Message integrity is verified by calculating a MAC with the symmetric encryption algorithms DES or triple DES in either the Cipher Block Chaining (CBC) mode or the Cipher Feedback (CFB) mode. These computations follow ANSI X9.19 (1986), which specifies methods for message authentication for use by financial retail institutions. They use the hashing function defined in ISO/IEC 10118-2 (1994) to produce a message digest of 16 octets (128 bits). The DES or triple DES algorithms are implemented in the merchant card, while currently, the client's card uses DES only. (Triple DES is planned for the future.)

The cardholder identifies himself or herself by using a secret code or PIN. The PIN is required to load the card with value but not to make a payment. In this way, payments do not have any relationship with customer's accounts. Anonymity is achieved with respect to the merchant but not with respect to the financial centers.

Each GeldKarte is personalized with an identifier and a symmetric encryption key. The parameters of the cardholder's card are the identifier CID and the key K_{RD}, while the corresponding merchant's parameters are MID and the master key K_{GM}. As mentioned earlier, with K_{GM}, the merchant card can derive the card K_{RD}, thereby avoiding the exchange of this secret key between both parties. The K_{GM} key is stored in the SAM of the merchant terminal (the merchant card).

During the personalization phase of the GeldKarte, the serial card number, the encryption keys, and the cardholder secret code are stored in the "secret zone" of the card. The same information is also stored in an encrypted file that will be stored under heavy security in the issuer bank.

Note that each party gives the transaction a unique number. This number is sent to the other party to include in its response. This mechanism builds a defense against replay attacks. The number that the merchant card defines is the transaction number TNo, and the number that the client's card assigns is the sequence number SNo.

Partial anonymity of the cardholder with respect to the merchant is possible, provided that the data stored in the card are not linked in any manner to the bank account of the cardholder. However, if the encrypted data are presented to the clearance center as proof, the identity of the cardholders can be revealed.

9.5 Mondex

The first patents for Mondex were filed on April 12, 1990, in the name of T. Jones and G. Higgins, two employees of the National Westminster Bank (Natwest) in the U.K. The objective was to replace physical money. In 1991, the project implementation started in partnership with several suppliers: Dai Nippon Printing Co. for the card and Hitachi Panasonic and Oki Electric Industry for the integrated circuits. In 1992, BT (formerly British Telecom) became interested in the project, while Natwest declared its intention to obtain the approval of the Bank of England to recognize Mondex as a new means of payment. In 1993, Midland Bank joined the project, and the two banks united to form Mondex UK, with equal shares from each. Finally, Mondex International was formed in the summer of 1996 as an independent company, with MasterCard as a majority owner, together with 17 major multinational corporations (Mayer, 1997).

Despite the shroud of secrecy that surrounds the technical characteristics of the project, some important elements were noted:

- The memory capacity of the chip is as follows: 16 K octets of ROM, 512 octets of RAM, and 8 K octets of EEPROM (Kirschner, 1998).
- The cardholder can store values in five different currencies.
- The transfer protocol allows the exchange of value among two Mondex cards, even remotely. This peer-to-peer function seems to be a unique capability of Mondex among electronic purse schemes.
- The Mondex exchanges use a new type of MIME e-mail messages.
- The Mondex protocol seems to be robust: following a failure, the transaction can continue from the point of failure; otherwise, traces of the aborted transaction will be recorded in an audit trail. In other words, Mondex transactions appear to be durable in the sense defined by Camp et al. (1995).

The Mondex protocols can be used to combine the functions of remote payment on the Internet with those of an electronic purse. The architecture of the resultant hybrid client is shown in Figure 9.4. The interface to the user offers two distinct protocol stacks that converge at the level of the purse. The first stack allows access to the Internet with the HTTP protocol on top of the TCP/IP layers, in addition to the physical and logical layers of the network. The second stack is for the functionalities of the electronic purse through a card reader. The intermediate layers between the user or the browser on one side and the TCP/IP layers on the other side are proprietary. Similarly, the protocols that control the Mondex purse are proprietary.

This architecture was used in the Smart Access project that allowed electronic transactions on the Internet from a GSM terminal. It can also be used to allow exchanges with the Proton and Chipper cards by adding the corresponding interfaces (Penrose, 1998a, 1998b).

9.5.1 Loading of Value

The card can be loaded with money in electronic form from specially adapted ATMs or from special telephone sets. Public phone sets need to be modified to recognize the Mondex protocols.

9.5.2 Payment

Payments are performed by inserting the Mondex card in the point-of-sale terminal or by filling out an HTTP form from a PC. The financial exchanges are done with a payment server that is responsible for the interface with the

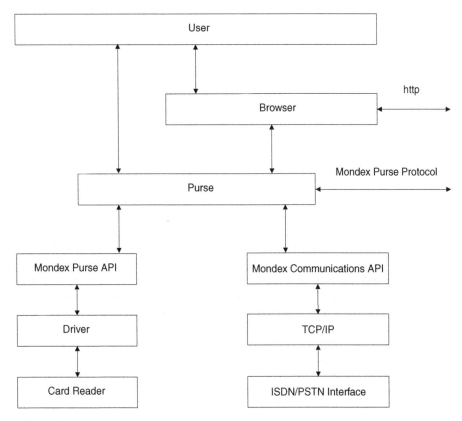

FIGURE 9.4
Configuration of a Mondex client.

banking network. Payment instructions are transmitted immediately from the card to the terminal and then from the terminal to the payment server, without the intervention of the verification centers; thus, the authorization is done offline. The payment server sends to the cardholder a receipt with the transaction references. This receipt is used to prove the validity of the transaction to the merchant server so that the buyer can receive the merchandise. In the case of the Internet, the merchandise is usually virtual information downloaded from the Internet.

9.5.3 Security

The details of the security protocols of Mondex are not known, which led to severe criticisms by some experts in security. It is known that public-key cryptography is used, but the key length was not announced. It was also suggested that the Hitachi chips used (S109 or S101) have technical vulnerabilities that cause "concern" (Dreifus and Monk, 1998, p. 147). Note, however, that the risks are limited to the amount stored in the card.

Mondex stores the details of the ten last transactions in the card; as a consequence, Mondex transactions are traceable, in contrast to physical cash. The Mondex card can be locked with a special four-digit code to prevent its use in the cases of theft or loss. Once locked, transfers from the card are not allowed, and the transactions stored cannot be displayed. This provides some security to the cardholder and provides limited confidentiality, until the code is broken.

9.5.4 Pilot Experiments

Starting in 1995, several trials were conducted at Swindon (a locality west of London) in the U.K.; in San Francisco and Manhattan in the U.S., and at Guelph in Canada. In Hong Kong, more than 45,000 cards were distributed with the participation of more than 400 merchants in three commercial malls.

At Swindon, Guelph, and Manhattan, the majority of the participants did not find any advantage in replacing coins and bills with an electronic purse (*Banking Technology*, 1998; Hansell, 1998). Merchants were unhappy with the difficulty of the operation and by the time it took to conduct a transaction (less than 5 seconds). The results gathered from the trials in Hong Kong and in New York City confirmed that, for the time being, the Mondex solution does not seem to be attractive to clients and merchants (Westland et al., 1998; Van Hove, 2001). At the same time, banks do not seem to welcome peer-to-peer exchanges, which are outside of their control, arguing that this functionality encourages money laundering and tax evasion. It seems, therefore, that in its current form, the Mondex card will experience many difficulties penetrating face-to-face commerce, and that it may be more suitable for remote micropayments.

9.6 Proton

In 1993, Banksys (an interbank society responsible for electronic payment in Belgium) initiated the Proton project. Introduced early in 1995, available over the whole of Belgium in 1996, the purse now shows considerable international success, even though it is second to GeldKarte in terms of number of units in circulation.

From an administrative viewpoint, Proton is the property of Proton World, which was a joint venture by American Express, Banksys, ERG, Interpay, and Visa International. In 2001, Proton World became wholly owned by the Australian-based ERG, which was behind the Octopus of Hong Kong.

The card used for Proton is supplied by CP8-Oberthur and Phillips. The chips are from SGS Thomson (ST16 601), Infineon (ex-Siemens) (SLE 44C40), or Motorola (SC46). Their capacities vary between 6 K and 16 K octets of ROM and from 1 K to 8 K octets of EEPROM (Kirschner, 1998).

The merchant terminals are equipped with a SAM to verify the cards and assure the mutual authentication of the card and the terminal. Security of the exchanges with the collection center is assured with the DES for confidentiality and RSA for authentication.

The Proton architecture is largely inspired by specifications that became the four-part CEN standard, EN 1546, in 1999. This standard defines the interfaces, the security architecture and the functionalities of electronic purses to ensure the interoperability of several applications. Beyond the generic payment application, Proton is planned for parking meters in Belgium. It can also be used for access control, which is already the case with Banksys or the French Hospital in Ganshoren.

9.6.1 Loading of Value

The loading of value in Proton can be done from special terminals using the PIN or from home using a special telephone set, the C-ZAM/phone, which allows phone purchasing as well. The C-ZAM/phone was adopted by the Rabobank in the Netherlands under the name Smartfone Rabo-bank.

The Belgian telephone operator Belgacom is studying the possibility of recharging the purse from public phones.

9.6.2 Payment

From the client side, Proton appears as an integrated-circuit card that does not require a PIN for usage (Dawirs, 1997); purchases with small amounts cause an automatic debit of the card. Because the verification of the purse is done offline, an electronic payment transaction takes about 2 to 3 seconds. Collection of transactions can also be done offline (for example, in the case of parking meters).

To make secure payments on the Internet, the client's computer must be equipped with a Proton card reader that has a chipcard security module (CSM). Banksys supplies both the card reader and the CSM. This application was developed in conjunction with the main cafeteria of Lueven Catholic University. The objective was to allow students to pay for their orders from various parts of the campus using the campus network.

The user must fill out a payment form in HTML on a Web page associated with a Java applet; this ensures platform independence. After placing the order, the client is invited to insert the Proton card in the reader. By clicking on the acceptance button, the client allows the Java applet to establish the connection between the Proton card and the CSM situated physically in the merchant server. The payment transaction continues without querying the authorization center (i.e., offline). The payments accumulated in the CSM are then transferred to Banksys to be credited to the cafeteria's bank account.

9.6.3 International Applications

In the Netherlands, the Interpay, the Dutch equivalent of Banksys, renamed the Proton technology Chipknip. The introduction of Chipknip set a ground-breaking record in the payment domain worldwide, because 12 million cards equipped with the security tools were put into circulation in 1997.

The Chipknip card is multifunctional. In addition to the purse function, it offers to its holders a traditional debit card, loyalty programs (management of discount coupons used in marketing), and the ability to pay for public transportation. The electronic purse function of Chipknip is for amounts less than 40 Dutch guilders (around U.S.$20). Beyond that amount, the function of debit card with PIN verification takes place.

The card has two suppliers: Bull and Phillips. The chips are from SGS Thomson. They vary in capacity from 6 K to 16 K octets for the ROM, 288 octets of RAM, and between 1 K and 8 K octets of EEPROM (Kirschner, 1998).

The Proton electronic purse is available in Germany and in Sweden; in Switzerland, its name is Cash. Proton is also present outside Europe: in Australia, in Hong Kong, and in New Zealand. American Express adopted it for its credit cards. In Canada, the purse is marketed under the name Exact. Finally, the introduction in Brazil was done in partnership with Banco do Brasil, one of the largest Brazilian banks, and Barra Shopping, the largest mall in Latin America, with more than 500 stores.

9.7 Harmonization of Electronic Purses

Table 9.1 shows a comparison of some technical and commercial features of the main electronic purses (Kirschner, 1998).

From this table, it is clear that electronic purse systems are often incompatible in terms of protocols or services. To access a richer set of applications, the holder must own several purses, particularly if there is a need for payment of cross-border exchanges through the Internet. Another approach is to leave the currency exchanges to an intermediary, under the supervision of a bank. Another value-added service that a nonbank could offer would be to facilitate the exchange of values in the same currencies among the incompatible purses.

The proliferation of purses and their lack of interoperability are discouraging the market. The lack of interoperability is an annoyance to users. It forces operational difficulties upon service providers and imposes additional costs on manufacturers. To bring such an unbridled development under control, some agreements are needed at the level of the protocols, the applications, and the terminals.

A new generation of mutliapplication electronic purses based on the EMV (EuroPay, MasterCard, Visa) specifications can help this harmonization.

TABLE 9.1

Comparison of the Principal Electronic Purses

Characteristics	Chipper	GeldKarte	Mondex	Proton
Country where it is used	The Netherlands	Germany, France	U.K., Australia, Canada, etc.	Belgium, Australia, Brazil, Sweden, etc.
Number of currencies	1	1	5	Several
Card manufacturer	Bull, Phillips	Gemplus, Giesecke & Devrient, ODS	Dai Nippon Printing	CP8-Oberthur, Phillips
Chip manufacturer	SGS Thomson	Infineon (ex-Siemens), Motorola	Hitachi	SGS Thomson, Infineon (ex-Siemens), Motorola
Size in octets				
ROM	8–16 K	12 K	16 K	6–16 K
RAM	288	256	512	—
EPROM, EEPROM	1–8 K	8 K	8 K	1–8 K
Security	RSA, triple DES, SAM	SAM, DES	Proprietary	SAM, triple DES, RSA
Anonymity	Yes	Yes	Yes	Yes

Europay (MasterCard), Visa, and ZKA (Zentraler Kreditausschuß) collaborated to define the Common Electronic Purse Specifications (CEPS) (CEPSO, 1999) for many functions. To illustrate, let us look at some examples in the areas of authentication of the purse by the issuer, loading of value, and point-of-sales payments.

9.7.1 Authentication of the Purse by the Issuer

Authentication of the purse by the issuing institution (which may be a telephone or a bus company, i.e., not necessarily a financial institution) can be done either online or offline.

Online authentication uses a secret key shared between the issuer of the purse and the smart-card of the purse. This keys allows the derivation of a common session key that will serve to compute the MAC to protect the integrity of the data exchanges and to encrypt the authentication data. The CEPS limits the key size on open networks to 8 octets (64 bits) but leaves open the choice of encryption and hashing algorithms.

Offline authentication uses public-key encryption using RSA for mutual authentication and for the subsequent exchange of a temporary session key between the chip card and the terminal. This session key will be used to encrypt the data according to triple DES.

9.7.2   Loading of Value

Loading the purse with value depends on if the purse is linked to a bank account.

If the purse is linked to a bank account, there is already some banking relationship between the issuer and the purse holder. Thus, the permission to convert the value to dematerialized money stocked in the purse is under direct control of the holder's bank. Authentication of the card, verification of the identity of the holder on the basis of a PIN, and authorization can be done in a single transaction. The exchanges involve the card, the load device, and the authorization server of the issuing bank.

The purse is not linked to a bank account when the line of credit is from a totally separate account or if it entails a revolving credit. In these cases, the risk of error or fraud increases, and the verification is more complex. The loading protocol must verify the integrity of the value transfer from the client's bank (or from that of the purse issuer) to the acquirer bank in addition to verifying the authenticity and the identity of the card holder and of the card that the holder presents. A Loading Secure Application Module (LSAM) provides security during the loading of value. The integrity of the communication is protected with signatures using the hashing algorithm SHA-1. Random numbers are included to protect against replay attacks.

9.7.3   Point-of-Sales Payments

The protocol for point-of-sales payments defines procedures for offline reciprocal authentication of the purse and the point-of-sales terminal. The protocol covers single transactions and a series of incremental transactions (such as payment for telephone calls).

This protocol conforms to the EMV specifications. Thus, the terminal can authenticate the public key of the card by using the public-key certificates of the card issuer and of the issuing bank. Similarly, the card authenticates the Purchase Secure Application Module (PSAM) of the terminal with the help of its certificate. The integrity of the exchanges is protected with MAC calculated by symmetric encryption. The terminal checks the validity of the PIN of the card holder.

The amount of the transaction is displayed on a screen, and the cardholder has to indicate his or her consent. Data relative to the transactions are exchanged between both sides of the transaction, and their integrity is ensured with the help of MACs. The amount of the value stored in the purse is updated. It should be noted that the use of screens raises some problems regarding their resistance to breaking as well as their reflectivity and energy consumption (Praca and Barral, 2001).

The specifications allow the possibility of canceling the last transaction, provided that the cancellation is done from the same PSAM that managed it in the first place and by using the same security parameters. The purpose of these restrictions is to discourage fraud.

Finally, an audit trail of the transactions is stored in the security module until it is collected by the authorized operator.

9.8 Summary

Several electronic purses were proposed for making micropayments. Reduction of the operational costs can be achieved using several practices, such as service prepayment, the reduction of computational load, and grouping of transactions before financial clearance.

The diffusion of these purses to the general population faces several hurdles. The more important among these obstacles are the lack of a network of charging points and the multiplicity of physical or logical interfaces. The transition to the EMV specifications offers the possibility of consolidation around a common format. Finally, a novel perspective seems to open with contactless payments in public transportation systems.

Questions

1. How can operational costs be reduced in micropayment systems?
2. What are the unique features of Mondex compared to other micropayment systems?
3. What are the main difficulties that Mondex promoters have faced? Compare with the case of Proton.
4. Speculate on the reasons behind the lack of electronic purse projects in the U.S.

10

Remote Micropayments

ABSTRACT

Between 1995 and 2003, products for remote micropayments passed through two generations. Those of the first generation dabbled with technical innovations and the subtleties of cryptography to produce new electronic monies, with different degrees of sophistication, without adequate care given to the practical aspects of their use. Once it became clear that the success of the majority of these systems will be questionable, they were quickly eclipsed in favor of a new generation more concerned with the actual needs of potential users than with technical prowess. Examples of these uses are the remote payment of occasional users or the security of payments among parties that do not know each other, such as for online auctions.

Whatever their generation, the different solutions address mostly the sale of nonmaterial content: information, newspaper archives, online games, horoscopes, job openings, pictures, music, videos, etc. The majority conform to a scheme whereby a trusted third party intervenes to facilitate the exchanges between the two parties of the transaction and to verify the level of security according to the interests at stake. The gains of the intermediary come from several sources, such as the interest on the floating value that accumulates in the intermediary's account until the payment is settled as well as from a commission on each transaction.

The bulk of this chapter covers a few representative systems from the first generation to underline their technical contributions. From the rich offer, we select the most outstanding examples, particularly:

- First Virtual, a system of security without encryption
- NetBill, which is based on a variant of public-key Kerberos
- Millicent, PayWord, and MicroMint, which are systems based on a jeton economy
- KLELine, where the payment operator combines several roles and covers bankcard payments in addition to micropayments

The common denominator of all of these systems is the use of a virtual purse or a virtual jeton holder to carry the various monetary values available to the user. This "virtual purse" is a bank account supplied with legal tender using a credit or debit card, a direct debit, or a fund transfer.

10.1 Security without Encryption: First Virtual

First Virtual, the first commercial offer to secure payment for digital information and services over the Internet, operated between 1994 and 1998 (New, 1995; Borenstein et al., 1996).

To ensure confidentiality and authentication without resorting to cryptography, the advanced solution was to use two independent networks to carry the exchanges in the clear, the Public Switched Telephone Network (PSTN) and the Internet. On the Internet, two different tools were mobilized: a browser and e-mail. Thus, use of First Virtual did not require any additional client software other than what is used normally for network usage.

10.1.1 Buyer's Subscription

To subscribe, the user established an account at First Virtual by filling out an application with indication of status, postal address, e-mail address, and a password of the customer's choice. Subscription could be done by postal mail, by telephone, by facsimile, or by the Internet. First Virtual would return by e-mail to the client an acknowledgment with a subscription number and the instructions to activate the account. Activation could be done by phone, by facsimile, or by postal mail. It was only after the account was activated that users had to give their banking coordinates. Once these references were verified, the First Virtual server would send to the customer a "Virtual Pin" that served as an access code to the payment server. In this manner, the transfer of the banking coordinates occurred only once, and never on the Internet. The subscription cost was U.S.$2.

For an annual subscription of U.S.$10, First Virtual supplied the merchants with the necessary software to run the merchant site.

10.1.2 Purchasing Protocol

The exchanges of the purchasing protocol are depicted in Figure 10.1:

1. The customer supplies the merchant's server with his or her Virtual Pin within a purchasing form that is completed using the browser. The purchaser can obtain the digital information without waiting

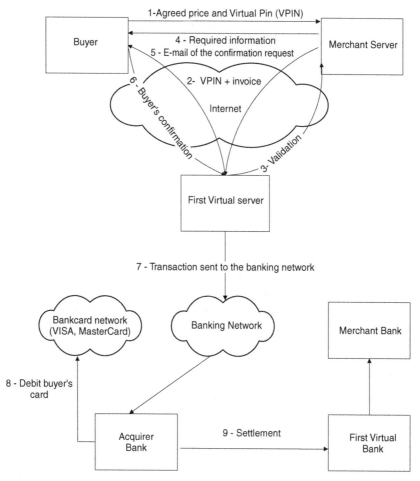

FIGURE 10.1

Message exchanges in the First Virtual system.

    for First Virtual's authorization, which, after being obtained, would compensate the merchant in case of fraud.

2. To verify with First Virtual that the account was valid, the merchant forwards the customer's identifier (Virtual Pin) together with the details of the invoice.

3. The First Virtual server checks the virtual PIN and responds to the merchant server.

4. If the responses were positive, the merchant server would continue to forward the required information to the customer.

5. Before initiating the settlement, the First Virtual server requests the customer's confirmation by e-mail, thereby closing the loop.

6. The customer responds to the First Virtual server. If the customer concurs, First Virtual submits the number of the customer's credit card to the acquirer bank (First USA) to debit the customer's credit account.

7. Depending on the merchant's preferences, the financial settlement takes place either through interbank exchanges or through the bank-card network.

10.1.3   Acquisition and Financial Settlement

The operations occurred in the following manner. Service providers would tally the transactions on their site and then send the details to the First Virtual server. That server would then debit a customer's account when the sum total of purchases exceeded U.S.$10. First USA would debit the client but, because the U.S. legislation allows repudiation of purchases by credit cards within 90 days, First USA would wait for this period before crediting the First Virtual bank account with the corresponding value, after deducting its commission (2% of the transaction amount). First Virtual would, in turn, fund the merchants after deducting a flat transaction fee of 29 cents per transaction.

10.1.4   Security

Correspondence between the identifiers and the banking coordinates of the users was kept under high security in a central computer uncoupled from the Internet.

Anonymity of the purchaser with respect to the merchant as well as the anonymity of both with regard to the respective banks of the customer and the merchant was assured. However, First Virtual was aware of all the exchanges, knowing not only the identities of the parties and their bank accounts but also the content of the orders and the invoices.

As explained above, the principle of the security model was to request the purchasers' confirmation by sending an e-mail to their addresses before debiting their accounts for any purchase. This measure prevented potential swindlers, even if they defrauded users from their identifier, from intercepting the confirmation request. The customer was further protected, because the supplier did not have the customer's banking coordinates. The customer could also refuse the invoice, even after receiving the information. Clearly, in this model, merchants assumed all the risks, because there could be long delays before the fraudulent use of an identifier was detected. Nevertheless, a vendor could choose not to satisfy a request until all banking verifications were conducted. First Virtual could also cancel the accounts in case of abuses, for example, when subscribers contested their bills frequently or when suppliers had many invoices challenged or had many complaints against them.

10.1.5   Evaluation

The main attraction of First Virtual was its extreme simplicity. A two-stage user-friendly procedure was able to bypass the problems of cryptography while allowing the online sale of images and texts. Nevertheless, this construct could not accommodate all types of applications, for example, the purchase of physical goods or payments among individuals, which did not make it suitable for online auctions.

10.2   NetBill

NetBill is a set of rules, protocols, and software designed for the delivery of text, images, and software through the Internet. It was developed by Carnegie Mellon University in partnership with Visa and Mellon Bank to address micropayments of the order of 1 cent (U.S.) (Cox et al., 1995; Sirbu and Tygar, 1995; Sirbu, 1997). The main characteristic of NetBill is that the customer is only billed after receiving the encrypted information; however, the decryption key that allows the customer to extract the information is only delivered after the bill is paid. This approach is particularly useful for repeated delivery of small quantities of information, such as subscription to news services or to information updates.

The NetBill server plays the role of a certification authority in addition to its function as a payment intermediary. It also manages the distribution of the pairs of RSA public/private keys as well as the session keys that are used to encrypt the exchanges between the customer and the merchant, as well as their exchanges with the NetBill server.

10.2.1   Registration and Loading of Value

Customers subscribe by giving the data of their credit card or the coordinates of a payment instrument encrypted with the help of a downloaded security module (Money Tool). In return, the customer receives from the NetBill server a user identification and a pair of RSA public and private keys, KP_C and KS_C, respectively.

Similarly, after registration, the merchant receives the program called Product Server as well as the pair of RSA public and private keys, KP_M and KS_M, respectively.

NetBill subscriber accounts are prepaid from bank checking accounts, preferably the bank of NetBill. The customer's software displays the amount available on the screen.

In a future phase, NetBill will consider authorizing postpayments.

10.2.2   Purchase

The basic purchasing protocol uses eight HTTP messages to cover the four main phases of the commercial and financial transaction: the negotiation, the order, the delivery, and the payment. These exchanges involve three parties: the customer, the merchant, and the payment intermediary (the NetBill server) called the *till*. The payment intermediary plays the roles of a notary and a trusted third party and communicates with the merchant directly. Through the merchant, it communicates indirectly with the customer. The exchanges are shown in Figure 10.2.

Before the establishment of communication channels, a phase of mutual recognition takes place to allow the partners to identify and authenticate themselves according to the public key Kerberos system (see Chapter 3). Thus, the merchant server authenticates the customer with the help of a session ticket and a certificate signed by the certification authority. The ticket is sealed (signed) by the customer private key and then encrypted with the merchant server public key. With the session ticket, the customer and the

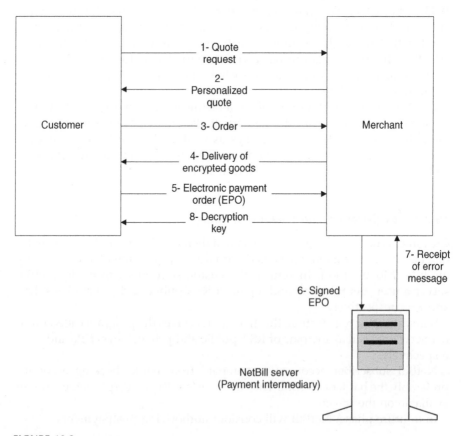

FIGURE 10.2
Messages exchanged in a NetBill purchase.

merchant will be able to share a session key to encrypt the exchanges that follow with a symmetric encryption algorithm. The session ticket also offers a defense against any penetration of the Kerberos server. Thus, the customer constructs a symmetric key $K_{challenge}$ and sends it to the merchant in the following message:

$$K_{one\_time} \{\text{customer name, merchant name, time stamp, } K_{challenge}\}, PK_M$$
$$\{K_{One\_time}\}, Sig_C$$

where $K\{X\}$ represents the encryption of message X with the key K. The first part of the message is encrypted with the symmetric key K_{one_time}, and the second is encrypted with the public key of the merchant, PK_M, using the public-key RSA algorithm. The merchant can recover the symmetric key by decrypting the second part with its private key SK_M. The variable Sig_C represents the signature (on the full or hashed message, depending on the case) with the private key SK_C of the customer to verify the message integrity upon reception. The time-stamp field may also include a nonce (a random number used only one time) that will be used to fend off replay attacks using past messages.

The merchant's server verifies the signature using the public key of the customer extracted from the certificate. It recovers the symmetric key K_{one_time} to obtain the key $K_{challenge}$. Next, the merchant produces the symmetric session key K_{CM} and constructs the session ticket τ_{CM}:

$$\tau_{CM} = \text{merchant's name, } KS_M \{\text{customer's name, customer's address, time stamp, expiration date, } K_{CM}\}$$

The ticket comprises two parts:

- The merchant's name in the clear
- The customer's name and address, a time stamp against replay attacks, the expiration date of the ticket, and the symmetric session key, K_{CM}, all encrypted with the merchant's private key. (Note that the expiration date is reconfigurable.)

The merchant's server encrypts τ_{CM} and K_{CM} with the help of the symmetric key $K_{challenge}$, then sends them to the customer in the encrypted message:

$$K_{challenge} \{\tau_{CM}, K_{CM}\}$$

Having generated $K_{challenge}$ and then sharing it exclusively with the merchant's server, the customer is sure that the message comes only from the merchant. The customer decrypts it to recover the session ticket τ_{CM} as well as the symmetric session key K_{CM}. Next, the customer verifies, with the help of the merchant's public key, that the session key is identical to the key that the session ticket contains.

10.2.2.1   Negotiation

During the negotiation phase, the customer requests the price of an item, and the merchant responds with a personalized quotation. This phase begins with the identification of the customer so that the merchant can customize the offer according to the customer's profile. The exchanges take place as follows (see Figure 10.2):

1. The customer presents the ticket τ_{CM} and then encrypts the request with the symmetric session key. This request is sent in Message 1:

    τ_{CM}, K_{CM} {customer credentials, data on requested product, request qualifiers, initial price, transaction identifier TID}

    The customer credentials are used to take advantage of special discounts offered to members of a given group. The data related to the merchandise can be composed in the request automatically by clicking on the reference to the goods. The request qualifiers are flags that indicate the customer's desires (for example, delivery options). Other data may be added as an option, such as the initial price and the transactional identifier, etc. The request is then sent to the merchant server.

2. Having identified the customer, the merchant server customizes the quotation according to the customer profile (for example, subscriber, privileged customer, volume reduction, etc.). The server then encrypts the offer as follows (Message 2 in Figure 10.2):

    K_{CM} {product description, proposed price, request qualifiers, transaction identifier TID}

    The product description is a text that will be printed on the customer's bill. The request qualifiers are flags that indicate what the merchant accepted in response to the customer's initial request. The presence of the transaction identifier in this message is mandatory.

3. Messages 1 and 2 will be exchanged as many times as necessary for the two sides to agree on the terms of the transaction.

10.2.2.2   Order

The customer indicates agreement by sending the following command (Message 3 in Figure 10.2):

$$\tau_{CM}, K_{CM} \{IDT\}$$

10.2.2.3   Delivery

After receiving the purchase order, the merchant sends the digital information encrypted with a symmetric encryption key K_{GOODS} without including this key in the message (Message 4 in Figure 10.2):

$$K_{GOODS}\{GOODS\},\ K_{CM}[SHA\ (K_{GOODS}\ \{GOODS\})],\ EPOID$$

where {GOODS} represents the encrypted information That file is stored on the customer's hard disk but will remain unusable without the decryption key; the client will receive this key after settling the bill. Nevertheless, the customer can verify the integrity of the received articles by applying the hashing algorithm SHA and comparing the results with the quantity $SHA(K_{GOODS}\{GOODS\})$ that is extracted from the message using the session key K_{CM}, which is a guarantee to the customer.

The merchant joins to the file an Electronic Payment Order Identifier (EPOID) to serve as an index in the NetBill registry of transactions. The EPOID is an exclusive identifier for each transaction that consists of three fields: a merchant identifier, the time stamp marking the end of goods delivery, and a unique serial number. The uniqueness of the EPOID prevents replay attacks, while the time stamp is used to date transactions.

10.2.2.4   Payment

The objective of this phase is to close the transaction by delivering the payment to the merchant and the decryption key to the customer. During this phase, the customer's software constructs an Electronic Payment Order (EPO) with the following fields:

1. Customer's identity
2. Product identifier
3. Negotiated price
4. Merchant identity
5. The digest of the encrypted goods $SHA(K_{GOODS}\{GOODS\})$
6. The digest of the product request data in Message 1, SHA (product request data) to forestall future disagreement over the details of the order
7. The digest of the customer's account number with an account verification nonce, SHA (customer's account number, nonce), which is used to verify the validity of the credentials sent to the merchant, for example, that the account number sent is the one that NetBill has given
8. EPOID, the transaction index to the NetBill registry
9. The ticket τ_{CN}
10. $K_{CN}\{$authorization ticket, customer's account number, account verification nonce, customer's memo field$\}$

Note that the authorization ticket is an access-control mechanism regulated by another Kerberos server functioning in the same way as the NetBill server but is under the control of another independent authority.

The EPO consists of two fragments. The first segment is composed of the elements 1 to 8 of the previous list and is sent in the clear. The second segment is addressed to the NetBill server and comprises elements 9 and 10.

The exchanges take place in the following manner:

1. The customer's software constructs an EPO. The customer signs with the signature SIG_C constructed with the help of the customer's private RSA key SK_C and sends it to the merchant in Message 5 of Figure 10.2. This message includes the ticket τ_{CM}, and the rest is encrypted with the symmetric session key K_{CM}:

$$\tau_{CM}, K_{CM}\{EPO, SIG_C\}$$

2. After verifying the integrity of the contents of the EPO, the merchant verifies the digital imprint of the file to make sure that the customer received the goods that were sent. If the merchant decides to continue the transaction, it adds the decryption key, endorses the EPO with its digital signature, and sends the new message to the NetBill server. This is Message 6 of Figure 10.2, which is sent in the following form:

$$\tau_{MN} K_{MN}\{(EPO, SIG_C), MAcct, MMemo, K_{GOODS}, SIG_M\}$$

where MAcct is the merchant account number, MMemo is a field reserved for the merchant's comments, and the signature is constructed with the help of the RSA private key of the merchant SK_M. By endorsing this message, the merchant confirms that the transaction took place in a regular fashion.

3. The NetBill server extracts the different fields for the message and checks the validity of the customer and merchant signatures, the uniqueness and the freshness of the EPOID, and determines the privileges of the customer. It verifies that the customer's account balance covers the sum requested, then debits the customer's account and credits the merchant account. In case the payment exceeds the balance, the server returns an error message inviting the customer to supply funds to the account for the transaction to conclude.

4. For a successful transaction, the NetBill server records the transaction, keeps a copy of the decryption key, K_{GOODS}, and digitally signs a receipt that contains the decryption key. The receipt is of the following form:

    Receipt = [result code, customer identity, price, product identity, merchant identity, K_{GOODS}, EPOID] SIG_N

    To increase the computation speed, this signature uses the DSA algorithm rather than the RSA. The message sent to the merchant is

Message 7 of Figure 10.2, which can be described in the following manner:

$$K_{MN}\{\text{receipt}\}, K_{CN}\{\text{EPOID, CAcct, Bal, Flags}\}$$

where CAcct, Bal, Flags are fields that contain details on the customer's account. These details are encrypted with the symmetric key K_{CN} to prevent the merchant's server from reading them.

5. The merchant's server extracts the receipt and encrypts it with the symmetric key K_{CM}, then adds the encrypted information on the customer's account to form Message 8 of Figure 10.2:

$$K_{CM}\{\text{receipt}\}, K_{CN}\{\text{EPOID, CAcct, Bal, Flags}\}$$

6. The customer's software applies the decryption key to retrieve the information bought and updates the balance of the funds available to the customer.

10.2.3   Financial Settlement

The funds to be credited to the merchants are accumulated and then deposited periodically through VisaNet in the bank accounts of the various merchants.

NetBill charges a commission varying between 2.5 cents for a 10-cent transaction and 7 cents for a U.S.$1 transaction (which is an overhead of between 25 and 7% of the transaction amount).

10.2.4   Evaluation

NetBill ensures the main security services (confidentiality and integrity of the messages, identification and authentication of the participants, as well as nonrepudiation). In addition, NetBill transactions satisfy the following requirements on monetary transactions (Camp et al., 1995):

- Atomicity, which means that the transaction must be executed as a whole for its effects to take place: The client is only billed after receiving the goods, and payment leads to access to the delivered goods. By separating the delivery of information from its decryption, customers can be assured that they received the information before settling the bill. At the same time, NetBill guarantees to the merchant that a transaction will be refused if the balance of the virtual purse is not sufficient to cover the purchase.

- Consistency, because multiple verifications are made throughout the purchase: Examination of the digital imprints ensures that the transaction reflects exactly the terms of the agreement between the two parties. The uniqueness of the EPOID prevents the reuse of ancient orders by unscrupulous merchants. Time stamping allows purging of stale transactions.

- Isolation, given that the transactions are independent of each other.

- Durability, because each party has a transaction record: In addition, the NetBill server has a record of the decryption key KGOODS that the customer can claim in case of a technical problem, irrespective of the origin.

The NetBill server plays the role of a trusted third party and an arbiter to resolve conflicts. Although the information relative to the customer's identity is dissociated from the goods exchanged, the EPO reveals the customer's identity to the merchant, and the transactions are traceable. Because the NetBill server intervenes in each exchange, this fact can be exploited to protect the customer's anonymity with respect to the merchant. The NetBill server intercepts all the customer messages to hide all information related to their origin before reexpediting them to the merchant. The server can ensure the merchant's anonymity in the same manner.

In short, NetBill presents interesting ideas (e.g., the electronic purchase order or the distinction of information delivery from information access), but the frequent use of digital signatures, particularly public-key signatures, tends to reduce the performances of commercial applications of a reasonable scale. In particular, because the intervention of the intermediary (NetBill server) is required for each transaction, the number of transactions that can be conducted concurrently is limited by the computational power available at the NetBill server.

10.3　KLELine

KLELine was comprised of a platform for secure payment, a virtual mall, and a payment system called Globe ID. The payment system was under the control of a bank, which allowed it to collect funds (Pays and de Comarmont, 1996; Bresse et al., 1997, pp. 56–59; Dragon et al., 1997, pp. 281–284; Pays, 1997; Romain, 1997; Sabatier, 1997, pp. 103–105). A succession of banks were associated with KLELine, La Compagnie Bancaire (which was acquired by Paribas) and the group BNP-Paribas, which favored the CyberComm solution and closed KLELine in January 2000.

KLELine covered a large range of payment instruments:

- A virtual purse of values less than 100 French francs (around U.S.$15): The virtual purse is supplied from a bank account to ensure that funds are always available.

- Access to the bankcard network for purchases exceeding 500 FF (around U.S.$5).

- Any of the two methods above for purchase amounts between 100 and 500 FF.

The KLELine accepted transactions in 183 currencies, with rates updated every 6 hours. KLELine charged a commission ranging from 10 to 20% for micropayments (payments that use the virtual purse) and from 2 to 4% for bankcard payments.

10.3.1 Registration

To subscribe to the service, the customer gives his or her banking coordinates, the mode of charging the customer's purse (debit or credit), an electronic address, and, optionally, a preference profile. The customer receives in return a personal identification number (PIN), a customer identifier (CID), and the customer's software, called either Klebox or PACK (Personal Authentication and Confirmation Kit). This software is a plug-in for the browser and gives access to the virtual purse. The customer uses the secret PIN code (that the customer can modify) for identification purposes with the server. This transaction is secured with a pair of RSA keys of 512 bits.

The primary function of the merchant's kit or SACK (Server Authentication and Certification Kit) is to secure the communications with the KLELine server through asymmetric encryption with certification. In addition, the SACK manages the back-office functions, such as personalization of the offer as a function of the customer's profile, issuance of sales slips, reception and recording of sales receipts, and update and management of currency exchange rates.

10.3.2 Purchase and Payment

The KLELine server was simultaneously an intermediary between the merchant and the customer and a gateway between the Internet and the banking network, a notary, a trusted third party, as well as a host of a virtual mall. It intervened to authenticate the parties present and to guarantee that the customer's bank data (account or bankcard number and PIN) were not exchanged in the clear on the open network (e.g., the Internet).

The CPTP (Customer Payment Server Transaction Protocol) describes the various phases in a purchase transaction, including payment. This protocol reproduces electronically the various phases that take place in a typical French department store (choice of the product, issuance of the sales slip, payment, issuance of sales receipt, pickup of the merchandise by showing the sales receipt). Thus, the transaction takes the following steps (see Figure 10.3):

1. The customer consults the online catalog with the browser.
2. The customer selects one or several items.
3. The merchant kit recognizes the number of the Klebox that is plugged into the navigator and identifies the customer.

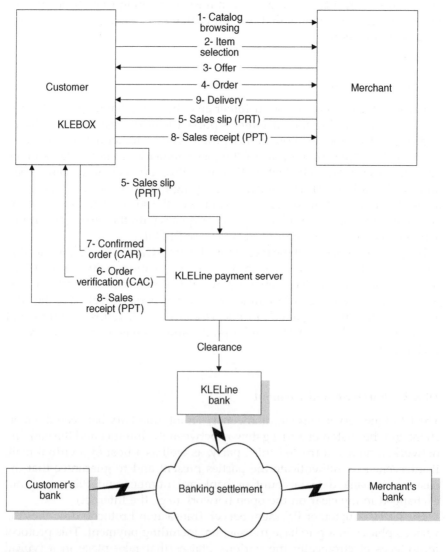

FIGURE 10.3
Message exchanges in a KLELine purchase.

4. The merchant server makes an offer to the customer.

5. The customer accepts the offer.

6. The merchant's server issues a signed sales slip and sends it to the KLELine payment server in the Payment Request Ticket (PRT) message, which is signed with the merchant's private RSA key. This message is sent first to the customer, who forwards it to the server.

7. The KLELine server processes the sales slip and uses the content of the PRT and the customer profile to compose the Confirmation and Authentication Challenge (CAC) message. The CAC requests the customer's confirmation on the order and contains a unique sequence number for this transaction and a time stamp to counter replay attacks. It presents the customer with a menu of payment instruments to select from and is signed using the private RSA key of the server.

8. The customer selects the payment instrument of choice, enters the PIN, and responds with the Confirmation and Authentication Response (CAR) message. The CAR, which confirms the order, consists of two parts: the first contains the PIN protected from attacks with a time stamp and a random number. This part is signed with the help of the public RSA key of the server, so that the server can verify its integrity. The second part contains a session key and is signed using the PIN and the customer identifier CID. The details of this operation are not known, except that it involves a message authentication code and that it uses the MD5 hashing function. This mechanism simultaneously ensures the integrity of the message content and the authentication of the sender. After this point, the transaction cannot be repudiated.

9. Once the authentication is assured, the KLELine server debits the customer's purse and credits the merchant's cash register.

10. The server sends to the customer the Payment Proof Ticket (PPT) message, which contains a sales receipt. This message is signed either with the session key in case of acceptance of the transaction or by the private key of the server in case of refusal.

11. The customer sends to the merchant the sales receipt included in the PPT message. If the transaction was successful, the sales receipt contains the order to be paid and a reference number of the transaction. (It is not clear how the merchant would be able to verify the integrity of the PPT message, in case of success, because the merchant does not know the confidential code of the customer.)

12. Following reception of the sales receipt, the merchant is supposed to deliver the selected article to the customer. Furthermore, KLELine also offers back-office services to help the merchant automate its internal procedures.

10.3.3 Financial Settlement

The clearance and settlement procedures used in banking networks are applied for bankcard purchases. For micropayments, the customer's virtual purse is debited for each transaction to the benefit of the merchant's virtual cash register. Then, 45 days after the purchase (to take into consideration delays in customer objections), KLELine empties the virtual cash register and distributes the amount collected among the merchant, the virtual mall, and itself.

10.3.4 Evaluation

The security operation depends on several features:

1. The payment server uses the CID to identify the buyer, which obviates the need to send banking references online.
2. The PIN does not travel in the clear through the merchant's server.
3. The PIN is not stored in the clear in the customer's workstation. As a consequence, in case of workstation failure, the software needs to be reinstalled.
4. Customers can always check the balance in their virtual purse and obtain a detailed statement of all transactions.
5. The sales receipt offers a proof of payment.
6. The merchants have in their possession the sales slip and the sales receipt that can be reconciled for bookkeeping purposes (these exchanges are not shown in Figure 10.3).
7. The transaction is protected against replay attacks with the help of a unique sequence number and a time-stamp.
8. KLELine keeps all customer and merchant information (card numbers, PINs, etc.) under tight security.
9. The merchant authentication and the verification of the integrity use RSA public-key encryption with a key size of 512 bits, MD5 hashing, and a nonspecified symmetric algorithm.
10. All transactions are recorded and the customer's identity is known by all participants, so there is no anonymity.

Given that the specifications of the CPTP protocol were never published, many details to conduct a complete evaluation are lacking. For example, the available documentation does not explain what is meant by a signature and it does not show which exchanges are encrypted with the help of the session key or which algorithm is employed (Pays and de Comarmont, 1996; Pays, 1997, p. 49).

10.3.5 Evaluation and Evolution

Banking support gave KLELine the significant advantage of relying on legal money and of recognizing 183 different currencies, which helped it internationally. At the time of its liquidation, the product was available to over 500 sites, with 85 in France. Unfortunately, its CPTP protocol remained proprietary, and many competitors were based on standardized protocols. This was one of the factors that favored the discontinuation of KLELine after the merger of BNP and Paribas.

From the ashes of KLEline was born a new virtual purse for payments with multicards, mutlicurrencies, and micropayments: Odysseo (http://www.odysseo.com). Contrary to KLELine, there is no client software. The monetary value is stored in a central server, and access to the purse can be done from any terminal (after a successful authentication). Merchant sites constitute a network of secured sites managed by Blue Line International (http://www.bluelineinternational.com). Access to the purse is through an SSL session secured with a symmetric encryption key of 128 bits.

The central server is managed by a specialized company, Ubizen (http://www.ubizen.com). This server carries the following tasks:

1. Merchant authentication with the help of a public-key infrastructure
2. Secure transmission with SSL
3. Nonrepudiation through time-stamping

This solution offers the user better protection than KLELine against hardware breakdowns and software failures at the client's side. For merchants, the offer includes management services for online catalogs and back-office operations (order taking, payment history, partial or total cancellation, accounting, banking interfaces, financial reporting, analysis of demographic data, etc.). Summarized in Table 10.1 are the principal differences between KLELine and Odysseo.

TABLE 10.1

Comparison between KLELine and Odysseo

Characteristic	KLELine	Odysseo
Client software	Klebox	—
Merchant authentication	RSA with 512 bits	Public-key infrastructure (key size unknown)
Security protocol	Proprietary (CPTP protocol)	SSL with 128 bits
Nonrepudiation	No	Yes

10.4 Millicent

Millicent constructs an economy of very small amounts (less than 1 cent) using a jeton currency called the *scrip*, with a value that varies from 0.1 cent to U.S.$50 (Digital Equipment Corp., 1995; Glassman et al., 1995; Manasse, 1995). The scrip represents an account established with a particular vendor for a limited period of time; this scrip is called vendor scrip.

A Millicent transaction involves three entities: the buyer, the vendor, and a broker. Brokers are payment intermediaries that relieve customers of the obligation of managing and maintaining several individual accounts with each vendor or service provider with which they wish to deal. Brokers can also offer their services to merchants by managing, in their place, and for a commission, the issuance and distribution of vendor scrip. In exchange, brokers get vendor scrips at discounted prices and supply them to their subscribers in the form of broker scrips. The responsibilities of a broker include issuing broker scrips, billing customers, collecting the scrips, and reimbursing vendors. They are also interfaces to the banking world on behalf of merchants, thereby alleviating their operational load by reducing the number of interfaces that they manage. Thus, long-term relationships are established between the consumers and the brokers on the one side, and between the brokers and the merchants on the other, instead of ephemeral relationships among vendors and buyers.

The holder of a scrip has a promise of service without a direct relationship with the banking system. Similar to the jetons stored in telephone cards, this is a preconsumption of service paid with the aid of traditional payment instruments. Just as for telephone cards, the Millicent scrip indicates the balance available after each purchase. One difference is that customers can close their accounts at any moment and ask the vendor for reimbursement in legal tender for the balance of the service that was not consumed. Also, because vendors know the balance of the scrip that they issued, they are able to detect double spending of a scrip that the vendor issued (using the scrip twice).

The broker scrip is valid for all the transactions conducted in the web of commercial relations centered around the broker. By getting a supply of scrip from a broker, a customer avoids having distinct scrip for each vendor. Furthermore, several brokers can establish a network among themselves to allow their customers to access vendors that are not in the same broker's network as their clients.

The function of the broker can be fulfilled by institutions known for their rigor and integrity, such as financial organizations or telecommunications operators. Some of the Millicent brokers in Europe included KLELine, Teledanemark, the Bank of Ireland, Visa, and MasterCard.

The cycle of scrip that underlies the Millicent model is presented in Figure 10.4.

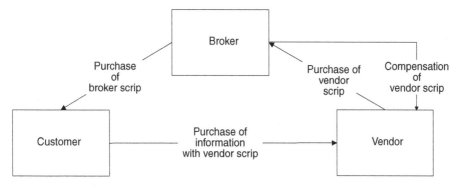

FIGURE 10.4
Cycle of the various scrips in the Millicent model.

TABLE 10.2

Management of Secrets in Millicent

Secret	Producer	User	Function
master_scrip_secret	Vendor or broker	Vendor	Verification of scrip authenticity
customer_secret	Vendor or broker	Vendor and buyer	Proof of ownership of the scrip
master_customer_secret	Vendor or broker	Vendor	Calculation of customer_secret

10.4.1 Secrets

The security mechanism in Millicent relies on three secrets for the production, validation, and spending of scrip. These secrets are the customer_secret, the master_customer_secret, and the master_scrip_secret (Glassman et al. 1995; O'Mahony et al. 1997, p. 205). The customer uses the customer_secret to prove legitimate ownership of the scrip. The vendor uses the master_customer_secret to derive the customer_secret from data in the scrip. Finally, the vendor uses the master_scrip_secret to prevent tampering or counterfeiting of scrip. The master_scrip_secret is known only to the vendor and to the broker, if the latter mints the scrip on behalf of the vendor. The master_customer_secret allows calculation of the customer_secret, on the basis of the data included in the scrip. Table 10.2 summarizes the management of secrets in Millicent.

10.4.2 Description of the Scrip

The vendor and the broker define the content of the scrip, but the structure remains as follows. A vendor scrip contains data fields reserved for the vendor: the expiration date, the balance, the customer, a code to verify the

TABLE 10.3

Fields in the Vendor Scrip

Field	Description
Vendor	Vendor identifier
Value	Balance value of the scrip, updated after each purchase
ID#	A unique serial number for the scrip, a part of which is used to define master_scrip_secret
Cust_ID#	Customer identifier that serves in the computation of a secret shared between the buyer and the vendor (customer_secret); customer_id is unique to each buyer
Expires	The date on which the scrip becomes invalid
Props (info)	Optional details describing the profile of a buyer to the vendor (age, address, fidelity program, etc.); definition of these details is up to each broker and the associated vendors
Certificate (Scrip-stamp)	Proof of the integrity of the scrip, i.e., that it was not altered; calculated by hashing the other fields with a keyed-hash function using the master_scrip_secret as a key

integrity of the scrip, and comments (Manasse, 1995; Millicent, 1995; O'Mahony et al., 1997, pp. 199–206). These fields are described in Table 10.3.

The field Certificate (also called Scrip_stamp) is computed using a hashing function H in the following manner:

$$\text{Certificate} = H(\text{Vendor} \parallel \text{Value} \parallel \text{ID\#} \parallel \text{Cust\_ID\#} \parallel \text{Expires} \parallel \text{Props} \parallel \text{master\_scrip\_secret})$$

where $\parallel$ represents the concatenation operation, and $H()$ is the function MD5 or SHA-1. This field is computed during the production of the scrip and allows the vendor or the broker to verify its authenticity when the customer presents it.

The vendor (or its broker) constructs the customer_secret with a hash using the customer identifier Cust_ID# and the master_customer_secret. This construction can be written in the following form:

$$\text{customer\_secret} = H(\text{Cust\_ID\#} \parallel \text{master\_customer\_secret})$$

It is the vendor (or the representing broker) that defines the correspondence between (Cust_ID# and the master_customer_secret). The vendor may also use several lists for the master_customer_secret that can be used alternately to repel falsifications and increase the level of security. In any event, knowledge of the master_customer_secret allows calculation of the customer secret on the basis of the data supplied in the scrip.

The customer receives the customer_secret when it gets the scrip originally; this can be done online if the connection is secure. The security of the links between the broker and the vendor on one side, and the customer on the other during the supply of the scrip is not strictly within the purview of Millicent. The vendor and the broker are free to choose the techniques to be used.

The convertibility of scrip from various vendors is possible at two levels. The first is for "scrip on us," i.e., scrips that originate from vendors that belong to the same broker's network. The second level of compatibility is for scrip outside the broker's network. In this case, the various brokers have to cooperate among themselves to assure this convertibility.

Note that the scrip can be made to correspond to any legal tender.

10.4.3 Registration and Loading of Value

At registration, the customer receives the customer's software (called Millicent Wallet) to be installed on the customer's computer and a customer ID (Cust_ID#). The customer chooses the broker from which to buy the broker scrip in exchange of fiduciary money. The broker defines the terms of the exchange, such as how to secure it and which algorithms to use.

The Millicent Wallet assists in the following functions:

- Purchase of broker scrip
- Exchange of broker scrip into vendor scrip, and vice versa
- Payment of purchases with vendor scrip
- Acceptance and storage of vendor scrip
- Acceptance of "change" from a vendor

The Millicent Wallet can be transported from one machine to another, because it keeps a record of past transactions and the balance of available scrip.

The customer obtains vendor scrip, either directly from the vendor or by exchanging broker scrip. Figure 10.5 depicts a simplified view of the exchanges to get a supply of scrip from the broker and to buy items from a vendor without any security mechanism.

In this example, the customer procures vendor scrip by exchanging some of the broker scrip with the broker. In exchange of broker scrip equivalent to U.S.$5.00, the broker returns $0.20 worth of vendor scrip, and the remainder ($4.80) in broker scrip. Thus, according to Millicent, the customer prepays for the service required using the vendor scrip, then receives the change back in the same currency.

When a purchase request is made, the vendor verifies if the scrip is its own. By comparing the number of the scrip with the numbers in an "approved list," the vendor is able to determine if the scrip was already spent. This is a protection against double spending that does not require the vendor to query a centralized database. The cost of the verification is that of a hash computation and a local database lookup. Thus, no direct link is necessary between the vendor and the broker, because all verifications are done locally.

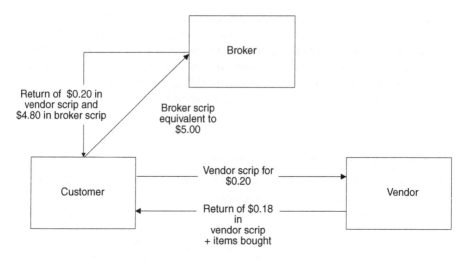

FIGURE 10.5
Purchase of vendor scrip and items (service) without security.

The Millicent protocol assumes that the loss of a portion of the vendor scrip (for example, the price of a page that was not downloaded for any reason) is negligible. In contrast, prepayment avoids the delays associated with multiple exchanges over the Internet.

This simplified view will be detailed by defining the management of security in Millicent.

10.4.4   Purchase

There are two ways to protect the communication channel between the customer and the vendor during a purchase. The first method consists of encrypting the exchanges using a symmetric encryption key constructed for each session, starting with the customer_secret that the customer has and that the vendor can compute. (The session key can also be identical to the customer_secret.) This key is denoted as $K_{customer\_secret}$.

Thus, the message sent to the vendor becomes

$$\text{Vendor, Cust\_ID\#, } K_{customer\_secret}\{\text{scrip, purchase request}\}$$

The fields Vendor and Cust_ID# are sent in the clear so that the vendor can recognize that it is the message recipient and to allow reconstruction of $K_{customer\_secret}$ to retrieve the scrip and the purchase request.

The vendor's response will include new scrip with the change, the required articles, and a copy of the initial Certificate field. This copy gives the buyer the possibility of verifying the authenticity of the response. These exchanges are depicted in Figure 10.6.

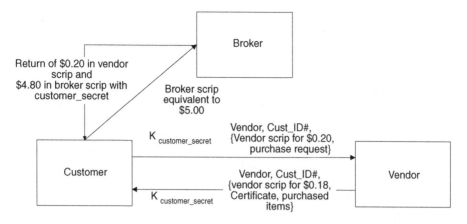

FIGURE 10.6
Purchase of vendor scrip and items (service) using an encrypted channel between the customer and the vendor.

This same purchase transaction can be done without assuring confidentiality but by furnishing a protection against the theft of scrip. In this case, the customer_secret and the purchase request are concatenated, and the result is hashed. The resulting digest is sent with the scrip. These exchanges are depicted in Figure 10.7.

The digest allows detection of any tampering with the scrip or with the purchase order. Even if an intruder succeeds in reading the exchange, the intruder will not be able to produce a purchase request and generate a new valid digest without the customer_secret. Accordingly, the integrity of the exchanges is protected.

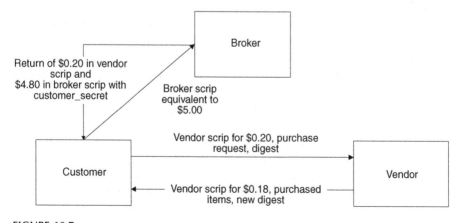

FIGURE 10.7
Purchase of vendor scrip and items (service) with protection against scrip tampering or falsification.

10.4.5   Evaluation

Millicent assumes that the trust relations among the customers, the merchants, and the brokers are asymmetric. The brokers are considered to be the most trustworthy, followed by the merchants, and, finally, the customers. The broker also acts as a trusted third party. The only time customers need to be trusted is when they evaluate the quality of the service. In an enterprise network, the broker may be a central directorate that manages access to various proprietary databases.

The Millicent broker intervenes each time a subscriber wants to buy vendor scrip for a new merchant within the broker's network. If the relations among customers and merchants are transient, the computational load on the broker's server may be excessive.

Millicent offers three levels of security. The highest level is that of "private and secure" exchanges; the intermediate level is that of "authentication without confidentiality," i.e., a cheater is prevented from benefiting from eavesdropping. Finally, the lowest level is when the scrip is sent in the clear. This can be justified when the amounts involved in each transaction are low, so that many illegal uses of the vendor scrip would be needed to acquire a substantial amount of money, which increases the probability of pinpointing the source of the fraud.

Thus, the security of the scrip is partially based on the smallness of the amounts involved per transaction. In addition, scrip is not easily reused because it can be recognized by a unique identifier, and it has an expiration date. Finally, the purchase is not anonymous; the vendor is aware of the buyer's identity and profile as recorded in the scrip.

The customer is able to detect an irregularity if the ordered items are not delivered or if the balance that the vendor indicates on the scrip does not correspond to the offered service. If the abuse is from the client, the cost to the merchants is that of detection of illegal scrip and denial of service. If the vendor is cheating, it will be the subject of many complaints, so that the broker can act accordingly and refuse to honor scrip from that vendor. Finally, if the broker is cheating, vendors will observe bad scrip coming from many customers, all originating from a single broker.

Note that the secrets are not sent in the clear; however, the customer_secret is stored on the customer's computer, which increases system vulnerability.

Millicent was able to reduce the cost per transaction with the following techniques:

- Prepayment in two ways: subscription by opening an account paid with legal money and prepayment of the service with vendor scrip before receiving the service. By assuming that users would not object to losing a micropayment, the purchase is done in one shot, without traffic going back and forth on the Internet for validation.

- Reduction of the cryptographic load: Authentication avoids public-key algorithms and relies on algorithms for symmetric encryption or hashing.

- The authorization for payments is done offline.

- The verification is done locally, which avoids online interrogation of the broker for each transaction. This is possible even when the broker issues the scrip in the vendor's name, because the verification lists are relatively small and can be updated daily on a local basis. Furthermore, the scrip contains the balance and a proof of its validity in the Certificate field. This saves the merchant the task of verifying the balance at each transaction and allows the client to keep track of expenses.

- Micropayments are grouped before financial settlement.

- Beyond the expiration date, the vendor scrip is not valid, which facilitates its management.

- Payments cannot be revoked or disputed, because no receipts are provided.

10.5 PayWord

PayWord is one of two variations on the theme of Millicent, the second being MicroMint. Both were proposed by Ronald L. Rivest and Adi Shamir, from the Laboratory for Computer Science of MIT (Massachusetts Institute of Technology) in the U.S. and the Weizmann Institute of Science in Israel, respectively (Rivest and Shamir, 1997; O'Mahony et al., 1997, pp. 213–220). In both cases, the authorization is done offline.

The economy of PayWord revolves around credit in jetons called paywords. Just like Millicent, the parties in a PayWord transaction are the buyer, the vendor, and the broker. Each broker authorizes users subscribing to its service to buy paywords that the broker produces to pay the vendors in the broker's network. At the same time, the broker agrees to reimburse the vendor in legal tender for the credits collected in jetons. Thus, PayWord is a postpayment system.

A chain of paywords is formed by applying a hash function recursively starting from an initial value, Wn. Each term is a coin of unique value that represents credit at the vender affiliated with the broker's network. The series of terms is represented in Figure 10.8.

The task of the vendor is to verify the signature on the initial value using a public-key algorithm. With this single signature verification, the vendor authenticates the whole chain of paywords. This verification is done offline, because the vendor is not required to contact the broker at each payment.

$$w_0 \xleftarrow{\quad h(w_1) \quad} w_1 \xleftarrow{\quad\quad\quad} \cdots \xleftarrow{\quad h(w_{n-1}) \quad} w_{n-1} \xleftarrow{\quad h(w_n) \quad} w_n$$

FIGURE 10.8
Formation of the chain of hash values in PayWord.

The local validation of paywords requires local storage of the revocation list (either in full or in a reduced form) and validation parameters.

The broker receives daily the last payword consumed by each user subscribing to its service, which allows the broker to update the list of current paywords. The broker constructs a blacklist of all abusive users or of those that declared loss or theft of their private key, SK_U.

The organization of the vendors in the broker's network as well as the clearance of accounts among the broker and the vendors are beyond the specifications of PayWord.

10.5.1   Registration and the Loading of Value

The user goes to a broker to open an account by communicating on a secured channel the coordinates of the user's bankcard or bank account. This channel can be postal mail or an Internet circuit secured with the help of SSL, for example. In response, the user receives the subscription card C_U in the following form:

$$C_U = \{B, U, A_U, PK_U, E, I_U\} SK_B$$

The card is signed with the private key of the broker, KS_B. It contains the names of the broker B and the user U, the user's address A_U (which may be a postal address, an IP address, or an e-mail address), the public key of the subscriber PK_U, an expiration date E, as well as some additional information I_U (for example, the number of the subscription card, the credit limit per vendor, the coordinates of the broker, the terms and conditions of the sale, etc).[1]

This subscription card authorizes the user to create chains of paywords with the sanction of the broker. By presenting this card to the vendor, the vendor can decipher the content using the public key of the broker PK_B. This assures the vendor that the broker promises to exchange the minted jetons with legal tender until the expiration date mentioned. As a precautionary measure, the goods sold will be delivered only to the address cited in the card.

Although the vendor knows the identity of the buyer, the system does not make any links between the sale of a particular item and the identity of the

---

[1] In this development, the term subscription card is used instead of certificate, which was used in the original description of PayWord, to avoid any confusion with the authentication certificates of X.509 type.

buyer, which offers a certain amount of privacy protection by reducing traceability. The subscriber has to protect, in particular, his or her private key, SK_U, although its storage on the customer's workstation is one of the weak points of the system. Nevertheless, the economy of PayWord can tolerate a certain rate of malfeasance. Given the smallness of the values carried by the jetons, the cost of sophisticated mechanisms to catch cheaters may exceed the values at risk. However, systematic and large-scale counterfeiting will be detected.

10.5.2 Purchase

The purchase takes place in two phases: commitment to a PayWord chain and delivery of the obligation.

10.5.2.1 *Commitment*

The commitment defines an association between a vendor and a subscriber for a limited duration. This association is similar to the one defined in a bank promissory note (Fay, 1997, p. 16). This is the promise of the subscriber to pay the broker the amount in legal tender that is equivalent to all the paywords $W_1, \ldots, W_{N-1}, W_N$ that the vender presents to the broker for reimbursement before the expiration date.

Each time a user wishes to contact a vendor, the user must construct a new chain of paywords $\{W_1, \ldots, W_N\}$ in the following manner:

1. Let N be the value in jetons of the credit needed to make the purchases. Select a random number denoted as W_N that will be the nth payword, with a value of one jeton.

2. Application of the hash function H (for example, MD5 or SHA-1) produces W_{N-1}. Thus,

$$W_{N-1} = H\ (W_N)$$

3. To produce W_{N-2} the hash function is applied a second time, i.e.,

$$W_{N-2} = H\ (W_{N-1}) = H\{H(W_N)\}$$

4. The chain $\{W_0, W_1, \ldots, W_{N-1}, W_N\}$ is constructed in this manner by applying the hashing function N times in sequence.

To link the chain to a given vendor, the user signs a commitment M to this chain in the following form:

$$M = \{V,\ C_U,\ W_0,\ D,\ I_M\}\ SK_U$$

where V represents the vendor, C_U the user's subscription card, W_0 the root of the payword chain, D the expiration date, and I_M any additional information (for example, the length of the chain or the value of a payword). The commitment M is signed with the user's private key, SK_U, before sending it to the vendor. This is represented by $\{\}\ SK_U$.

The commitment represents the largest amount of computation for the user, because it implies a signature with the RSA algorithm.

By presenting the commitment as a sign of authenticity to the vendor and joining the subscription card, the vendor, with the help of the public keys of the user and the broker, respectively PK_U and PK_B, can verify the user's signature on the commitment M and the broker signature on the user's subscription card C_B. Next, the vendor can be sure that the received values of the expiration dates D and E as well as the value of the last value of the payword chain, W_0, are correct.

10.5.2.2   Delivery

The user spends the paywords in an ascending order: W_1 before W_2, etc. If the price of an item is i paywords, the payment P of the user U to the vendor V is defined by

$$P = (W_i, i)$$

Thus, the payment is not signed.

The vendor must carry i hashing on the root of the chain W_0 (indicated in the commitment) to verify the payment validity. However, the number of operations can be reduced by exploiting the properties resulting from the way that the chain $\{W_0, W_1, \ldots, W_{N-1}, W_N\}$ was constructed. If W_j paywords were already spent (where $j < i$), the verification of P requires $(i - j)$ hash operations on W_j instead of i hash operations on W_0. Because the verification of W_j already required j operations on W_0, the vendor will save computations by keeping track of the index of the last value utilized for each user.

Note the following:

- Only the commitment has to be signed to guarantee the integrity of the root W_0. The payment is not signed.
- The authenticity of the user is checked using the subscription card, which is the reason it can be called a certificate.
- Verification is local, and authorization is offline, i.e., the vendor does not have to contact the broker for each payment.
- The payment does not mention the item to be purchased, which prevents tracing of the transactions. Although this characteristic protects the user's privacy somewhat, it does not protect from negligent or bad-intentioned vendors.

- The broker plays the role of a small-claims judge in addition to the functions of trusted third party.

- The vendor can keep a record of all paywords (even those that were spent and redeemed) until their expiration date to protect against replay attacks.

10.5.3 Financial Settlement

A vendor does not have to be affiliated with a broker to be able to authenticate the user's subscription card C_U, because the vendor only needs the broker's public key, KP_B. However, redemption of the paywords with legal tender requires a formal relationship between the vendor and the broker.

On a periodic basis (for example, on a daily basis), the vendor sends to the broker a reimbursement request for each subscriber of the broker's network. This message contains the commitment M of the user and the last payword received.

The broker then verifies the validity of each commitment using the public key PK_U of each user, given that the broker recognizes the subscription cards that it issued and their expiration dates. To verify the validity of the last payword per user, for example, W_K, the broker will have to do K hashing operations on W_0 of that particular user.

The broker groups small amounts before charging them to a credit card account. The interface to the bankcard account is controlled by the rules and regulations of the network of bankcards from the point of view of security, financial settlement, etc. Thus, the subscribers are responsible with respect to two authorities: the bank that issued their credit card and their PayWord broker.

Financial settlement with legal tender is done through the banking circuits and is outside the specifications of PayWord.

10.5.4 Computational Load

10.5.4.1 Load on the Broker

For each user member of its network, the broker carries the following operations:

- Periodically (monthly):
  — Renews the subscription card, which requires a public-key signature
- Daily:
  — Verifies the commitment of each user who made a purchase, which implies verifying a public-key signature.
  — Verifies the payment by computing successive hashes

These computations do not need to be performed in real time.

10.5.4.2   Load on the User

The user performs the following calculations:

- Periodically (monthly):
  - Verifies the subscription card and the certificate, which requires verification of a public-key signature
- Daily (for each vendor):
  - Signs a new commitment that requires generation of a public-key signature
  - Computes the hash of each payword used
  - Records the various commitments, the various payword chains, and the last payword employed

The user is the only entity that needs to sign the commitment (i.e., generate a public-key signature) and calculate hashes online.

10.5.4.3   Load on the Vendor

The vendor must do the following:

- Verify daily the subscription cards and the commitments received from customers; these computations are online but are less intensive than those required from the customers
- Compute online the necessary hashes for the paywords received
- Keep a record of all received commitments until their expiration dates as well as the last valid payword received during the day

TABLE 10.4

Comparison between Millicent and PayWord

Characteristics	Millicent	PayWord
Verification	Offline	Offline
Nature of the system of payment	Prepayment	Credit (postpayment)
Representation of the monetary value	Counter	Digital function (hash)
Security	Three levels to choose from	Prevention of double spending
Storing of value	In counters	Number of hash operations
Multiplicity of currencies	A set of counter registers per currency	A root W_0 per currency
Anonymity	No	No
Traceability	Yes	No

10.5.5 Evaluation

As an evaluation, Table 10.4 contains a comparison between the approaches of Millicent and PayWord.

10.6 MicroMint

MicroMint is the second scheme that Rivest and Shamir developed for micropayments (Rivest and Shamir, 1997; O'Mahony et al., 1997, pp. 228–236).

The MicroMint economy is based on jetons called MicroMint coins. These include a sequence of bits whose validity can be easily checked but whose production is extremely expensive. Just as in the case of minting metallic coins, the per-unit cost of fabrication decreases as production increases because of the economies of scale. At the same time, small-scale forgery is not economical. One difference from PayWord is that MicroMint avoids public key encryption to reduce the computational load.

The cycle of coins in the MicroMint economy is shown in Figure 10.9. Vendors exchange the coins collected for legal tender daily. The duration of validity of new coins is typically 1 month. Beyond that delay, unused coins are returned to the broker in exchange for legal tender or other MicroMint coins.

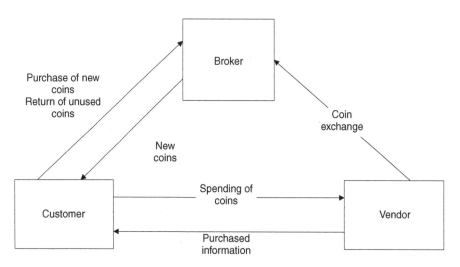

FIGURE 10.9
Cycle of coins in MicroMint.

10.6.1   Registration and Loading of Value

Purchasing of coins from the broker is done by means of bank payments (credit cards, checks, etc.). The broker has to keep inventory of all the coins purchased by each subscriber to the broker's network. MicroMint does not prescribe the relations between the broker and its clients

If the charging of value is done through the Internet, the subscriber and the broker may resort to symmetric cryptography to fend off any attempts to steal coins during the charging. This is possible because the relations between the broker and the users that are affiliated with the broker network are generally long term.

10.6.2   Purchase

Purchase is done by exchanging the coins for a service or information between the customer and the vendor. The vendor has to verify the validity of the coins, but this verification does not mean that the coins are not being reused. It is up to the broker to detect later if coins were double-spent and then trace the double spending to the user or to the vendor suspected to be the source of the fraud.

10.6.3   Financial Settlement

Every 24 hours, the vendors present the received coins to the broker. After verification of the regularity of the coins and verification that they were not previously exchanged, the broker reimburses the vendor for the equivalent amount in legal tender, less the broker's commission. The broker is free to accept or reject coins that were already exchanged.

10.6.4   Security

It is assumed that forgers have no interest in cheating to gain a negligible amount. The security mechanisms are installed to discourage systematic fraud, such as large-scale counterfeit, the theft of coins, or sustained double spending.

In MicroMint, a coin is minted following k collisions of the hashing function H: $x \rightarrow y$, where x and y are vectors of dimensions n and m, respectively $(m < n)$.

A collision is said to occur when the hash of x_1 and x_2 is the same vector y, i.e.,

$$H\,(x_1) = H(x_2) = y$$

Thus, to have k collisions, the following condition must be satisfied:

$$H\ (x_1) = H(x_2) = \ldots = H\ (x_{k-1}) = H(x_k) = y$$

where the x_i vectors are distinct.

The coin C is formed from these vectors x_i:

$$C = \{x_1, x_2, \ldots, x_{k-1}, x_k\}$$

Verification consists of assuring that the hash of all of these vectors x_i are equal. For example, assume that $n = 52$ and $k = 4$. In the case of the standard hash functions, such as MD5 or SHA-1, the length of the hash is 128 and 160 bits, respectively. To find the required 52 bits, the lowest-order 52 bits are retained. However, the verification of the collision does not detect whether the coin is a forgery or was previously spent.

10.6.4.1 Protection against Forgery

Because small-scale forgery is not attractive, the defensive measures are oriented toward industrial counterfeit. Some suggested measures are as follows:

- Monthly renewal for criteria for validity, for example:
    - Some high-order bits of the hash correspond to a certain mask.
    - The minted coins belong to a certain subset whose hash has a given structure, which allows for the separation of fake pieces based on the examination of a hash.
    - The various x_i must satisfy some conditions.
- The broker can augment the computation time by starting to mint the coins several months in advance, which will make it more difficult for counterfeiters to be ready at the required time.
- An extreme measure in case the MicroMint server is penetrated is to recall all the coins in circulation and replace them with a new issue.

10.6.4.2 Protection against Coin Theft

Encryption can protect the coins from theft during their collection by the broker. Use of public-key cryptography is possible because the relations between the vendors and the broker are relatively stable and long term. Another approach, which avoids encryption, is to personalize the coins of each subscriber to the broker's network. In return for sacrificing anonymity, the usage of the coin will require explicit permission of the owner, which makes them closer to electronic checks. On the other hand, if the coins are specific to a given vendor, the stolen coins would not have any value, because the vendor would put a stop on them. Anonymity can be restored, but the universality of the coins will be sacrificed.

10.6.4.3   Protection against Double Spending

The broker can detect double spending because it controls an inventory of coins purchased by each subscriber in its network. The broker is capable of identifying the vendors who returned these coins for redemption. With the cooperation of the merchants, and by analyzing the data they collected, the broker would be able to trace the subscribers that may have used them. Because MicroMint does not use digital signatures, the identity of a forger can be repudiated. Nevertheless, the broker may refuse to supply the suspected forgers with new coins.

All this means that in the MicroMint scheme, detection of the double spenders will not be easy. However, given the small magnitude of the amounts in play, the risks should be acceptable.

10.6.5   Evaluation

Table 10.5 compares the properties of PayWord with those of MicroMint.

10.7   eCoin

The practical application of virtual jeton holders in micropayments is illustrated by a defunct commercial application — eCoin.

TABLE 10.5

Comparison between PayWord and Micromint

Characteristic	PayWord	Micromint
Security	Public-key encryption; verification against a blacklist	No encryption; no easy protection against double spending
Nature of the jeton	Specific to a vendor	General for any vendor but can be personalized for one or several users and restricted to a vendor
Financing in jetons	Credit in paywords	Debit in coins
Nominal value of the jeton	1 payword = 1 cent	1 coin = 1 cent
Verification by the payment intermediary (the broker)	Offline	Offline
Storage of value	In the number of hashes	In the number of hashes
Multiplicity of currencies or denominations	One root W_0 per currency or denomination	One condition on the x_i per currency or denomination
Anonymity	No	No
Traceability	No	Yes

All exchanges in eCoin passed by the broker. The broker distributed the jetons to the clients, verified them at purchase time, and refunded the merchants for all purchases that were authorized. It was imperative to link the traffic between the client terminal and the eCoin server, because the channel was used for the distribution of jetons for payments.

Each eCoin jeton consisted of 15 octets, five of which were reserved for a time stamp and for the expiration date, while eight were for identification of the jeton. The value was stored in a virtual jeton holder called eCoin Wallet Manager, a plug-in to the client browser.

The eCoin server (the broker) maintained a copy of the jetons distributed and a list linking the jetons with the user. Transactions were thus traceable to identify dishonest users who would try to use the same jeton twice. However, it was not possible to identify them irrefutably without digital signatures.

At the moment of payment, the traffic was diverted toward the broker (the eCoin server) to verify the jetons spent. After verification, the server would return the control to the merchant server to complete the purchase. The exchanges during a transaction with eCoin are illustrated in Figure 10.10.

10.8 Comparison of the Different First-Generation Remote Micropayment Systems

The various first-generation remote micropayment systems that were discussed are compared in Table 10.6.

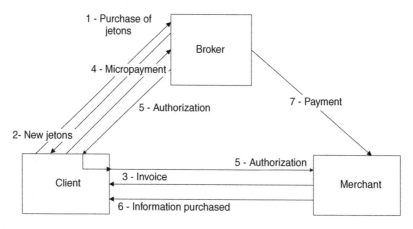

FIGURE 10.10
Exchanges in eCoin.

TABLE 10.6
Comparison among the Systems of Remote Micropayment

Characteristic	NetBill	KLELine	Millicent	PayWord	Micromint
Services offered	Payment system	Commercial mall; banking gateway; payment intermediary; EDI	Payment system	Payment system	Payment system
Product available	Payment software	Payment software, hosting software	Payment software	Trusted third party, notary	—
Authorization	Online	Online	Offline	Offline	Offline
Role of the intermediary	Trusted third party, notary	Trusted third party, notary	Trusted third party, notary	Notary	Notary
Security protocols	Public-key Kerberos	Proprietary (CPTP) for micropayments	Symmetric-key cryptography, hashing	Public-key cryptography, hashing	No encryption, hashing; no protection against double spending
Storage of the secrets by the customers	The payment intermediary keeps a copy of the decryption key of the items; the session keys are stored on the client machines	PIN to be memorized	The customer_secret is stored on the customer's computer	Private key of the customer is stored on the customer's workstation	—
Instruments for loading value	Credit card, direct debit, fund transfer	Under direct control of a bank	Credit card, direct debit, fund transfer	Credit card, checks	Credit card, checks
Nature of money	Legal tender	Legal tender	Legal tender	Jeton	Jeton
Subscription mode	Prepayment	Prepayment	Prepayment	Credit	Credit
Minimum payment in legal tender	1 cent	<1/6 cent	<0.1 cent	0.1 cent	—
Financing of the internal economy	N/A	N/A	Debit (prepayment) in scrip	Credit in paywords	Debit in coins
Storage of value	In a counter	In a counter	In a counter	In the number of hashes	In the number of hashes
Currencies	U.S. dollars	183 currencies; rate updated every 6 hours	Variable	One root W_0 per currency and per denomination	One condition on the x_i per currency and per denomination
Revocability of the payment	Possible until the electronic payment order is signed	Up to 45 days from the payment	Irrevocable and without any possibility for challenge	Irrevocable and without any possibility for challenge	Irrevocable and without any possibility for challenge
Billing	Per transaction	Grouped transactions	Grouped transactions	Grouped transactions	Grouped transactions

10.9 Second-Generation Systems

Remote micropayment systems of the second generation can be grouped into three categories: prepaid cards systems, systems based on e-mail, and Minitel-like systems. Let us consider each category.

10.9.1 Prepaid Cards Systems

In telephony, one of the advantages of prepaid cards is to avoid long-term subscriptions, which helps occasional or traveling users. As a consequence, major actors in the area of prepaid telephone cards are attempting to penetrate the sphere of remote micropayments by building on their recognized know-how and by exploiting their presence at a large number of points of sale. While with prepaid cards, the installation of a special client software can be avoided, the merchant site must be equipped to manage the authorization, credit verification (or redirection of requests to the authorization centers), statistics collection, and the payment to content suppliers.

Prepaid cards can be rechargeable or expendable. The Smartcodes card from Sep-Tech, for example, is a virtual purse managed by a trusted third party (the payment operator). To reduce costs to a level suitable for micropayments, the security infrastructure should be as light as possible. Thus, the card includes just the circuitry to produce codes of 128 bits, each of which is valid for only a single transaction. The ephemeral code is sent to the merchant without encryption because its value is timed. The payment is validated by cross-referencing the ephemeral code with the card identification number and the amount of the transaction.

Users receive the Smartcodes cards from their banks or from their Internet access providers. A secret code allows activation of the card before inputting the amount of the transaction.

In contrast, the Easycode card is expendable. In this case, the codes are written on the back of the card and are revealed gradually by scratching specific spots. These codes must be used in the order of their positions on the back of the card. There is an upper limit on the value stored; also, the value stored in a given card can be transferred to another card in case of loss, theft, or nonconsumption.

10.9.2 Systems Based on Electronic Mail

In systems based on e-mail, both parties must have an e-mail address. Among the most renowned members of this family are PayPal® (www.paypal.com) and c2its[SM] (www.c2it.com) of Citibank.

The offers can be distinguished on the bases of technical or organizational aspects. The technical elements include the following factors:

- The recognized means of payment
- The access network — mobile or terrestrial
- The exchange protocols, such as SMS (Short Message System) or WAP (Wireless Application Protocol), etc.
- The means for verifying and securing the exchanges, such as certification authorities or encryption techniques, over various connections; in general, the communications among the hosts and the servers for the authorizing centers of banks are encrypted with symmetric keys of relatively large size, for example, 1024 bits
- The traceability of the exchanged documents

The organizational and managerial aspects relate to the following:

- Conditions for enrollment and methods for recruitment, for example, whether it is necessary to open an account to participate in the network or what is needed to open an account
- Target market segments
- Registration fees and a graduated schedule of costs
- Customer-facing operator, e.g., Internet service provider or telephone operator (whether mobile or terrestrial)
- Delay before the exchange funds are available
- Possibility of canceling a payment
- Efforts to reduce operational costs (offline verification, grouping of transactions before financial settlement, etc.)
- Role of the intermediary in the negotiation — it can be at the crossroad from the beginning of the negotiation or the parties can be left to agree on the terms of the transaction and then inform the intermediary by e-mail of the amount of the transaction and the coordinates of the two parties

Typical exchange for a payment by bankcard is depicted in Figure 10.11.

10.9.2.1   PayPal®

Let us consider PayPal to illustrate the presentation. PayPal was formed at the initiative of Nokia, Deutsche Bank, and some investors from the Silicon Valley. PayPal completely succeeded in replacing its competitor Billpoint as the system for exchanging valuables on the Internet using the auction site eBay®. Competitors such as eMoneyMail (Bank One) and Billpoint had to withdraw, respectively, in 2002 and 2003, despite their considerable financial backings.

To settle a purchase by PayPal, the user sends to the payment operator a request with the e-mail address of the recipient and the amount in question.

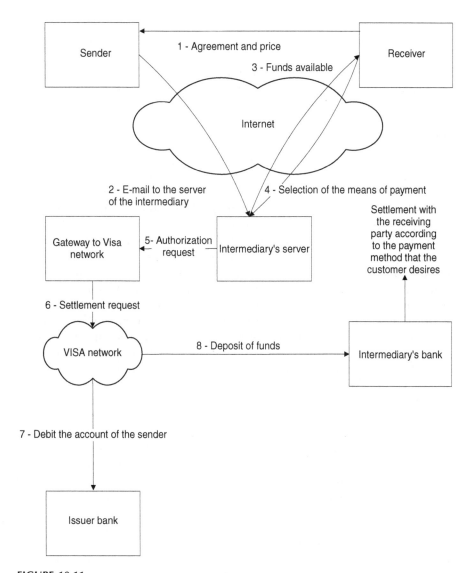

FIGURE 10.11

Typical exchanges in a system of remote micropayments with bankcards based on e-mail.

This message of remote payment is secured through SSL. User identification relies on a secret code of 20 digits.

The recipient receives a message from PayPal signaling the availability of funds from the sender. To receive the funds, recipients must register with PayPal (if they are not already part of the network) and specify their preferred means of payment (transfer of funds to a bank account or payment by a check withdrawn on a PayPal account). Users can also deposit their dues in a subaccount of PayPal's account. Clearly, transactions are not anonymous.

The main advantage of this system is its ease of use; the tools needed are e-mail and a browser — not special software. It relies on a well-known payment mechanism that does not require the user to learn new ways or to mobilize a special amount for prepayment. It is exposed, however, to credit card fraud, especially whenever smart cards are not used.

10.9.3 Minitel-like Systems

These systems inherited from the Minitel the specific role of the provider of access to the Internet or to the telephone network and the billing structure (per transaction or by duration) in addition to the grouping of micropayments before settlement. The distribution of the tasks of billing, collecting payments, and refunding the merchants can follow two distinct schemes.

In the first case, the access provider bills its own subscribers, while the payment intermediary is responsible for accounting for the online purchases of subscribers from merchant sites within the network. The intermediary starts the financial compensation, receives the value, and reimburses the merchant and the telecommunications operator. This was the scheme adopted by a payment intermediary that ceased operation — 1ClickCharge.

In the second approach, the access provider centralizes billing and accounting. The payment service provider operates the payment platform and supplies the access provider with the necessary tools to monitor, secure, and record client transactions. Thus, the total amount of purchases for each subscriber is added and included in the bill of the access provider. The access provider collects the monies, reimburses the merchants, and gives commission to the payment intermediary. This model, which seems to be more manageable than the previous one, was adopted by ClickShare (www.clickshare.com), WISP (www.trivnet.com), and iPIN (www.ipin.com).

The financial settlement is carried out in one of two ways. The payment provider sends a daily or monthly statement to the associated access providers with a record of the purchases from each of their subscribers. Each access provider includes the amounts as line items in their billing statements, deducts its own part, and then gives the rest to the payment operator, which then distributes the sums to the various content providers after taking its commission. This is the situation for iPIN

ClickShare and WISP use an alternative method. Here, the payment operation gives the access provider a record of the transactions and debits the latter of the sum total corresponding to all the purchases of their subscribers plus a commission. Access providers are, thus, responsible for collecting from their subscribers. They also reconcile the records of buyers and vendors on a daily basis by downloading the data collected by the payment platform.

10.10 Summary

The evolution of Internet sites toward efficient payment models requires a means of payment for small amounts, without adding undue complications to users and requiring them to reveal their financial information for each purchase. The first generation of micropayment tools was too ambitious technically, which turned out to be nonrealistic in practice. Since then, the new generation has learned a lesson and is reincorporating existing and well-mastered techniques of e-mail, prepayment cards, and the Minitel.

Questions

1. It is sometimes said that the first generation of remote micropayment systems consisted of technical solutions in search of a problem. Do you agree?

2. List the hurdles that the developers of Millicent had to face and overcome.

3. Why is the role of the access provider important in second-generation remote micropayment systems?

4. Analyze the position of PayPal within the financial environment using Porter's model (presented in Chapter 1).

11

Digital Money

ABSTRACT

This chapter presents the most ambitious solution for online payment systems. In effect, digital money is a new instrument that matches the speed and the ubiquity of informatics networks, while respecting the properties of classical fiduciary money (privacy, anonymity, and difficulty in counterfeiting). What distinguishes digital from classical money is that the support of the money is "virtual," because the value is stored in computer memory, on the hard disk of the user, or in a microprocessor card in the form of algorithms.

The bank of the system exchanges digital money against physical money after checking a verification database. This database can be centralized or distributed. The exchange of value among the parties is done in real time via the network, but financial settlement with the bank can be either in real time or in nonreal time.

One of the particularities of digital money systems compared with other systems is the possibility of making the transactions totally anonymous, i.e., dissociating the payment note from the holder's identity, just as in the case of physical money. The disruptive effect of this new invention is that the dematerialization of money can lead to the formation of new universal money that is independent of the actual monetary system. The economic and political implications of such a proposition are enormous and may lurk behind technical or legalistic debates.

The theoretical literature is replete with systems of digital money (Brands, 1993; Eng and Okamoto, 1994; Okamoto, 1995; Chan et al., 1996), and Wayner's book gives some simplified examples (1997). The focus of this chapter will be on two systems that passed from the study phase either to commercialization (DigiCash) or to an advanced prototype (NetCash).

11.1 Building Blocks

Money in the form of paper notes or metallic coins has no link with the nominal identity of the holder (buyer or seller) and of their banking coordinates. This attribute can be reproduced in some face-to-face systems (see Chapter 9). Anonymity in a remote transaction depends on two factors: the ability to communicate in an anonymous manner and the capability to make an anonymous payment. Clearly, an anonymous communication is a *sine qua non* condition for anonymous payments; once the source of a communication is identified, the most sophisticated scrambling strategies will not be able to mask the identity of the intervening party or the station of origin (Simon, 1996). Indeed, identification of the line on the basis of the calling number is the authentication method used in Minitel.

Systems for digital money attempt to reproduce the same properties by strengthening the confidentiality of a transaction with anonymity and untraceability (Chaum, 1989, undated Web citation). Anonymity protects the identity of the actor (buyer or seller) by dissociating it from the completion of the transaction. Untraceability means that two payments from the same individual cannot be linked to each other by any means (Sabatier, 1997, p. 99). As a consequence, the system operator (the bank) must be able to update the account balance for each actor without knowing the actor's identity or linking the various transactions that the actor may have performed with others. Nevertheless, each actor can authorize the system operator to track the exchanges to guarantee, for example, payment or the delivery of merchandise or to arbitrate disagreements.

The operation of the system is based on blind signatures, the principle of which was explained in Chapter 3. This mechanism is an extension of the authentication of a message with the public-key encryption RSA algorithm. It allows the payer to mint the digital coin and the bank to seal it without the bank having access to the coin's serial number. Of course, the message exchanges are secured with the typical mechanisms of confidentiality and authentication. To avoid cluttering the presentation with secondary details, these mechanisms will not be emphasized.

11.1.1 Case of Debtor Untraceability

Consider a debtor that would like the bank to sign a digital note blindly. The bank has a public verification key e, a private encryption key d, and an RSA modulus N ($N = pq$, where p and q are prime odd numbers) of length 512 or 1024 bits.

11.1.1.1 Loading of Value

To request the bank's authorization without disclosing the serial number of the means of payment, s, the debtor chooses a random number r uniformly distributed between 1 and $(N-1)$

$$1 \leq r \leq (N-1)$$

The sizes of the numbers s and r are 200 bits. Only the debtor has this number r, which is called the blinding factor. Its role is to hide the serial number s of the payment note. This operation consists of encrypting the blinding factor with the bank public key, then sending it to the bank for digital signing (see Figure 11.1):

$$s\, r^e \bmod N$$

The bank then seals the note by signing it with its private key:

$$(s\, r^e)^d \bmod N = s^d\, r \bmod N$$

and returns it to the debtor. The bank debits the client's bank account with the amount that corresponds to the digital note.

In general, the message sent to the bank contains several fields, including the serial number s. These fields indicate, among other items, the type of banking operation and the debtor's account number, the whole signed by the debtor's private key. This is represented by

$$(\text{"withdrawal," "debtor's account number," } s\, r^e)^c$$

where c is the debtor's (client) private key. The corresponding public key is included in a certificate signed by the competent certification authority.

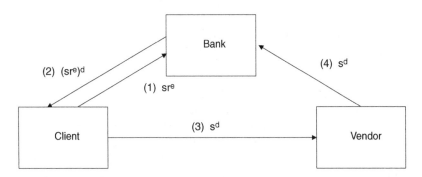

FIGURE 11.1
Simple system ensuring the buyer's untraceability.

The note is anonymous because it includes only the serial number encrypted with the bank secret key. The bank knows only the account number and, because the serial number of the note was blinded, the bank will not be able to reconcile the merchant's request for financial settlement with the note that it blindly signed.

11.1.1.2 Purchase

The debtor obtains the signed note by dividing what arrives from the bank by r to get

$$s^d \bmod N$$

Using the bank's public key, the debtor verifies that the received note is what was sent, because

$$(s^d)^e \bmod N \equiv s \bmod N$$

The buyer then sends the signed payment note to the creditor (the merchant).

11.1.1.3 Deposit and Settlement

The merchant deposits the note in the merchant's bank account by sending the following message to the bank:

$$\text{"deposit," "merchant bank account," } s^d$$

After verification that the serial number of the note has been previously deposited, the bank credits the merchant account with the equivalent sum and withdraws the note from circulation. The bank may also send a receipt to the merchant in the following form:

$$f(\text{"deposit," "merchant bank account," } s^d)^d$$

This protection against double spending obligates the bank to keep a list of all the serial numbers for the notes that were deposited.

11.1.1.4 Improvement of Protection

In some circumstances, clients may be interested in forfeiting untraceability to protect their rights (such as in cases of litigation with the merchant or the theft of notes, etc.). The construction of the random number r is then done by applying a one-way function (if not a hashing function) to a number r' that the client keeps hidden. This allows the debtor to prove, if necessary, that the debtor is the originator of the note, which the bank signed blindly,

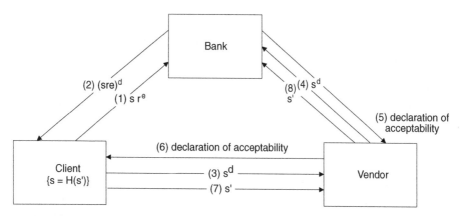

FIGURE 11.2
Protection of the buyer through the use of precursors.

and it does this by presenting r' to an arbitrator. In contrast, a merchant that refuses to fulfill its obligations or a thief that stole the note would not be able to do so.

Similarly, the serial number s can be formed by applying a special algorithm that allows the client to prove to an arbiter that it is the originator of the note. If s' is a random number, s can be constituted as

$$s = \{[s' \oplus H(s')] \parallel H(s')\} \equiv g(s')$$

where $H()$ is a hash function, $\oplus$ is the "exclusive OR," and $g()$ is the total operation on the precursor s'.

The debtor must then supply the precursor to the arbiter to prove the authenticity of the note that the merchant deposited.

The use of the precursors allows the client to confirm to the bank that the client spent the note that the merchant is depositing. These exchanges are shown in Figure 11.2.

In this configuration, the bank verifies that the serial number of the payment note was not previously used, and provides the merchant with a statement of acceptability. The merchant forwards the statement to the buyer. This statement includes an expiration date and can be signed with the bank's private key so that the merchant and the client can ascertain its integrity and its origin. If the client is convinced of the statement's authenticity, it sends the precursor s' to the bank through the vendor. The debtor verifies the precursor and forwards it to the bank, which confirms also that it is the entity that produced the serial number of the note. At this point, the bank gives its definitive approval, records the note as deposited, retains a copy of the precursor, and credits the merchant's account with the corresponding amount.

11.1.2   Case of Creditor Untraceability

To achieve creditor untraceability, the merchant issues the note, and then sends it to the client, who, in turn, sends it to the bank for a blind signature and for withdrawal of the corresponding amount from the client's account. When the creditor receives the signed note, it will use the bank's public key to verify that the note was derived from the note that was sent to the bank, as follows:

$$(s^d\ r)^e \bmod N = s\ r_e \bmod N$$

The client sends the note to the merchant, who, in turn, sends it to the bank after removing the blinding factor, as shown in Figure 11.3. The bank then credits the merchant's account with the corresponding amount. Compared with the preceding case, the exchange consists of five messages instead of four.

Notice that the message the bank receives is identical to the one in the preceding case. Thus, the bank will not be able to tell which of the two parties is untraceable, the client or the vendor.

11.1.3   Mutual Untraceablity

Mutual untraceablity is constructed through the fusion of the two preceding cases. The debtor, after receiving the note the creditor issued and blinded, adds another blinding factor by choosing a random number x, uniformly distributed between 1 and $(N-1)$. The note is sent to the bank for a blind signature and becomes

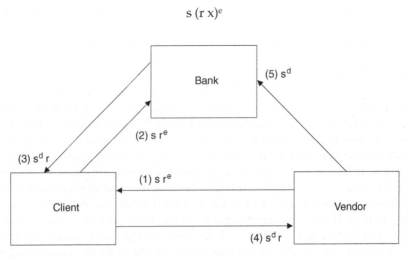

FIGURE 11.3
Merchant (creditor) untraceability.

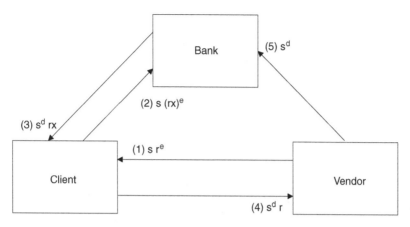

FIGURE 11.4
Mutual untraceability of the debtor and the creditor.

This technique is based on double-blinding and corresponds to the exchanges shown in Figure 11.4.

Another way to achieve mutual untraceability is to combine the two simple cases of creditor untraceability and debtor untraceability and to ask a trusted third party to be the intermediary among the parties. The client pays the third party using the debtor untraceability protocol. From its side, the trusted third party retransmits the payment to the merchant according to the rule of creditor untraceability. The bank can be this trusted third party, because even in that role, it cannot trace the payer or the payee.

11.1.4 Description of Digital Denominations

The representation of several denominations of digital money utilizes the binary representation of the value of a note; a denomination is assigned to each bit that is set to 1. In this way, j denominations of digital money can represent all the values between 1 and $(2j - 1)$ monetary values. Each of these denominations corresponds to one of the prime numbers in the bank's public exponent of an RSA system, as shown in Figure 11.5.

Thus, to represent the value of 1, the public exponent of the bank will be 3. Similarly, the value of 2 decimal (or 10 in binary) will correspond to the public exponent of 5, the value of 4 decimal (or 100 in binary) to 7, the value of 8 decimal (or 1000 in binary) to the public exponent of 11, and so on. Thus, with the public exponent of $11 \times 7 \times 5 \times 3$, a note with four denominations will be capable of representing all decimal values between 1 and 15 monetary units (in binary between 0001 and 1111). The following example shows how payments made with the denominations constructed in the manner indicated solve the problem of change (Chaum, no date).

Consider a digital money with four denominations. Assume that a customer would like to buy two digital notes, each representing 15 monetary

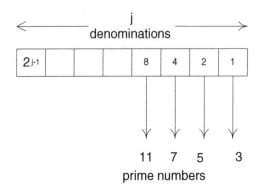

FIGURE 11.5
Representation of the various denominations with prime numbers.

units. Each note has its own serial number, say, s_1 and s_2, respectively. The corresponding blinding factors are r_1 and r_2, respectively:

The bank public exponent that is used to represent the full value note is therefore

$$h = 3 \times 5 \times 7 \times 11$$

To load the value, the client mints the two notes, hides them with the corresponding blinding factors, and sends them to the bank for its blind signature. The message sent to the bank will be the concatenation of two requests:

$$s_1 \times r_1{}^h \parallel s_2 \times r_2{}^h$$

The bank debits the client's account with the stated amount, signs the notes without knowing the serial number of the notes, and then returns the composite message:

$$s_1{}^{1/h} \times r_1 \parallel s_2{}^{1/h} \times r_2$$

The inverse $1/h$ is the bank private exponent and is calculated with the following formula:

$$h \times (1/h) = 1 \bmod [(p-1)(q-1)]$$

The exponent also indicates the total value of the note.

To pay with the first note for an item valued at 10 units (with a binary representation of 1010), the client utilizes the exponent 11×5 in its purchase request. Therefore, the client composes a message to be sent to the merchant in the following form:

$$s_1^{1/h* \, (5 \times 11)} = s_1^{1/(3 \times 7)}$$

To obtain the residue change, the client becomes an untraceable creditor of the bank. Following the rules of the protocol of creditor untraceability. The client forms a "cookie jar" in which to hold the residues and hides it with a blinding factor r_a. This second note takes the form:

$$j \, r_a^{(5 \times 11)}$$

The message sent to the merchant to buy items with a total value of 10 units with a note of 15 units is composed of the concatenation of the two messages:

$$s_1^{1/(3 \times 7)} \| j \, r_a^{(5 \times 11)}$$

The bank returns via the merchant the following message to indicate the change:

$$j^{\, 1/(5 \times 11)} \, r_a$$

The residue change corresponds to the following exponents

$$(3 \times 5 \times 7 \times 11)/(5 \times 11) = 3 \times 7$$

i.e., the binary value 0101 or 5 monetary units. These exchanges are associated with the first payment and are shown in Figure 11.6.

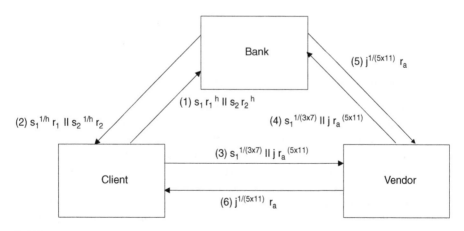

FIGURE 11.6
Exchanges with multiple denominations and returns of change (first payment).

Assume that the second payment is for a value of 12 units (binary representation of 1100); the public exponent is 7×11. Therefore, the composite message (using a new blinding factor r_b for the residual in the cookie jar) is

$$s_2^{1/h* \, (7 \times 11)} \, \| \, j^{\, 1/(5 \times 11)} \, r_b^{\, (7 \times 11)}$$

$$s_2^{\, 1/(3 \times 5)} \, \| \, j^{\, 1/(5 \times 11)} \, r_b^{\, (7 \times 11)}$$

In this case, the bank returns a message through the merchant that has the total residual change in the cookie jar:

$$j^{1/\{(5 \times 11) \, (7 \times 11)\}} r_b$$

The change for the second payment corresponds to the following exponents:

$$(3 \times 5 \times 7 \times 11)/(7 \times 11) = 3 \times 5$$

i.e., the binary value 0011 or 3 monetary units. Thus, the total residual change is $5 + 3 = 8$ units. This corresponds to the content of the cookie jar.

This approach depends on the withdrawal of notes of fixed value with the unspent parts later credited to the payer during a refund transaction. Other schemes were devised to allow change to be spent without an intervening withdrawal transaction (Chaum, undated).

In the general case, the message sent to the bank for blind signature can take the following form:

$$f(s) \times r^h \, (\text{mod } n)$$

where $f()$ is a one-way function.

11.1.5   Detection of Counterfeit (Multiple Spending)

The fabrication of false money is the principal source of fraud for paper bills. To prevent counterfeiting, the support is perfected (paper quality, insertion of a magnetic wire, presence of watermarks, etc.) to make the creation of equivalent notes more difficult. In contrast, the duplication of digital money is much easier; it is sufficient to reproduce the file that contains the monetary data. This introduces the danger of reusing the same note, which is called "double spending." Even if the bank can distinguish between good notes and reused notes on the basis of the serial number, the identity of the counterfeiter remains hidden.

The following algorithm, due to David Chaum and his team, allows the detection of cheaters without divulging the identity of honest debtors (Chaum et al., 1990).

11.1.5.1 Loading of Value

Consider a debtor that owns a bank account u to which the bank associates a counter v:

1. The debtor forms k blinded messages of the following format:

$$B_i = s_i \times r_i^e \, 1 \leq i \leq k$$

In this case, s_i is formed by $f(x_i, y_i)$, with

$$x_i = g(a_i, c_i)$$

and

$$y_i = g\{a_i \oplus (u \,\|\, (v + i)), w_i\}$$

The variables a_i, c_i, and w_i are independent and uniformly distributed over the interval $(1, N-1)$, where N is the modulus for computations. The k messages are sent to the bank as candidates to make the digital notes. The functions $f()$ and $g()$ have two arguments that are without collisions, so each result corresponds to a single combination of the two inputs.

2. The bank chooses at random $k/2$ candidates for verification and asks the debtor to produce the corresponding variables a_i, c_i, r_i, and w_i. Let R be the set of these $k/2$ candidates.

3. Once the bank has finished all the checks, and if it did not discover any irregularities, the bank sends the product of the $k/2$ candidates signed with its private key d:

$$\prod_{i \notin R} B_i^d = \prod_{1 \leq i \leq k/2} B_i^d$$

4. The bank withdraws the corresponding amounts from the debtor's account.

As in the preceding cases, the debtor removes the blinding factors r_i from the $k/2$ candidates ($1 \leq i \leq k/2$) to form the digital note as the product of all the serial numbers raised to the power of the private key of the bank:

$$\prod_{1 \leq i \leq k/2} s_i^d$$

The probability for detecting fraud clearly depends on the number k and the frequency of cheating. For a given value ε, if the proportion of irregular candidates among those that the bank has checked exceeds ε, the probability of catching the cheat is $[1 - \exp(-c\varepsilon k)]$ for a constant c.

11.1.5.2   Purchasing

When the debtor sends the merchant the digital note, the merchant starts a check procedure to catch cheats. The procedure is as follows:

1. The merchant selects a random binary string $\{z_1, z_2, \ldots, z_{k/2}\}$ and stores it in the vector $\mathbf{Z}$ of dimension $k/2 \times 1$
2. For all the elements z_i, $1 \leq i \leq k/2$:
    a. If $z_i = 1$, the merchant asks the client to send a_i, c_i, and

    $$y_i = g\{a_i \oplus [u \,\|\, (v + i)], w_i\}$$

    b. If $z_i = 0$, the merchant asks the client to send

    $$x_i = g(a_i, c_i), a_i \oplus [u \,\|\, (v + i)]$$

    c. and w_i
3. The merchant verifies that the digital note is of the correct form, and then sends the note, the vector $\mathbf{Z}$, and the client responses to the bank with a request for financial settlement.

11.1.5.3   Financial Settlement and Verification

The bank stores the note of digital money, the vectors of the tests, and the responses. If the note is being reused, the probability of receiving two complementary values for two different merchants and for the same bit z_i is high. In other words, it is highly probable that the bank would have received both a_i and $a_i \oplus [u \,\|\, (v + i)]$. With these two values, the bank can extract the debtor's account number u and then identify the debtor.

11.1.5.4   Proof of Double Spending

One of the difficulties of the mechanism just presented is that the discovery of the cheater's identity depends on the bank's accuracy or honesty. To protect the client from any foul play, it is necessary that a signature be used.

Several approaches are possible. It is not the intention here to review all the proposals but to highlight their ideas. For example, s_i can be formed with $f(x_i, y_i)$ and then inscribed in each note using the following equations:

$$x_i = g(a_i, c_i)$$

and

$$y_i = g\{a_i \oplus [(u \parallel z_{i'} \parallel z_{i''}) \parallel (v + i)], w_i\}$$

where $(z_{i'}, z_{i''})$ are random variables.

At the time of loading the value, the debtor will have to supply the signature on the factors $g(z_{i'} \parallel z_{i''})$ for each of the $k/2$ notes that the bank chooses to check. During the purchase, the client will also have to supply the signature of the factor $g(z_{i'} \parallel z_{i''})$ in the used note. In this way, the bank can prove the identity of the cheat without any doubt, if it can break at least $1 + k/2$ of the elements $(z_{i'}, z_{i''})$.

11.2 DigiCash (Ecash)

DigiCash is a payment system that was conceived within the framework of the European informatics project ESPRIT. The purpose was to create a digital currency based on the work of David Chaum and his team. DigiCash uses a digital currency called Ecash to replace coins and notes and to retain anonymity and untraceability; in addition, it adds security services for communication on open networks.

In 1990, David Chaum formed DigiCash in the Netherlands and in the U.S. to commercialize the system. The Mark Twain Bank of St. Louis, Missouri, agreed to issue this digital currency in U.S. dollars. It managed a special account for DigiCash to collect the various amounts circulating in the electronic currency before depositing them in the creditor's bank accounts. About 5000 clients registered to use the currency with 300 merchants. For a while, Deutsche Bank was commercializing its digital money in Germany. In 1999, however, the Mercantile Bank, which bought the Mark Twain Bank, judged the results to be insufficient and withdrew its support. Ecash technologies is now the owner of all DigiCash patents.

The system works as follows: the user mints the Ecash coins and then sends them to the bank for signature. The coins that the bank returns are then stored on the user's hard disk. They can be used to pay for purchases made at merchants that subscribe to the DigiCash system. The bank can add the role of a trusted third party to its role of a payment intermediary and a gateway to banking networks. Presented in Figure 11.7 are the relations that DigiCash establishes among the various parties of a transaction. In the general case, there is no guarantee to the client that the merchandise will be delivered.

DigiCash uses the same structure for the exchange of value among two peer entities, such as from one user to another.

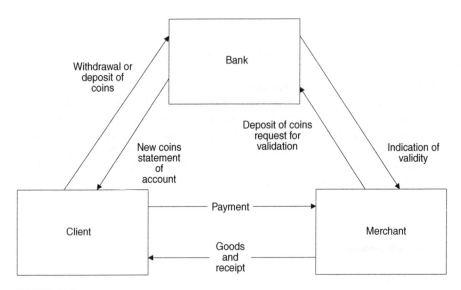

FIGURE 11.7
Relations among the various entities according to DigiCash.

11.2.1   Registration

The user downloads a plug-in to the browser called Cyberwallet (that Digi-Cash issues), to store the client's coins, track expenditures, and verify expiration dates. At registration time, the client receives all the public keys that will be used for minting and verifying the various denominations. Similarly, the merchant registers and downloads the merchant's software to manage customers, payments, and reimbursements from the bank. The merchant and the clients must have their accounts in the same bank, because interbanking relations are not yet supported.

11.2.2   Loading of Value

To charge the digital purse with monetary value, the client software generates new coins without value, each with a unique serial number. The serial numbers are chosen at random and are large enough (100 digits) to minimize collisions with the choices of other clients. After blinding the coins, the software sends them to the bank to be signed with a blind signature. The bank then debits the client's account for the corresponding amount (and the relevant commission) and puts its signature without uncovering the serial number of the minted coin. When they are returned to the users, the coins will be charged with the requested values and signed by the bank.

The note that the bank signs with its private key SK can be represented as follows:

$$\{\text{Serial number}\}SK$$

It is seen that the note is anonymous because it depends only on the serial number and the bank's private key. The exchanges are also anonymous because they depend only on the client's identity, and only the client knows the blinding factor. The only transaction that is perhaps not anonymous is the debit of the client's account. However, the client's anonymity is only protected if the communication channel between the client and the bank does not allow a third party to guess the client's identity, for example, by revealing the terminal address or location.

The bank associates a pair of private and public keys for the same modulo N to each coin denomination; this pair of keys is called the coin denomination key. The coin is of the following form:

$$\text{Key\_type, } \{\text{Serial number}\}SK_{\text{coin\_denomination\_key}}$$

where Key_type indicates the coin denomination key and other related information (for example, the currency and the expiration date).

To ensure the integrity of the exchanges, the client uses the client's private key to sign the loading value request. To ensure confidentiality, the client uses a secret session key that will be encrypted with the bank's public key. The bank's response, which includes the coins that were signed blindly, is signed with the bank's private key and then encrypted with the session key.

To protect customers against failures in the operation (terminal failures or network failures), the bank can store the trail for the last n loading transactions (for example, $n = 16$) of each customer. This facilitates error recovery and the reconstitution of the value stored in their software.

11.2.3 Purchase

The merchant's software sends a payment request to the client's software in the clear. It is of the following form:

Payment request = {currency, amount, time stamp, merchant's account, description}

Because the message is sent in the clear, although it contains details about the order amount, the currency to be used, and some description of the order, an eavesdropper may be able to see what is ordered and for how much.

Note that if several banks are allowed to sign electronic money blindly, another field would be added to identify the emitter bank.

The payment message contains details of the transaction and the monetary notes encrypted with the bank's public key PK_{Bank}:

$$\text{payment} = PK_{\text{Bank}}[\text{payment\_info; } \{\text{notes, } H(\text{payment\_info})\}]$$

The merchant has to join this message with the financial settlement request, so that the bank can verify that the merchant and the client accepted the same sales conditions. The payment_info field can be expressed as follows:

{Bank identification, amount, currency, number of notes, time stamp, merchant identifier, H(description), H(client_code)}

H(client_code) is the hash of a secret code that the customer's software selects. It could be used to show a proof of payment to the bank, if the customer decides to reveal its identity.

11.2.4   Financial Settlement

To ask for reimbursement, the merchant signs the payment with its private key SK_M and then encrypts it with the bank's public key:

$$\text{Settlement request} = PK_{Bank} \{(\text{payment}) \, SK_M \}$$

In this way, the bank can verify that the client and the merchant agree on the terms of the transaction. The bank must also verify that the notes used for purchase were not previously used. Once all the verifications are made, the bank credits the merchant's account with the amount of the transaction and informs the merchant of the result with the message that it signs with its private key:

$$\text{Receipt} = \{\text{result, amount}\} \, SK_{Bank}$$

[Note that the customer will behave in a similar manner if the customer desires to return the notes that were not spent, either for reimbursement with legal tender or for exchange against new notes at the time of expiration of the old notes.]

Protection against double spending of digital money is the bank's function; the bank has to verify the serial number of each digital coin or note to be deposited. After each settlement, it records in a large database the serial numbers of all coins that were spent. Nevertheless, a cheater can use the same coins without detection in between settlement requests.

11.2.5   Delivery

The merchant is obligated to deliver the purchased items to the client. In case this does not happen, the client can ask the bank to intervene by revealing the secret code client_code whose hash value it has already sent, thereby giving up anonymity to trace the flow of money.

11.2.6   Evaluation

DigiCash uses blind stamping to verify the validity of digital coins and notes without knowing their details. Furthermore, there is no relation between the serial number of the digital money and the identity of the holder. Once a coin is deposited with the bank, it is withdrawn from circulation.

The major drawback of this system is the radical change it introduces into the banking system. The magnitude of the changes is responsible for the bank's lack of enthusiasm in adopting this means of payment.

From the point of view of transactional properties, the following are observed:

- The transactions are not atomic, because the client is not necessarily aware of the current status of the payment. In the case of a network failure, the status of the money sent is undetermined (Camp et al., 1995).
- Double spending can occur undetected.
- The bank can be a bottleneck to the whole system because it has to verify every coin to be deposited.
- The larger the size of the database that contains the serial numbers of the deposited coins, the longer the response time of the bank.

11.3   NetCash

NetCash is an online payment system with digital coins. It was designed at the Information Science Institute (ISI) of the University of Southern California (USC) in collaboration with the Massachusetts Institute of Technology (MIT) (Mevinsky and Neuman, 1993; Neuman and Mevinksy, 1995; O'Mahony et al., 1997, pp. 168–181). It uses the Asynchronous Reliable Delivery Protocol (ARDP), which resides on top of User Datagram Protocol (UDP) to provide reliable delivery of short messages without the connection setup and teardown overhead of Transmission Control Protocol (TCP) (Salehi et al., 1999). The system can be modified to work with group payments and operate offline.

The bank associated with NetCash is called NetBank. It was given permission by the U.S. authorities to deliver digital money, provided that it keeps sufficient funds to ensure payments with legal tender. Thus, the NetCash server has the dual role of a gateway and of a traditional bank. The NetCash designers asked that the U.S. federal government establish a new federal institution, the Federal Insurance Corporation (FIC), to keep in escrow the public keys used in the system.

11.3.1   Registration and Value Purchase

First, the user must open a bank account at the NetBank by filling out an electronic form with the necessary information and sending it to the e-mail address netbank@agents.com. The bank has to authenticate the user before proceeding.

The user purchases coins with money in this account. The purchase request (or electronic check) is sent, together with a symmetric session key K_{Client}, and both are encrypted with the public key of NetBank PK_N:

$$PK_N\{\text{Electronic check, } K_{Client}\}$$

This encryption protects the confidentiality of the exchange. NetBank processes the check and, if there are no errors and the electronic check clears, sends back the new coins encrypted with the symmetric key K_{client}. The user stores the coins in the hard disk of the user's terminal.

Each digital coin has a unique serial number whose integrity can be verified with NetBank's digital signature. Thus, each coin takes the following form:

$$\text{Coin} = \{\text{Expiration date, serial number, value}\}\ SK_N$$

where SK_N is the private key of the NetCash server. To reduce the computational load, the signature can be applied to the hash of the above fields instead of to the fields themselves.

Similarly, to obtain change from the server, the user forms a change request message that includes the digital coins to be divided and a secret session key and then encrypts the message with the server's public key. The server then decrypts the message with its private key, extracts the coins and removes them from circulation, issues the new coins according to the user's request, and then sends them to the user encrypted with the session key that the user sent.

11.3.2   Purchase

A typical purchase transaction takes place in the following form:

1. The customer and the merchant agree on a price.
2. The buyer sends to the vendor a message encrypted with the merchant's public key PK_M, which contains the payment in digital coins, the identifiers of the items to be purchased, a secret session key K_1, a transaction identifier, and the buyer's digital signature:

$$PK_M\ \{\text{Payment, items identifiers, } K_1, \text{transaction identifier, } Sig_{Buyer}\}$$

3. The merchant decrypts the message with the merchant's private key, extracts the session key, and verifies the integrity of the message by comparing the received signature with the one recomputed. It then forwards the coins to the server by adding another session key K_2 and then encrypting the whole message with the NetCash server's public key PK_N:

$$PK_N \{Payment, K_2, transaction\ identifier\}$$

4. The NetCash server verifies that the serial numbers of the coins are valid, i.e., that the coins were not withdrawn from circulation. In this case, it replaces them with new coins that are sent to the merchant encrypted with the private key K_2:

$$K_2 \{New\ coins\}$$

5. The merchant decrypts the message, extracts the new coins, and then sends to the buyer a receipt and signs the message with the help of the session key K_1:

$$\{Receipt, transaction\ identifier\}\ K_1$$

The receipt has the following format:

$$Receipt = \{Amount, transaction\ identifier, date\}\ KS_M$$

i.e., it is signed with the merchant's private key KS_M.

Figure 11.8 summarizes the exchanges that take place during a purchase.

11.3.3 Extensions of NetCash

Some ideas have been suggested to increase the robustness of the basic protocol (O'Mahony et al., 1997, pp. 178–181). For example, the basic mechanism in NetCash uses prepayment but does not offer a remedy if the merchant does not deliver the purchased items after receiving the payment.

To prevent this eventuality, the minted coin is divided into three distinct parts, one part can be spent only for purchases, the second can be called to make claims against the merchant, and the third is a generic piece that allows the coin to be used if the first two parts are not used within their time windows. Thus, the digital coin can be written as follows:

$$Digital\ coin = \{generic\_coupon, purchase\_coupon, claims\_coupon\}$$

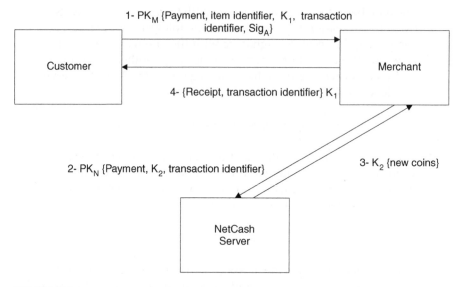

FIGURE 11.8
Exchanges during a purchase using NetCash.

The generic coupon is identical to the digital coupon previously described. Each of the new parts contains a field associated with the buyer or merchant, respectively. All parts are signed with the NetCash server's private key. Thus, we have the following:

generic_coupon = {Server's name, server's IP address, date of issue, date of expiration, serial number, value} SK_N

purchase_coupon = {Server's name, server's IP address, merchant's information, date of issue, date of expiration, serial number, value} SK_N

claims_coupon = {Server's name, server's IP address, client's information, date of issue, date of expiration, serial number, value} SK_N

The various "windows" of validity should be distinct. They can be called the validity window, the purchase window, and the claims window. The data relative to the merchant include its public key, PK_M. Before attempting to verify the purchase coupon that the merchant presents, the NetCash server will ask the merchant to prove ownership of the private key SK_M. NetCash will also verify that the purchase coupon did not expire. The data related to the merchant are constructed in a similar manner.

The operation takes place as follows:

1. The buyer purchases the digital coin from the server by specifying the client's identity and public key, as well as the merchant's identity and public key.

2. The purchase transaction follows the same steps as above but with the purchase coupon.

3. If the buyer does not receive a receipt or if the purchased items are not delivered, the customer can claim the money by sending the claims coupon to the server during the claims window.

4. The server checks if the purchase coupon was deposited. If this is the case, it informs the customer of the depositing merchant's identity, the public key, and the value of the payment, and it signs the whole answer with the server's private key SK_N:

$$\{Merchant's\ identity,\ PK_M,\ amount\}SK_N$$

This constitutes a proof that the merchant received and cashed the money that the client paid.

5. Otherwise, if the purchase coupon was not spent, the server refunds the customer for the amount spent.

6. If both the purchase coupon and the claim coupon have expired, the generic coupon can be used according to the normal procedure but without protection.

The additional protection that this modification introduces imposes the following constraints:

- The customer has to plan the purchase in advance and to know the merchant's name and public key, which eliminates impulsive purchases.

- The server will have to make additional encryptions using its secret key for signing each minted coin.

However, because the purchase coupon is specific to a given merchant, local verification on the basis of the serial numbers and the purchase windows is possible without queries to the central server. Merchants can also group payments and request financial settlements by one single transaction.

Finally, the direct transfer of value between two individuals can be done after establishing a symmetric key to secure the exchanges. The key distribution can be done with the Diffie–Hellman method for key exchange or by encrypting the session key with the public key of the recipient, if this key is known.

11.3.4 Evaluation

For the time being, the digital coins can be converted only to U.S. dollars.

The use of public-key encryption increases the computational load, which delays the system's response and limits the transactions to those that exceed the threshold of US$0.25 to $0.30.

Anonymity is not possible in the case of a unique server. This server has to identify the buyer and then register the serial number of each digital coin that it gives the buyer. Furthermore, the server intervenes to authorize each payment (unless the merchant groups the transactions). Thus, it is possible for the server to trace the flow of money and then tie payments to payers.

To gain some kind of partial anonymity, several servers have to be connected in a network. Each server mints its own money, and each digital coin contains fields for the name and address of the minting server or issuer. Thus, a coin has the following form:

Coin = {Name of the minting server, IP address of the minting server, expiration date, serial number, value} SK_S

where { }SK_S represents the signature using the private key of the server. The use of networked servers allows the separation of the sale of digital coins from the approval of purchases. Thus, by making the tracing of transactions more difficult, the holder's privacy can be better protected. For example, a customer may buy the coins from one server, exchange them with the coins that a second server mints, and then make payments that a third server authorizes. Without the collusion of all these servers, it will be impossible to completely trace the flow of digital coins from the moment of issuance to the moment of exchange with legal tender.

Given the multiplicity of servers, and to reduce the risks of encountering fake servers, any server has to authenticate itself by presenting a certificate issued by a competent certification authority that guarantees its public key PK_S. In the U.S., the proposed new federal authority FIC would play that role.

The certificate would contain several fields, including a unique certificate identifier, Certificate_ID, the server's name, the dates of issue and of expiration, and the public key of the server PK_S, and would be signed with the private key of the FIC, Sig_{FIC}:

Certificate = {Certificate_ID, name of the server, PK_S, date of issue, expiration date} Sig_{FIC}

However, the distribution of certificates to merchants and to clients is not part of the NetCash specifications.

Figure 11.9 shows the evolution of a purchase transaction during which two NetCash servers intervene — the issuer server and the acquirer server. The customer buys from the issuer server the coins encrypted with the server's public key and utilizes them to pay for goods. The merchant is

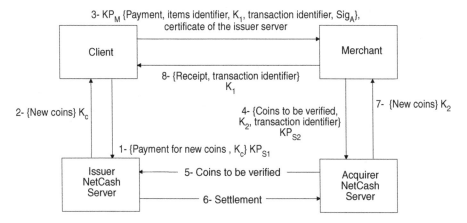

FIGURE 11.9
Exchanges during a transaction with two NetCash servers.

reimbursed by presenting the coins to the acquirer server. The public keys of the respective servers are PK_{S1} and PK_{S2}.

Because of the presence of multiple servers, each of which mints separate money, independently of the others, construction of a new banking system for settlement and regulation of funds is needed. This new infrastructure is based on the utilization of the electronic checks of NetCheque, which will be discussed in Chapter 12.

11.4 Summary

Digital money is similar to cash in the sense that it is anonymous and untraceable. It is different from traditional cash in two aspects. First, the support is electronic (user's hard disk or a microprocessor's memory). Second, confidentiality of the transaction and authentication of the payment instrument can be added at the cost of some computational complexities. The value is stored as the output of an algorithm or in a digital record. Digital money is so radically different from other instruments that it raises many legal and financial issues that the financial industry and regulators have not yet been able to address. This is the main reason as to why digital money has had limited success.

Questions

1. Name some of the legal challenges that digital money systems have to face.
2. Consider inheritance. How can untraceability and inheritance be made compatible within the scope of digital money?

12

Dematerialized Checks

ABSTRACT

The data from the Bank for International Settlements (BIS), which were partially reproduced in Chapter 2, underscore the differences among countries in the usage of checks. It was seen that, according to the data from 2000, the annual rate of check utilization in the U.S. was 49.6 billion checks, which corresponds to 58.3% of the volume of scriptural transactions. This was still the highest usage among the G-10, even though it decreased significantly from 1996, when the corresponding data were 63 billion checks or 74.7% of the volume of scriptural transactions. The second highest rate was in France, where the portion of checks in the volume of scriptural transactions is 43.6% (4.5 billion checks) compared to about 1.3% in Switzerland and 0.2% in Sweden. Clearly, replacement of classical paper-based checks with dematerialized checks only interests those in areas where checks remain a significant payment instrument.

The use of an electronic format for all or some of the check-processing chain promises to increase the efficiency of operation, to reduce costs, and to augment the security of financial transactions. Nevertheless, the problems associated with check dematerialization are not only technical. They require complete reengineering of the various steps for processing check payments, and of the ways cash flow and float values are computed. In addition, the personnel currently involved in the manual processing of checks will have to be retrained. In other words, check dematerialization is a comprehensive socioeconomic project.

This chapter begins with an overview of the classical treatment of paper-based checks and then describes ways to dematerialize this treatment. Next are presented the following projects for virtual checks: NetCheque, Bank Internet Payment System (BIPS), and eCheck.

12.1   Classical Processing of Paper Checks

The focus of this section is on the U.S. and France, which are the two G-10 [1] countries mostly involved in the dematerialized treatment of checks. Although the precise details vary from one country to the other, a general view helps in appreciating the efforts exerted to automate the treatment.

12.1.1   Checkbook Delivery

In the U.S., a checking account must be paid for, which has led to the exclusion of one fourth to one third of the population (Hawke, 1997; Mayer, 1997, p. 451). The holder must explicitly order a new checkbook with 50 checks, and several checkbooks can be ordered at once. Checkbooks are delivered by mail. For the past several years, the printing of checks has no longer been part of the banking monopoly.

In France, a checking account is free of charge. The issuance of a new checkbook is automatic, based on tracking of the number of checks used. The new checkbook contains 25 or 50 checks. The user can also request a new checkbook directly at the bank branch or by Minitel and pick it up in person. Or, the checkbook can be sent by registered mail at the user's expense; however, the user has to be present in person to receive the checkbook when it arrives.

12.1.2   Check Processing

The classical processing of paper-based checks comprises three phases: the send phase, the clearance phase, and the return phase (Dragon et al., 1997, pp. 112–124). Illustrated in Figure 12.1 is the case of a beneficiary that receives checks from users U_1 and U_2 drawn on Bank A, and users U_3, U_4, and U_5 drawn on Bank B.

The send phase begins when the payer issues a check by filling in the amount and the date and then sign it, thereby transforming the paper-based check into a payment instrument. The issuer sends the check to the beneficiary, which then endorses the check and deposits it in the beneficiary's bank. By endorsing each check, the payees are mandating their banks to cash the amounts indicated from the various payers' banks.

The payee's bank prepares for check clearance by sorting the checks and classifying them according to the paying banks, as follows. Checks destined to the banks that belong to the same clearinghouse as the beneficiary's bank

---

[1] As previously indicated in Chapter 2, the "G-10" or "Group of 10" includes the following countries: Belgium, Canada, France, Germany, Italy, Japan, the Netherlands, Sweden, Switzerland, the U.K., and the U.S.

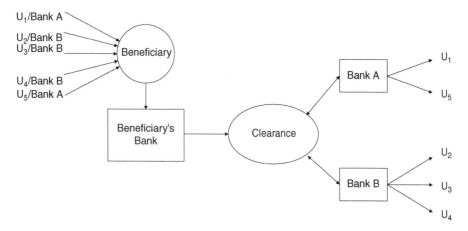

FIGURE 12.1
Classical processing of paper checks.

are put in one group, while the other checks are grouped according to their clearinghouses and then routed to their destinations.

The physical exchange of checks takes place during the clearance phase. In the case of manual clearing, representatives of the various banks meet every day at a given hour in the clearinghouse, where they bring the checks to be exchanged. The checks are transported by cars, trains, trucks, or planes. A limited number of banks send their own representatives, whereas the remaining banks rely on those attending to represent them.

Finally, in the return phase, the paying bank verifies the signature, archives the checks (for example, by microfilming the check's front and back sides), and then debits the drawer. During exceptional conditions, such as when checks are to be tracked or are written for large amounts, additional verifications are performed. Rejected checks are treated separately; these checks are drawn, for example, on accounts with insufficient funds, on closed accounts, on invalid accounts, on blocked accounts (accounts for businesses under bankruptcy or under financial reorganization), or on accounts with overdraft protection that was exceeded. In the U.S., most banks return cashed checks to the drawer, which entails new costs.

12.2 Dematerialized Processing of Paper-Based Checks

The techniques for dematerialized processing of paper-based checks fall into two main categories, depending on the degree to which the end-to-end treatment eliminated paper. For electronic check presentment, the beneficiary's bank sends the drawer's bank only the information that allows identification of the check. In the case of check imaging, the beneficiary's bank sends a scanned image of both sides of the check.

12.2.1 Electronic Check Presentment

Electronic check presentment is the electronic representation of the data contained in the paper-based check on an immaterial support. By eliminating the paper medium after the presenting bank, which is usually the beneficiary's, received the check, payment data are put in an electronic format and then transmitted to the clearing network. The presenting bank archives the paper-based checks. Thus, the use of check presentment aims to avoid the material transport of checks among the various banks to reduce the cost of check clearing and to increase the security of interbank settlements.

In France, the exchange of check images, ECI (Échange d'Images Chèques), were operational since 1990 through the Creic (Centres Régionaux d'Échanges d'Images Chèques — Regional Centers for the Exchange of Check Images). The information presented, called the check image, is the CMC7 (Caractères Magnétiques Codés à 7 Bâtonnets — Magnetic Characters Coded with 7 Links) line that appears on the check as a line of alphanumeric magnetic characters defined by an AFNOR (Association Française de Normalisation) standard. It indicates the check number, the amount drawn, the code of the issuer bank, the branch code, the account number, the coordinates of the account holder, etc. However, this information does not include the signature of the payer.

Since June 30, 2002, the clearinghouses have been closed, and French banks are obliged to send and return check images via the electronic exchange SIT (Système Interbancaire de Télécompensation), either directly or through one of the 17 direct participants. The payee's bank is responsible for archiving the checks instead of the payer's bank, in contrast to the former manual system. The elimination of physical exchange should eliminate the risks of check loss, damage, or routing errors. One settlement date is uniformly applied: $T + 1$. All the checks returned before 6 p.m. of a working day are cleared and settled the next day. Thus, the delays for "out of area" checks are divided by 2 or 3.

The U.S. version of the check image is called ECP (Electronic Check Presentment). It was introduced in 1996 by the New York Clearing House (NYCH) Association as check truncation. Here, the line corresponding to the CMC7 is called the MICR (Magnetic Ink Character Recognition), which consists of characters printed with magnetic ink. This line is processed with high-speed magnetic recognition equipment to construct the file of payment data as defined in ANSI X9.37 (1994). The verification of the MICR line before the physical exchange of checks gives the payer's bank the chance to detect irregularities within the legal delay for clearing a check.

The objective of ECP is not so much cost reduction as it is the prevention of abuses by users that take advantage of the time delay between the physical exchanges of checks and the availability of funds to depositors. In September 1990, the Expedited Funds Availability Act obliged banks in the U.S. to make the funds of deposited checks available to the depositors within a certain time interval defined by law, even though the checks may not have been physically exchanged. This has led to a dramatic increase in the frequency of check fraud.

In check truncation, the paying bank has to reconcile the electronic payment data with the exchanged paper checks that will be returned later to the payer's bank and the payer. This adds some additional processing steps. However, if the paying bank does not return the paper-based checks to the payer, ECP becomes the equivalent of the method of check images used in France.

One hurdle that the generalized use of ECP must surmount is the variability of the MICR formats used by U.S. banks. Nine banks, among them the Bank of America, established the Electronic Check Clearing House Organization (ECCHO) in 1990 to promote the use of ECP by deriving a uniform format. Today, this association has about 70 members.

12.2.2 Point-of-Sale Check Approval

A pilot service called SafeCheck was developed in the U.S. by Small Value Payments Co. (SVPCo) by combining ECP and debit-card processing. It builds on the infrastructure for electronic funds transfer to provide real-time authorization of checks presented at the point of sale. A reader terminal captures the MICR data, while the cashier enters the purchase amount manually. The data are transmitted to the issuer bank (the customer's bank) to verify the availability of funds and to authorize the transaction. The approved check is canceled and returned to the customer.

The advantage of this approach is that it can leverage the existing infrastructure for debit transactions to facilitate the use of checks of purchases. The difficulty arises from the wide variety of MICR formats and the need to upgrade the data-processing and network equipment.

12.2.3 Check Imaging

The check image is the digitized picture of both sides of the check to replace the physical check completely in banking networks. The digital picture can be transmitted over a computer network, stored in the various banks, and sent to the user instead of the cashed checks. At the destination, the image can be displayed on-screen or printed on a printer. The success of such a scheme requires using image compression algorithms, in addition to securing the exchanges over the computer network. In this scheme, the presenting bank receives the beneficiary's check and transforms it into a check image to then be transmitted to the clearing computer. The presenting bank is also responsible for archiving the image. It has to include in the image file the indexing references to allow location of the images for access and retrieval of the archived images.

In 1992, the FSTC (Financial Services Technology Consortium, online a), which is concerned with the applications of new technologies for financial services, started a U.S. project, PACES (Paperless Automated Check Exchange & Settlement), to study the feasibility of establishing a national system for the exchange of check images (Financial Services Technology Consortium, no date). A protocol called CIIP (Check Image Interchange Protocol) was developed to transmit the image file on command. The format of the image file is specified in ANSI X9.46 (1997) for TCP/IP networks. In parallel to the check image transmission, data extracted from the MICR line are collected and processed by electronic means. In 2000, SVPCo took over the PACES project, and in 2001, Viewpointe Archive Services announced the first commercially available platform to support the exchange of check images.

The compressed image in black and white or in color can be presented in one of several formats, such as the algorithms defined in the ITU-T Recommendations T.4 or T.6, which are used for the transmission of Group III and Group IV facsimile or the JPEG (Joint Photographic Expert Group) algorithm. The presentation format can be TIFF (Tagged Image File Format) or COF (Common Output Format) of the Federal Reserve.

12.3 NetCheque

NetCheque is an experimental system for virtual checks, from the Information Science Institute (ISI) of the University of Southern California (USC) (Neuman, 1993). Originally, it was conceived for the distributed management of access to computation resources in a university computation center. These resources could be, for example, the amount of memory blocks allocated, the number of processing cycles, or the number of pages to be printed, etc. This concept was extended to means of payment through virtual checks. A prototype is available at http://www.netcheque.org.

Kerberos is used for authentication and for the production and distribution of session keys. Three session keys are needed: the key K_{CB1} between the client and its bank (B1), the key K_{MB2} between the merchant and its bank (B2), and the session key K_{BB} between the two banks. In general, the Kerberos server is distinct from the two banks. The configuration is depicted in Figure 12.2.

According to Kerberos terminology, the client's credentials with respect to the client's bank are a session ticket that the Kerberos server supplies and an authenticator AuthC that the client constructs and encrypts with the help of the relevant session key.

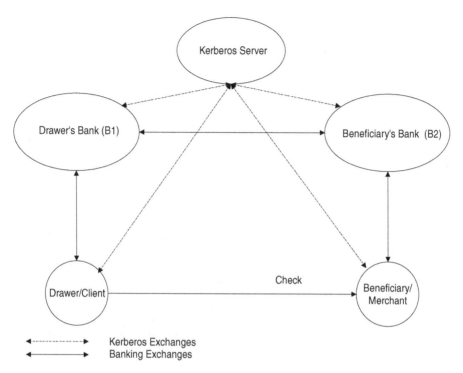

FIGURE 12.2
Use of Kerberos in NetCheque.

12.3.1 Registration

Users register at their banks by opening accounts and then obtaining virtual checks. Each account is defined by the identifier of the bank server, the user's name, a list of access permits, and the set formed by the account balance for each of the currencies that can be used. NetCheque virtual checks include the following fields:

- The amount
- The currency type
- The check date
- The drawer's account number
- The beneficiary's name
- The digital signature of the account holder
- The endorsement of the beneficiary (the merchant) and of the banks used

The drawer's (the client's) bank verifies the last two fields, while the remaining fields are sent in the clear. These clear fields are the NetCheque restrictions.

12.3.2 Payment and Financial Settlement

The drawer draws a check, giving to the beneficiary (the merchant) the right to transfer funds from the drawer's account. Customers use a Kerberos ticket, T_{CB}, to authenticate themselves to their bank B1; similarly, the merchant needs the T_{MB} ticket to authenticate itself next to its bank B2. Finally, the ticket T_{BB} allows the bank B2 to authenticate itself to the other bank B1.

The session ticket T_{CB} includes, among other information, the identity S of the Kerberos server, the customer's name C, the coordinates of its bank B1, and the session key K_{CB}. The ticket content, except for the server's identity, is encrypted with the long-term key K_{SB1} between the Kerberos server and the customer's bank B1. Thus,

$$T_{CB} = \{S, K_{SB1}(C, B1, K_{CB})\}$$

Similarly, the ticket T_{MB} contains the equivalent information for the merchant:

$$T_{MB} = \{S, K_{SB2}(M, B2, K_{MB})\}$$

Finally, the ticket T_{BB} is given by

$$T_{BB} = \{S, K_{SB1}(B2, B1, K_{BB})\}$$

The customer constructs the authenticator, AuthC, containing the customer's identity C, the customer's account number No_C, and the digest of the check, and all of this information is encrypted with the session key K_{CB}:

$$AuthC = K_{CB}\{H(Check)\}$$

The encrypted authenticator and the session key are concatenated to the check, and the message sent to the merchant becomes

$$Check \parallel AuthC \parallel T_{CB}$$

In a similar manner, the merchant establishes a secure communications channel with its bank using the session ticket T_{MB}. To endorse the check, it constructs an authenticator AuthM:

$$AuthM = K_{MB}\{H(Check)\}$$

The merchant concatenates its authenticator to the message received by the client. The message E1 sent to the merchant's bank takes the following form:

$$E1 = \text{Check} \parallel \text{AuthC} \parallel T_{CB} \parallel \text{AuthM} \parallel T_{MB}$$

The merchant's bank endorses the check by constructing the authenticator AuthB:

$$\text{AuthB} = K_{BB}\{H(\text{Check})\}$$

and joining the session ticket T_{BB}. The message E2 sent to the client's bank is as follows:

$$E2 = \text{Check} \parallel \text{AuthC} \parallel T_{CB} \parallel \text{AuthB} \parallel T_{BB}$$

The customer's bank verifies all the data included in the tickets and the authenticators to confirm that the bank B2 and the customer are authentic and then decrypts each authenticator with the corresponding key K_{CB} or K_{BB}. The exchanges are shown in Figure 12.3.

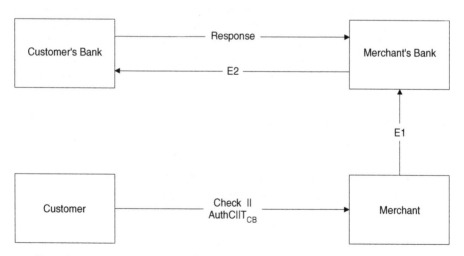

FIGURE 12.3
Exchanges during a payment using a virtual check according to NetCheque.

12.4 Bank Internet Payment System (BIPS)

BIPS and eCheck are among several projects for virtual checks that the FSTC has promoted. The electronic equivalent of the checkbook would be a tamper-resistant microprocessor card or a file on a workstation. The account numbers used for the electronic checkbook are such that the debtor's bank will be the only entity capable of establishing correspondence with the real bank account.

BIPS aims at the design of a remote payment system using the Internet as the communication support. The program targets large users of checks, such as professionals, businesses, and their banks. Thus, the system covers salary payment, payment for services, and refunds from one enterprise or a government agency to individuals or to other enterprises. The informatics exchanges can be interactive through the Web or be by e-mail.

The participants already using X12 or EDIFACT messages will have to change their architectures and adopt the Network Payment Protocol (NPP). Alphanumeric EDI messages must be restructured according to XML (Extensible Markup Language). Another ambition of BIPS is to completely replace the mechanisms for interbanking communications at the worldwide level, which puts it in competition with existing networks such as SWIFT (Society for Worldwide Interbank Financial Telecommunications). However, the restrictions that the U.S. federal government imposes on the export of encryption software may thwart these ambitions.

BIPS has two other competitors. The first is OFX (Open Finance Exchange), jointly developed by Microsoft, Intuit, and CheckFree. CheckFree merged with Transpoint, a joint society of Microsoft and First Data Corporation, to strengthen the offer for online bill presentment. The second is the Gold standard proposed by Integrion. The OFX aims to provide personalized banking services, such as home banking, starting with SGML (Standard Generalized Markup Language). OFX is not compatible with XML or with SWIFT.

12.4.1 Types of Transactions

BIPS recognizes the following banking transactions (FSTC, 1998b):

- **Direct debits**. Direct debit is a payment instrument that the creditor initiates, once the debtor agreed to it on paper. It can be used, for example, when a corporation would like to collect the funds available in the accounts of its various subsidiaries. It can also be used by private or public institutions that collect regular payments, such as rents, taxes, contributions, and telephone, gas, or electricity bills.

- **Fund transfer**. This is an instrument uniquely suited for business-to-business payments as well as regular payments to individuals made by commercial and industrial corporations and government agencies. Common examples include payroll payments, social security benefits and investment dividends, payments by insurance companies, etc.

- **Payments authorized remotely through a computer connection.** This is similar to the télé-Tip that was introduced in France in 1995 for the payment of gas, electricity, and telephone bills through the Minitel.

The BIPS designers renamed direct debit transactions "pull transactions" and the funds transfers as well as the computer payments as "push transactions." To accommodate current commercial practices, the agreement between the creditor and the debtor is always done on paper. In the same manner, BIPS does not mandate that the interactions between the creditor and its bank necessarily be conducted on the Internet.

Figure 12.4 illustrates the exchanges that take place during a pull transaction as specified in BIPS. The exchanges in a push transaction for BIPS are presented in Figure 12.5.

The creditor, usually an enterprise, initiates the direct debit according to the agreement with the debtor and its bank. However, the creditor and the debtor do not have to use the same bank; therefore, the direct debit will call for an authentication according to the procedures of Recommendation X.509.

The financial network shown in Figure 12.4 is, in reality, composed of several specialized networks, depending on whether the payment is a large-value payment, a large-volume payment, or a payment by bankcard, as was previously indicated in Chapter 2.

12.4.2   BIPS Service Architecture

The BIPS server resides in the secure financial network and includes the following components:

- An e-mail server or a Web server to receive client requests by e-mail or through the Web
- A front-end to interpret the BIPS messages and relay them to the electronic payments handler (EPH)
- An EPH whose primary role is to validate the BIPS message, the digital signatures, and the certificates (examination of the certificates relies on a certification infrastructure, as explained in Chapter 3)
- A payment system interface (PSI) to translate the BIPS into the format of the selected payment mechanism (this element is responsible for the interface with the other banking networks, particularly the ACH [Automated Clearinghouse] network for large-volume payments, the wire transfer network FedWire, and CHIPS [Clearing House Interbank Payment System] for large and medium amounts, the bankcard network, and other banking networks)

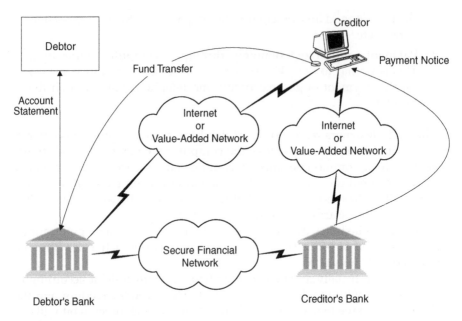

FIGURE 12.4
Exchanges in a pull transaction of BIPS.

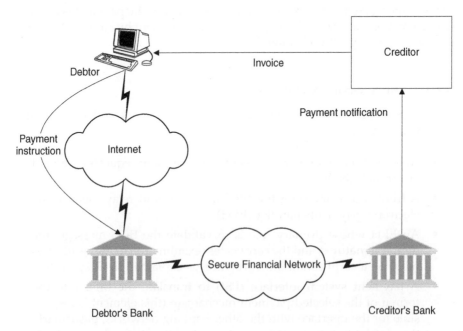

FIGURE 12.5
Exchanges in a push transaction of BIPS.

- An event log for recording an audit trail of all events for auditing purposes and for error recovery (the list of recorded events includes receipt of messages, their forwarding to other entities, successful and failed authentication, etc.)

- A repository to maintain a copy of all transactions for processing and audit purposes

The new NPP defines the structure of the dialogues for the different transactions with messages in XML.

The certification system used to authenticate the customers is external to BIPS. It conforms to Version 3 of X.509 and is responsible for maintaining the revocation lists. Encryption algorithms are used to verify message integrity and to protect confidentiality. The security of e-mail is achieved with S/MIME (Secure Multipurpose Internet Mail Extensions), PGP (Pretty Good Privacy), or Open PGP. Web exchanges are secured with SSL (Secure Sockets Layer) or its standardized version TLS (Transport Layer Security) or, strictly for HTTP (HyperText Transfer Protocol) exchanges, with S-HTTP (Secure-HTTP).

As an example, the organization of these elements in the BIPS server of the debtor's bank is shown in Figure 12.6.

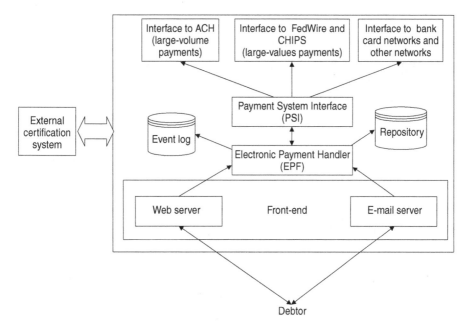

FIGURE 12.6
Components of the banking server of BIPS.

12.5   eCheck

The eCheck is a virtual check sponsored by the FSTC and the U.S. Treasury (Jaffe and Landry, 1997). Currently, it is the sole electronic payment mechanism recognized by the U.S. Treasury and was used in a pilot trial for 50 subcontractors of the Department of Defense for payments up to U.S.$100,000. The eCheck technology was licensed to Clareon. Clareon incorporated it into its PayMode secure electronic payment engine.

12.5.1   Payment and Settlement

The aim of this new instrument is to permit a debtor, whether an institution or an individual, to initiate a fund transfer using a standard message sent by e-mail (FSTC, online b). From the exchanges in Figure 12.7, it is seen that the circuit of paper-based checks is reproduced with a dematerialized base for the check and for the control message. Clearance and interbank settlement of the virtual checks use ANSI X9.46 (1997) and X9.37 (1994). This scenario is called "deposit and clear."

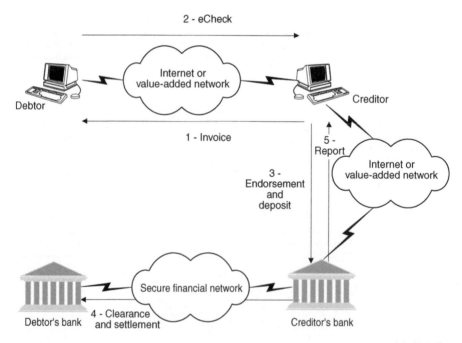

FIGURE 12.7
Exchanges in a payment by eChecks (Deposit and Clear).

The flow of exchanges for the deposit-and-clear scenario is similar to the case of the traditional check system. The payer sends a signed eCheck to the payee. The payee then deposits it in this bank. Clearance and settlement go through the traditional financial network. Experience has shown that it takes less than 15 minutes for the debtor to send a virtual check and for the creditor to endorse it and deposit it in the creditor's bank. Because banking clearance takes the usual electronic path, payment to the creditor does not occur before the debtor's bank completes all the necessary verifications.

One of the most critical aspects of payment with virtual checks is the reconciliation of the data that the customer's computer or microprocessor card records with the data that the customer financial application records and with the bank statement.

One of the advantages of virtual checks is that they provide a means for attaching some free-format commentary files or messages to the payment to give needed information regarding billing statements or purchase orders. This attachment will be removed before depositing the virtual check in the creditor's bank. Thus, attachments can be separated from a digitally signed package of electronic documents while retaining the original authorizing signatures. This explains why, in 1999, the management of the eCheck initiative was transferred from FSTC to CommerceNet in order to extend the technology to applications outside of banking.

A variation of the preceding circuit can be useful if the creditor's bank is not equipped to process electronic checks, while the debtor's bank is so equipped. In this case, called "cash and transfer," the debtor sends the check to the creditor, who endorses it and then sends it to the debtor's bank. After making the prerequisite verifications, the debtor sends the necessary funds through the banking network, as shown in Figure 12.8.

Certified virtual checks can be treated in one of two ways. Debtors send their checks to their banks to verify the availability of funds, and then the banks block them before the debtors sign the checks. The banks can then return the checks to the debtors (Option 1) or send them directly to the creditors (Option 2). The corresponding flows are represented in Figure 12.9, where the network between the creditor, the debtor, and the debtor's bank is not shown so as to avoid unnecessary cluttering of the diagram.

Let us now consider two more payment scenarios. Direct fund transfers are initiated when the payer sends the eCheck to his or her bank. The actual fund transfers take place through the usual interbank circuits. The lockbox scenario is of interest to large organizations that receive a large number of checks, because it allows them to leave the verification of the eChecks that they received to their banks, provided that the payers send the eChecks to a special account. The bank will then initiate the clearance and settlement through financial networks for all the checks that were proven to be valid. In Figure 12.10, the corresponding exchanges are illustrated.

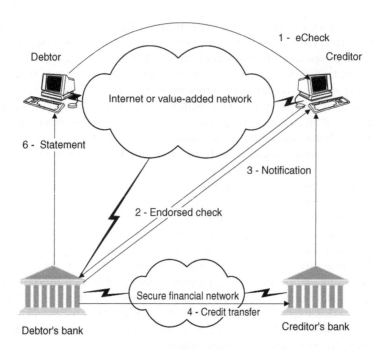

FIGURE 12.8
Exchanges when the creditor's bank cannot process eChecks (Cash and Tranfer).

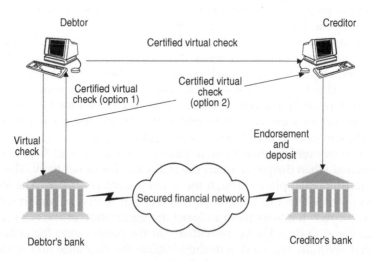

FIGURE 12.9
Exchanges for certified virtual checks with eCheck.

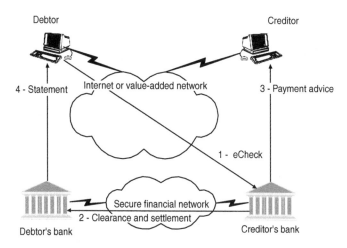

FIGURE 12.10
Exchanges for the lockbox scenario with eCheck.

12.5.2   Representation of eChecks

The representation of the eCheck uses the Financial Services Markup Language (FSML) (FSTC, 1998a). This language is defined within the framework of SDML (Signed Document Markup Language), which specifies how to sign digital documents (Kravitz, 1998). Although SDML, in turn, is inspired by SGML of ISO 8879 (1986), it is not compatible with XML (Liu, 1998). The designers of FSML were aiming at financial applications of microprocessor cards and could not postulate that the bandwidth would be free and memory abundant. It is unfortunate that two projects for the FSTC, namely, eCheck and BIPS, use incompatible languages.

FSML describes the checks using a sequence of blocks to represent the data related to the check. These blocks can be nested and are signed with the help of public-key cryptography algorithms and hashing algorithms, by encrypting the message digest with the private key of the sender. The FSML represents the alphanumeric characters that are part of the check data using 7-bit ASCII, which limits the alphabet (for example, no accents are used).

There are two categories of blocks in an eCheck. The first comprises the action, the check data, the debtor's account number, any attached documentation, and the invoice. It is signed with the private key of the sender, whose corresponding public key is included in the user's certificate. The second category comprises the account number and the debtor's certificate; it is signed with the private key of the bank. Presented in Figure 12.11 are the various blocks in the check.

Endorsement of an eCheck means adding an endorsement block, with the identity of the endorser, its banking coordinates, its certificate, and the certificate of its bank. The endorser and its bank sign the block indicated in

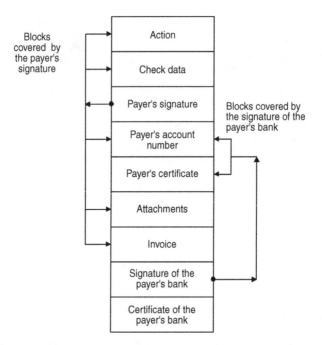

FIGURE 12.11
Representation of virtual checks in eCheck.

Figure 12.12. Typical sizes of eCheck files are shown in Table 12.1 (U.S. Department of Treasury, no date).

Two algorithms are currently used for signature calculations. These algorithms are MD5 with RSA, and SHA-1 with the DSA algorithm. The X.509 Version 1 signed by the U.S. Treasury Department is utilized for authentication of banks; banks, in turn, certify their clients. The e-mail protocol is MIME.

The check presentment file in the banking circuit conforms to ANSI X9.46 (1997), which is also used for presenting check images. A future convergence among the settlement procedures for these two means of payment cannot be excluded.

12.6 Comparison of Virtual Checks with Bankcards

The main features that distinguish virtual checks from remote payments with bankcards based on SSL/TLS or SET are as follows:

1. SSL/TLS and SET are for interactive transactions on the Web, while virtual checks can be used either on the Web or by e-mail.

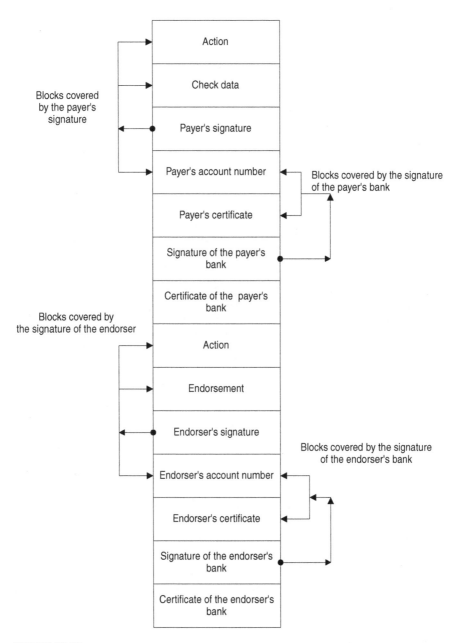

FIGURE 12.12
Representation of an endorsed virtual check in eCheck.

2. SSL/TLS and SET use real-time authentication and online verification of the certificates. In contrast, authentication in the case of virtual checks is not necessarily in real time, which reduces the constraints imposed on the system.

TABLE 12.1

Size of eCheck Files

Description	Total Size (octets)	Compressed Size (octets)
Single check	4762	2259
Single check with remittance	7052	3135
Single check with remittance and endorsement	1443	4839

TABLE 12.2

Comparison between Electronic Payments by Virtual Checks and by SET

Characteristics	Virtual Check	SET
Payment model reproduced	Paper-based checks	Bankcards (credit and debit)
Interaction with the user	E-mail, Web	Web
Authentication	Offline	Online
Authorization of the payer	Sent jointly with the payment instructions	Preauthorization at subscription time
Trusted third party	Not available	Mandatory
Data on the payer's account	Available to the merchant	Hidden from the merchant

3. In SET, encryption of the transaction data hides the information on the user's account from the merchant's server but not from the payment gateway.

4. In SET, the payment gateway plays the role of a trusted third party (arbiter and small-claims judge), without knowing the details of the transaction. Thus, it verifies the agreement between the buyer and the merchant using the appropriate digests. There are no trusted third parties in the case of virtual checks.

Summarized in Table 12.2 are the differences between electronic payments by virtual checks and by bankcards through SET.

12.7 Summary

The dematerialized check is a form of the télé-Tip adapted to the Internet that can coexist and then substitute paper-based checks for remote payment of bills (Dragon et al., 1997, pp. 128–129, 234). It can also be associated with other solutions based on microprocessor cards. In this case, a three-party dialogue has to be established between the card reader, the debtor's "electronic checkbook," and the bank server, according to a secure payment protocol. For example, the cardholder will have to enter a PIN before using the smart card to make a payment.

In principle, the legal status of dematerialized checks in the U.S. is identical to paper-based checks, and all the laws regulating the use of paper-based checks would be readily applicable to electronic checks.

Questions

1. What are the advantages of dematerialized checks?
2. What are the differences between electronic check presentment and check imaging?
3. Discuss the factors favoring or opposing the use of check imaging.
4. What would be needed to help popularize eCheck?

13

Security of Integrated Circuit Cards

ABSTRACT

The smart card (or integrated-circuit card) is the culmination of a technological evolution starting from cards with bar codes and magnetic strip cards. Even though their rates of penetration vary among countries, the interest that they raise is undeniable, because they offer storage and signal-processing capabilities that allow complex decisions. Nowadays, smart cards or microprocessors can be found in payment applications, in electronic purses, in client-fidelity programs, as well as in access controls to physical sites, to computer networks, and to pay TV.

In this chapter, we examine the ways to protect smart cards and to secure the transactions that are carried out, and we also discuss the limits of these protections. After an overview of smart cards and their applications, we describe the various aspects of security within the life cycle of a typical card. We introduce next the multiapplication smart cards and the operating systems that control resource sharing among several applications, in particular, the ISO standards and the EMV (EuroPay[1], MasterCard, Visa) specifications. After a review of the techniques to integrate smart cards with personal computers, this chapter ends with a list of several ways to bypass the security of smart cards, including when biometrics are used.

13.1 Overview

Although the first patents for integrated-circuit cards were awarded in the 1970s in the U.S., Japan, and France, large-scale commercial development started in Japan in the 1970s and in France in the 1980s (Ugon, no date). At present, microprocessor cards are gradually replacing magnetic-strip cards in electronic commerce (e-commerce) applications that require large capacity

---

[1] In 2002, EuroPay became MasterCard Europe.

for storage, for the processing of information, and for security (McCrindle, 1990, pp. 15–20; Lindley, 1997, pp. 122–123; Dreifus and Monk, 1998). The main reason is the spectacular increase in fraud, particularly due to online transactions. In 2000, for example, the loss due to fraud experienced by banks in the European Union increased by 48% to reach 400 million euros, which is about 0.07% of the turnover (Austin, 2001). In the U.S., about half of the contested transactions are for online purchases, although they constitute only between 1 and 2% of the total volume of transactions.

While principally remaining a European phenomenon, the worldwide market for smart cards is expected to increase by 50% between 2001 and 2004. Depicted in Figure 13.1 is this evolution and the changes in the geographical distribution of card issuance within this period (Parsons, 2002).

13.1.1   Classification of Smart Cards and Their Applications

Smart cards can be classified according to several criteria, for example

- Usage in a closed system or an open system
- Duration of card usage, which distinguishes disposable cards from rechargeable cards
- Intelligence of the card, which ranges from simple memory for information storage to wired logic and a programmable microprocessor
- Necessity for direct contact with the card reader and the way power is supplied, which distinguishes contact cards from contactless cards: contact cards have a data transfer rate of 9.6 kbit/sec, whereas the rate for contactless cards can reach 106 kbit/sec. Because power

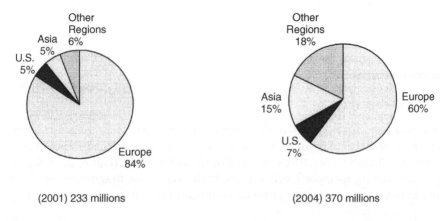

FIGURE 13.1
Projected number and geographic distribution of smart cards issued annually in 2001 and in 2004. (From Celent Communications and Parsons, M., *Red Herring*, 109, 72, January 2002).

supply in these cards can be done by induction, they seem to attract the interest of railroad and bus companies as a way for customers to pay fares

- Characteristic application of the card, which differentiates monoapplication cards from multiapplication cards

The first applications of microprocessor cards were in closed systems with in-house specifications or from a single manufacturer. This remains the case for telephone cards, or for cards used for payments to vending machines (parking meters, public transportation, etc.). Integrated-circuit cards can be classified according to their computational capacities in the following categories:

- *Memory cards.* These cards include a microprocessor of reduced capacity that can stock data and even offer minimal protection to the data; the processing power of this type of card is adapted to prepayments, such as for telephone cards.
- *Wired-logic cards.* These cards are used to offer minimum control to regulate access to encrypted television channels; their cost is about U.S.$1.
- *Microprocessor cards, known also as smart cards or integrated-circuit cards.* These are programmable machines capable of making complex computations. The price of microprocessor cards remains between U.S.$5 and $20, which is relatively high, and the market remains fragmented among several incompatible operating systems and programming languages (although this is changing).

Telephone cards are usually contact cards, monoapplication and disposable; their cost is around 50 cents (U.S.). The total value of telephone units corresponds to an equal number of memory cells. Cells are progressively erased as the card is used, until all units are depleted. Their storage capacities are on the order of 8 to 256 kbits but can reach hundreds of megabits.

Microprocessor cards contain, within the thickness of their plastic (0.76 mm), a miniature computer (operating system, microprocessor, memory, and integrated circuits). These cards can carry out advanced methods of encryption to authenticate participants, to guarantee the integrity of data, and to ensure their confidentiality. Payment authorizations can also be done offline, which is not the case for magnetic-strip cards. These cards can also be used for physical access control (enterprises, hotels, etc.) or logical control (software, confidential data repositories, etc.). In this case, the card takes the shape of a small calculator that uses dynamic passwords with which to authorize the holder's access.

Multiapplication cards can even offer several personalized services, such as an electronic pass, an identification card, support for portable electronic folders (health record, school or professional transcripts), electronic purse, etc. Such cards are already used at several European universities (Martres and Sabatier, 1987, pp. 105–106; Lindley, 1997, pp. 36–37, 95–111). Mass applications have been in preparation for more than one decade in Belgium, France, and Germany. The cards Sésame-Vitale in France and Versicherten-Karte in Germany will be visible interfaces to a whole informatics system for the acquisition and storage of administrative and vital data on cardholders (McCrindle, 1990, pp. 143–146). Finally, the Hemacard program in Belgium targets at-risk populations, such as pregnant women, patients with cancer, or those with chronic diseases. However, the coexistence of several applications on the same card requires isolation of the space used by each of them. If access to the card depends on a personal identification number (PIN), then communication with each application will have to be controlled by a second set of secret keys established between the supplier of the application and the cardholder.

13.1.2   Integrated-Circuit Cards with Contacts

In integrated-circuit cards with contacts, the clock rate varies between 3.5 and 5 MHz. The power supply is between 3 to 5 V. In the circuits of the previous generation, the programmable memory was supplied by a different power source from 5 to 15 V or even 21 V. Microelectronics advances eliminated the need for this additional source, and the required voltage is generated through internal circuitry. This reduces the chances of attacks through voltage fluctuations; furthermore, the freed contact can be used to increase the rate of data exchange through duplex transmission.

An integrated-circuit card has eight contacts that must resist wear and corrosion due to various factors (abrasion, corrosion, sweat, pollution, chemical substances, etc.) (see Figure 13.2). The standards define the locations of the contacts between the card and the card readers to ensure their compatibility. Manufacturers, while respecting the standards constraints, have been able to use the arrangement of the contacts as a distinctive means for identification.

Card readers are generally equipped with mechanisms to grab the card automatically as soon as it is inserted in the reader slot. This setup enhances the reliability of the reading and strengthens the card reader's resistance to users' abuse and to vandalism, although at the cost of increased complexity and operational and maintenance costs.

13.1.3   Contactless Integrated-Circuit Cards

Contactless cards communicate with the reader through inductive or capacitive coupling. As for the power supply, it is provided by batteries

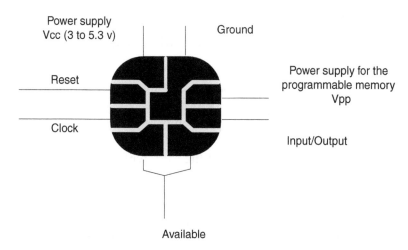

FIGURE 13.2
The eight contacts of an integrated-circuit card with contacts.

or, in a passive way, through inductive or passive coupling (Sherman et al., 1994; Thorigné and Reiter, 1994). Embedded batteries are from lithium or carbon and manganese or use the photovoltaic effect for recharging (Praca and Barral, 2001). Data exchanges with the reader occur at a distance of 10 to 20 cm, with rates to 19.2 kbit/sec. The frequencies used are in the band reserved for industrial applications (less than 150 kHz) or in the band of scientific and medical applications (mostly around 6.78 MHz).

Contactless cards have several advantages. They reduce the maintenance cost by avoiding the wear in the mechanical system to pull in tickets with magnetic strips. They offer a means of payment that can accommodate persons moving at a small speed, such as in the case of public transportation or at toll booths. In Singapore, the contactless electronic purse Cash-Card has been in use since 1996 for micropayments (tolls, parking meters, etc.). In the Parisian region, the metro and railways organizations, respectively, the RATP and the SNCF, are planning to replace paper metro or train ticket payments with an electronic purse by 2004 (Navigo project). This will be achieved by equipping the gates with card readers (called validators) that can receive payment data at a distance of 10 cm. The antenna is integrated in the surface of the card and transmits a carrier frequency of 13.56 MHz. The data-exchange protocols must be robust enough to resolve contentions among several cards trying to access the reader at the same time. At any station, a server will supervise all the gates equipped with readers and will update the blacklist daily (for buses, the validation lists will be present on each bus).

13.2 Description of Integrated-Circuit Cards

13.2.1 Memory Types

An integrated-circuit card has three types of memory: a read-only memory (ROM); a random access memory (RAM); and a programmable read-only memory (PROM), also called nonvolatile memory (NVM).

- The ROM contains the mask of the card, which is fixed in several kilobits during a masking operation. The mask includes the card's operating system. The security algorithm mostly used in single-application cards is part of the M4 mask. The configuration of this mask changes according to the domain of application. The mask adapted for banks is called B0, while the one used for telephone cards (publiphone) is B03 (Guillou et al., 2001).

- The RAM is programmable and can contain intermediary results as long as there is a power supply, which explains its other name, "volatile memory." The content of static RAM (SRAM) remains stable, while the content of dynamic RAM (DRAM) must be refreshed periodically to prevent leakage.

- The ROM keeps its content in the absence of a power supply. It has two principal forms: electrically programmable read-only memory (EPROM) and electrically erasable programmable read-only memory (EEPROM). The EPROM's content is erased only after exposure to ultraviolet radiation. This type of memory is often found in disposable cards. Programmable or secure cards use EEPROM memories to store encryption keys, updates, or bug corrections. This category of memory requires two clock cycles to write an octet of data, the first to erase the existing data and the second to record the new octet. New memory technologies, such as flash or Ferrite RAM (FeRAM), can reduce the write time and the power consumption.

A typical smart card uses an 8-bit processor, a ROM of 16 K octets, a RAM of 256 to 512 or even 1024 octets, and a PROM of 3 K up to 32 K octets. For example, the smart card used in the Navigo project operates with an 8-bit processor, an EEPROM of 2 K octets, and a proprietary operating system. In a top-of-the-line card, the size of the ROM can be 128 K octets, the RAM 4 K octets, and the EEPROM 64 K octets (Borst et al., 2001; Praca and Barral, 2001).

From a functional viewpoint, the memory of a secured integrated-circuit card is organized in the form of a hierarchy of zones (Martres and Sabatier, 1987, pp. 91–93):

- The fabrication zone is the part of the memory recorded before personalization. It includes the lot number of the wafer, the name of the manufacturer, its serial number, the serial number of the card, and the identity of the card supplier.

- The secret zone locks the cardholder's confidential information, such as the PIN or the secret cryptographic keys and the personal data files.

- The transactions zone, or working zone, stores the temporary confidential information related to individual transactions, such as the amount, the balance of the stored value, and the details related to the vendor. This part is shared among the various applications in multiapplication cards.

- The access control zone records, under the control of the microprocessor, all successful accesses or access attempts to the secret zone or to the transactions zone (if this zone is protected). Thus, it is possible to block access to the card after three unsuccessful attempts.

- The free reading zone, or the open zone, is where nonconfidential information, such as the names of the holder and the issuer, and the expiration date, is stored. This zone is accessible to the different applications of a multiapplication card.

13.2.2   Operating Systems

Often, the operating system — which resides in ROM — is specific to a given manufacturer. The system mostly used for monoapplication cards is part of the proprietary mask M4. For multiapplication cards, there are three competing contenders. MULTOS™, used in the Mondex electronic purse, is the candidate of the Maosco consortium, led by Mastercard, and which includes Hitachi, Gemplus, and Siemens as members. The MULTOS technical specifications cover the operating system, the MEL (MULTOS executable language) assembler language, the API, and the chip. On the opposing side, Visa International, with the assistance of Sun Microsystems, Schlumberger, Motorola, and Gemplus, is spearheading the JavaCard Forum. This consortium, founded in 1997, proposes a specification for smart cards based on the Java virtual machine: the JavaCard (see Section 13.5.3.3). The third candidate is Windows for Smart Card from Microsoft, which utilizes the Visual Basic programming language. It has a library of cryptographic commands with which to facilitate the production of various kinds of electronic signatures as well as the management of certificates in public-key infrastructures. Finally, the German company ZeitControl proposed the BasicCard, with a secure proprietary system and cryptographic functions such as elliptic curve cryptography. The advantages are less costs and an easier way to program them (Husemann, 2001).

13.3 Standards for Integrated-Circuit Cards

Standards for smart cards are needed for two main reasons: they increase the scale factor, which reduces the unit production costs; and they facilitate the networking of applications on a worldwide level through harmonized interfaces. Standardization has advanced on the physical aspects of the cards; however, the lack of standardization at the logical level is hampering financial applications on a large scale. In this section, the standards for cards with contacts and for contactless cards are presented, as are the EMV specifications for multiapplication cards.

13.3.1 ISO Standards

The standard ISO/IEC 7816 ensures physical compatibility between integrated-circuit cards with contacts and card readers. The first specifications naturally focused on the physical dimensions of the card, the positions of the contacts, the power supplies, the shapes and durations of the electric signals, and the protocols for communication between the card and the terminal. With the increase in commercial and telecommunications applications of smart cards, other parts were added, so that the standard consists today of six parts, with more to come.

- ISO/IEC 7816-1 (1998), the oldest part, specifies the physical characteristics of the card, the dimensions of the integrated circuit, the resistance to static electricity and electromagnetic radiation, the flexibility of the support, and finally, the location of the integrated circuit on the card.
- ISO 7816-2 (1988) defines the dimensions and the positions of the metallic contacts on the card.
- ISO/IEC 7816-3 (1997) describes the electric signals (polarity, voltage, duration, etc.), the transmission protocols between the card and the terminal, and the card's response to a reset originating from the terminal. Four protocols are currently defined:
  - A character-oriented half-duplex protocol chosen by the value $T = 0$
  - A block-oriented half-duplex protocol identified by the value $T = 1$
  - A block-oriented full-duplex protocol identified by the value $T = 2$; this mode, however, is rarely used
  - The value $T = 14$ indicates the use of proprietary protocols. It is used to support applications that were already planned be-

fore the standard in France and in Germany in the health field, such as the card Sésame-Vitale in France

- The values $T = 3$ to $T = 13$ are reserved for future use.

- ISO/IEC 7816-4 (1995) defines the logical organization of the data stored in the card and the framework for secure access to these data, in particular:
  - Cardholder authentication using a PIN
  - Authentication of an external entity using a secret key; this method is used to authenticate the terminal or the bank
  - Verification of the data integrity using a cryptogram that is often a message authentication code
  - Encryption of the data

- ISO/IEC 7816-4 commands fall into three categories: administrative commands, security commands, and communications management commands. In general, card manufacturers prefer to pick and choose from the list of commands so that most of the cards that are commercially available provide only a subset of the ISO/IEC 7816-4 commands. Additional proprietary commands will be added to the facility file and data management.

- ISO/IEC 7816-5 (1994) defines the procedure by which to register the applications to obtain a worldwide application identifier AID.

- ISO/IEC 7816-6 (1996) defines the interindustry data elements.

The EMV specifications build on the ISO standards to define multiapplication cards for financial transactions. The European Telecommunications Standards Institute (ETSI) also published specifications for multiapplication telephone card.

Finally, the ISO series of standards ISO 10536 defines the physical and mechanical aspects of contactless cards.

13.3.2 EMV (EuroPay, MasterCard, Visa)

The EMV specifications sprang from a collaboration of Europay, MasterCard, and Visa to define the architectures of smart cards capable of accommodating several applications, while keeping conformance with norms derived from ISO/IEC 7816. They were initially published in 1996 and then revised in 1998 and in 2000 (EMV'96, 1996; EMV2000, 2001).

Since February 1999, the maintenance of these specifications has been the responsibility of EMVCo, an organization registered by the state of New York and the U.S. As a consequence, the international use of the EMV specifications is subject to the restrictions on export, particularly for Cuba, North Korea, Iran, Iraq, Libya, the Sudan, and Syria. The list of laboratories that can verify the conformance of the terminals to the EMV specifications is available at http://www.emvco.com.

The latest revision of EMV from December 2000 (Version 4.0) clarifies ambiguities in the penultimate version (Version 3.1.1) or fills the gaps. For example, the location of the contacts and the areas of metal contacts were clarified to ensure compatibility with existing French cards. The first implementations based on EMV took place in January 2002.

13.3.2.1 Properties of Encryption Keys

Each EMV terminal contains the public key of the certification authority of each known application. The module of the keys of the certifying authorities must be less than or equal to $(2^{1984} - 1)$, while the module of the issuers should not exceed that of the corresponding certifying authorities. To speed the encryption, the exponent of the public key is either 3 or $(2^{16} + 1)$ (Menezes et al., 1997, p. 291). The expiration of these keys depends on their length, according to the schedule in Table 13.1.

13.3.2.2 Migration to EMV

MasterCard and Visa plan to complete the transition from magnetic strip cards to EMV smart cards by January 2005, at the latest. Beyond this date, banks that would not have made the transition would be considered liable for frauds recorded on payments made with cards they issued. Visa has also asked banks to upgrade their automated teller machines (ATMs) to accept EMV cards, starting from October 2001. American Express is marketing a smart card conforming to EMV under the name of Blue Card, with three different functions: debit/credit, fidelity, and access control. A simplified version, the Green Card, offers the debit/credit function only.

Migration to EMV requires a switch in the whole operating mode, with respect to processing of payments, and imposes a renewal of the ATMs and merchant terminal as well as the distribution of new cards. This migration

TABLE 13.1

Schedule of the Expiration of the Public EMV Keys of the Certified Authorities

Length of the Keys (bits)	Expiration Date
768 $(2^9 + 2^8)$	December 31, 2002
896 $(2^9 + 2^8 + 2^7)$	December 31, 2004
1024 (2^{10})	December 31, 2008
1152[a] $(2^{10} + 2^7)$	December 31, 2010

[a] Was introduced on December 31, 2002.

is more advanced in England than in France, because the rate of loss due to fraud in the former is five times the rate in the latter (Buliard, 2001).

Before the transition to EMV completes, some provisional means can contribute to the reduction of fraud, for example, examination of the cards with ultraviolet rays in the case of face-to-face commerce. For online sales, the verification of the authenticity of the card can be improved with additional numbers on the back side of the card. Similarly, the authentication of the card holder can be enhanced by comparing the address of the cardholder with that given for the delivery of the items bought.

13.4 Security of Microprocessor Cards

Security of the integrated-circuit card covers the secret data stored in the card and the access rights to the service. The objectives of the security process are: to prevent forgeries at all stages of production; to prevent theft of the firmware for applications and security; to protect the stored information; and to detect and prevent any illegal or abusive use. The protection must be given during the production as well as during the utilization.

13.4.1 Security during Production

The fabrication, personalization, and distribution of microprocessor cards take place in seven steps (Daaboul, 1998; Service Central de la Sécurité des Systèmes d'Information, 1997):

1. Design and development of the integrated circuits
2. Design and development of the card firmware
3. Fabrication of the silicon wafers
4. Insertion of the firmware, packaging of the integrated circuit, and final testing
5. Prepersonalization, with the addition of the programs related to the final use of the card and verifications of their correct operations
6. Personalization of the integrated circuit by adding the names of the issuer organization and of the holder and with the addition of the application software
7. Issuance of the smart card on the plastic support, with embossing, imprinting of the logos, and distribution of the card

Summarized in Table 13.2 are the access conditions to the memory of the smart card during the phases of the life cycle of the card.

TABLE 13.2

Access Conditions to the Memory of a Smart Card throughout Its Life Cycle

Phase in the Life Cycle	Fabrication	Prepersonalization	Personalization	Utilization	Invalidation
Access Mode	Physical addressing		Logical addressing		
Operating System	Not accessible				
Fabrication Data	Read, write, update	Read, but can be blocked			
Directories	Read, write, update			Not accessible in most cards	
Data	Read, write, update				
PIN	Read, write, update				

The production of a smart card requires the intervention of the following participants:

- Designers of the integrated circuits and the security software
- Manufacturers of the integrated circuits and producers of the security software
- Certification authorities
- Developers of the application software
- Card producers (art designers, embossers, printers, etc.)
- Card issuers (which are the entities legally responsible for the card content and for the delivery of each card to its intended user)

The security of a smart card has to take into account all the phases of production and transportation of the card among the various participants. A sufficient level of security must be ensured for each phase of the production cycle before passing into the next phase.

During the design phase, security relates to the protection of documents containing the requirements on the integrated circuit, the design documents, the software, the prepersonalization procedures, and the security of the production environment. The security policy has to establish an inventory of all possible threats as well as available countermeasures during fabrication and transport. Control of the production environment includes the material and tools used, the good and defective products, and the inventories. Among the potential threats are disclosure of the requirements, modification or theft of goods and materials, and modification of the software, including the microcode of the integrated circuit and the operating system.

Protection of the silicon wafers during their production is the work of the manufacturers of integrated circuits. During this phase, the chips are probed individually while they remain in the silicon wafers to verify the functioning of the microprocessor. The chips declared to be valid are blocked using symmetric encryption with a foundry key of 8 to 16 bits; the fabrication lot number as well as the number of the manufacturer are also engraved. The fabrication key is generated from a master key by using a diversification algorithm. Such an algorithm allows the derivation of the fabrication key from the master key using the card serial number. This method allows a central authority to authenticate all the lots without having to stock all the keys. At the end of the phase, the bad ICs are extracted from the wafer, and the verified ICs are delivered to the card manufacturer.

During prepersonalization, the card supplier cuts the wafer to separate the individual chips. These are then tested again before being molded under pressure into the milled plastic card. The logo of the application supplier is also put on the back of the card. Once it is installed, the chip is tested to verify the correct operation. The supplier unlocks the microprocessor using the foundry key, adds the operating system, etches the serial number of the card, and then blocks writing through direct memory access. From this point, any communication with the memory will have to be through logical addressing to protect all the stored data from modification or unauthorized access. Access to the card is then blocked again with the help of a prepersonalization key or transport key that is associated with the issuer, before delivering the integrated circuit to the application supplier.

During personalization of the smart card, the application supplier records files related to the application on the card, in addition to the personal data of the cardholder, such as the cardholder's identity and PIN, as well as the unblocking keys. At the end of this phase, the cards are distributed to their holders. Depending on the commercial offer, the holder may be able to modify some of the personal parameters, such as the PIN. Details of the security during this phase will be presented later in this chapter.

Most smart cards keep a record of unsuccessful access attempts. The counter is reset to 0 when a valid entry is made, unless a threshold is reached. Once the number of attempts exceeds a threshold, access to the card or to a specific file is blocked. Some cards allow the user to choose the thresholds; in others, the threshold is predefined to a number from 3 to 7 (Ugon, 1989).

A card may be invalidated for several reasons:

- The expiration date of the card is reached. In a monoapplication card, the expiration date of the card is the same as that of the application. However, for a multiapplication card, the card's expiration date is that of the master file. When an application expires, the operating system blocks all write and update operations, but read operations remain possible for analysis purposes.
- All available places in the memory zone reserved for the data of new transactions were exhausted. It is still possible to read the card when the correct PIN is entered, but it is impossible to write new data. To avoid service interruption, a new card is necessary.

- The card can be invalidated by the issuing institution following fraudulent usage or a declaration of theft, which leads to a simultaneous blocking of the smart card PIN and of the unblocking key. A partial blockage, for example, blocking a specific PIN, will only affect the applications that use that PIN and can be unlocked by the owner of the application with the related unblocking key.

13.4.2 Physical Security of the Card during Usage

The security architecture of cards used in financial transactions was standardized in ISO 10202 (Parts 1–7), while ANSI X9.17(1985)/ISO 8732 specified the key management. Physically, the rectangular plastic support encloses elements for identification of the issuer of for personalization of the card for the cardholder. The location of some of these elements may be country dependent.

The face of the card includes the following:

- The contacts for the microprocessor of the smart card
- Drawings of the logo of the card issuer, the financial institution, and the bankcard schema operator
- Embossing (in relief) of the card serial number, the name of the cardholder, the date for limit of validity
- A hologram to increase security and make counterfeiting more difficult (The location of this hologram is the same in all countries.)

The reverse of the card contains a place for the cardholder's signature, the return address in case of loss, and, in North America, the list of associated bank networks. The elements of identification and verification of the PIN, the expiration date, as well as codes describing the functionalities allowed to the user are recorded on magnetic tracks on the reverse side.

The card number consists of 10 to 19 characters divided into groups of four digits. The first digit of the first group identifies the network of the bankcard schema (4 for Visa, 5 for MasterCard, etc.). Following are codes that identify the country, the representative of the bankcard schema in that country, and the bankcard. The last four digits form a verification code called "Luhn's key."

In general, a smart card includes tamper-resistant circuits that inhibit output operations when a physical attack is detected. A dielectric layer offers passive protection of the integrated circuits from impurities, dust, and radiation. When this passive layer is violated, the integrated circuit may react to light, temperature, voltage, or frequency differentials.

Physical protection of the memory cells can be employed to prevent a selective erasure or to distribute the storage of sequential words in noncontiguous memory cells. Finally, there are special fuses with which to deactivate the test circuits that are used before the distribution of the cards.

13.4.3 Logical Security of the Card during Usage

Several measures assure logical security during usage. In face-to-face commerce, the identification of the card is done by the merchant, using physical means (identity card of the holder, signature, etc.), or by calling an authorization server or using the holder's PIN, which is the electronic signature for cash withdrawal or for payments. In contrast, online authorization systems rely on cryptographic procedures to authenticate the participants (card holder, card, and merchant terminal). The process consists of two phases: reciprocal authentication of the cardholder and the card; and reciprocal authentication of the card and the network terminal. In case of offline verification, for example, in the South African system for paying electric consumption, the card contains the credit awarded to the cardholder, encrypted with a symmetric algorithm (Anderson and Bezuidenhoudt, 1996).

Let us consider in more detail the protection of online transactions. The first line of defense consists of the authentication of the legitimate user and the card before establishing, if needed, a secure logical channel between the smart card and the host system through the reader. The establishment of this channel requires the reciprocal authentication of the card and the authorization server on the network. Time-stamping of the transactions ensures nonrepudiation, which requires a precise clock with power supply ensured in all conditions. A second series of measures includes recording details of the transaction in an audit file and counting unsuccessful attempts to access the card, with blockage when the number exceeds a predetermined threshold. Finally, there is a period of validity of cards that is limited to reduce the possibility of cloning or attacks by replaying old messages.

The procedures for cryptographic authentication rely on algorithms for symmetric or asymmetric encryption. While most systems use proprietary protocols, the EMV consortium published its own specifications for financial transactions. Also, for mobile telephony, GSM (Groupe Spécial Mobile — Global System for Mobile Communication) employs standardized remote authentication procedures for mobile users. Let us review cryptographic authentication in more detail.

13.4.3.1 Authentication with Symmetric Encryption

The advantage of this mode of authentication is that adding a cryptographic coprocessor can be avoided and, consequently, the cost of the card can be reduced. The authentication exchanges take place in the following fashion (Hamman et al., 2001):

1. After insertion of the card in the slot of the machine, the card reader generates a random number and sends it to the card.

2. The card computes the message authentication code (MAC) of the concatenation of the random number and the card identification number (CID). The derivation of CID depends on the specifications of the system and depends on the chip serial number, the account number, a secret code, and the expiration date. The card sends the MAC and the CID to the authentication server.

3. Using CID, the server derives the card encryption key from the master key. It performs the inverse computations of that of the card and compares the results with the numbers received.

4. The result of the comparison defines the success or failure of the authentication.

The GeldKarte described in Chapter 9 uses this technique for authentication.

13.4.3.2 Authentication with Public-Key Encryption

Authentication with asymmetric encryption can be static or dynamic. In static authentication, the data exchanged are fixed during the fabrication of the card. In contrast, the exchanges during dynamic authentication do not allow an intruder to deduce the secrets of the card.

13.4.3.2.1 Static Authentication

The signature of the card ID computed with the RSA public-key algorithm and stored in the chip identifies the card at each payment. However, this method is vulnerable to cloning by copying the cryptogram of a valid card into a fake card.

13.4.3.2.2 Dynamic Authentication

Shown in Figure 13.3 are the exchanges in dynamic authentication using public-key encryption. The exchanges occur in the following steps:

1. The terminal sends a random number, RAND, to the card.

2. The card concatenates its identification number CID and the random number and calculates the digital signature of the concatenation with its private key SK_C. This signature is sent to the terminal accompanied by the card certificate signed by the corresponding certification authority.

3. To ensure the card authenticity, the terminal decrypts the signature using the card's public key of the card, PK_C, as extracted from the certificate. It compares the results to the hash obtained directly with the hash function $H()$.

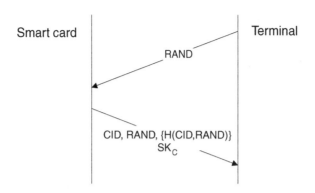

FIGURE 13.3
Dynamic authentication.

This deterministic authentication method usually requires a cryptographic coprocessor to offload the main microprocessor and speed the calculations. This would increase the cost of smart cards. To overcome this constraint, Guillou, Davio, and Quisquater proposed in 1988 a method for probabilistic authentication using zero-knowledge techniques. In this scheme, the authentication is interactive and consists of three exchanges: a commitment by the claimant, a challenge by the verifier, and a response by the verifier. The counterpart to the reduction in the computational load is an increase in the number of exchanges and the presence of a residual error. However, the verifier can reduce the error probability by requesting additional iterations.

The starting point is at the following equation (Guillou et al., 1988):

$$G \times Q^e \equiv 1 \bmod n$$

where G is a public number calculated from the card CID; Q is the signature of CID as computer by a banking authority with the RSA algorithm and its private key; (e, n) constitute the public key of the verifier; and d is the corresponding private key.

The exchanges take place as follows:

1. The card sends to the verifier (the server or the card reader) the commitment t:

$$t = R^e \bmod n$$

   with R a random number between 1 and $(n - 1)$.
2. The verifier responds with a random challenge V between 0 and $(e - 1)$.
3. The card responds with T calculated as follows:

$$T = R \, Q^V \bmod n$$

4. The verifier can now verify the authenticity of the card by reconstituting the commitment as follows:

$$G^V\, T^e \bmod n = G^V\, R^e\, (Q^V)^e \bmod n$$
$$= (G\, Q^e)^V\, R^e \bmod n$$
$$= R^e \bmod n$$

The exchanges are illustrated in Figure 13.4.

It should be noted that any legitimate claimant can finish each iteration with success without revealing the secret code Q. All that the verifier can derive is that the claimant has the necessary credential, without being able to reconstitute its value. It can be shown that an impostor has only one chance in $(e - 1)$ to guess the response. In the case of $e = 2^{16} + 1$, there is one possibility of cheating in 65,536 attempts, which seems to be sufficient in banking applications (Guillou et al., 2001).

13.4.4   Examples of Authentication

13.4.4.1   *Memory Card Reader for the Minitel*

The memory card reader LECAM® (Lecteur de Cartes à Mémoire) was conceived as an adjunct of the Minitel to secure transactions with smart cards. The security algorithm used in the card can be represented in the following form (Coutrot and Pommier, 1989; Ugon, 1989):

$$R = f(E,S,M)$$

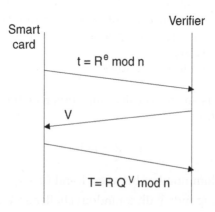

FIGURE 13.4
Interactive authentication with a zero-knowledge proof.

where E = a random number supplied by an external application residing on the host computer of the network that is managing
the transaction;

S = a secret session key previously stored in the card's secret's
zone;

M = the message to be sent;

R = the result calculated with a one-way function $f()$. The function $f()$ depends on the security system as well as the cryptographic algorithms used.

13.4.4.2   Smart Card of French Banks

Until recently, the procedure used for verification in French smart cards of
the BO' mask was static authentication. This authentication consists in verifying that the cubic power modulo n (where n is a 320-bit public key of the
banking authority) of an integer A gives another integer J called the *redundant
identity of the card*. This is written as follows:

$$A^3 \bmod n = J$$

Both A and J are 320 bits long. A is attributed during the personalization
phase of the smart card, while J is derived from the card identity CID
according to

$$J = (1 + 2^{160})\, CID$$

The length of CID is 160 bits, and it contains the following parameters in
the clear: the serial number of the chip, the account number, the name of the
cardholder, the PIN, and the expiration date (Guillou et al., 1988).

The card sends the values of CID and A to the terminal, which then verifies
that A^3 and J conform to the relation indicated above. The protection comes
from the difficulty of getting the cubic root (modulo n) which, according to
number theory, requires factoring n into its prime factors. The complexity
increases with the length of n.

Nevertheless, because each authentication reuses the same values, it is
possible to deduce the value of the public key n, as well as the correspondence between CID and card parameters, by observing the exchanges
between an authentic card, even when it is expired, and a payment terminal.
After succeeding in factoring n and discovering the relationship between the
card parameters and the value CID, one can select any arbitrary 16-digit
value to be the card number as well as other parameters from which the
values CID and J can be computed. The cubic root of J modulo n will yield
the value of A that will be sent to the payment terminal for verification. The
counterfeit will fool a local verifier and will remain unnoticed as long as the
authorization is carried out locally. Yet, French banks use semionline

verification, i.e., the authorization server is called once every 24 hours or when the transaction amount exceeds a surveillance threshold.

In 1998, Serge Humpich was able to penetrate the system defenses and fake a French smart card called the "Yescard," thus demonstrating the possibility of overcoming the static authentication procedure. Following this affair, French banks decided to speed their migration from the BO' mask to a standard derived from the EMV specifications called CB 5.2. The two environments BO' and EMV will coexist until July 1, 2003. After this date, all cards will have to be switched to the EMV environment. Simultaneously, temporary security measures were taken to increase the length of the encryption key to 768 bits.

Information on the Humpich affair is available on the somewhat polemical site run by Laurent Pélé (http://www.parodie.com/monétique).

13.4.4.3   EMV Card

An EMV session includes the following phases:

1. Application selection
2. Authentication of the application data
3. Evaluation of the processing restrictions due to the incompatibilities among the versions of the card and the server, due to geographical restrictions, or due to the limits that the service provider imposes
4. Cardholder authentication
5. Risk management (request for authorization, security verification, audit management, etc.)
6. If needed, online authorization request, by calling the server of the verification center
7. Update of the data on the card

EMV recognizes three authentication modes:

1. Static authentication for offline transactions
2. Dynamic authentication for offline and online transactions
3. Dynamic authentication with production of an application cryptogram (This mode was added in Version 4.0 for both offline and online transactions.)

13.4.4.3.1   Static Authentication

For each application, the card contains the digital signal of critical data computed with the private key of the issuer as well as the certificate of the issuer from the certification authority. The digital signal is cast during the personalization of the card. To ensure a smooth evolution without service disruption, each terminal must be able to recognize the public keys of six different authorities per registered application.

As indicated earlier, this authentication mode cannot protect against counterfeited cards.

13.4.4.3.2   Dynamic Authentication

This authentication follows the classic deterministic scheme. The card receives from the terminal a random number that it will concatenate with its card identity to calculate a MAC that will be returned to the terminal. The card must present to the terminal its certificate from the issuer as well as the latter's certificate signed by the certification authority. The terminal verifies the validity of both certificates and the integrity of the received signature before confirming that the card is authentic. In this way, this mode allows authentication of the card and verification of the legitimacy of the data exchanged while blocking possible counterfeiting.

As in the previous case, each terminal must be capable of recognizing the public key of six certification authorities per registered application.

13.4.4.3.3   Dynamic Authentication with Application Cryptogram Generation

This optional mode was introduced in EMV 2000 and includes two sets of data exchanges. The first is similar to the case of dynamic authentication, as it is related to the production and transmission of the card digital signature computed with its private signature key on the concatenation of its identity and the random number from the terminal. The second is associated with the production card digital signature on the transaction data and its transmission to the terminal.

To illustrate the operation of this mode, consider an offline authentication and an online authentication.

13.4.4.3.3.1   Offline Authentication

Figure 13.5 shows the main exchanges of EMV. Explained in Table 13.3 are the meanings of the symbols used in that figure.

After receiving the READ_RECORD command from the terminal, the card returns the identifier of the certification authority CA, the public-key certificates of the issuer and of the integrated-circuit card, the account number associated with the card, and the PIN stored in the card. The card response indicates whether the cardholder's verification will be conducted online, which will necessitate the bank's intervention, or offline, thereby using the terminal alone. Both certificates are needed for the terminal to authenticate the card's public key in CERT_C. To execute the challenge-response authentication, the terminal issues an INTERNAL_AUTHENTICATE command with the authentication-related data ARD. These data include the date and time of the transaction, the card identifier, the associated account number (PAN, primary account number), the terminal identifier, and a nonce1 to

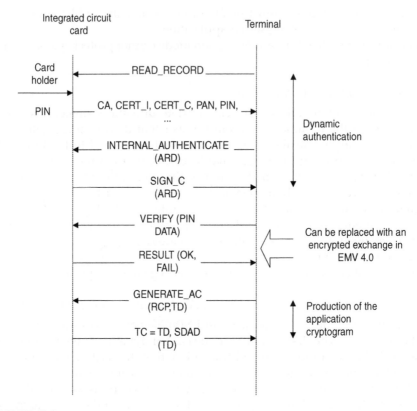

FIGURE 13.5
Exchanges in an EMV transaction with dynamic authentication and generation of the application cryptogram (offline case).

avoid replay attacks. The card responds with a signature over these data using its private signature key.

These exchanges correspond to the dynamic authentication. Next, the terminal issues the VERIFY command to verify the cardholder's PIN. The card's response indicates success or failure after comparing the code received from the terminal and the code stored in the card. In principle, the exchanges are not encrypted or authenticated, because EMV is for face-to-face commerce only.

The GENERATE_AC command includes the reference control parameter (RCP) and the transaction data (TD). The value of RCP in this case indicates that the transaction takes place without the intervention of the bank (offline). According to the EMV 4.0 specification, the TD must include a random number. The card responds with the transaction certificate (TC), which contains the TD and the application cryptogram computed using the card's private key for signature. The signature is calculated over all the data exchanges, including the random number received from the terminal and

TABLE 13.3

Symbols Used in EMV Exchanges

Variable	Name	Comments
ARD	Authentication related data	Includes date, time, card identifier, PAN, terminal identifier, nonce1
ARQC	Authorization request cryptogram	
CA	Identifier of the certification authority	
CERT_C	Public key certificate of the integrated circuit card	As certified by the issuer bank
CERT_I	Issuer public key certificate	As certified by the certification authority
IAD	Issuer authentication data	
PAN	Private account number	User's account inscribed on the card
RCP	Reference control parameter	Includes user preferences and whether the transaction is online or offline
SIGN_C(ARD)	Signature of the card on ARD	Uses the card's private key
TC	Transaction certificate	
TD	Transaction data	Amount, currency, date, time, PAN, terminal identifier, nonce2

the command TC. This command TC indicates that the card is allowing the transaction to proceed and will be forwarded later to the bank for auditing purposes.

For offline authentication, EMV Version 4.0 (2002) allows the possibility of checking the integrity of the PIN that the terminal received by encryption with a pair of keys reserved for that purpose. With the command GET CHALLENGE, the terminal requests that the card send a random number RAND1 that it will concatenate to the PIN and another random number RAND2 that the terminal generates. Then, the whole will be encrypted with the card public key PK_C and sent in the VERIFY command. The card will then decode the cipher with its secret key SK_C to verify that the values of the PIN and the random number RAND1 are the same that it already sent. These exchanges are illustrated in Figure 13.6.

13.4.4.3.3.2 Online Authentication

Figure 13.7 depicts the exchanges during an online dynamic authentication with generation of the application cryptogram.

The behavior is the same as in the preceding case, until the terminal sends the command GENERATE_AC with the parameter RCP indicating an online transaction. To respond, the card produces an authorization request cryptogram (ARQC) that the terminal passes to the bank in online authorization requests. The ARQC contains the MAC of the TD that the card computes using the session key that it shares with the acquirer bank

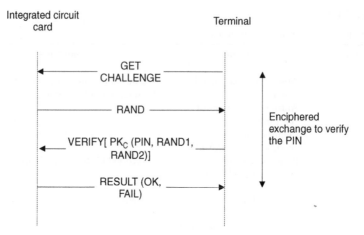

FIGURE 13.6
Exchanges to verify the PIN in EMV 4.0 (offline case).

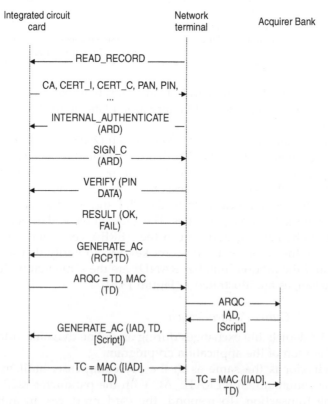

FIGURE 13.7
The EMV exchanges for card authentication and cardholder verification (optional fields are enclosed in square brackets [ ]). (Originally published in Van Herreweghen, E. and Wille, U., *Proc. USENIX Workshop Smartcard Technol.*, (Smartcard '99). ©1999 USENIX Association.)

and that is derived from a shared master key. Thus, the bank can verify the integrity of the request and authenticate its origin (because of symmetric encryption). If the verification is successful, the bank returns an authorization response message that contains issuer authentication data (IAD). It may also contain an optional command script to be delivered to the card. The terminal generates a second GENERATE_AC command that contains the IAD and the script. The card responds with a message that contains the transaction certificate TC and its MAC to indicate approval of the transaction. The bank can later verify the integrity of the TC and the authenticity of its origin. Note, however, that the terminal does not verify the response, because it does not have the session key that is shared between the card and the bank.

13.4.5   Evaluation

The EMV specifications were developed for point-of-sale applications, where the EMV terminal is under the control of the merchant or the issuer bank. In these situations, the transaction is face-to-face, and the terminal and the bank communicate over a secure channel. This has the following implications for the protocol:

- Because the purchase is face-to-face and the goods are delivered to the purchaser immediately, the protocol does not address merchant authentication. Also, it does not include a description of the goods or guarantee delivery of the goods before debiting the customer's account. Given that the physical presence of the card can be visually ascertained, the protocol does not link different parts of the transaction to verify that the same card is used throughout. There is no provision to authenticate the terminal to the card explicitly, and the terminal does not explicitly authorize or sign any part of the transaction. Thus, if the purchase is offline, in which case the bank is not involved, the purchaser does not have formal proof of having conducted the transaction.

- Because a secure channel is assumed between the card and the terminal, no tools are provided to secure the messages they exchange. Once it authenticates the card, the terminal does not check the computations that the card performs. Consequently, and in the offline case, the merchant is not able to verify the transaction data.

- Because the merchant and the bank trust each other, the merchant terminal does not verify the data that it receives from the bank, and the bank trusts the terminal to deliver the messages to the card.

Based on the above considerations, the EMV specifications are not suitable for remote transactions over nonsecure networks, such as the Internet (Van Herreweghen and Wille, 1999).

13.5 Multiapplication Smart Cards

Multiapplication smart cards make it possible to have several applications on the same card, provided that the card's resources can be shared without compromising security. The definition of standardized interfaces helps to facilitate the development of these applications independently as well as porting them from one card to another, even if they are made by different suppliers. The ISO/IEC 7816-4 (1995) standard is the preferred starting point for achieving that goal, not only in the case of open specifications such as EMV and Java Card, but also for proprietary solutions. This section begins with a presentation of the file system that ISO/IEC 7816-4 defined.

13.5.1   File System of ISO/IEC 7816-4

The file system defined in ISO/IEC 7816-4 supports two categories of files: dedicated files (DF) and elementary files (EF). Each file has an identifier coded on 2 octets in hexadecimal notation. The way the files are arranged is illustrated in Figure 13.8.

The Master File (MF) is at the root of the tree and is always identified by the file identifier 3F 00. The file identifier of the first DF is 01 00, and the last is 3E 00. Thus, a card cannot contain more than 62 dedicated files in addition to the MF. Each DF is associated with a given application and may contain one or more elementary files. Application selection may be through the SELECT FILE command with the AID as an argument or indirectly with the help of the special elementary DIR (Directory) or ATR (Answer to Reset).

The EFs contain the data. Each EF is identified by its position in the tree, i.e., by the path leading back to the master file. The identifier is also coded on 2 octets and takes the form $xx\ yy$, where xx is the identifier of the DF to which the EF belongs, and yy is a sequential number of the EF in that particular directory. Thus, xx is 3F if the file depends directly on the master file and the number of elementary files in a directory cannot exceed 63.

The elementary files 2F 00 and 2F 01 under the MF have special indexing functions. The first is called DIR and the other is ATR (see §9.4 of ISO/IEC 7816-4). Roughly speaking, the file DIR contains elements that allow the identification of the applications, while the file ATR specifies how the card can find the applications or the various objects.

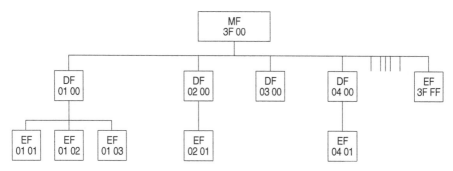

FIGURE 13.8
Tree structure of ISO/IEC 7816-4 file system.

The maximum number of elementary files in a card is thus, $63 \times 63 = 3969$ files. Clearly, this structure is rigid and does not suit dynamic situations where files can be added or deleted corresponding to the addition or removal of applications. In fact, ISO/IEC 7816-4 does not allow the creation of new files; therefore, the various suppliers of integrated-circuit cards had to define proprietary commands for file management.

The ISO/IEC 7816-4 distinguishes between two types of elementary files: internal EFs and working EFs. The latter contain data for the exclusive use of entities external to the card. The internal EF contains data that the card uses during its operation. For example, in a monetary application, the following files can be present:

- Key files for the storage of the keys that will be used to derive a session key as specified by the payment protocol employed — Given the sensitivity of banking transactions, the applications that use purses will probably need several keys, one for each action, such as for certification, for debit, for credit, or for electronic signature. Each key will be associated with an individual file.

- PIN files to stock the PINs that control access to the application file. The application files and the access conditions are irrevocably defined during the personalization phase.

- Purse files — For each purse, the file indicates the maximum balance, the maximum payment for each transaction, the current balance, and a backup balance to recover the previous value in case of a failure.

- Certificate files, in the case of public-key encryption.

- Application usage files.

13.5.2 The Swedish Electronic Identity Card

The example of the Swedish electronic identity card (EID) can illustrate the preceding presentation. This card is defined by the Swedish Standard SS 61 43 30 on the basis of the work done by SEIS (Secured Electronic Information in Society). The SEIS is a Swedish nonprofit organization of about 50 firms and organizations in the financial, industrial, and public administrative sectors in Sweden, with the aim of taking advantage of new networking technologies.

In this application, the master file contains the following elementary files (SEIS, 1998):

- The file EF_{PAN} defines the account embossed on the card PAN.
- The file EF_{PIN} defines the master PIN that all applications on the card may use.
- The DIR file (2F 00), in conformance with the specifications of ISO/IEC 7816-4, holds the application identifier for the electronic identity card.
- The dedicated file of the application includes three elementary files:
  - The application usage file (AUF) EFAUF
  - The file for the private RSA key EFPrK
  - The certificate file EFCERT

Figure 13.9 shows the logical organization of the files for the Swedish electronic identity card (SEIS, 1998, Annex B).

13.5.3 Management of Applications in Multiapplication Cards

There are three possible cases depending on the type of relationship among the applications that coexist on the same multiapplication card:

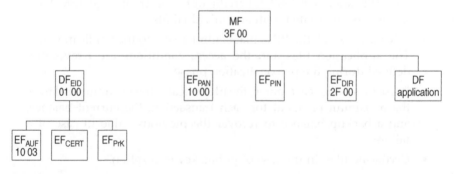

FIGURE 13.9
Logical organization of the Swedish Electronic Identity Card.

- A primary application may control other secondary applications.
- Several applications may be federated under the control of a central authority.
- All installed applications are independent.

13.5.3.1   Secondary Applications Controlled by the Primary Application

This situation requires perfect coordination among the providers of the various applications to share harmoniously the resources of the card. In fact, the current operating systems for smart cards are not multitasked and are closer to the systems used by the computers of the 1970s.

In the distribution and organization of the data files, the secondary applications will be considered logical subsets of the primary application, as shown in Figure 13.10. The index file (DIR or ATR) that points to the dedicated files for the secondary applications is defined during the personalization of the card. Information sharing is solely between each secondary application and the primary application. Security relies on the primary application authenticating the secondary applications.

Several telephone companies in Germany and the Netherlands already offer cards that are capable of managing several applications. Their commercial success, however, is uncertain, given the lack of standardization.

13.5.3.2   Federation of Several Applications under a Central Authority

In this case, the applications share common data, for example, the personal data of the cardholder. This sharing is done under the auspices of a central authority that controls the master file, as indicated in the diagram of Figure 13.11. This authority is usually the card supplier.

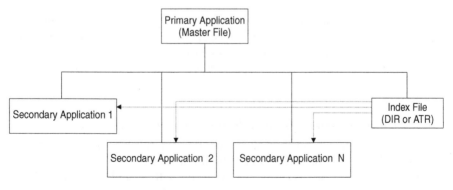

FIGURE 13.10

Logical representation of the configuration of the dedicated files for a primary application controlling secondary applications.

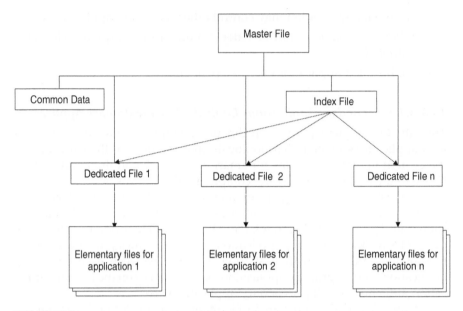

FIGURE 13.11
Several applications under the control of a central authority.

The central authority attributes a unique application identifier to each application and stores it in the index file. This file points to the dedicated file for each application. Any application has the right to read this file, but the write permissions are reserved to the central authority.

Before reactivating itself, an application has to authenticate itself to a common security module. Similarly, when an application A requests permission to access the data of another application B, the common security module has to authenticate it. This module next sends the authentication request to the application to determine whether application A owns the access privileges.

In this configuration, mechanisms have to be provided to allow multiple applications to share digital credentials, such as keys and certificates, effectively. Some of the important initiatives in this area are the format defined in PKCS #15, but many other aspects have to be standardized (Nyström, 1999).

13.5.3.3   *Independent Multiapplications on the Same Card*

In this case, there are no central authorities. The application identifiers are assigned in the chronological order of their addition and are maintained in an index file, whose access is protected, typically with a public-key algorithm.

Each application supplier creates new applications according to the specifications of the card API, then distributes its product to the user directly or indirectly through an intermediary. One of the currently popular methods for managing the various applications is shown in Figure 13.12. It consists

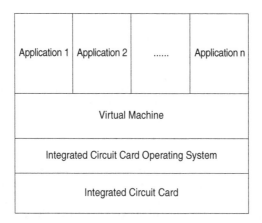

FIGURE 13.12
Management of applications with a virtual machine.

of superimposing a virtual machine on the card's operating system. This virtual machine is responsible for the control of interactions among the applications as well as the communications of each one with the smart card.

The mechanism is the foundation of the JavaCard architecture, used by Visa and American Express (in its Blue Cash offer). The JavaCard specifications simplify the Java virtual machine to define a subset of the Java language suitable for integrated-circuit cards. This feature facilitates the development of portable applications independent of the chip or of the underlying operating system (Sun Microsystems, 1999). Exchanges with the peripheral equipment conform to ISO/IEC 7816. However, the secure sharing of objects and resources among the applications of the JavaCard remains under study (Montgomery and Krishna, 1999; Oestreicher and Krishna, 1999).

13.6 Integration of Smart Cards with Computer Systems

Support of integrated-circuit cards on computer systems faces several levels of fragmentation, particularly the operating systems of cards and computers, the peripherals (card readers, setup boxes, etc.), communication protocols, data formats, etc. To face this proliferation, the solution generally considered is to offer the applications a single interface despite the underlining heterogeneity. The Java-oriented OpenCard Framework (OCF) and the PC/SC toolbox for Windows are the principle examples. Note also that there are experimental systems that adapt UNIX to integrated-circuit cards (Itoi et al., 1999) and that MUSCLE (Movement for the Use of Smart Cards in a Linux Environment) works on transposing the PC/SC to a Linux environment.

13.6.1 OpenCard Framework

OCF Version 1.2 (2000), comes from the OpenCard Consortium, which includes IBM and Sun Microsystems, as well as others. The activities of this consortium are devoted exclusively to the Java language to establish a common programming interface for all Web applications independent of the access peripherals or the applications residing in the card (Hermann et al., 1998). The framework of the interactions as established by the OpenCard specifications is depicted in Figure 13.13.

The OCF services shield the Java applications from the specifics of each smart card or access terminal (ATM terminal, computer, card reader, etc.) through the abstract services CardService for cards and CardTerminal for peripherals. The OCF is capable of handling several simultaneous requests for access, which is useful for multiple slot readers. Thus, the OCF supports authentication systems that rely on reading several cards at once. These systems are encountered in some financial or health-care applications, where a card is needed to open the access of another card that belongs to a customer or a patient (Kaiserswerth, no date).

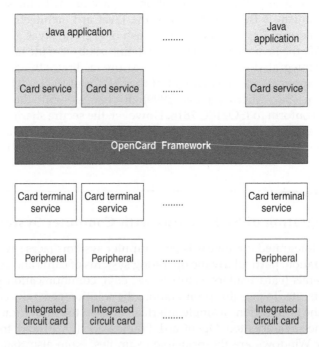

FIGURE 13.13
Framework of operation of OpenCard.

13.6.2   PC/SC

The PC/SC is shorthand notation for the "Interoperability Specification for Integrated Circuit Cards and Personal Computer Systems." This specification was produced by a work group formed by leading integrated-circuit cards and personal computer vendors such as Bull CP8, GemPlus, Schlumberger, Hewlett-Packard, IBM, Sun Microsystems, and Microsoft. It focuses on the Windows operating system and supports multiple languages (PC/SC, 1997). Microsoft also published its own cryptographic library, CryptoAPI, which can be used in the Windows environment to access the cryptographic functions and the certificates stored in a smart card. CryptoAPI also provides secure messaging and certificate management. However, this effort is hampered by U.S. export restrictions on cryptographic software. Figure 13.14 shows the architecture of PC/SC.

The resource manager of the integrated-circuit card is the hub of PC/SC exchanges; it controls the device drivers for the peripherals as well as all the exchanges with the applications through service providers. These maintain

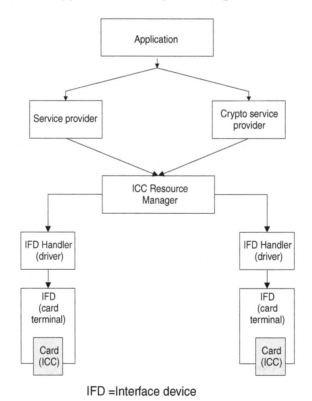

IFD =Interface device

FIGURE 13.14
Framework of operation of PC/SC.

the various card registers and follow the communication protocols. A specific provider, the cryptographic service provider, takes care of all the functions of authentication and security of the exchanges.

13.7 Limits on Security

Fraud comes mostly from the use of inauthentic cards or the theft of card numbers and PINs. In general, integrated-circuit cards offer greater security than magnetic strip cards. These have a magnetic strip on which is recorded the holder's identification data and data to verify the PIN. In face-to-face transactions, the card holder signs an agreement by writing next to a card imprint using a hand unit. One possible source of fraud is the use of a false imprint with a fake signature, in cases where the authorization center is not called directly at the time of purchase. A false card can be made by recording on the magnetic strip the data from a valid card (collected, for example, through a doctored terminal). The corresponding PIN can then be stolen by spying the user while the code is entered, either through direct observation or with a hidden camera.

In the Internet era, new methods were invented. A common swindle consists in producing card numbers using a generator program and retaining only those numbers that are attributed to cardholders. To do so, it is sufficient to attempt a subscription of one of the online services using a given number; if the subscription is accepted, this indicates that the number is valid. The impostor can now make purchases with this card number until the real cardholder discovers the theft by reviewing the statement and contesting the transactions.

Although these attacks are more difficult with smart cards, a number of physical and logical attacks on the security of smart cards are recorded in the literature (Anderson and Kuhn, 1996; Kömmerling and Kuhn, 1999). These attacks were made by amateurs, technical experts, and organizations specialized in reverse engineering. There are four main categories of attacks: logical (noninvasive) attacks, physical (destructive) attacks, attacks exploiting the developers' negligence, and attacks against the exchanges between cards and card readers.

13.7.1 Logical (Noninvasive) Attacks

Noninvasive attacks are active or passive. Active attacks modify the environmental conditions to perturb the functioning of the integrated circuits to compromise the hidden information. Write operations into the EEPROM memory can be affected by changing the ambient temperature, by imposing an instantaneous surcharge on the power supply, or by applying shorter clock pulses to perturb the operation of the microcontroller. Given that the encryption keys and the security software are stored in this memory, security is vulnerable to this type of attack, called glitch attacks, because they can prevent the execution of verification instructions. Take, for example, the following (Anderson and Kuhn, 1996):

- The functioning of the random number generation can be perturbed to furnish a fixed number if the voltage is sufficiently decreased.
- For PIC16C84 microcontrollers, the security bit can be set to zero with erasure of the memory by augmenting the power supply from Vcc to Vpp-0.5 volts.
- For DS5000 controllers produced by Dallas Semiconductor, a brief reduction in the power supply can disable the security mechanism without erasing the memory content.

Some secured processors are so sensitive to modifications in their environments that they declare many false alarms.

In contrast, passive attacks concentrate on eavesdropping and observing the operation of the card to detect variations in the current supply or leakages in radiation. This is because each instruction has a specific signature, which allows the distinction, for example, of branching instructions or operations that involve a coprocessor. Among the passive methods is differential power analysis (DPA), which is based on the principle that power consumption depends on the bits involved (Messerges et al., 1999).

13.7.2 Physical (Destructive) Attacks

Destructive techniques start with the extraction of the integrated circuit from the plastic support. First, a cut is made in the plastic around the chip module, until the epoxy resin becomes visible. This resin is then treated with fuming nitric acid ($HNO_3 > 98\%$) and washed in acetone until the silicon surface is fully exposed. Once the chip is uncovered, it is possible to probe the behavior of the various components and recover the cryptographic keys embedded in the cards by trial and error. The use of laser probes or focused ion beams allows the exploration of the states of the microcontroller to extract the necessary information. However, the cost of carrying out this type of attack is relatively high, which puts these attacks outside the realm of hobbyists or typical hackers.

13.7.3 Attacks due to Negligence in the Implementation

As explained previously, symmetric encryption is used in smart card protocols for verification and authentication. Yet, although many card manufacturers indicate their inclusion of the DES algorithm in the CBC mode in their cards, a recent study underlined several deficiencies in the implementation of this algorithm (Itoi and Honeyman, 1999):

- In the Cryptoflex card from Schlumberger, the DES algorithm is available by special order only.
- In the Multiflex card from the same supplier, the algorithm is available for the internal authentication command only and returns 48 bits only instead of 64 to be used as an encryption key.
- In the multifunction card (MFC) from IBM that is used for multiple applications, it is not possible to define the key directly.
- In the GPK card from Gemplus, the size of the key is limited to 40 bits to satisfy the French legislation on cryptography at the time of its design.
- In certain cases, the internal test circuits of the chip that should have been deactivated before distribution in the marketplace are reactivated. This allows potential attackers to access the card's operational circuits from a limited number of probe points, which considerably reduces the number of combinations to be examined (Kömmerling and Kuhn, 1999).

13.7.4 Attacks against the Chip-Reader Communication Channel

The focus with this category of attacks is on the link between the integrated circuit and the reader. To illustrate the discussion, let us return to user authentication through biometric identification that was presented in Chapter 3. In these systems, the processing of the biometric signal involves four stages:

1. The acquisition, with the aid of sensors, of the biometric imprint of the individual to be identified
2. The extraction of the distinguishing features
3. The comparison of the features with the template stored in the card
4. The evaluation of the similarity between the two using a similarity score and a confidence value

The decision depends on the degree of similarity evaluated in light of the security risks and the risk policy. It is transmitted next to the payment applications within the card before proceeding with the transaction. Given that the operations for data acquisition, feature extraction, comparison, and decision can be executed by the integrated circuit or by a processor within the reader, several combinations are possible, as shown in Table 13.4 (Hachez et al., 2000; Rila and Mitchell, 2002).

TABLE 13.4

Possible Organizations of the Acquisition, Feature Extraction, Comparison, and Decision Functions

Configuration	Acquisition of the Imprint	Feature Extraction	Comparison and Decision
I	Reader	Reader	Reader
II	Integrated circuit of the card	Reader	Reader
III	Reader	Reader	Integrated circuit of the card
IV	Integrated circuit of the card	Reader	Integrated circuit of the card
V	Integrated circuit of the card	Integrated circuit of the card	Integrated circuit of the card

In configuration V of Table 13.4, all functions reside in the same integrated circuit, which offers the best protection against any attack on the exchanges at the price of larger complexity and cost. In contrast, the risks of vandalism or falsification increase as the reader performs additional functions. The exchanges associated with the other configurations are depicted in Figure 13.15. It is seen that the complexity of the card in Configuration I is the lowest, contrary to Configuration IV, where the card carries out the functions of data collection, comparison, and decision making. We also notice that there are three exchanges in Configuration II and two in Configuration III, which makes the latter less vulnerable to attacks on the communication link.

To protect the exchanges shown above from being tapped or modified, cryptographic means must be provided to assure their confidentiality and integrity. Replay attacks will be foiled by inclusion of random numbers, time-stamping, sequence numbers, etc. in the protected messages (Hachez et al., 2000). Physical protection of the reader should be provided to prevent manipulations or its replacement by a doctored unit. To abort a transaction if a false reader or a false card are detected, thereby preventing revelation of the system's secrets, any exchange of values must be preceded by a reciprocal authentication phase between the card and the card reader

13.8 Summary

Compared to cards with bar codes or with magnetic strips, smart (or integrated circuit) cards have computational capacities that allow them to make complex decisions. The applications of smart cards are primarily in banking, mobile telephony, and pay TV; however, their use as badges for access control is increasing. Smart cards can also be used in fidelity applications to record qualifying purchases and available rewards.

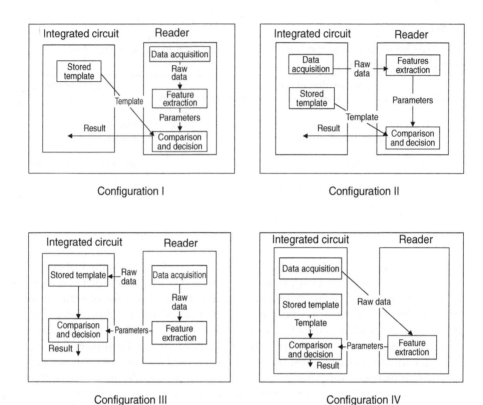

Configuration I   Configuration II

Configuration III   Configuration IV

FIGURE 13.15
Distribution of the operations of data acquisition, feature extraction, comparison, and decision between the integrated circuit of the card and the card reader.

There are several categories of smart cards: rechargeable or not, with or without contacts with the reader, and single application or multiple applications. The security of smart cards covers the whole life cycle. Protection measures are aimed at the data recorded in the chip as well as at the exchanges in which it participates. Measures of logic security include authentication of the holder, of the card, and of the card reader and securing all the communication channels involving the triplet smart cards, card readers, and host system. Authentication can use symmetric or asymmetric cryptography and can be online or offline. In banking applications, most systems are proprietary; however, migration to the EMV specification and the generalized use of conforming cards should radically change the picture.

The EMV uses three modes for authentication: static, dynamic, and dynamic with generation of the application cryptogram. The sharing of resources in multiapplication cards among the applications they support must be strictly controlled to maintain the suitable security level in each case. The integration of integrated-circuit cards with computers depends on the operating system used in the computer system: OpenCard for Java and

PC/SC for Windows. There are limits to the security of smart cards at the physical or logical level and also due to sloppiness in the implementation. Finally, identity control using biometric characteristics stored in a template within the smart card requires protection of the communication channel between the chip and the reader.

Questions

1. What are the main constraints on the security of smart cards?
2. Compare the OpenCard Framework with PC/SC.
3. What are the main points of vulnerability of biometric authentication with templates in a smart card?
4. Summarize the main limitations of ISO/IEC 7816-4.
5. What is a major drawback of static authentication?
6. What is a major advantage of dynamic authentication over static authentication?
7. Compare and contrast deterministic dynamic authentication and probabilistic dynamic authentication.
8. List some of the advantages and disadvantages of contactless smart cards.

14

Systems of Electronic Commerce

ABSTRACT

In this chapter, several representative systems used to improve the informatics infrastructure of electronic commerce (e-commerce) are discussed. We begin with the European programs SEMPER (Secure Electronic Marketplace for Europe) and CAFE (Conditional Access for Europe), as well as the common project of the World Wide Web Consortium (W3C) and CommerceNet, JEPI (Joint Electronic Payment Initiative). We present next real-life applications: the platforms PICS™ (Platform for Internet Content Selection), P3P (Platform for Privacy Preference), and the analysis of user profiles and fidelity cards. Finally, we give some configurations to guarantee the quality of the offered services.

14.1 SEMPER

SEMPER was a project sponsored by the Directorate General XIII of the European Commission between 1995 and 1998 as part of the ACTS (Advanced Communication Technologies and Services) project. It was conducted by a consortium of about 20 companies, financial institutions, and European universities, with project management by IBM France and the technical leadership of IBM research laboratories in Zurich (Waidner, 1996; Asokan et al., 1997; Lacoste, 1997; Abad Peiro et al., 1998; Lacoste et al., 2000).

SEMPER is a security platform with programmable modules that operates in a coherent framework. It takes into account all aspects of e-commerce over open and insecure networks, be they legal, commercial, social, or technical, with the goal of establishing a platform common to all commercial offers. The "openness" of the architecture in this context not only means that it is nonproprietary, but also that it can evolve smoothly with the addition of new components in the form of plug-ins. From the operator's viewpoint, this means independence from the network architecture and other informatics properties (choice of operating system or application software) and from the peripheral equipment (for example, smart card or computer). From the user's viewpoint, the operating principle is to adapt to the selected means of payment by hiding the particularities of the various methods of electronic payment through specific applications programming interfaces (APIs), thereby giving the end user the ability to make the desired payment in a transparent manner. Finally, a *deal* in SEMPER considers all aspects of a transaction, including quality of service and security.

14.1.1   SEMPER Architecture

The designers of SEMPER selected an architecture in four layers, as follows (and as shown in Figure 14.1):

- The commerce layer handles the details of the business sessions, such as ordering items, payment instructions, signatures, etc. It integrates different payment systems, maintains the status of a deal, and records its progress with the aim of maintaining its coherence. For example, it verifies that the price in an invoice matches that in the offer; otherwise, it requests a new invoice. The protocols used in this layer depend on the type of business exchange and the means of payment.

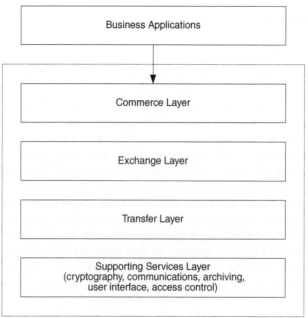

FIGURE 14.1
SEMPER architecture.

- The exchange layer is responsible for establishing dialogues among the various entities: the debtor, the creditor, the issuing and acquiring institutions, and, if needed, the arbiter, the notary, or the trusted third party. The underpinning of SEMPER is fair exchange. Each party must be ensured that the transaction is reciprocal, i.e., that each party will receive something in return for a value that is sent, as specified in the agreement reached. The exchange and transfer layers work together to ensure that no payment is without transfer and no transfer is without payment.

- The transfer layer uses containers to transmit and receive information (signed documents or payments) originating from, or destined to, higher layers on the basis of associated security attributes. This layer consists of three distinct blocks for the functions of certification, verification of the statements (such as signature), and payment. These blocks are called, respectively, the certification block, statement block, and payment block (Figure 14.2). Separation of the various roles permits smooth adaptation of the architecture to the chosen method of payment, without disturbing the whole structure. The transfer manager is in charge of the syntactic analysis of the message arriving from the exchange layer before directing it to the corresponding block. In the reverse direction, it fills the message headers before sending them to the appropriate network (the Internet or the bank network). The certification block verifies and authenticates the parameters of the client or of the financial institution. The SEMPER distinguishes among three types of certificates:

  - Identification certificates to establish a link between a physical entity and the public key contained in the certificate

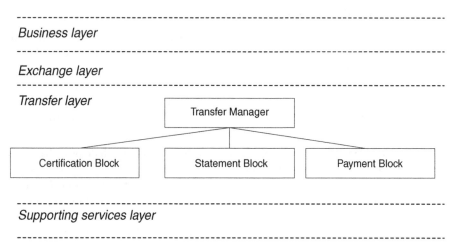

FIGURE 14.2
The transfer layer of SEMPER.

- Authorization certificates to associate a given attribute to a physical entity

- Hybrid certificates that combine the characteristics of the two preceding certificates

- The certification authority for the SEMPER project was the Society for Mathematics and Data Processing GMD (Gesellschaft für Mathematik und Datenverarbeitung) in Darmstadt. The payment block establishes the correspondence between the payment messages on the Internet side and those on the banking network side.

- Finally, the supporting services layer offers to all the other layers cryptographic services (key generation, encryption, signature, hashing, handling of certificates, etc.), communication services, bookkeeping (archival) services, access control, time-stamping, etc. It also offers a trustworthy user interface leading to other services in a uniform manner. The communication secures the exchanges, including anonymous communication with normal or pseudonymous addresses. The archival services are for the exchanged messages, the certificates, and the pairs of encryption keys for public-key algorithms. They allow the verification of the correct functioning of the whole system, for example, detection of double spending of digital money, facilitation of recovery following a crash, or the meeting of legal requirements. Access control services control the access to the resources. Finally, time-stamping is a tool for nonrepudiation services or for retransmitting delayed or lost messages.

14.1.2 Payment Terminology in SEMPER

Not only are the terms used in the architecture specific to SEMPER, but SEMPER divides the payment instruments into two categories (Abad Peiro et al., 1996, 1998):

- Cash like instruments, such as integrated-circuit cards storing values in legal monies (electronic purse systems like Chipper or payments with digital money, like Ecash or DigiCash)

- Account-based systems, such as virtual checks or electronic fund transfers; according to the relationship between the time the debtor makes the payment and the time the value is actually taken from the payer's account, there are pay-now systems (in which case, the debtor is instantaneously debited) and pay-later systems (in which case, the merchant's account is credited before the buyer's account is debited)

Debit cards are usually of the pay-now type, unless the debit is deferred, but credit cards are of the pay-later type. SEMPER architects call direct debits *indirect pull* and credit transfers *indirect push*.

The payment gateway forms the junction between the existing banking systems and the new instruments used over the Internet, for example, SET (Secure Electronic Transaction), Chipper, or the digital money systems by DigiCash (eCash) (SEMPER Consortium, 1996). This is a generic gateway with a function to hide the details of the payment system from the banking circuits. Therefore, in the case of credit payments, the interface of the gateway to the acquiring bank has to conform to ISO 8583 (1993), which is the international standard that describes the exchanges between the issuer and the acquirer banks. Similarly, in the case of credit transfers, the gateway uses the EDIFACT messages of ISO 9735 (1988).

In SEMPER, a purse is the end point of a value transfer that a payment instrument initiates. Thus, a purse is an abstraction of an instance of a payment system that is available to the user. Accordingly, each bankcard that the payer owns is a purse for online payments, if it follows the procedures outlined in one of the protocols for electronic payments, such as SET or EMV (EuroPay, MasterCard, Visa). SEMPER also recognizes the electronic purses that conform to Chipper, eCash, or any other purse specifications. The same purse can be involved in several simultaneous business transactions under the control of a payment manager.

14.1.3   The Payment Manager

The payment manager has two functions: to administer the selection of the purse that fits a given transaction and to keep track of available purses and of the information related to active and past transactions. In particular, the payment manager gathers the information needed to recover following a crash.

Illustrated in Figure 14.3 are the components of the payment manager and the internal and external interfaces. Each purse communicates with the instance of the selected payment systems through specialized adapters. Each adapter provides the junction between the payment model and the external instance of the payment system (Abad Piero et al., 1996).

14.2   CAFE

CAFE was one of the programs of the European research project ESPRIT (Boly et al., 1994; O'Mahony et al., 1997, pp. 158–168). The consortium that undertook the research comprised market analysis groups (Cardware, Institut für

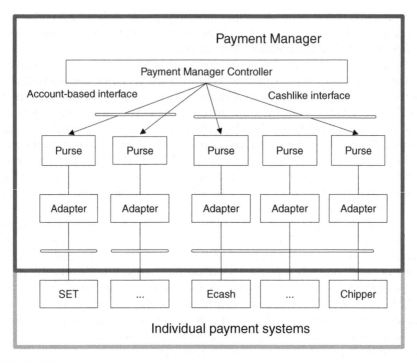

FIGURE 14.3
Organization of the SEMPER payment manager.

Sozialforschung), suppliers of banking software and terminals (Gemplus, Ingenico, Siemens), and designers of cryptographic software (DigiCash, CWI Amsterdam, and the Dutch telephone company KPN).

The objective of the CAFE program was to develop a multiapplication integrated-circuit card to serve as the electronic equivalent of a traditional wallet. This electronic wallet would contain an electronic purse for face-to-face commerce, such as making payments for public transportation and telephone usage. In addition, it would include personal documents, such as a driver's license, a passport, or a health card. As shown in Figure 14.4, the electronic wallet takes the form of a hand calculator, with a screen and a keyboard, and communicates by infrared with other similar wallets as well as with card readers and recharging terminals. The wallet contains an electronic chip with which to perform the necessary cryptographic operations to secure the transactions. Gemplus designed the first wallets, and the cryptographic chips were implemented in encryption processors from Siemens.

The CAFE payment architecture is that of a system of prepayments with offline authorization, which makes it particularly well adapted to payment by the public of the fares for transportation and for telephone calls. An important restriction is that the payments are made with fixed-denomination coins, which creates the problem of change.

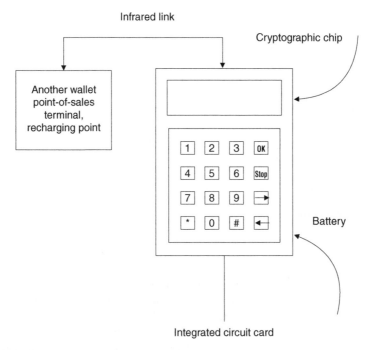

Infrared link

Cryptographic chip

Another wallet point-of-sales terminal, recharging point

Battery

Integrated circuit card

FIGURE 14.4
Presentation of the CAFE electronic wallet.

The electronic money is supplied online using specialized distribution points. Two entities can also transfer money from one purse to another directly, without the intervention of a third party. The issuer of the money keeps a replica of the electronic money issued to the user. Thus, if the user loses the usage of the purse, or if the purse is broken into or stolen, the replica can be used to show how much money was spent and how much value remains available. However, the issuer cannot credit the holder for the money that would have been used after the loss (or theft) of the purse.

The first line of defense is based on the resistance of the wallet to physical attacks. The parameters of the encryption algorithm must reside in tamper-resistant devices. The logical security of CAFE payments is based on the algorithms for digital money presented in Chapter 11. Blind signature is used to make the payer nontraceable. Because the identities of the various actors are hidden, the authentication of the intervening parties can only be achieved with the help of a trusted third party whose role is also to prevent the repeated use of the same digital coin (double spending). This trusted third party is called the "guardian" in the CAFE vocabulary and is implemented in the cryptographic chip. Agreement of the guardian ensures the legitimacy of the money used for payment, because the guardian gives its approval only once for a given money. Honest merchants are supposed to refuse any coin that a guardian has not certified. This guarantees that the money is not reused.

An application can invoke additional defenses, such as secret codes associated with the various functions, spending limits, limits on the recharging of value, etc. The CAFE communication protocols allow for the recovery of the initial state, if the transaction is interrupted for any reason, such as human error, faulty communications, or malfunctioning terminals.

The CAFE originated in 1992, which makes it one of the oldest programs for electronic wallets in Europe. A field trial with several hundreds of wallets and tens of recharging points took place between October 1995 and February 1996 in two buildings of the European Commission. So far, however, no commercially viable product has emerged.

14.3 JEPI

JEPI was a joint initiative of the W3C and the CommerceNet consortium. Started in 1995, its objective was the automation of the negotiations between the buyer and the e-commerce server to choose the instrument of payment. Two protocols were added on top of HTTP (HyperText Transfer Protocol):

- The upper layer is a specialized protocol UPP (Universal Payment Preamble) with which to negotiate the means of payment. This layer represents the main contribution of JEPI.
- The lower layer is a general protocol PEP (Protocol Extension Protocol) with which to negotiate the mode of operation of the selected instrument of payment.

This solution did not get a suitable response, and the JEPI project was terminated in 1996.

14.4 PICS and P3P

PICS™ is a method defined by the W3C to associate labels to the content on the Internet to control access to it. These labels are put in two additional HTTP headers that precede the content. The assignment of labels can be manual or automatic, with programs for content analysis. The specifications define the vocabulary used. In contrast, the classification criteria and the category boundaries are left for choice. Also, the integrity and the authenticity of the labels are protected.

P3P is another specification from the W3C that gives users some control over the collection and the commercial use of the personal data that commercial Web sites gather. This is done by automatic analysis of the policies of the Web sites about the use and the protection of these data (Cranor, 2002).

Accordingly, each site describes the potential uses of the information they collect from users and the measures taken for its protection, while users indicate their preferences. The site policy is coded with XML. The automatic comparison of the site policies with the user's preferences reveals any potential discrepancy so that users make their decisions knowingly; in other cases, the browser's response is already programmed.

The P3P specifications describe a standard framework for presentation, including a vocabulary and special data elements to describe the site policy about personal information as well as an exchange protocol using HTTP. According to the client-server architecture, the P3P platform comprises two parts: the user's preferences at the client station and the site policy at the server. In the case of cookies, compact policies are used. These are short summaries of the full P3P policy that describes how cookies are used when a site is accessed.

According to P3P, the term *user agent* indicates the software that performs the access function that can be incorporated in a browser, in an electronic wallet, or in stand-alone applications. Currently, P3P is not widely accepted or deployed over the majority of commercial Web sites. As a consequence, users with P3P browsers can experience some difficulties when accessing specific sites. For example, when the user visits a site that uses cookies but did not declare a P3P policy, while the browser (Internet Explorer 6.0 from Microsoft or Netscape 7 from AOL Time Warner) is configured to refuse cookies unless the site conforms to P3P, the site will be blocked, or at least some functions will be disturbed. Note that APPEL (A P3P Preference Exchange Language) is intended to modify the default configuration of the user agent. The APPEL files are encoded in XML.

14.5 Analysis of User Behavior

The analysis of user behavior takes advantage of tools for searching large databases or data mining to extract some key indicators and levers on which an enterprise or a government can act. The various techniques for knowledge extraction (trends, segmentation of the population, classification, etc.) are described in the literature and will not be treated here.

The origins of the analyzed data are diverse. Online forms are one direct source. The analysis of server access logs allows for reconstitution of the route that the visitors took to the site and for evaluation of the effectiveness of references or affiliations. In addition, the loading time, the volume of visitors, the mean time of a visit, and the characteristics of the accessed pages (duration of access, number of downloads, etc.) contribute to improve site administration and organization. Finally, commercial data related to the items sold, the after-sales calls, and the rate of complaints or returns give guidance on the evolution of the offer.

It should be noted, however, that the prediction of individual purchases as well as the mailing of promotional offers to customers based on a profile derived from the previous data poses ethical problems about the protection of privacy. The legal aspects of privacy protection will be reviewed in the next chapter.

14.6 Fidelity Cards

Initially, accreditation or merchant cards in commerce were not targeted for a universal use. This is changing for two main reasons. The decentralization of banking services is encouraging banks to look for new points of access to the banking network, while merchants are seeking ways to enrich their private cards with financial features. Thus, point-of-sale terminals (POSTs) or cash registers in stores may function as ATMs by connecting to the banking network. The combined service offer allows payment for purchases, cash withdrawal and deposit to bank accounts at supermarket checkout counters, in addition to the fidelity functions. Furthermore, the collection system must integrate the monetary functions for payment by cash, by cards, or by checks (query of blacklists, authorization requests, transmission of payment data) with the traditional function of cash registers. Consider, for example, the following:

- Reading of the optical code on the article to identify it and to determine the price in an electronic file
- Tracking of sales activity to make the link with the store inventory
- Tracking of sellers and cashiers

Supermarket chains are associated with banks so as to add financial services to commercial offers. Consider, for example, the association of the British chains Tesco and Sainsbury's with the Royal Bank of Scotland or the partnerships established in France between the big department stores and the institutions offering consumer credit with the Cofinoga, Aurore, or Cofidis cards.

Every Penny Counts, Inc. (EPC) is the U.S. reincarnation of the same principle. Founded in New Jersey, in 1998, EPC proposes an electronic "piggy bank" to use to save the cash returned at the cashier's register. The collected amounts are deposited in a bank account that EPC manages. Each participant will be able to request a reimbursement either in cash or in the form of additional services, for example, as donations given to a charity of choice. Another variation called AutoGive operates on the principle of an automatic donation of 2% on all the purchases that the EPC cardholders pay at the regular price. EPC collects the rebates, and the amounts accumulated are then transferred, after deducting a commission, to the bank account of the organization that the cardholder designated. However, it is the account manager (EPC) that collects the interest on the amounts stored in the bank, while the merchants get a commission of 3% of the amounts deposited from their cash registers. Note the risk to the user's privacy, given that all the grocery purchases are documented in computer databases and related to the user's identity.

EPC owns a patent on the computerization of the collection and recording of contributions from payments at the cashier's register (Burke, 1997).

14.7 Quality of Service Considerations

Dynamic variations in the load affect the administrative policy of sites. In a simple site, such as that depicted in Figure 14.5 with a Web server, an application server, and a query server, the offered quality varies depending on the total load. An increase in the number of users or the incident of requests may saturate the capacity of the system and reduce the available throughput.

When the data reside in several databases, either because they were copied to ensure their safety or increase the reliability, or because they are from distinct sources (case of portals), it is often useful to divide the request into several subrequests and then group the subrequests to the same database. The configuration becomes that shown in Figure 14.6 (Bochmann et al., 2001).

The architecture with full replication and distributed queries, shown in Figure 14.7, improves the response time in case of congestion by balancing the loading of each server. A class of users can then receive priority treatment, and it is possible to accept or refuse a request depending on the availability of resources. Note, however, that the replication of identical databases raises the problem of synchronization of their respective contents.

In reality, as shown in Figure 14.8, users connect to the different servers through the network; consequently, the quality of service that they will experience depends on the quality of the network administration. The complexity comes from the fact that the Internet does not have a centralized administration, and the routing of traffic from one end of the network to the other requires the collaboration of several network operators.

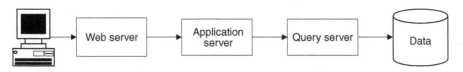

FIGURE 14.5
Simple architecture of electronic commerce.

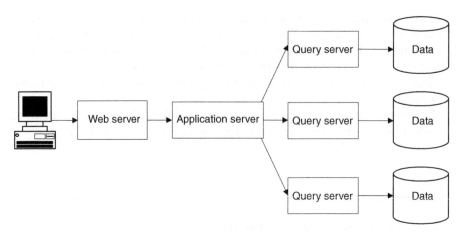

FIGURE 14.6
Architecture for electronic commerce with distributed data.

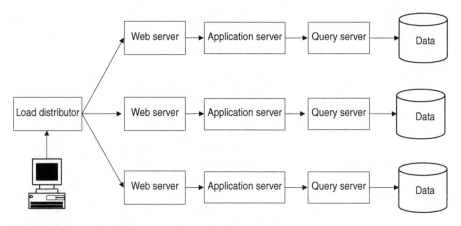

FIGURE 14.7
Architecture for electronic commerce with full replication and load balancing.

14.8 Summary

In this chapter, we reviewed several approaches to improve the informatics structure of e-commerce. The European research program SEMPER considered an architecture with four layers that is open and evolving and is capable of accommodating a large number of proprietary components. The specifications of PICS and P3P are more practical; nevertheless, they are struggling for acceptance. The PICS focuses on content labeling, while P3P is responsible for the protection of personal data. With or without the necessary precautions, mining of the data-accumulated transactions and the fidelity

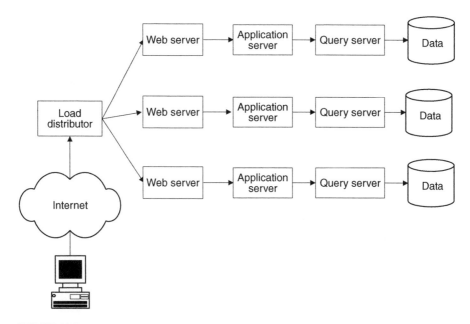

FIGURE 14.8

Architecture for electronic commerce with full replication, load distribution, and connection to the Internet.

programs offer sales possibilities, if not client retention, which is becoming vital because of the slowing of the new economy. Finally, several configurations allow better load distribution on the servers as well higher reliability so that the service quality can match the agreements with users.

Questions

1. Starting with SEMPER, list some of the ways that the word *purse* is used in e-commerce.
2. Speculate on the reasons that have prevented a commercially viable product based on CAFE.
3. What are the different uses of the word "wallet" in e-commerce systems?
4. Why is the use of cookies in e-commerce systems problematic?
5. How is data mining used in e-commerce systems?
6. List some of the ways an e-commerce site can improve server response time.
7. What are the possible causes for the slow uptake of P3P?

15

Electronic Commerce in Society

ABSTRACT

The experience acquired during several decades shows that electronic commerce (e-commerce) transforms the organization of tasks, modifies the balances within administrations or enterprises, and brings about a reevaluation of the roles played by economic agents. Success does not depend only on technical mastery but also on good management of sociological and cultural factors, particularly when integration of the processes used in several independent entities is at stake.

The Minitel, a French product of the 1980s, was the first success in the consumer arena. With the Internet, there is the possibility, for the first time in history, to blanket the whole planet with the whole value chain from the producer to the end user. Initially, the Internet allowed the growth of a worldwide economy of donations, which produced new forms of social interactions. Participants in this economy formed new types of remote collaborative networks, bypassing such earlier structures as the postal service and the telephone network. The spirit of sharing in the Internet has taken the form of an economy based on programs available at no cost and free of commercial constraints, the freeware, and of advice given without a fee on the newsgroups.

The monetary economy established after the commercialization of the Internet is of a different nature. First, a complete infrastructure dedicated to e-commerce for the general public or for businesses is costly: it calls for complex information systems to secure electronic payments. These systems require sophisticated maintenance. Second, the Internet matured in a context where it is physically impossible to sustain ambitious growth rates because of the saturation of the rich markets, the progressive depletion of natural resources, and the risks of pollution. In this sense, postmodernity can be conceived of as an attempt to accommodate the cultural and ideological necessity of growth, given the limits of the physical world (Haesler, 1995, pp. 315–316, 325). The theme of e-commerce would thus give a meaning, a design, and a collective ambition to the utopia of this end of the century, a

virtual economy without borders in a huge, worldwide market totally deregulated but nevertheless dominant. The production of wealth would not be
considered as the addition of material value but due to the movement of
information in the form of simple monetary transactions. This virtual economy expresses the emergence of a new social class with wealth that does
not arise from physical properties but from the power to manipulate symbols
(Lash, 1994).

In the following, we will consider certain aspects regarding the wide-scale
use of e-commerce, in particular, the following:

1. The communication infrastructure
2. The harmonization of architectures and the standardization of specifications
3. The issuance of electronic money
4. The protection of intellectual property
5. The risks that electronic surveillance puts on privacy
6. Content filtering and censorship
7. Taxation of e-commerce
8. Fraud prevention
9. Dematerialization of records

15.1 Communication Infrastructure

E-commerce relies on a high-quality telecommunication infrastructure with
worldwide coverage. Conversely, without ubiquitous and reliable transmission pipes, e-commerce cannot be generally considered. Nevertheless, applications in South Africa showed that in places that lack an adequate
telecommunications network, prepaid cards can compensate for deficiencies
in the infrastructure to collect bills and at the same time reduce operating
costs.

The deregulation of telecommunication services increases the risks of
weakening the foundations of such commerce. Instead of the vertical integration in the classical telecommunications infrastructure, the current infrastructure includes several physical and virtual networks, overlayed or
associated and managed by distinct administrative authorities. A series of
participants must coordinate their interventions from one end of the connection to the other to ensure successful exchanges. As shown in Figure 15.1,
they include the following:

- One or several bandwidth providers whose role is to establish the
  physical pipes between the two communication end points

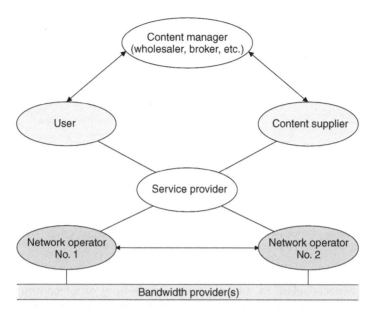

FIGURE 15.1
Contemporary organization of telecommunication services.

- One or more network operators responsible for the operation and maintenance of the switching and transport network; this network, which is overlayed on the physical network of links and trunks, usually includes equipment from several manufacturers, implementing equivalent techniques that should be interoperable

- Service providers (e.g., trusted third party, certification authority, payment intermediary, etc.) responsible for installation and management of the message exchanges; security may be incumbent on several certified providers to manage access using certificates provided by recognized certification authorities; other providers may host security applications, merchant sites, give guarantee labels, etc.

- Suppliers for the content sold online (databases, information, games, etc.)

- Retailers or brokers between content suppliers or payment services and the users

In such a fractured architecture, trouble localization and repair is a delicate operation. There is a wide variety of network element management systems, of trouble tickets, and of diagnostic procedures. In Figure 15.2, a virtual private network is illustrated, where each of several operators is responsible for one part of the ensemble but does not have a global view. At the same time, each operator will keep the data concerning their parts. As a consequence, the user (usually an enterprise or a financial institution) will not be able to assess the performance of its virtual private network (VPN).

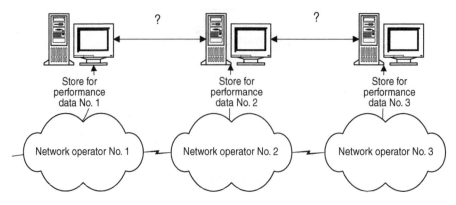

FIGURE 15.2

Collection of data performance in the distributed management of virtual private networks. (The links among the stores of the various operators are not defined.)

These aspects relate to the traditional know-how of telecommunication networks operators, as opposed to those that relate to end-user applications or to payment instruments. Clearly, efforts should be expended in organizations for telecommunication standardization to harmonize all the systems involved and to ensure seamless and secure arrival of the data related to the performance of the VPN to the rightful owners of that data.

15.2 Harmonization and Standardization

One of the objectives of the automation and dematerialization of monetary exchanges is to improve productivity and reduce the operational costs of financial transactions. Accordingly, e-commerce extends beyond the confines of telecommunications and information technology to the reorganization of operational processes. The interfaces that must have a minimum set of interoperability include the following:

- Interfaces among the same category of applications, such as micro-payment systems
- User interfaces to applications
- Operating systems of terminals (mobile telephones, computer, personal digital assistants, smart cards, etc.)
- Protocols for charging with value electronic or virtual purse or jeton holders
- Certification and certificate revocation procedures

Clearly, standardization for e-commerce covers a large area, given the many ramifications of the subject and the wide diversity of the participants. In addition to specialized organizations, such as the W3C (World Wide Web Consortium), the IETF (Internet Engineering Task Force), the ITU (International Telecommunication Union), and the ISO (International Organization for Standardization), the subject is already on the agenda of several other international organizations, such as the OECD (Organization for Economic Cooperation and Development), the WOIP (World Organization for Intellectual Property), the UN/ECE (U.N. Economic Commission for Europe), etc.

Harmonization may be difficult because of the proliferation of interested parties. The overabundance of standards, and sometimes their competition, may constitute a barrier to the spread of e-commerce. However, standardization may be a trap, should it lead to a monopoly position of one manufacturer, giving that manufacturer the possibility of imposing its products, irrespective of their quality. Another danger of the standardization of files or the use of a universal language is that it facilitates the preying on privacy by reducing the cost of data collection and interpretation.

15.3 Issuance of Electronic Money

The modern monetary system was freed of material support by replacing it with a generalized system of trust based on continuous growth. In the past, monetary creation was absolutely limited by the scarcity of materials to be exchanged or of precious metals for minting coins. On the contrary, money liberated from a material basis, in particular, electronic money, gives the impression of a creation *ex nihilo* of money that can multiply autonomously without any physical work.

The right to issue money is a privilege that benefits the state, because the cost of manufacture of currency is much less than the face value of the money. The difference represents a loan without interest that the holder of the currency gives to the issuer. *Seignorage* (from the French *seigneuriage*) is the government revenue due to this difference. For example, the large circulation of dollars outside the U.S. (which amounts to 50 to 70% of the currency issued) finances the budgetary deficit by about $7.2 to $19.2 billion (Bottelier, 2002).

"Trust" in a payment instrument is a subtle mixture of objective and subjective data. Acceptance of means of payment cannot be decreed but depends on the confluence of economic, sociological, political, and certainly, human factors. Technological promises should not make us forget that cultural factors will govern the political choices, thus affecting the future of each offer of e-commerce. In Chapter 2, it was shown how the preference for individual instruments of payment varied according to the country. Similarly, all surveys confirmed that the Internet is better received in the countries that Hall calls low-context countries (the Anglo-Saxon and the Nordic countries) than in high-context countries (France, Italy, and Spain) (*Les Échos*, 1998). Designers of e-commerce systems will have to take into consideration these national and geographical preferences.

Many private monies are used alongside legal monies in daily life. These are often promises of service in the form of coupons, phone jetons, fidelity points, preferential stocks, etc. A company, Zhone Technologies, paid for the construction of its headquarters in the city of Oakland, California, with a promise of 1500 jobs and $6 million in stock options for the city. The telephone unit is also capable of playing the role of virtual money for microtransactions, particularly for mobile telephony (Pedersen, 1997). In principle, the area of action of these private monies could be expanded, provided that their issuers mutually recognize their respective instruments. This would have two important consequences: banks would be excluded from the circuits of financial exchanges; and private monies would benefit private entities at the expense of the public treasury, thereby increasing the deficit of state budgets. Nevertheless, the security of these private currencies would have to be greatly enhanced for them to play this subversive role, given that they already face the same challenges that state monies have to overcome, such as counterfeiting (Norris, 2001).

Last, the generalized use of legal electronic monies raises the problem of escheat, i.e., the reversion of personal tangible abandoned property to the state after a period of inactivity. The situation becomes more complicated in the case of electronic or virtual purses, because several cases would have to be considered: residual values, full values in cards purchased as collectibles, anonymous cards, lost cards with access control, etc. (Lorenz, 1997). The determination of dormancy is also problematic for online systems, because this would require the tracking of all transactions, which may be cumbersome beyond a certain number of participants and would not apply in the case of person-to-person exchanges.

15.4 Protection of Intellectual Property

With globalization and digitization, the protection of intellectual property requires rules that are followed worldwide regarding the reproduction and the broadcast of virtual works, such as music recordings. The widespread use of digital optical disks and the spread of the Internet facilitate the storing, copying, editing, and distributing of all kinds of digital contents. Hypertext links on Web sites can lead the users to content that is protected by copyright.

Signed in 1996 under the auspices of the World Intellectual Property Organization (WIPO) and coming into force in 2002, two treaties — called the Internet treaties — update copyrights and related rights (rights of performers, publishers, and broadcasters) to fit the era of e-commerce. These are the

WIPO Copyright Treaty (WCT) and the WIPO Performance and Phonogram Treaty (WPTT) that allow authors, artists, and performers to control the broadcast of their works on digital networks such as the Internet. The most controversial aspects of these two treaties are the limits they put on the ability to copy digital productions and the prohibition of any action to intentionally circumvent the technical means to protect copyrights and related rights, in particular, the manufacture, distribution, importation with the intent to distribute, and use of any equipment for circumvention.

The Digital Millennium Copyright Act of 1998 integrated the stipulations of these treaties into U.S. law. However, the excessive zeal in the application of this law led to grotesque situations, such as the case of Professor Edward Felten of Princeton University. A consortium of musical publishers started the Secure Digital Music Initiative (SDMI) to prevent piracy by developing a file format that includes coding and watermark techniques to protect the content by controlling the playback, storage, and broadcast of digital music. To prevent access without prior authorization, the rights to be exercised are recorded in an indelible manner in the music file. This allows playback control through *a posteriori* verification that the owner is legitimate; were it not the case, playback would be prevented. In September 2001, SDMI announced a public challenge to encourage attempts to break certain algorithms proposed for content protection. A team of researchers from Princeton and Rice Universities, led by Professor Felten, concentrated on the task and was able to reveal the mechanisms of certain algorithms. In a panic mode, the SDMI tried to block publication of the results on the basis of the previously mentioned law and did not change its mind until the intervention of the Electronic Frontier Foundation (EFF) (Craver et al., 2001).

From a technical viewpoint, the methods considered to protect the digital content are based on encryption or on watermark algorithms.

15.5 Electronic Surveillance and Privacy

Browsers pass to e-commerce sites several pieces of information concerning the user: the IP network address, operating systems, and origination of the connection. These data, as well as those in the forms filled at checkout, allow merchants to collect personal data. In addition, cookies or spyware, located in the hard disk of client stations, allow merchant sites to record and remember information about the users. Merchant sites can take advantage of cookies to personalize their interfaces and to adapt their offers to the client's preferences and past choices. Another advantage is that users can avoid reentering their identifiers (login and password) at each visit. The integrated circuits of bankcards, of mobile phones, of automatic payment for tolls or parking, and the access terminals to buildings or offices are all means by which to track movements and profile habits that are available to authorities or hackers. The same can be said regarding digital money or content (books, music, films, etc.), which, once dematerialized, become simple information that can be tracked and recorded.

There are more things to come. Identifiers embedded in the processor or the operating systems of terminals allow the localization of individuals at any moment without their knowledge (Gray, 1999). Of course, these techniques are the basis of localization services that are so useful for mobility; nevertheless, they enable the tracking of individuals. Finally, personal data contained in X.509 certificates can be decrypted if the encryption algorithm is not strong enough or if the directory of the certification authority is accessible (Renfro, 2002). The information threat is multiplied if the accumulated data are diverted from their original purposes so as to establish links with files accumulated in several unrelated domains — private or public, commercial or government, national or international — without verifying the consistency of the criteria used in the collection or the determination of content accuracy.

Efforts to establish some order in this area aim at regulating access to personal data collected during e-commerce transactions by defining the following:

- The explicit purpose for data collection
- The right of an individual to verify the accuracy of the data and to correct any possible errors
- The protection accorded to the collected data
- The conditions under which the data can be released to third parties

The concrete forms that the regulation takes vary according to the aims of the legislator: to stimulate electronic transactions or to protect individual rights against potential aggressions. Japan, for example, started in August 2002 to use a national system for access control to local government services using an 11-digit unique identifier, before voting for a law for privacy protection.

In the U.S., laws for the protection of privacy were voted following revelations on the spying of different security agencies against political opposition (for example, the Cointelpro). The Federal Privacy Act of 1974 forbids government institutions from collecting information on the political or social behaviors of individuals without their knowledge. The Internet was explicitly considered in 1997 in the Federal Internet Privacy Protection Act, which focused on the information in medical, financial, or work records. Finally, the e-Government Act of December 2002 forces the government to evaluate the impact of technologies on private life before their implementations. It is seen that these laws do not consider the problem of privacy protection in its entirety, because they do not apply to commercial entities. In the commercial sector, self-regulation prevails, with the exception of the financial sector, where the laws are stricter (e.g., the so-called Graham–Leach–Bliley Act).

Yet, for many enterprises, while declaring their respect for privacy, data collected from individuals are sources of income and their sale is negotiable. Has not the online vendor Amazon declared them *transferable assets*? Nevertheless, the Children Online Privacy Protection Act (COPPA) requires that the operators of Web sites directed at children under the age of 13 obtain parental consent before collecting certain personal information.

The frailty of protections for private life jumps to the eye; they are floundering under the simultaneous assaults of merchants that would like to profit from the data they collected on their customers as well as the state's repressive and security-oriented policies. Even when employers promised not to spy on their employees or provided them with the necessary tools to ensure the confidentiality of their personal data, the U.S. courts upheld the right of employers to monitor all the exchanges on their networks (Waldmeir, 2001). Courts found that cookies and other spyware are consistent with federal laws. Many constitutional safeguards were simply disregarded by invoking moral (for example, battle against pedophilia) or political considerations (war on drugs or on terror, fight against hooligans or organized crime, etc.). In the U.K., for example, video surveillance of public areas is widespread. More than a third of the U.S. employers monitor the online activities of their employees, including the content of electronic exchanges. The Echelon network, established by the intelligence services of Anglo-Saxon countries (Australia, Canada, New Zealand, the U.K., and the U.S.) monitors telephone calls, facsimile, and data transmissions through nodes spread throughout the world. Next, with the Patriot Act, the U.S. government claimed the right to monitor the files related to all kinds of electronic transactions and to mobilize, for that effect, numerous service providers: banks, travel agencies, car rental agencies, casinos, etc. Libraries and bookstores are ordered to indicate to the authorities who borrowed or bought what, with the obligation of hiding the inquiry from the targeted person.

The goal of establishing a "panoramic" surveillance network of the civilian population is being further defined with the Total Information Awareness project. The aim is to establish a digital signature of each resident by constituting extra-large databases from banking, financial, or medical records, voice or written communications, library records, electronic payments, trips, etc., of the whole population, in order to derive individual profiles. These will then be cross-checked with the information available from intelligence and police agencies to uncover potential terrorists. Clearly, the development of e-commerce can bring about a society of control and surveillance.

In Canada, plans for the federal government to link the elements that it has on every person with the data from the provincial or territorial governments were aborted. The fear came from seeing all information on the person (education, civil status, language, nationality, ethnic origin, physical capacities, revenues, professional work, etc.), their parents or spouse, as well as their family members, under single control in a super file, the *Longitudinal Labour Force File*. Faced with criticism in favor of a better protection to private life, the project was withdrawn in May 2000.

In France, the first attempt to link all administrative files using the social security number as a single identifier — the Safari (Système Automatisé pour les Fichiers Administratifs et le Répertoire des Individus — Automated System for Administrative Files and Individuals Directory) project — provoked an uproar that led to the enactment of the law on Informatics and Freedoms (Informatique et Libertés) of 1978 (revised in 2002) and the establishment of the National Commission on Information and Freedoms (Commission Nationale de l'Information et des Libertés [CNIL]). Recently, the commission studied threats to personal privacy due to e-commerce and the Internet and defined a code of conduct with respect to use of the Internet at the workplace. According to its recommendations, personal use is allowed as much as it does not affect the work and does not threaten "public order and good moral standards." Nevertheless, an enterprise can monitor this usage provided that the employees are informed of such surveillance, and it can prohibit downloading of software or access to specific sites. The Court of Cessation ruled in October 2001 that enterprises cannot open the personal e-mail of employees, even when internal regulations prohibit the use of e-mail for personal messages, i.e., the secrecy of personal correspondence is upheld at all times, including with electronic media. Despite all these principles, the CNIL accepted in 1998 the establishment of a megafile containing a large amount of personal information collected from transcripts of police reports, for tickets, civil offenses, or crimes, irrespective of whether the individual was involved as a suspect, witness, victim, or merely implicated, rightly or wrongly, during the investigation. This file can be consulted during administrative investigations, for example, within the context of fighting tax evasion (Bonnaure, 1999).

The insertion of cookies to personalize Web pages and service offers is allowed, provided that the user was informed of the attempted installation, that means to block the insertion were provided, and that the purpose of data collection was clearly explained. The creation of digital files on individuals, particularly nongovernmental files, requires an authorization from the CNIL that depends on the content (sensitive data, genetic data, social status) and its final use (for example, identity or attribute verification to exercise rights). The CNIL considers that the simplification of administrative procedures justifies the interconnection of individual files owned by different departments. To protect privacy, however, each interconnection will be restricted to a "sphere" of activities (social, fiscal, justice and law enforcement, culture and education, etc.) instead of a generalized interconnection using a single identifier. Thus, a moving portal would be the window to all necessary formalities for a change of address (heating, electricity, water, telephone, taxation, etc.). Similarly, a card for daily life (CVQ) (Carte de Vie Quotidienne) would be coupled with the Moneo electronic purse to serve for all local government procedures (registration for schools, libraries, and sport activities; payments for public transportation, etc.). Finally, to secure personal medical data captured with the health insurance smart card Sésame-Vitale, transmissions and back-office stores must be protected. There are even procedures to destroy the files totally, in case of national emergency, such as an invasion by foreign troops.

At the European level, the European Directive 95/46 harmonizes the legislation in various European countries by establishing a general framework for the processing of personal data collected from clients. The aim is to distinguish sensitive data from common data. In the first case, the data cannot be diverted for the purpose indicated at the time of collection. For common data, users will have to indicate their permission or their refusal to have them shared with other applications or entities. In 1998, another European directive established privacy as a fundamental human right. Each member state of the European Union is obliged to integrate this directive in its legislation and to forbid, for example, the inappropriate use of client's financial data or purchasing profiles. Furthermore, the transmission of personal data outside the European community is strictly forbidden if the countries of destination do not offer the same levels of protection. The strict application of this directive, especially in Germany, prevented outsourced processing of human resources files outside the European Union.

In November 2001, the European Parliament voted to forbid the use of a computer network to collect information on client stations. This prohibition covers spying software ("spyware") that is used to monitor user activity secretly as well as identification or tracking mechanisms hidden in the client station. The protection of data collected by mobile telephony operators came into force in all the member states on January 16, 2002. The only data that can be gathered are data necessary to localize the user. Later use of the data is forbidden without the explicit approval of subscribers. In addition, subscribers must be able to delete these data or request that they be processed anonymously. Finally, the operators must install the mechanisms to ensure data confidentiality. Nevertheless, this protection was considerably weakened in May 2002 with an amendment voted in the war against terrorism. In addition, in March 2002, the European Commission agreed, in principle, without parliamentary authorization, to a request from the U.S. to share the data that European airlines would have on their passengers with the U.S., including bank card numbers and dietary restrictions.

15.6 Filtering and Censorship

Content or traffic filtering raises an interesting problem because it questions a founding myth of the Internet concerning the free circulation of information. It imposes significant technical hardship as well as legal responsibilities on computer professionals (network administrators and access providers).

In the case of business-to-business commerce, professional services and value-added services providers have several roles: consulting, training, installing, and maintaining. These intermediaries can also offer security services: certification of parties, notarization of exchanges, time-stamping of transactions, maintaining archives or information to satisfy legal requirements.

In the case of consumer commerce, many countries impose filtering of exchanges to prevent the broadcast of contents deemed illegal. Focusing only on Western countries, we note that in 1998, the German subsidiary of CompuServe had some problems with the law for not blocking access to child pornography sites and for not filtering pictures with a pedophilic nature. Following a lawsuit started in 1999 by the Jewish Student Union of France (UEJF, Union des Etudiants Juifs de France) and the International League against Racism and anti-Semitism (LICRA, Ligue Internationale Contre le Racisme et l'Antisémitisme), a French court ordered Yahoo! to block access to its virtual auction site for Nazi memorabilia from France and its territories, even though the incriminated site was intended for U.S. audiences only. These regional variations of what is prohibited and what is not are clearly a hardship for managing online e-commerce on a worldwide basis.

Another subject of concern is the *spam* that comes from publicity campaigns destined for the largest number of Internet users without their prior consent and without considering any embarrassment that the content may cause them. For telemarketers, spam is attractive because messages are easy to reproduce and quick to spread, thereby reducing the cost of the campaign. In contrast, the resulting traffic may congest the network and overload the mailbox of the recipients. This electronic pollution ends up costing the enterprises in terms of bandwidth, storage, and time to read and to delete.

"Spammers" construct their lists by conducting systematic collections of e-mail addresses from mailing lists, Web pages, newsgroups, files purchased from other online merchants, etc. In the middle of Februray 2003, AOL was able to filter 1 billion messages in 24 hours, which corresponds to 28 undesirable e-mails per subscriber (Décision Micros & Réseaux, 2003). It is believed that techniques for advanced filtering and for adaptive learning by mail agents has reduced the proportion of spam to less than 10% of total Internet traffic.

15.7 Taxation of Electronic Commerce

The fiscal consequences of e-commerce are more important than sales by mail-to-order because of the ubiquitous and permanent nature of the cyberspace. This is why the growth of e-commerce exchanges poses the issue of taxing the associated transactions. On the one side, tax exemption of e-commerce favors online enterprises at the expense of traditional stores and reduces the income of all the governments involved. On the other side, taxation may put a brake on the expansion of electronic markets. This is the context in which the divergent viewpoints of the European Union and the U.S. are expressed on that issue.

In the U.S., the Internet Tax Freedom Act of 1998 aims at promoting e-commerce through fiscal advantages, by imposing a moratorium for 3 years for additional taxes on "virtual goods." This law was renewed in 2001, but some states, such as California, are moving toward taxing Internet sales. A 2001 study by the University of Tennessee suggested that loss of revenues from untaxed Internet sales would reach $45 billion by 2006 and $55 billion by 2011 (Morrison, 2003). A directive from the European Union requires a value-added tax (VAT) on digital products delivered online (sales of software, remote maintenance, etc.), even when the merchants are outside Europe. In other words, U.S. vendors of online articles or digital services are obliged to establish a physical presence in one of the countries of the European Union and apply the VAT according to the rules of the countries where the buyer resides.

15.8 Fraud Prevention

Fraud in e-commerce has multiple sources: fraudulent descriptions of articles sold online, rigged auctions, nondelivery of purchased items, false payments, propagation of malicious rumors, etc. In addition to the technical means that were presented in this book, the repression of fraud usually takes two approaches, the legislative approach and self-regulation. Thus, legislation defines the overall responsibility of the participants so that the party with grievances can ask the authorities to intervene to redress torts. Industry self-regulation relies on private institutions to accord "approval seals" to sites that promise to follow a specific line of conduct with respect to fraud prevention and the protection of data collected from the clients. In Table 15.1, a partial list of such labeling organizations for online commerce is presented.

15.9 Archives Dematerialization

Management of electronic documents includes data archival in addition to day-to-day management. This is a critical issue whenever the archives must survive and remain accessible over long periods (30 to 50 years). In case of a tax audit, for example, companies are obligated to present all records that contributed, whether directly or indirectly, to the consolidated statement, including accounting, payroll, inventory, billing, etc. Should the file be in a proprietary format, the application programs may have to be available in the version of that time. What is needed, therefore, is a guarantee for the

TABLE 15.1

Partial List of Labeling Organizations for Online Commerce

Label	Description	URL
WebTrust	Site of the organization formed through the collaboration of the American Institute of Certified Public Accountants and the Canadian Institute of Chartered Accountants	http://www.cpawebtrust.org
TRUSTe	Founded by the Electronic Frontier Foundation and CommerceNet, this organization gives its seal to sites that follow the recommendations of the Federal Trade Commission with respect to the protection of private and confidential data and which agree to be monitored by TRUSTe and to follow its mechanisms to fix problems; a special seal is accorded to sites that conform to COPPA for the protection of minors less than 13 years of age	http://www.truste.org
BBBonline	Online version of the Better Business Bureau, a U.S. nonprofit organization financed by the member merchants that gives information to the public	http://bbbonline.org
AECE	Site of the Spanish Association of Electronic Commerce (Asociación Española de Comercio Electrónico)	http://www.aece.org

readability of the archives by guaranteeing their integrity, their longevity, and the possibility of restoring them at will. Some support media, however, are precarious. Magnetic supports are not recommended because they can be reused, and hence, the data that they contain can be falsified. In contrast, traditional storage media, such as paper or punched cards, are only used once. Throwaway integrated-circuit cards offer the required guarantees, because there is a progressive destruction of the memory cells as data are recorded. In contrast, the rechargeable microprocessor card calls for a public-key infrastructure to ensure nonrepudiation.

Nevertheless, the readsbility of the archives does not depend solely on the physical condition of the support media but also on the availability of peripherals and software for reading. A requirement for archiving materials for 30 years may be out of the question given the speed with which technological developments are taking place. How easy it is today, for example, to find readers for $5\frac{1}{4}$-inch floppy disks — less than a decade ago the most prevalent storage medium — to access the reams of data that these disks contain? As a consequence, it is important to ensure the presence of adequate reading terminals, and if the format of the files is proprietary, to preserve a copy of the application software that was used to produce them.

15.10 Summary

E-commerce forces a complete revision of the whole value chain: relations between production and demand, the power relations within the enterprise, the collaboration among partners, modalities for the dissemination of the information, etc. These changes go beyond the technical domain and modify financial or legal aspects of the conduct of business, thereby posing questions of the nature of society. These challenges should occupy engineers, computer scientists, jurists, lawyers, marketers, sociologists, buyers, investors, bankers, politicians, etc. in the years to come.

Questions

1. What is the role of standards in e-commerce? Discuss their advantages and disadvantages.
2. What is escheat? How is it affected by the various types of e-commerce?
3. Can copyright protection clash with the free flow of information for e-commerce?
4. What are the similarities between U.S. and European positions regarding the protection afforded to data collected from users in e-commerce?
5. What are the difficulties regarding taxation of e-commerce?

Web Sites

General

http://www.nic.fr
Site that contains data on Internet usage in Europe

http://www.ustreas.gov/press/releases
Site of the U.S. Treasury Department

http://www.bis.org
Site of the Bank of International Settlements, which contains data on the various transactions with the different scriptural means of payments for the G-10 countries

http://www.disa.org/apps/acroglos
Glossary of acronyms of EDI

Standards

http://web.ansi.org
ANSI (American National Standards Institute) standards

http://www.iata.org
IATA (International Air Transport Association) standards

http://www.isss/Workshop/eBES/Default.htm
ebXML standards

http://www.unece.org/trade/untdid/Welcome.htm
EDIFACT standards

http://www.ietf.org
IETF standards

http://www.rsasecurity.com
Information on the PKCS industry standards

http://www.w3.org
W3C standards for electronic payments

Encryption

http://www.rsasecurity.com/rsalabs/

Site with an excellent FAQ on the theory and practice of cryptographic systems

http://www.counterpane.com/

Site that gives information on the practice of cryptography

http://csrc.nist.gov/pki

Site that gives information on the PKI program of NIST

http://www.freeswan.org

Site that has open-source implementations for Linux of IPSec, IKE, and other protocols

Kerberos

http://web.mit.edu/kerberos/www/index.html

The official Web page for Kerberos

http://www.cybersafe.com

Site for a commercial version of the free version of Kerberos (Heidmal). The free version was written by Johan Danielsson and Assar Westerlund from the Swedish Institute of Computer Science. It is not restricted by the U.S. laws on exporting encryption.

Certification

http://www.pkiforum.com

Site that gives the latest news in the certification field concerning authorities, suppliers, products, fairs, etc.

http://www.verisign.com

Site of VeriSign, the renowned certification authority

http://www.identrus.com

Site of Identrus, established by banks in 2000 to manage, on their behalf, the issuance and distribution of digital certificates to their partners and clients

http://www.cartes-bancaires.com/html/grpmnt/comm1.html

Site of the Global Trust Authority (GTA), a nonprofit organization established by banks for the worldwide management of public-key certificates stored in integrated circuits (smart) cards

http://csrc.nist.gov/pki/program/Welcome.html

Site with information on the PKI activities at NIST

Biometrics

General

http://www.biometric.freeserve.co.uk

Bilingual site (English and French) maintained by Julian Ashbourn, with general information on the techniques, commercial offers, and associations active in this area

http://www.biometricgroup.com

http://www.biometrics.org/REPORTS/BioRef.html

List of 544 references on the subject prepared in 1995 by Roger Johnson of Los Alamos National Laboratories

http://www.cs.rug.nl/~peterkr

Site prepared by Peter Kruizinga, from the University of Groningen in the Netherlands, with current information on research in biometrics

http://vismod.www.media.mit.edu/vismod/demos/facerec/feret_res.html

Site with the results of the tests sponsored in 1996 by the U.S. Army on face recognition

Standards Organizations

http://www.bioapi.org
BioAPI Consortium

http://www.ibia.org
Site of the International Biometric Industry Association (IBIA)

http://www.afb.org.uk
Site of the Association for Biometrics

http://www.bsi.bund.de
Site of the Bundesamt für Sicherheit in der Informationtechnik (BSI — Federal Information Security Agency)

Products

Face Recognition

http://www.identrix.com
http://www.miros.com
http://www.viisage.com

Fingerprints

http://www.infineon.com
http://www.secugen.com
http://www.authentec.com
http://www.veridicom.com
http://www.identix.com
http://www.whovision.com
http://www.tcs.thomson-csf.com

Iris Scan

http://www.iriscan.com

Hand Geometry

http://www.biomet.ch
http://www.recogsys.com

Keyboard Recognition

http://www.biopassword.com

Retinal Scan

http://www.eye-dentify.com

Speech Recognition

http://www.trintech.com

EDIFACT

http://lexmercatoria.org

Site that publishes UNCITRAL Model Law on Electronic Commerce of 1996

http://netman.cit.buffulo.edu/WG/ByBds/EWOS/ovw.html

Site for the EWOS (European Workshop for Open Systems) documentation. The EWOS was formed by the European industry with the support of the European Commission to encourage the compatibility of software solutions and to guide the standardization work at the European level as well as worldwide. In 1997, EWOS became ISSS (Information Society Standardization System) within the European Committee for Standardization (CEN), focusing on the integration of XML and EDI in Europe.

http://www.unece.org/cefact

Site of the UN/ECE (U.N. Economic Commission for Europe), which hosts the documents of the CEFACT, including those that relate to ebXML

XML

XML and its derivatives are widely covered over the Web. Some of the sites are as follows:

http://www.w3.org/XML

http://www.w3.org/Encryption
Reference sites for the W3C responsible for the development of XML and its encryption

http://www.xmlfr.org
Site publishing detailed information on XML, with daily updates

http://www.mutu-xml.org
Site dedicated to the exchange of information among XML developers

Integration XML/EDIFACT

http://www.oasis-open.org/home/index.php
Site of the OASIS consortium

http://www.commerce.net
CommerceNet is an industrial nonprofit consortium for the promotion of global solutions for electronic commerce

http://www.rosettanet.org
RosettaNet is another U.S. nonprofit consortium working to facilitate the electronic exchanges among the various entities of the supply chain

http://www.unece.org/cefact
Site of the CEFACT of the United Nations that collaborates with the OASIS consortium to integrate EDIFACT and XML in ebXML

http://www.uddi.org
Site of the organization responsible for establishing the directory of services available on the Web

http://msdn.microsoft.com/xml

Site that presents the solutions that Microsoft proposes for EDI/XML integration

---

SSL/TLS/WTLS

http://www2.psy.uq.edu.au/~ftp/Crypto/

Site with information on SSL and SSLeay. In particular, some information on bugs in commercial software is available at the following address:

http://www2.psy.uq.edu.au/~ftp/Crypto/ssleay/vendor-bugs.html

http://www.openssl.org

Site of the project OpenSSL

http://www.wapforum.org

Site of WAP Forum that contains the WAP specifications, including those of WTLS

SET

http://www.mastercard.com;

http://www.visa.com

SET specifications can be downloaded from one of the above two sites

---

Purses

http://www.jrc.es/cfapp/leodb/recent.cfm

Leo Van Hove, together with volunteer collaborators, maintains free of charge a bibliographical site on the various electronic purses

http://www.ecml.org

Site of the Electronic Commerce Modeling Language (ECML) alliance http://epso.jrc.es/newsletter

The newsletter of the Electronic Payment Systems Observatory (EPSO)

Micropayments

System	Site
ClickShare	http://www.clickshare.com
DigiCash	http://www.ecashtechnologies.com
iPIN	http://www.ipin.com
NetBill	http://www.ini.cmu.edu/netbill
NetCash	http://www.isi.edu
NetCheque	http://www.isi.edu
WISP	http://www.trivnet.com

Smart (Microprocessor) Cards

http://www.cardeurope.demon.co.uk

Card Europe is an association to promote the applications of smart cards.

http://www.geocities.com/ResearchTriangle/Lab/1578/smart.htm

A site maintained by Bo Lavare on smart card securities. This site contains information on GSM cards and TV cards for cable or satellite.

The newsgroup *alt.technology.smartcards* publishes a FAQ with the latest details.

http://www.visa.com

http://www.mastercard.com

Sites that publishes the EMV specifications on the use of smart cards for payment systems

http://www.opencard.org

Site managed by the OpenCard Consortium, which contains the technical details of the OpenCard Framework

http://www.pcscworkgroup.com

Site that contains the PC/SC specifications on the interoperability of smart cards and PCs under Windows

http://java.sun.com/products/javacard

Site that gives the JavaCard specifications

http://www.linuxnet.com

Site of MUSCLE, a movement to adapt integrated-circuit cards to the Linux environment

http://www.seis.se

Site of SEIS (Secured Electronic Information in Society), a nonprofit Swedish association that publishes the specifications of the electronic identity card that was later adopted as a Swedish Standard

http://www.parodie.com/monetique

Site maintained by Laurent Pélé, which gives the details of the Humpich affair

---

Electronic and Virtual Checks

http://www.ftsc.org

Site of FSTC, which gives information on the U.S. project on dematerialized checks

http://www.echeck.org

Site that gives the BIPS specifications and the eCheck project

http://www.netcheque.org

Site presenting the prototype of the NetCheque system

SEMPER

http://www.semper.org

Site that publishes the SEMPER documentation

Labeling Organizations

http://www.cpawebtrust.org

Site of the CPAWebTrust run by the American Institute of Certified Public Accountants and the Canadian Institute of Chartered Accountants

http://www.truste.org

Site of TRUSTe, an organization run by the Electronic Frontier Foundation and CommerceNet, which awards its privacy seal to sites that comply with TRUSTe oversight and conflict resolution procedures. TRUSTe also offers a special seal for sites addressed to children to indicate their compliance with the COPPA requirements.

http://bbbonline.org

Online version of the Better Business Bureau

http://www.aece.org

Web site of the Spanish Association of Electronic Commerce (Asociación Española de Comercio Electrónico)

http://www.ftc.gov/bcp/conline/edcams/kidsprivacy

Site of the Federal Trade Commission of the United States, with extensive details on COPPA and the protection of the privacy of children less than 13 years of age

Organizations

http://www.eff.org

Site of the Electronic Frontier Foundation, which was founded in 1990 to defend freedom of expression, privacy, and the openness of the information society

http://www.ircm.qc.ca/bioethique/english/telehealth/observations.html

Site of the Center of Bioethics, Clinical Research Institute of Montreal, which can be consulted on issues related to the confidentiality of electronic medical records

Organizations

Acronyms

3D-Set	Thre-Domain SET
3D-SSL	Three-Domain SSL
ACH	Automated Clearing House
ACS	Access Control Server
ACTS	Advanced Communication Technologies and Services
ADSL	Asymmetric Digital Subscriber Line
AECE	Asociación Española de Comercio Electrónico (Spanish Association for Electronic Commerce)
AES	Advanced Encryption Standard
AFCEE	Association Française pour le Commerce et les Échanges Électroniques (French Association for Electronic Commerce and Exchange)
AFNOR	Association Française de Normalisation — French Association for Standardization
AH	Authentication Header
AIAG	Automotive Industry Action Group
AID	Application Identifier
AIR-IMP	AIR Interline Message Procedures
ANSI	American National Standards Institute
ANX®	Automotive Network eXchange
API	Application Programming Interface
APPEL	A P3P Preference Exchange Language
ARDP	Asynchronous Reliable Delivery Protocol
ARQC	Authorization Request Cryptogram
AS	Authorization Server
ASC	Accredited Standards Committee
ASN.1	Abstract Syntax Notation 1
ATM	Asynchronous Transfer Mode
ATM	Automated Teller Machine
ATR	Answer to Reset
BACS	Banker's Automated Clearing Service
BC	Biometric Consortium

BER	Basic Encoding Rules
BIN	Bank Identification Number
BIPS	Bank Internet Payment System
BIS	Bank for International Settlements
BSI	Bundesamt für Sicherheit in der Information-technik (Germany: Federal Information Security Agency)
BSP	Bank Settlement Payment
BTX	Bildschirmtext
CAC	Confirmation and Authentication Challenge
CAFE	Conditional Access for Europe
CALS	Computer-Aided Acquisition and Logistics Support, became Continuous Acquisition and Life-Cycle Support, then became Commerce at Light Speed
CAN	Customer Account Number
CAPI	Cryptographic Application Programming Interface
CAR	Confirmation and Authentication Response
CARGO-IMP	CARGO Interchange Message Procedures
CAS	Channel-Associated Signaling
CASE	Computer-Aided Systems Engineering
CAVV	Cardholder Authentication Verification Value
CBC	Cipher Block Chaining
CBEFF	Common Biometric Exchange File Format
CCD	Cash Concentration and Disbursement
CCD	Charge-Coupled Device
CCITT	Comité Consultatif International Télégraphique et Téléphonique
CCS	Common Channel Signaling
CDT	Committee for Development of Trade
CEFACT	Center for Trade Facilitation and Electronic Business
CEFIC	Conseil Européen des Fédérations de l'Industrie Chimique — European Council of Industrial Chemistry Federations
CEN	Comité Européen de Normalisation — European Committee for Standardization
CEPS	Common Electronic Purse Specifications
CFB	Cipher Feedback
CFONB	Comité Français d'Organisation et de Normalisation Bancaires
CGI	Common Gateway Interface

CGM	Computer Graphics Metafile
CHAPS	Clearing House Automated Payment System
CHIPS	Clearing House Interbank Payment System
CID	Cardholder ID
CID	Confidential Identity
CIDX	Chemical Industry Document Exchange
CIIP	Check Image Interchange Protocol
CMC7	Caractères Magnétiques Codés à 7 Bâtonnets — Magnetic Characters Coded with Seven Links
CMP	Certificate Management Protocol
CMS	Cryptographic Message Syntax
CNIL	Commission Nationale de l'Information et des Libertés — National Commission on Information and Freedoms
COF	Common Output Format
COPPA	Children Online Privacy Protection Act
COST	Computer Security Technologies
CPRQ	Condensed Payer Authentication Request
CPRS	Condensed Payer Authentication Request Response
CPS	Certification Practice Statement
CPTP	Customer Payment Server Transaction Protocol
Creic	Centres Régionaux d'Échanges d'Images Chèques — Regional Centers for the Exchange of Check Images
CRI	Centrale des Règlements Interbancaires — Exchange for Interbanking Payments
CRL	Certification Revocation List
C-SET	Chip-Secured Electronic Transaction
CSM	Chipcard Security Module
CTI	Computer Telephony Integration
CTP	Corporate Trade Payments
CTX	Corporate Trade Exchange
CVQ	Carte de Vie Quotidienne — Card for Daily Life
CVV2	Customer Verification Value
DAP	Directory Access Protocol
DDOS	Distributed Denial of Service
DEC	Digital Equipment Corporation
DEDICA	Directory-Based EDI Certificate Access and Management
DER	Distinguished Encoding Rules
DES	Data Encryption Standard
DF	Dedicated File

DGI	Direction Générale des Impôts — General Directorate for Taxation
DIN	Deutsches Institüt für Normung e.V. — German Institute for Standardization
DIR	Directory
DIS	Draft International Standard
DISA	Data Interchange Standards Association
DISP	Directory Information Shadowing Protocol
DNS	Domain Name Service
DOD	Department of Defense
DOI	Domain of Interpretation
DPA	Differential Power Analysis
DRAM	Dynamic Random Access Memory
DSA	Digital Signature Algorithm
DSA	Directory System Agent
DSL	Digital Subscriber Line
DSP	Directory System Protocol
DSS	Digital Signature Standard
DSSSL	Document Style Semantics and Specification Language
DTD	Document Type Definition
DUA	Directory User Agent
EAI	Enterprise Application Integration
EAN	European Article Numbering
EBES	European Board for EDI Standardization
EBPP	Electronic Bill Payment and Presentment
EBS	Elektronik Banking Systems GmBH
ebXML	Electronic Business XML
ECB	Electronic Code Book
ECC	Elliptic Curve Cryptography
ECCHO	Electronic Check Clearing House Organization
ECDH	Elliptic Curve Diffie–Hellman
ECDSA	Elliptic Curve Digital Signature Algorithm
ECI	Échange d'Images Chèques — Exchange of Check Images
ECML	Electronic Commerce Modeling Language
ECP	Electronic Check Presentment
EDE	Encryption-Decryption-Encryption
EDI	Electronic Data Interchange
EDIFACT	Electronic Data Interchange for Administration, Commerce and Transport
EDIINT	EDI Internet Integration

EEG	EBES Expert Group
EEPROM	Electrically Erasable Programmable Read-Only Memory
EF	Elementary File
EFF	Electronic Frontier Foundation
EFT	Electronic Funds Transfer
EFTA	European Free Trade Association
EIC	Échange d'Images Chèques — Exchange of Truncated Checks
EID	Electronic Identity Card (in Sweden)
EIPP	Electronic Invoice Payment and Presentation
EMV	EuroPay, MasterCard, Visa
EPC	Every Penny Counts, Inc.
EPH	Electronic Payments Handler
EPO	Electronic Payment Order
EPOID	Electronic Payment Order Identifier
EPROM	Electrically Programmable Read-Only Memory
ERCIM	European Research Consortium for Informatics and Mathematics
ESP	Encapsulating Security Payload
ETEBAC	Échange Télématique entre les Banques et Leurs Clients — Telematic Exchange among Banks and Their Clients
ETSI	European Telecommunications Standards Institute
ETSO	European Science and Technology Observatory
EWG	EDIFACT Work Group
FACNET	Federal Acquisition Computer Network
FAQ	Frequently Asked Questions
FeRAM	Ferrite Random Access Memory
FIC	Federal Insurance Corporation
FinXML	Fixed Income Markup Language
FIXML	Financial Information Exchange Markup Language
FpML	Financial Products Markup Language
FSML	Financial Services Markup Language
FSTC	Financial Services Technology Consortium
FTP	File Transfer Protocol
GALIA	Groupement pour l'Amélioration des Liens dans l'Industrie Automobile — Group for the Improvement of Ties in the Automobile Industry
GDS	Goppinger Datenservice
Gie	Groupement d'Intérêt Économique

GMD	Gesellschaft für Mathematik und Datenverarbeitung
GMT	Greenwich Mean Time
GOCPKI	Government of Canada Public Key Infrastructure
GPRS	General Packet Radio Service
GSM	Groupe Spécial Mobile — Global System for Mobile Communication
GTA	Global Trust Authority
GTDI	General-Purpose Trade Data Interchange
HA-API	Human Authentication-Application Program Interface
HEDIC	Healthcare EDI Coalition
HHA	Handheld Authenticator
HIBCC	Health Industry Business Communications Council
HMAC	Hashed Message Authentication Code
HTML	HyperText Markup Language
HTTP	HyperText Transfer Protocol
HyTime	Hypermedia/Time-Based Document Structuring Language
IAD	Issuer Authentication Data
IADF	Internal Application Data File
IANA	Internet Assigned Numbers Authority
IATA	International Air Transport Association
IBA	Italian Banking Association
IBIA	International Biometric Industry Association
ICMP	Internet Control Message Protocol
IDEA	International Data Encryption Algorithm
IEC	International Electrotechnical Commission
IETF	Internet Engineering Task Force
IETM	Interactive Electronic Technical Manuals
IFTM	International Forwarding and Transport Message
IFX	Interactive Financial Exchange
IKE	Internet Key Exchange
IMAP	Internet Message Access Protocol
INCITS	InterNational Committee for Information Technology Standards
INRIA	Institut National de Recherche en Informatique et en Automatique
InterNIC	Internet Network Information Center
IP	Internet Protocol

IPSEC	Internet Protocol Security
IRC	Internet Relay Chat
IRML	Investment Research Markup Language
ISAKMP	Internet Security Association and Key Management Protocol
ISDN	Integrated Services Digital Network
ISI	IBM Smartcard Identification (protocol)
ISI	Information Science Institute
ISITC	Industry Standardization for Institutional Trade Communications
ISO	International Organization for Standardization
ISP	Internet Service Provider
ITAR	International Traffic in Arms Regulation
ITLS	Integrated Transport Layer Security
ITU	International Telecommunication Union
ITU-T	International Telecommunication Union — Telecommunication Standardization Sector
JEPI	Joint Electronic Payment Initiative
JPEG	Joint Photographic Expert Group
JRT	Joint Rapporteurs Team
KEA	Key Exchange Algorithm
L2TP	Layer 2 Tunneling Protocol
LACES	London Airport Cargo EDP Scheme
LDAP	X.500 Lightweight Directory Access Protocol
LETS	Local Exchange Trading System
LICRA	Ligue Internationale Contre le Racisme et l'Antisémitisme — International League against Racism and Anti-Semitism
LSAM	Loading Secure Application Module
LVMH	Louis Vuitton–Moët–Hennesy
MAC	Message Authentication Code
MD	Message Digest
MDDL	Market Data Definition Language
MDG	Message Development Group
MEL	MULTOS Executable Language
MEMS	Microelectromechanical System
MF	Master File
MFC	Multifunction Card
MIA	Merchant-Initiated Authorization
MIA	Mortgage Industry Architecture
MIC	Message Integrity Check
MICR	Magnetic Ink Character Recognition

MIME	Multipurpose Internet Mail Extensions
MISPC	Minimum Interoperability Specification for PKI Components
MIT	Massachusetts Institute of Technology
MITL	Multi-Industries Transport Label
MOSET	Merchant-Originated SET
MPI	Merchant Server Plug-in
MRO	Maintenance, Repair, and Operations
MTA	Message Transfer Agent (X.400 messaging)
MUSCLE	Movement for the Use of Smart Cards in a Linux Environment
NACHA	National Automated Clearing House Association
NAETEA	Network-Assisted End-To-End Authentication
NAS	Network Access Server
NASP	National Association of State Purchasing Officials
NFS	Network File System
NIST	National Institute of Standards and Technology
NMAC	Nested Message Authentication Code
NMDS	Narrowband Multiservice Delivery System
NNTP	Network News Transfer Protocol
NPP	Network Payment Protocol
NSA	National Security Agency
NTM	Network Trade Model
NVM	Nonvolatile Memory
NWDA	National Wholesale Druggists Association
NYCH	New York Clearing House
OAEP	Optimal Asymmetric Encryption Padding
OASIS	Organization for the Advancement of Structured Information Standards
OBI	Open Buying on the Internet
OCF	Open Card Framework
OCSP	Online Certificate Status Protocol
ODA	Open Document Architecture
ODETTE	Organisation des Données Échangées par Télétransmission en Europe — Organisation for Data Exchange and Teletransmission in Europe
OECD	Organization for Economic Cooperation and Development
OFB	Output Feedback
OFTP	ODETTE File Transfer Protocol
OFX	Open Finance Exchange
OI	Order Information

OSI	Open Systems Interconnection
OSPF	Open Short Path First
OTP	Open Trading Protocol
P3P	Platform for Privacy Preference
PACES	Paperless Automated Check Exchange and Settlement
PACK	Personal Authentication and Confirmation Kit
PAN	Primary Account Number
PC	Personal Computer
PCA	Primary Certification Authority
PEDI	Protocol EDI
PEP	Protocol Extension Protocol
PESIT	Protocole de Transfert de Fichier pour le Système Interbancaire de Télécompensation — File Transfer Protocol for the Interbanking System for Remote Clearance and Settlement
PGP	Pretty Good Privacy
PI	Payment Instructions
PICS™	Platform for Internet Content Selection
PIN	Personal Identification Number
PIP	Partner Interface Processes
PKCS	Public Key Cryptography Standards
PKI/PKIX	Public Key Infrastructure
PKP	Public Key Partners
PMI	Privilege Management Infrastructure
PNNI	Private Network-to-Network Interface
PNS	Paris Net Settlement
POP	Post Office Protocol
POSA/R	Point-of-Sale Activation and Recharge
POST	Point-of-Sale Terminal
PPP	Point-to-Point Protocol
PPT	Payment Proof Ticket
PROM	Programmable Read-Only Memory
PRT	Payment Request Ticket
PSAM	Purchase Secure Application Module
PSI	Payment System Interface
PSTN	Public Switched Telephone Network
RA	Root Authority
RADIUS	Remote Authentication Dial-in User Service
RADSL	Rate-Adaptive Digital Subscriber Line
RAM	Random Access Memory
RBAC	Role-Based Access Control

RCP	Reference Control Parameter
RFC	Request for Comment
ROM	Read-Only Memory
RPC	Remote-Procedure Call
RPPS	Remote Payment and Presentment Service
RRES	Réseaux Récriproques d'Échange de Savoirs
RTGS	Real-Time Gross Settlement
RTP	Real-Time Protocol
S/MIME	Secure Multipurpose Internet Mail Extensions (Secure MIME)
S/WAN	Secure Wide Area Network
SACK	Server Authentication and Certification Kit
SAFARI	Système Automatisé pour les Fichiers Adminis-tratifs et le Répertoire des Individus — Auto-mated System for Administrative Files and Individuals Directory
SAGITTAIRE	Système Automatique de Gestion Intégrée par Télétransmission de Transactions avec Imputa-tion de Règlements Étrangers
SAIC	Science Applications International Corporation
SAM	Security Application Module
SAML	Security Assertion Markup Language
SAP	Systems, Applications, Products
SASL	Simple Authentication and Security Layer
SCSSI	Service Central de la Sécurité des Systèmes d'In-formation
SDMI	Secure Digital Music Initiative
SDML	Signed Document Markup Language
SDSI	Simple Distributed Security Infrastructure
SEIS	Secured Electronic Information in Society
SEL	Systèmes d'Échange Locaux
SEMPER	Secure Electronic Marketplace for Europe
SET	Secure Electronic Transaction
SET SCCA	SET Compliance Certification Authority
SETREF	SET Reference Implementation
SGML	Standard Generalized Markup Language
S-HTTP	Secure HyperText Transfer Protocol
SHA	Secure Hash Algorithm
SIA	Security Industry Association
SIC	Swiss Interbank Clearing

SIMPRORANCE	Comité Français pour la Simplification des Procédures du Commerce Internationale — French Committee for the Simplification of Procedures of International Commerce
SIPS	Service Internet de Paiement Sécurisé — Secure Internet Payment Service
SIT	Système Interbancaire de Télécompensation — Interbanking Clearance and Settlement System
SITA	Société Internationale de Télécommunications Aéronautiques — International Society for Aeronautical Telecommunications
SITPRO	Simplification of International Trade Procedures
SKIP	Simple Key Management for Internet Protocols
SMS	Short Message Service
SMTP	Simple Mail Transfer Protocol
SNMP	Simple Network Management Protocol
SNNTP	Secure Network News Transfer Protocol
SNP	Système Net Protégé
SOA	Source of Authority
SOAP	Simple Object Access Protocol
SRAM	Static Random Access Memory
SSB	Società per i Servizi Banacari
SSC	Serial Shipment Container Code
SSH®	Secure Shell®
SSL	Secure Sockets Layer
SSO	Single Sign-On
STP	Straight-through Processing
STPEML	Straight-through Processing Extensible Markup Language
SWIFT	Society for Worldwide Interbank Financial Telecommunications
SwiftML	Society for Worldwide Interbank Financial Telecommunication Markup Language
TACACS	Terminal Access Controller Access System
TARGET	Trans-European Automated Real-Time Gross Settlement Express Transfer System
TBF	Transferts Banque de France
TC	Transaction Certificate
TCP	Transmission Control Protocol
TD	Transaction Data
TDCC	Transportation Data Coordinating Committee
TDFC	Transfer de Données Fiscales et Comptables — Transfer of Fiscal and Accounting Data

TDI	Trade Data Interchange
TEDIS	Trade Electronic Data Interchange System
TEK	Token Encryption Key
TEP	Terminal for Electronic Payment
Tep	Titre Électronique de Paiement — Electronic Payment Title
TFM	Transaction File Manager
TGS	Ticket Granting Server
TID	Transaction ID
TIFF	Tagged Image File Format
Tip	Titre Interbancaire de Paiement — Interbank Payment Title
TLS	Transport Layer Security
TMN	Telecommunications Management Network
TTC	Terminal Transaction Counter
UBL	Universal Business Language
UCC	Uniform Code Council
UCC	Uniform Commercial Code
UCS	Uniform Communication Standards
UDDI	Universal Description, Discovery, and Integration
UDEF	Universal Data Element Framework
UDP	User Datagram Protocol
UEJF	Union des Étudiants Juifs de France — Jewish Student Union of France
UMTS	Universal Mobile Telecommunication System
UN/ECE	United Nations Economic Commission for Europe
UNCID	United Nations Rules of Conduct for Interchange of Trade Data by Teletransmission
UNCITRAL	United Nations Commission on International Trade Law
UNCL	United Nations Code List
UN-JEDI	United Nations Joint Electronic Data Interchange
UN-TDI	United Nations Trade Data Interchange
UNI	User Network Interface
UPP	Universal Payment Preamble
URL	Uniform Resource Locator
USC	University of Southern California
VAN	Value-Added Network
VAT	Value-Added Tax
VDSL	Very High bit Rate Digital Subscriber Line
VLSI	Very Large-Scale Integration

VPN	Virtual Private Network
W3C	World Wide Web Consortium
WAN	Wide Area Network
WAP	Wireless Application Protocol
WCT	WIPO Copyright Treaty
WDP	Wireless Datagram Protocol
WEEB	West European EDIFACT Board
WIM	Wireless Identification Module
WINS	Warehouse Information Network Standard
WIPO	World Intellectual Property Organization
WML	Wireless Markup Language
WOIP	World Organization for Intellectual Property
WPTT	WIPO Performance and Phonogram Treaty
WSDL	Web Services Description Language
WTLS	Wireless Transport Layer Security
WTP	Wireless Transaction Protocol
xBRL	Extensible Business Reporting Language
XDR	External Data Representation
XFRML	Extensible Financial Reporting Markup Language
XHTML	Extensible HypertText Markup Language
XHTMLMP	XHTML Mobile Profile
X-KISS	XML Key Information Service Specification
XKMS	XML Key Management Specification
X-KRSS	XML Key Registration Service Specification
XML	Extensible Markup Language
XML-DSIG	XML Digital Signature
XOR	Exclusive OR
ZKA	Zentraler Kreditausschuß

References

3-D Secure, 3-D Secure™ — System Overview V.1.0.2, September 26; 3-D Secure™ — Protocol Specifications, Core Functions, V.1.02, July 16; 3-D Secure™ — Functional Requirements: Access Control Server, V.1.02, July 16; 3-D Secure™ — Functional Requirements: Merchant Server Plug-in V.1.02, July 16a. Available at http://www.visaeu.com. 2002

Abad Peiro, J.L., Asokan, N., Steiner, N., and Waidner, M., Designing a Generic Payment Service, IBM Research Report RZ 2891, September, 1996.

Abad Peiro, J.L., Asokan, N., Steiner, M., and Waidner, M., Designing a generic payment service, *IBM Syst. J.*, 37, 1, 72–88, 1998.

Abadi, M. and Needham, R., Prudent engineering practice for cryptographic protocols, *IEEE Trans. Software Eng.*, 22, 1, 6–15, 1996.

Adams, J., Friends or foes? *Eur. Card Rev.*, 5, 2, 15, 1998.

Adams, J., On the right lines? *Eur. Card Rev.*, 9, 1, 12–17, 2002.

AF-SEC-0179.000, Methods for securely managing ATM network elements — Implementation agreement, Version 1.0, ATM Forum Technical Committee, April, 2002.

Agnew, G.B., Cryptography, data security, and applications to e-commerce, in *Electronic Commerce Technology Trends: Challenges and Opportunities*, Kou, W. and Yesha, Y., Eds., IBM Press, Toronto, 2000, pp. 69–85.

Althen, B., Enste, G., and Nebelung, B., Innovative secure payments on the Internet using the German electronic purse, in *Proc. 12th Annu. Comput. Security Appl. Conf.*, December 9–13, 1996, pp. 88–93.

Anderson, R.J. and Bezuidenhoudt, S.J., On the reliability of electronic payment systems, *IEEE Trans. on Software Eng.*, 22, 5, 294–301, 1996.

Anderson, R. and Kuhn, M., Tamper resistance — A cautionary note, in *Proc. 2nd USENIX Workshop on Electron. Commerce*, Oakland, CA, 1996, 1–11.

Anderson, R.J., Cox, I.J., Low, S.H., and Maxemchuck, N.F., Eds., Copyright and privacy protection, *IEEE J. Selected Areas Commun.*, 16, 4, 1998.

ANSI INCITS 358, Information Technology — BioAPI Specifications (Version 1.1), 2002.

ANSI X3.92, American National Standard — Data Encryption Algorithm (DEA), 1981.

ANSI X3.105, American National Standard — Data Link Encryption, 1983.

ANSI X3.106, American National Standard — Data Encryption Algorithm, Modes of Operations, 1983.

ANSI X9.9 (revised), American National Standard — Financial Institution Message Authentication (Wholesale), American Bankers Association, (replaces X9.9-1982), 1986.

ANSI X9.17-1985, American National Standard — Financial Institution Key Management (Wholesale), American Bankers Association, 1985. This is the basis for ISO 8732.

ANSI X9.19, American National Standard — Financial Institution Retail Message Authentication, American Bankers Association, 1986.

ANSI X9.30:1-1997, American National Standard — Public Key Cryptography for the Financial Services Industry: Part 1: The Digital Signature Algorithm (DSA), (Revision of X9.30:1-1995), American Bankers Association, 1997.

ANSI X9.30:2-1993, American National Standard — Public Key Cryptography for the Financial Services Industry: Part 2: The Secure Hash Algorithm 1 (SHA-1), (Revised), American Bankers Association, 1993.

ANSI X9.37-1994, American National Standard — Specification for Electronic Check Exchange, 1994.

ANSI X9.46-1997, American National Standard — Financial Image Interchange: Architecture, Overview, and System Design Specification, 1997.

ANSI X9.52-1998, American National Standard — Triple Data Encryption Algorithm Modes of Operation, 1998.

ANSI X9.57-1997, American National Standard — Public Key Cryptography for the Financial Services Industry: Certificate Management, (Revision of X9.57-1995), 1997.

ANSI X9.62-1998, American National Standard for Financial Services — Public Key Cryptography for the Financial Services Industry: The Elliptic Curve Digital Signature Algorithm (ECDSA), American Bankers Association, 1998.

ANSI X9.68:2-2001, American National Standard — Digital Certificates for Mobile/Wireless and High Transaction Volume Financial Systems: Part 2: Domain Certificate Syntax, 2001.

ANSI X9.84-2001, American National Standard — Biometric Information Management and Security, 2001.

Asokan, N., Janson, P.A., Steiner, M., and Waidner, M., The state of the art in electronic payment systems, *IEEE Comput.*, 30, 9, 28–35, 1997.

Austin, D., Chips with everything, *Banking Technol.*, 18, 4, 53, 2001.

Baldwin, R.W. and Chang, C.V., Locking the e-safe, *IEEE Spectrum*, 34, 2, 40–46, 1997.

Banes, J. and Harrington, R., *56-bit Export Cipher Suites for TLS*, Internet draft, July 19, in *Proc. 51st IETF*, Muñoz, J., and Syracuse, N., Eds., Corporation for National Research Initiatives, Reston, VA, 2001. Available at http://www.ietf.org/proceedings/01aug/i-d/draft-ietf-tls-56-bit-ciphersuites-01.txt.

Bank for International Settlements, Committee on Payment and Settlement Systems, *Statistics on Payment Systems in the Group of Ten Countries, Figures for 1995*, Basel, December, 1996. Available at http://www.bis.org.

Bank for International Settlements, Committee on Payment and Settlement Systems, *Statistics on Payment Systems in the Group of Ten Countries, Figures for 1996*, Basel, December, 1997. Available at http://www.bis.org.

Bank for International Settlements, Committee on Payment and Settlement Systems, *Statistics on Payment Systems in the Group of Ten Countries, Figures for 1998*, Basel, February, 2000. Available at http://www.bis.org.

Bank for International Settlements, Committee on Payment and Settlement Systems, *Statistics on Payment Systems in the Group of Ten Countries, Figures for 1999*, Basel, March, 2001. Available at http://www.bis.org.

Bank for International Settlements, Committee on Payment and Settlement Systems, *Statistics on Payment Systems in Selected Countries, Figures for 2000*, Basel, July, 2002. Available at http://www.bis.org.

Banking Technology, UK business slow on e-commerce, *Banking Technol.*, 15, 4, 14, 1998.

Barbaux, A., Quel moyen de paiement pour le Net? *Internet Prof.*, 24, 72–79, 1998.

Bellare, M. and Rogaway, P., Optimal asymmetric encryption — how to encrypt with RSA, in *Adv. Cryptology — Eurocrypt'94, Workshop on the Theory and Application of Cryptographic Techniques*, Perugia, Italy, 1994, DeSantis, A., Ed., *Lect. Notes Comput. Sci.*, Springer-Verlag, Berlin, 1995, pp. 92–111.

Bellare, M., Canetti, R., and Krawczyk, H., Keying hash functions for message authentication, in *Adv. Cryptology — CRYPTO'96 Proc.*, Springer-Verlag, Heidelberg, 1996, pp. 1–15.

Berget, P. and Icard, A., *La Monnaie et ses Mécanismes, Que sais-je?* 12th ed., No. 1217, Presses Universitaires de France, Paris, 1997.

Berman, D., Card sharps, Business *Week E-Biz*, April 3, 2000, EB68–76.

Bidgood, A., VADs interworking: a cloud on the EDI horizon, in *The EDI Handbook*, Gikins, M. and Hitchcock, D., Eds., Blenheim Online, London, 1988, pp. 211–261.

Billaut, J.-M., Les intermédiaires sont-ils condamnés? in *l'Internet et la Vente*, Aumetti, J.-P., Ed., Les Éditions d'Organisation, Paris, 1997, pp. 63–73.

Bochmann, G.V., Kerhervé, B., Lutfiyya, H., Salem, M.-V.M., and Ye, H., Introducing QoS to electronic commerce applications, in *Electronic Commerce Technologies, Proc. 2nd Int. Symp.*, ISEC 2001, Hong Kong, China, Kou, W., Yesha, T., and Tan, C.J., Eds., Lect. Notes Comput. Sci., 2040, Springer-Verlag, Heidelberg, 2001, pp. 137–478.

Böhle, K., Rader, M., and Riehm, U., Eds., *Electronic Payment Systems in European Countries — Country Synthesis Report — Final Report*, Forschungszentrum Karlsruhe GmBH, Karlsruhe, 1999.

Boly, J.-P., Bosselaers, A., Cramer, R., Michelsen, R., Mjølsnes, S., Muller, F., Pedersen, T., Pfitzmann, B., de Rooij, P., Schoenmakers, N., Schunter, M., Vallée, L., and Waidner, M., The ESPRIT project CAFE — High security digital payment systems, in *ESORICS'94*, Lect. Notes Comput. Sci., 875, Springer-Verlag, Heidelberg, 1994, pp. 217–230.

Boneh, D. and Shacham, H., Fast variants of RSA, *CryptoBytes*, 5, 1, 1–9, 2002. Available at http://www.rsa.com/rsalabs/pubs/cryptobytes.html.

Bonnaure, P., Big Brother, acte II, *Futuribles*, 244, 131–135, 1999.

Bontoux, S., Accélerez le chiffrement SSL de votre serveur, *Internet Prof.*, 64, 60–63, 2002.

Borchers, A. and Demski, M., The value of Coin networks: the case of Automotive Network Exchange®, in *Organizational Achievement and Failure in Information Technology Management*, Khosrowpour, M., Ed., IDEA Group Publishing, Hershey, PA, 2000, pp. 109–123.

Borenstein, N.S. et al., Perils and pitfalls of practical cybercommerce, *Commun. ACM*, 39, 6, 37–44, 1996.

Borst, J., Preneel, B., and Rijmen, V., Cryptography on smart cards, *Comput. Networks*, 36, 423–435, 2001.

Bottelier, P., US dollars circulating abroad put aid in another perspective, *Financial Times*, April 3, 14, 2002.

Boucher, X., Le m-commerce brise ses chaînes, *Internet Prof.*, 52, 48–51, April 2001.

Bouillant, O., *Messageries électroniques*, Eyrolles, Paris, 1998.

Box, D., Ehnebuske, D., Kakivaya, G., Layman, A., Mendelsohn, N., Nielsen, H.F., Thatte, S., and Winer, D., Simple Object Access Protocol (SOAP) 1.1, W3C Note, May 8, 2000. Available at http://www.w3org/TR/SOAP.

Brands, S., Untraceable off-line electronic cash in wallets with observers, in *Adv. Cryptology — Crypto'93, Proc.*, Springer-Verlag, Heidelberg, 1993, pp. 302–318.

Brayshaw, E., A chip off the old fraud, *Banking Technol.*, 18, 3, 10, 2001.

Bresse, P., Beaure d'Augères, G., and Thuillier, S., *Paiement Numérique sur Internet*, Thomson Publishing France, Paris, 1997.

Breton, T., *Les Téléservices en France. Quels Marchés pour les Autoroutes de l'Information?*, La documentation française, Paris, 1994.

Bryan, M., Ed., Guidelines for Using XML for Electronic Data Interchange, Version 0.05, January 25, 1998. Available at http://www.geocities.com/WallStreet/Floor/5915/guide.html.

Buliard, F., French to migrate cards, *Banking Technol.*, 18, 3, 8, April, 2001.

Burke, B.V., *Automatic Philanthropic Contribution System*, U.S. Patent 5621 640, April 15, 1997.

Burr, W., Dodson, D., Nazaria, N., and Polk, W.T., MISPC — Minimum Interoperability Specifications for PKI Components, Version 1, September 3, 1997, NIST Special Publication 800-15.

Cabinet FG Associés [online], Message Électronique et Droit de la Preuve: Un Concubinage Conflictuel. Available at http://www/fgassocies.com/m1/internet/a12.html.

Cafiero, W.G., International standards for EDI/EFT, *EDI Forum*, 4, 2, 74–77, 1991.

Camp, L.J., Sirbu, M., and Tygar, J.D., Token and notational money in electronic commerce, in *Proc. 1st USENIX Workshop Electron. Commerce*, New York, 1995, pp. 1–12.

Le Canard enchaîné, Le minitel est "plus branché que jamais," April 25, 2001.

Carasik, A., Secure Shell FAQ, revision 1.4, http://www.tigerlair.com/ssh/faq, February, 2001.

Catinat, M., Internet: la politique européenne, *COMMUN. STRATEGIES*, 35, 199–205, 1999.

CEN EN 1546-1, European Committee for Standardization, Identification Card Systems — Inter-sector Electronic Purse — Part 1: Definitions, Concepts and Structures, 1999.

CEN EN 1546-2, European Committee for Standardization, Identification Card Systems — Inter-sector Electronic Purse — Part 2: Security Architecture, 1999.

CEN EN 1546-3, European Committee for Standardization, Identification Card Systems — Inter-sector Electronic Purse — Part 3: Data Elements and Interchanges, 1999.

CEN EN 1546-4, European Committee for Standardization, Identification Card Systems — Inter-sector Electronic Purse — Part 4: Data Objects, 1999.

CEPSO, Common Electronic Purse Specifications — Functional Requirements, Version 6.3, September, 1999. Available at http://www.visa.com/pd/cash/ceps.html.

CERT/CC CA-2001-19, CERT® Advisory CA-2001-19 — "Code Red" worm exploiting buffer overflow in IIS indexing service DLL, last revised January 17, 2002. Available at http://www.cert.org/advisories/CA-2001-19.html.

Chadwick, D.W., and Ottenko, A., RBAC policies in XML for X.509 based privilege management, in *Security in the Information Society: Visions and Perspectives*, Ghonaimy, M.A., El-Hadidi, M.T., and Aslan, H.K., Eds., Kluwer, Dordrecht, 2002, pp. 39–53.

Chan, A., Frankel, Y., MacKenzie, P., and Tsiounis, Y., Mis-representation of identities in E-cash schemes and how to prevent it, in *Adv. Cryptology — ASIACRYPT'96*, Springer-Verlag, Heidelberg, 1996, pp. 276–285.

Chang, R.K.C., Defending against flooding-based distributed denial-of-service attack: a tutorial, *IEEE Commun. Mag.*, 40, 10, 42–51, 2002.

Charmot, C., EDI, L'Échange de Données Informatisé: Définition et Enjeux, *Télécom Interview*, 33, 8–10, 1997a.

Charmot, C., *L'Echange de Données Informatisé (EDI)*, *Que Sais-je*? No. 3321, Presses Universitaires de France, Paris, 1997b.

Chaum, D., Blind signatures for untraceable payments, in *Crypto 82*, Plenum Press, New York, 1983, pp. 199–203.

Chaum, D., Privacy protected payments: unconditional payer and/or payee untraceability, in *Smart Card 2000*, Chaum, D. and Schaumüller-Bichl, I., Eds., Elsevier/North Holland, Amsterdam, 1989, pp. 69–93.

Chaum, D. [online], On line cash checks. Available at http://www.digicash.com/publish/online.html.

Chaum, D., Fiat, A., and Naor, M., Untraceable electronic cash, in *Adv. Cryptology — CRYPTO'88*, Springer-Verlag, Heidelberg, 1990, pp. 319–327.

Chavanne, R. and Paris, L., L'alternative des logiciels libres, *Internet Prof.*, 24, 80–83, 1998.

Cheng, P.-C., An architecture for the Internet Key Exchange Protocol, *IBM Syst. J.*, 40, 3, 721–746, 2001.

Cho, S., Customer-focused Internet commerce at Cisco Systems, *IEEE Commun. Mag.*, 37, 9, 61–63, 1999.

Christensen, E., Curbera, F., Merideth, G., and Weerawarana, S., Web Services Description Language (WSDL) 1.1, W3C Note, March 15, 2001. Available at http://www.w3.org/TR/wsdl.

Clapaud, A., Faut-il investir dans l'ebXML? *Internet Prof.*, 62, 40–42, 2002.

Comer, D.E., *Interworking with TCP/IP Vol. I: Principles, Protocols, and Architecture*, 3rd ed., Prentice Hall, Englewood Cliffs, NJ, 1995.

CommerceNet, Lessons learned from implementing MIME-based secure EDI over Internet, August 31, 1997.

Coutrot, F. and Pommier, J.P., Network security using smart card and Lecam, in *Smart Card 2000, Proc. IFIP WG 11.6 Int. Conf. Smart Card 2000: The Future of IC Cards*, Laxenburg, Austria, October, 1987, Chaum, D. and Schaumüller-Bichl, I., Eds., Elsevier/North Holland, Amsterdam, 1989, pp. 187–199.

Cox, B., Tygar, J.D., and Sirbu, M., NetBill security and transaction protocol, in *Proc. 1st USENIX Workshop Electron. Commerce*, New York, 1995, pp. 77–78.

Cranor, L.F., *Web Privacy with P3P*, O'Reilly and Associates, Sebastopol, CA, 2002.

Craver, S.A., Wu, M., Liu, B., Stubblefield, A., Swartzlander, B., Wallach, D.S., Dean, D., and Felten, E.W., Reading between the lines: Lessons from the SMDI challenge, *Proc. 10th USENIX Security Symposium*, Washington, D.C., August 13–17, 2001.

Cuijpers, E.P.E., The ISI-Protocol v2.0 Implementation at the Server of the R*u*G, Version 1.1, Eindhoven University of Technology, Department of Mathematics and Computing Science, 1997. Available at http://www.iscit.surfnet.nl/team/Erik/masterh/isi2implt.htm.

Daaboul, T., Spécification d'une Plate-Forme de Personnalisation de Carte à Puce, rapport de thèse professionnelle, ENST, Paris, 1998.

Dabbish, E.A. and Sloan, R.H., Investigations of power analysis attacks on smartcard, in *Proc. USENIX Workshop on Smartcard Technol.*, Chicago, IL, 1999, pp. 151–161.

Datapro Information Services, SITA Group Network Services, March 1998.

Daugman, J.G., *Recognizing Persons by Their Iris Patterns*, U.S. Patent 5 291 560, 1994.

Daugman, J.G., Biometric personal identification system based on iris analysis, in *Biometrics: Personal Identification in Networked Society*, Jain, A., Bolle, R., and Pankati, S., Eds., Kluwer, Dordrecht, 1999, pp. 104–121.

Dawirs, M., Porte-monnaie électronique: le procotole normalisé par le CEN et l'implantation Proton, *6ème Colloque Francophone sur l'Ingénieurie des Protocoles*, Liège, Belgium, 1997.

Dawkins, P., Open communications standards: their role in EDI, in *The EDI Handbook*, Gikins, M. and Hitchcock, D., Eds., Blenheim Online, London, 1988, pp. 56–73.

de Galzain, P., Electronic Data Interchange (EDI) in the automotive industry, in *EuroComm 88, Proc. Int. Congr. Bus., Public and Home Commun.*, Amsterdam, Schuringa, T.M., Ed., Elsevier/North Holland, Amsterdam, 1989, pp. 259–264.

de Lacy, J., France, the sexy computer, *Atlantic Monthly*, 18–26, July 1987.

Dean, D. and Stubblefield, A., Using client puzzles to protect TLS, in *Proc. 10th USENIX Security Symposium*, Washington, D.C., August 13–17, 2001.

Décision Micro & Réseaux, Le chiffre de la semaine: 1 milliard de spams en 24 h, 542, 8, 17 March 2003.

del Pilar Barea Martinez, M., EDI: des organismes fédérateurs pour le commerce international, *Télécom Interview*, 33, 12–14, 1997.

Diffie, W. and Hellman, M.E., New directions in cryptography, *IEEE Trans. Info. Theory*, 22, 6, 644–654, 1976.

Digital Equipment Corporation, The Millicent Microcommerce System: Defining a New Internet Business Model, 1995. Available at http://www.millicent.com/html/executive-overview.html.

Dobbertin, H., Bosselaers, A., and Preneel, B., RIPEMD-160: a strengthened version of RIPEMD, in *Proc. Fast Software Encryption Workshop*, 1996, pp. 71–82.

Doraswamy, N. and Harkins, D., *IPSec: The New Security Standard for the Internet, Intranets, and Virtual Private Networks*, Prentice Hall PTR, Upper Saddle River, NJ, 1999.

Dowland, P.S., Furnell, S.M., and Papadaki, M., Keystroke analysis as a method of advanced user authentication and response, in *Security in the Information Society: Visions and Perspectives*, Ghonaimy, M.A., El-Hadidi, M.T., and Aslan, H.K., Eds., Kluwer, Dordrecht, 2002, pp. 215–226.

Dragon, C., Geiben, D., Kaplan, D., and Nallard, G., *Les Moyens de Paiement: Des Espèces à la Monnaie Électronique*, Banque Editeur, Paris, 1997.

Dreifus, H. and Monk, J.T., *Smart Cards*, John Wiley & Sons, New York, 1998.

Du Pré Gauntt, J., Digital currency and public networks: so what if it is secure, is it money? in *USENIX Workshop Electro. Commerce*, Oakland, CA, 1996, pp. 77–86.

Dupoirier, G., *Technologie de la GED: Technique et Management des Documents Électroniques*, 2nd ed., Hermès, Paris, 1995.

Eaglen, C., The relationship between EDI and electronic funds transfer, in *The EDI Handbook*, Gikins, M. and Hitchcock, D., Eds., Blenheim Online, London, 1988, pp. 85–103.

ElGamal, T., A public-key cryptosystem and a signature scheme based on discrete logarithms, in *Adv. Cryptology — CRYPTO'84 Proc.*, Springer-Verlag, Heidelberg, 1985, pp. 10–18.

Emmelhainz, M.A., *EDI: A Total Management Guide*, 2nd ed., Van Nostrand Reinhold, New York, 1993.

EMV'96, Integrated Circuit Card Specification for Payment Systems, Version 3.1.1, May, 1998. Available at http://www.visa.com or http://www.mastercard.com.

EMV2000, Integrated Circuit Card Specification for Payment Systems, Version 4.0, December, 2000. Available at http://www.visa.com or http://www.master-card.com.

Eng, T. and Okamoto, T., Single-term divisible electronic coins, in *Adv. Cryptology — Eurocrypt'94, Proc.*, Springer-Verlag, Heidelberg, 1994, pp. 306–319.

Enoki, K., Concept of i-mode service — new communication infrastructure for the 21st century, *NTT DoCoMo Tech. J.*, 1, 1, 4–9, 1999.

European Commission, Recommendation 94/820 on October 19, publication L 339, December 28, 1994.

European Telecommunications Standards Institute — ETSI STC TE9 1993, Terminal Equipment: Requirements for IC Card and Terminals for Telecommunication Use. Part 3: Application Independent Card Requirements, prEN726-3 Version 14, September 1993.

European Telecommunications Standards Institute — ETS 301 141-1-1997, Signalling Protocols and Switching: Narrowband Multi-Service Delivery System (NMDS). Part 1: NMDS Interface Specification, 1997.

Even, S. and Goldreich, O., Electronic wallet, in *Adv. Cryptology, Proc. Crypto'83*, Chaum, D., Ed., 1983, pp. 383–386.

Fallon, T. and Welch, B., BACS — practical control issues, in *Electronic Banking and Security*, Welch, B., Ed., Blackwell Scientific, Oxford, 1994, pp. 31–42.

Fay, A., *Dico Banque*, La Maison du Dictionnaire, Paris, 1997.

Feigenbaum, J., Towards an infrastructure for authorization, in *Invited Talks on Public Key Infrastructure, 3rd USENIX Workshop Electron. Commerce*, 1998, pp. 15–19.

Financial Services Technology Consortium (FTSC) [online a], Check Image Exchange Project. Available at http://www.ftsc.org/projects/imaging/index.html.

Financial Services Technology Consortium (FTSC) [online b], Echeck Q and A. Available at http://www.echeck.org/kitprint/q&at/q-a.htm.

Financial Services Technology Consortium (FTSC), FSML Version 1.17, 1998a.

Financial Services Technology Consortium (FTSC), BIPS Specification, V.1.0, 1998b. Available at http://www.fstc.org/projects/index.html.

Flom, L. and Safir, A., Iris Recognition System, U.S. Patent 4 641 349, 1987.

Ford, W. and Baum, M.S., *Secure Electronic Commerce: Building the Infrastructure*, Prentice-Hall, Englewood Cliffs, NJ, 1997.

Ford, W. and O'Higgins, B., Public-key cryptography and open systems interconnection, *IEEE Commun. Mag.*, 39, 7, 30–35, 1992.

Forrester, S.E., Palmer, M.J., McGlaughlin, D.C., and Robinson, M.J., Security in data networks, *BT Technol. J.*, 16, 1, 52–75, 1998.

Foucault, G., Alain Grangé Cabane: "Internet passe de la foire au marché," *le Figaro*, September 9, 1996.

France Télécom, *La Lettre de Télétel & Audiotel*, 13, 7–9, June, 1995.

France Télécom, *La Lettre des Services en Ligne Audiotel, Minitel et Internet*, 43, 6–8, 1997.

Freier, A.O., Karlton, P., and Kocher, P.C., The SSL Protocol Version 3.0, 1996. Available at http://www.netscape.com/PROD/eng/ssl3/ssl-toc.html.

Froissard, F., À quoi sert le Net? *Décision Micro & Réseaux*, 494, 6, February 4, 2002.

FT-IT Review, European Webranking 2002, *Financial Times*, November 6, VIII, 2002.

Fu, K., Sit, E., Smith, K., and Feamster, N., Dos and don'ts of client authentication on the Web. Available at http://cookies.lcs.mit.edu. A shorter version was published in the *Proc. 10th USENIX Security Symp.*, Washington, D.C., August 2001.

Fumer, W. and Landrock, P., Principles of key management, *IEEE J. Selected Areas Commun.*, 11, 5, 785–793, 1993.

Futuribles, Disparités des pratiques d'information en Europe, 45, 69, 1999.

Garfinkel, S.L., *PGP: Pretty Good Privacy*, O'Reilly and Associates, Sebastopol, CA, 1995.

GeldKarte, *Schnittstellenspezifikation für die GeldKarte mit Chip*, debis Systemhaus GEI, August 9, 1995.

George, N., SEB refocuses from internet to branches, *Financial Times*, October 25, 19, 2001.

George, T., On a virtual roll, *Banking Technol.*, 15, 4, 48, 1998.

Girolle, M. and Guerin, D., La cryptographie et son utilisation dans le commerce électronique sur Internet, *Télécom*, 111, 37–42, spring 1997.

Glassman, S., Manasse, M., Abadi, M., Gauthier, P., and Sobalvarro, P., The Millicent protocol for inexpensive electronic commerce, in *Proc. 4th Int. World Wide Web Conf.*, 1995. Available at http://www.milicent.com/works/details/papers/millicent-w3c4/millicent.html.

Goralski, W., *ADSL and DSL technologies*, McGraw-Hill, New York, 1998.

Granet, J., Procédure de transfert de données fiscales et comptables, *Télécom Interview*, 33, 15–18, 1997.

Gray, B., Privacy, *login:*, 24, 6, 74–76, 1999.

Groupement des Cartes Bancaires, Chip-Secured Electronic Transaction (C-SET) Security Architecture, Version 1.0, January 29, 1997. Available at http://www.imaginet.fr/~cb-mail.

Guillou, L.C., Davio, M., and Quisquater, J.-J., L'état de l'art en matière de techniques à clé publique: Hasard et redondance, *Ann. Télécommun.*, 43, 9–10, 489–505, 1988.

Guillou, L.C., Ugon, M., and Quisquater, J.-J., Cryptographic authentication protocols for smart cards, *Comput. Networks*, 36, 437–451, 2001.

Hachez, G., Koeune, F., and Quisquater, J.-J., Biometrics, access control, smart cards: A not so simple combination, in *Proc. 4th IFIP WG8.8 Working Conference on Smart Card Research and Advanced Applications (CARDIS 2000)*, Bristol, England, U.K., Domingo-Ferrer, J., Chan, D., and Watson, A., Eds., Kluwer, Dordrecht, 2000, pp. 273–288.

Haesler, A.J., *Sociologie de l'Argent et Postmodernité*, Droz, Genève-Paris, 1995.

Hall, E.T. and Hall, M.R., *Understanding Cultural Differences*, Intercultural Press, Yarmouth, ME, 1990.

Hamman, E.-M., Henn, H., Schäck, T., and Seliger, F., Securing e-business applications using smart cards, *IBM Syst. J.*, 40, 3, 635–647, 2001.

Hansell, S., Got a dime? Citibank and Chase end test of electronic cash, *New York Times*, November 4, C1, 1998.

Hawke, J.D., Jr., Testimony of the Treasury Under Secretary for Domestic Finance, House Government Reform and Oversight, Subcommittee on Government Management, Information and Technology, June 18, RR-1768, 1997. Available at htttp://www.ustreas.gov/press/releases/pr1768.htm.

He, L.-S., Zhang, N., and He, L.-R., A new end-to-end authentication protocol for mobile users to access Internet services, in *Security in the Information Society: Visions and Perspectives*, Ghonaimy, M.A., El-Hadidi, M.T., and Aslan, H.K., Eds. Kluwer, Dordrecht, 2002, pp. 239–250.

Hendry, M., *Implementing EDI*, Artech House, Boston, 1993.

Hermann, R., Husemann, D., and Trommler, P., OpenCard Framework 1.1, October, 1998. Available at http://www.opencard.org/docs.

Hill, J., An analysis of the RADIUS authentication protocol, 2001. Available at http://www.untruth.org/~josh/security/radisu/radius-auth.html.

Hill, R., Minitel: an example of electronic commerce compared to the World Wide Web, *EDI Forum*, 9, 2, 89–96, 1996.

Hill, R., Retina identification, in *Biometrics: Personal Identification in Networked Society*, Jain, A., Bolle, R., and Pankati, S., Eds., Kluwer, Dordrecht, 1999, pp. 123–124.

Huitema, C., *IPv6: The New Internet Protocol*, Prentice-Hall, Englewood Cliffs, NJ, 1996.

Husemann, D., Standards in the smart card world, *Comput. Networks*, 36, 473–487, 2001.

IETF RFC 821, Simple Mail Transfer Protocol, Postel, J., August 1982.

IETF RFC 822, Standard for the Format of Internet Text Messages, Crocker, D., 1982.

IETF RFC 1320, The MD4 Message Digest Algorithm, Rivest, R., April 1992.

IETF RFC 1321, The MD5 Message Digest Algorithm, Rivest, R., April 1992.

IETF RFC 1487, X.500 Lightweight Directory Access Protocol, Yeong, W., Howes, T., and Kille, S., July 1993.

IETF RFC 1492, An Access Control Protocol, sometimes Called TACACS, Finseth, C., July 1993.

IETF RFC 1510, Public-Key Based Ticket Granting Service in Kerberos, Sirbu, M. and Chuang, J., May 6, 1996.

IETF RFC 1661, The Point-to-Point Protocol (PPP), Simpson, W., July 1994.

IETF RFC 1767, MIME Encapsulation of EDI Objects, Crocker, D., March 1995.

IETF RFC 1847, Security Multiparts for MIME. — Multipart/Signed and Multipart/Encrypted, Galvin, J., Murphy, S., Crocker, S., and Freed, N., October 1995.

IETF RFC 1991, PGP Message Exchange Formats, Atkins, D., Stallings, W., and Zimmerman, P., August 1996.

IETF RFC 2015, MIME Security with Pretty Good Privacy (PGP), Elkins, M., September 1996.

IETF RFC 2045, MIME (Multipurpose Internet Mail Extensions) Part One: Mechanisms for Specifying and Describing the Format of Internet Message Bodies, Freed, N. and Borenstein, N., November 1996.

IETF RFC 2046, Multipurpose Internet Mail Extensions (MIME) Part Two: Media Types, Freed, N. and Borenstein, N., November 1996.

IETF RFC 2104, HMAC: Keyed-Hashing for Message Authentication, Krawczyk, H., Bellare, M., and Canetti, R., February 1997.

IETF RFC 2109, HTTP State Management Mechanism, Kristol, D. and Montulli, L., February 1997.

IETF RFC 2222, Simple Authentication and Security Layer (SASL), Myers, J., October 1997.

IETF RFC 2246, The TLS Protocol Version 1.0, Dierks, T. and Allen, C., January, 1999.

IETF RFC 2251, Lightweight Directory Access Protocol v3, Whal, M., Howes, T., and Kille, S., December 1997.

IETF RFC 2311, S/MIME Version 2 Message Specification, Dusse, S., Hoffman, P., Ramsdell, B., Lundblad, L., and Repka, L., March 1998.

IETF RFC 2315, PKCS #7: Cryptographic Message Syntax Version 1.5, Kaliski, B., March 1998.

IETF RFC 2401, Security Architecture for the Internet Protocol, Kent, S. and Atkinson, R., November 1998.

IETF RFC 2402, IP Authentication Header, Kent, S. and Atkinson, R., November 1998.

IETF RFC 2406, IP Encapsulating Security Payload (ESP), Kent, S. and Atkinson, R., November 1998.

IETF RFC 2407, The Internet IP Security Domain of Interpretation for ISAKMP, Piper, D., November 1998.

IETF RFC 2408, Internet Security Association and Key Management Protocol (ISAK-MP), Maughan, D., Schertler, M., Schneider, M., and Turner, J., November 1998.

IETF RFC 2409, The Internet Key Exchange (IKE), Harkins, D. and Carrel, D., November 1998.

IETF RFC 2437, PKCS #1: RSA Cryptography Specifications Version 2.0, Kaliski, B. and Staddon, J., October 1998.

IETF RFC 2440, OpenPGP Message Format, Callas, J., Donnerhacke, L., Finney, H. and Thayer, R., November 1998.

IETF RFC 2510, Internet X.509 Public Key Infrastructure — Certificate Management Protocols, Adams, C. and Farrell, S., March 1999.

IETF RFC 2527, Internet X.509 Public Key Infrastructure — Certificate Policy and Certification Practices Framework, Chokani, S. and Ford, W., March 1999.

IETF RFC 2560, X.509 Internet Public Key Infrastructure — Online Certificate Status Protocol (OCSP), Myers, M., Ankney, R., Malpani, A., Galperin, S., and Adams, C., June 1999.

IETF RFC 2585, Internet X.509 Public Key Infrastructure — Operational Protocols: FTP and HTTP, Housley, R. and Hoffman, P., May 1999.

IETF RFC 2660, The Secure HyperText Transfer Protocol, Rescorla, E. and Schiffman, A., August 1999.

IETF RFC 2661, Layer Two Tunneling Protocol L2TP, Townsley, W., Valencia, A., Rubens, A., Pall, G., Zorn, G., and Palter, B., August 1999.

IETF RFC 2773, Encryption using KEA and SKIPJACK, Housley, R., Yee, P., and Nace, W., February 2000.

IETF RFC 2829, Authentication Methods for LDAP, Wahl, M., Alvestrand, H., Hodges, J., and Morgan, R., May 2000.

IETF RFC 2865, Remote Authentication Dial In User Service (RADIUS), Rigney, C., Willens, S., Rubens, A., and Simpson, W., June 2000.

IETF RFC 2951, TELNET Authentication using KEA and SKIPJAK, Housley, R., Horting, T., and Yee, P., September 2000.

IETF RFC 2986, PKCS #10: Certification Request Syntax Specification, Version 1.7, Nystrom, M. and Kaliski, B., November 2000.

IETF RFC 3106, ECML v1.1: Field Specifications for E-Commerce, Eastlake, D. and Goldstein, T., April 2001.

IETF RFC 3156, MIME Security with OpenPGP, Elkins, M., Del Torto, D., Levien, R. and Roessler, T., August 2001.

IETF RFC 3193, Securing L2TP using IPsec, Patel, B., Adoba, B., Dixon, W., Zorn, G., and Booth, S., November 2001.

IETF RFC 3335, MIME-Based Secure Peer-to-Peer Business Data Interchange over the Internet, Harding, T., Drummond, R., and Shih, C., September 2002.

Industry Canada, Electronic Commerce Task Force, Electronic Commerce in Canada — Priorities for action, July 1998.

International Telecommunication Union — Recommendation X.209, Specifications of Basic Encoding Rules for Abstract Syntax Notation One (ASN.1), 1988.

International Telecommunication Union — Recommendation X.500 (ISO/IEC 9594-1), Information Technology — Open Systems Interconnection — The Directory: Overview of Concepts, Models and Services, 2001.

International Telecommunication Union — Recommendation X.501 (ISO/IEC 9594-2), Information Technology — Open Systems Interconnection — The Directory: Models, 2001.

International Telecommunication Union — Recommendation X.509 (ISO/IEC 9594-8), Information Technology — Open Systems Interconnection — The Directory: Public-Key and Attribute Certificate Frameworks, 2001.

International Telecommunication Union — Recommendation X.511 (ISO/IEC 9594-3), Information Technology — Open Systems Interconnection — The Directory: Overview of Concepts, Models and Services, 2001.

International Telecommunication Union — Recommendation X.518 (ISO/IEC 9594-4), Information Technology — Open Systems Interconnection — The Directory: Procedures for Distributed Operation, 2001.

International Telecommunication Union — Recommendation X.519 (ISO/IEC 9594-4), Information Technology — Open Systems Interconnection — The Directory: Protocol Specifications, 2001.

International Telecommunication Union — Recommendation X.520 (ISO/IEC 9594-6), Information Technology — Open Systems Interconnection — The Directory: Selected Attribute Types, 2001.

International Telecommunication Union — Recommendation X.521 (ISO/IEC 9594-7), Information Technology — Open Systems Interconnection — The Directory: Selected Object Classes, 2001.

International Telecommunication Union — Recommendation X.525 (ISO/IEC 9594-9), Information Technology — Open Systems Interconnection — The Directory: Replication, 2001.

International Telecommunication Union — Recommendation X.500, 1993.

International Telecommunication Union — Recommendation X.800, Security Architecture for Open Systems Interconnection for CCITT Applications, 1991.

International Telecommunication Union — Recommendation X.811, Security Architecture for Open Systems: Authentication Framework, November 1995.

International Telecommunication Union — Recommendation X.812, Information Technology — Open Systems Interconnection Security Frameworks for Open Systems: Access Control Framework, November 1995.

International Telecommunication Union — Recommendation X.813, Information Technology — Open Systems Interconnection Security Frameworks for Open Systems: Non-repudiation Framework, October 1996.

International Telecommunication Union, Yearbook of Statistics, February 2001.

Internet Professionel, 65–66, 13, June–July 2002.

Inza, J., Banesto Easy SET project, July 6, 2000. Available at http://www.banesto.es or http://www.inza.co/setfacil/easyset.ppt

ISO Contribution N1737, Mechanisms for Bridging EDI with SGML, ISO/IEC JTC1/SC18/WG8, August 29, 1994.

ISO 7372, Trade Data Interchange — Trade Data Elements Directory (Endorsement of UNECE/TDED, Volume 1), 1993.

ISO/IEC 7498-1, Information Technology — Open Systems Interconnection — Basic Reference Model: The Basic Model, 1994.

ISO 7498-2, Information Technology — Open Systems Interconnection — Basic Reference Model — Part 2: Security Architecture, 1989.

ISO/IEC 7498-3, Information Technology — Open Systems Interconnection — Basic Reference Model: Naming and Addressing, 1997.

ISO/IEC 7498-4, Information Technology — Open Systems Interconnection — Basic Reference Model — Part 4: Management Framework, 1989.

ISO/IEC 7816-1, Identification Cards — Integrated Circuit(s) Cards with Contacts — Part 1: Physical Characteristics, 1998.

ISO 7816-2, Identification Cards — Integrated Circuit(s) Cards with Contacts — Part 2: Dimensions and Location of the Contacts, 1988.

ISO/IEC 7816-3, Identification Cards — Integrated Circuit(s) Cards with Contacts — Part 3: Electronic Signals and Transmission Protocols, 1997.

ISO/IEC 7816-4, Identification Cards — Integrated Circuit(s) Cards with Contacts — Part 4: Interindustry Commands for Interchange, 1995.

ISO/IEC 7816-5, Identification Cards — Integrated Circuit(s) Cards with Contacts — Part 5: Numbering System and Registration Procedure for Application Identifiers, 1994.

ISO/IEC 7816-6, Identification Cards — Integrated Circuit(s) Cards with Contacts — Part 6: Interindustry Data Elements, 1996.

ISO 8731-1, Banking — Approved Algorithms for Message Authentication — Part 1: DEA, 1987.

ISO 8731-2, Banking — Approved Algorithms for Message Authentication — Part 2: Message Authenticator Algorithm, 1992.

ISO 8372, Information Processing — Modes of Operation for a 64-bit Block Cipher Algorithm, 1987.

ISO/IEC 8824-1, Information Technology — Abstract Syntax Notation One (ASN.1): Specification of Basic Notation, 1998.

ISO/IEC 8824-2, Information Technology — Abstract Syntax Notation One (ASN.1): Information Object Specification, 1998.

ISO/IEC 8824-3, Information Technology — Abstract Syntax Notation One (ASN.1): Constraint Specification, 1998.

ISO/IEC 8824-4, Information Technology — Abstract Syntax Notation One (ASN.1): Parameterization of ASN.1 Specifications, 1998.

ISO/IEC 8825-1, Information Technology — ASN.1 Encoding Rules: Specification of Basic Encoding Rules (BER), Canonical Encoding Rules (CER) and Distinguished Encoding Rules (DER), 1998.

ISO 8583, Financial Transaction Card Originated Messages — Interchange Message Specifications, 1993.

ISO 8583-2, Financial Transaction Card Originated Messages — Interchange Message Specifications — Part 2: Application and Registration Procedures for Institution Identification Codes (IIC), 1998.

ISO 8583-3, Financial Transaction Card Originated Messages — Interchange Message Specifications — Part 3: Maintenance Procedures for Codes, 1998.

ISO 8732/ANSI X9.17 — Financial Institution Key Management — Wholesale.

ISO/IEC 8859-1, Information Technology — 8-Bit Single-Byte Coded Graphic Characters Sets — Part 1: Latin Alphabet No. 1, 1998.

ISO/IEC 8859-2, Information Technology — 8-Bit Single-Byte Coded Graphic Characters Sets — Part 2: Latin Alphabet No. 2, 1987.

ISO/IEC 8859-3, Information Technology — 8-Bit Single-Byte Coded Graphic Characters Sets — Part 3: Latin Alphabet No. 3, 1988.

ISO/IEC 8859-4, Information Technology — 8-Bit Single-Byte Coded Graphic Characters Sets — Part 4: Latin Alphabet No. 4, 1998.

ISO/IEC 8859-5, Information Technology — 8-Bit Single-Byte Coded Graphic Characters Sets — Part 5: Latin/Cyrillic Alphabet, 1999.

ISO/IEC 8859-6, Information Technology — 8-Bit Single-Byte Coded Graphic Characters Sets — Part 6: Latin/Arabic Alphabet, 1999.

ISO/IEC 8859-7, Information Technology — 8-Bit Single-Byte Coded Graphic Characters Sets — Part 7: Latin/Greek Alphabet, 1999.

ISO/IEC 8859-8, Information Technology — 8-Bit Single-Byte Coded Graphic Characters Sets — Part 8: Latin/Hebrew Alphabet, 1999.

ISO/IEC 8859-9, Information Technology — 8-Bit Single-Byte Coded Graphic Characters Sets — Part 9: Latin Alphabet No. 5, 1999.

ISO/IEC 8859-10, Information Technology — 8-Bit Single-Byte Coded Graphic Characters Sets — Part 10: Latin Alphabet No. 6, 1998.

ISO/IEC 8859-13, Information Technology — 8-Bit Single-Byte Coded Graphic Characters Sets — Part 13: Latin Alphabet No. 7, 1998.

ISO/IEC 8859-14, Information Technology — 8-Bit Single-Byte Coded Graphic Characters Sets — Part 14: Latin Alphabet No. 8 (Celtic), 1998.

ISO/IEC DIS 8859-15, Information Technology — 8-Bit Single-Byte Coded Graphic Characters Sets — Part 15: Latin Alphabet No. 9.

ISO 8879, Information Processing — Text and Office Systems — Standard Generalized Markup Language (SGML), 1986.

ISO 9735, Electronic Data Interchange for Administration, Commerce and Transport (EDIFACT) — Application Level Syntax Rules, 1988.

ISO 9735-1, Electronic Data Interchange for Administration, Commerce and Transport (EDIFACT) — Application Level Syntax Rules (Syntax Version Number: 4, Syntax Release Number: 1) — Part 1: Syntax Rules Common to All Parts, 2002.

ISO 9735-2, Electronic Data Interchange for Administration, Commerce and Transport (EDIFACT) — Application Level Syntax Rules (Syntax Version Number: 4, Syntax Release Number: 1) — Part 2: Syntax Rules Specific to Batch EDI, 2002.

ISO 9735-3, Electronic Data Interchange for Administration, Commerce and Transport (EDIFACT) — Application Level Syntax Rules (Syntax Version Number: 4, Syntax Release Number: 1) — Part 3: Syntax Rules Specific to Interactive EDI, 2002.

ISO 9735-4, Electronic Data Interchange for Administration, Commerce and Transport (EDIFACT) — Application Level Syntax Rules (Syntax Version Number: 4, Syntax Release Number: 1) — Part 4: Syntax and Service Report Message for Batch EDI (Message Type — CONTRL), 2002.

ISO 9735-5, Electronic Data Interchange for Administration, Commerce and Transport (EDIFACT) — Application Level Syntax Rules (Syntax Version Number: 4, Syntax Release Number: 1) — Part 5: Security Rules for Batch EDI (Authenticity, Integrity and Non-repudiation of Origin), 2002.

ISO 9735-6, Electronic Data Interchange for Administration, Commerce and Transport (EDIFACT) — Application Level Syntax Rules (Syntax Version Number: 4, Syntax Release Number: 1) —Part 6: Secure Authentication and Acknowledgment Message (Message Type — AUTACK), 2002.

ISO 9735-8, Electronic Data Interchange for Administration, Commerce and Transport (EDIFACT) — Application Level Syntax Rules (Syntax Version Number: 4, Syntax Release Number: 1) — Part 8: Associated Data in EDI, 2002.

ISO 9735-9, Electronic Data Interchange for Administration, Commerce and Transport (EDIFACT) — Application Level Syntax Rules (Syntax Version Number: 4, Syntax Release Number: 1) — Part 9: Security Key and Certificate Management Message (Message Type — KEYMAN), 2002.

ISO/IEC 9797-1, Information Technology — Security Techniques — Message Authentication Codes (MAC), Part 1: Mechanism Using a Block Cipher, 1999.

ISO/IEC 9797-2, Information Technology — Security Techniques — Message Authentication Codes (MAC), Part 2: Mechanism Using a Hash-Function, 2002.

ISO/IEC 10116, Information Technology — Security Techniques — Mode of Operation for an n-bit Block Cipher Algorithm, 1997.

ISO/IEC 10118-1, Information Technology — Security Techniques — Hash-Functions — Part 1: General, 1994.

ISO/IEC 10118-2, Information Technology — Security Techniques — Hash-Functions — Part 2: Hash-Functions Using an n-bit Block Cipher Algorithm, 1994.

ISO/IEC 10118-3, Information Technology — Security Techniques — Hash-Functions — Part 3: Dedicated Hash-Functions, 1998.

ISO/IEC 10118-4, Information Technology — Security Techniques — Hash-Functions — Part 4: Hash-Functions Using Modular arithmetic, 1998.

ISO/IEC 10179, Information Technology — Processing Languages — Document Style Semantics and Specification Language (DSSSL), 1996.

ISO 10202-1, Financial Transaction Cards — Security Architecture of Financial Transaction Systems Using Integrated Circuit Cards — Part 1: Card Life Cycle, 1991.

ISO 10202-2, Financial Transaction Cards — Security Architecture of Financial Transaction Systems Using Integrated Circuit Cards — Part 2: Transaction Process, 1996.

ISO 10202-3, Financial Transaction Cards — Security Architecture of Financial Transaction Systems Using Integrated Circuit Cards — Part 3: Cryptographic Key Relationships, 1998.

ISO 10202-4, Financial Transaction Cards — Security Architecture of Financial Transaction Systems Using Integrated Circuit Cards — Part 4: Security Application Modules, 1996.

ISO 10202-5, Financial Transaction Cards — Security Architecture of Financial Transaction Systems Using Integrated Circuit Cards — Part 5: Use of Algorithms, 1998.

ISO 10202-6, Financial Transaction Cards — Security Architecture of Financial Transaction Systems Using Integrated Circuit Cards — Part 6: Cardholder Verification, 1994.

ISO 10202-7, Financial Transaction Cards — Security Architecture of Financial Transaction Systems Using Integrated Circuit Cards — Part 7: Key Management, 1998.

ISO 10202-8, Financial Transaction Cards — Security Architecture of Financial Transaction Systems Using Integrated Circuit Cards — Part 8: General Principles and Overview, 1998.

ISO/IEC 10536-1, Identification Cards — Contactless Integrated Circuit(s) Cards — Part 1: Physical Characteristics, 1992.

ISO/IEC 10536-2, Identification Cards — Contactless Integrated Circuit(s) Cards — Part 2: Dimensions and Locations of Coupling Elements, 1995.

ISO/IEC 10536-3, Identification Cards — Contactless Integrated Circuit(s) Cards — Part 3: Electric Signals and Reset Procedures, 1996.

ISO/IEC 10744, Information Processing — Hypermedia/Time-Based Structuring Language (HyTime), 1992.

ISO/IEC 11179-1, Information Technology — Specification and Standardization of Data Elements — Part 1: Framework for the Specification and Standardization of Data Elements, 1999.

ISO/IEC 11179-2, Information Technology — Specification and Standardization of Data Elements — Part 2: Classification for Data Elements, 2000.

ISO/IEC 11179-3, Information Technology — Metadata Registries (MDR) -- Part 3: Registry Metamodel and Basic attributes, 2003.

ISO/IEC 11179-4, Information Technology — Specification and Standardization of Data Elements— Part 4: Rules and Guidelines for the Formulation of Data Definitions, 1995.

ISO/IEC 11179-5, Information Technology — Specification and Standardization of Data Elements — Part 5: Naming and Identification Principles for Data Elements, 1995.

ISO/IEC 11179-6, Information Technology — Specification and Standardization of Data Elements — Part 6: Registration of Data Elements, 1997.

ISO/IEC TR 13335-5, Information Technology — Guidelines for the Management of IT Security — Part 5: Management Guidance on Network Security, 2001.

ISO/IEC 14662, Information Technology — Open-EDI Reference Model, 1997.

Itoi, N. and Honeyman, P., Smartcard integration with Kerberos V5, in *Proc. USENIX Workshop on Smartcard Technol.*, Chicago, IL, 1999, pp. 51–61.

Itoi, N., Honeyman, P., and Rees, J., SCFS: A UNIX filesystem for smartcards, in *Proc. USENIX Workshop Smartcard Technol.*, Chicago, IL, 1999, pp. 107–117.

Jackson, D., Preparing the organization for EDI, in *The EDI Handbook*, Gikins, M. and Hitchcock, D., Eds., Blenheim Online, London, 1988, pp. 149–155.

Jaffe, F. and Landry, S., Electronic checks: the best of both worlds, *Electron. Commerce World*, July, 1997. Available at http://www.echeck.org/kitprint/article.htm.

Juels, A. and Brainard, J., A cryptographic countermeasure against connection depletion attacks, in *Proc. Internet Soc. Symp. Network Distributed System Security*, Kent, S., Ed., IEEE Computer Society Press, Washington (also available online at www.rsasecurity.com), 1999, pp. 151–165.

Jupiter Communications, cited in Nelson, S.A., Internet traders swap Beanies and trust, *USA Today*, July 20, 15A, 1998.

Kaiserswerth, M., The OpenCard Framework and PC/SC. Available at http://www.opencard.org/docs/ocfpcsc.pdf.

Kaplan, S. and Sawhney, M., E-Hubs: the new B2B marketplaces, *Harvard Business Review*, May–June, 97–101, 2000.

Kelly, E.W., The future of electronic money: a regulator's perspective, *IEEE Spectrum*, 34, 2, 21–22, 1997.

Kimberley, P., *Electronic Data Interchange*, McGraw-Hill, New York, 1991.

Kirschner, L., The battle of the electronic purses: who will win? *Card Forum Int.*, 2, 3, 33–41, 1998.

Kömmerling, O. and Kuhn, M.G., Design principles for tamper-resistant smart card processors, in *Proc. USENIX Workshop Smartcard Technol.*, Chicago, IL, 1999, pp. 9–20.

Kravitz, J., SDML-Signed Document Markup Language, Version 2.0, April 1998. Available at http://www.fstc.org/projects/sdml/sdml_det.html.

Krawczyk, H,. SKEME: a versatile secure key exchange mechanism for Internet, in *Proc. Internet Soc. Symp. Network and Distributed Syst. Security*, IEEE Computer Society Press, Washington, 1996, pp. 114–127.

Kwon, E.-K., Cho, Y.-G., and Chae, K.-J., Integrated transport layer security: end-to-end security model between WTLS and TLS, in *Proc. 15th Int. Conf. Inf. Networking*, 2001, pp. 65–71.

Kwon, T., Yoon, M., Kang, J., Song, J., and, Kang, C.-G., A modeling of security management system for electronic data interchange, in *Proc. 2nd IEEE Symp. Comp. Commun. (ISCC)*, IEEE Computer Society Press, Washington, 1997, pp. 518–522.

Lacoste, G., SEMPER: a security framework for the global electronic marketplace, IBM France, August 1997. Available at http://www.semper.org.

Lacoste, G., Pfitzmann, B., Steiner, M., and Waidner, M. (Eds.), *SEMPER — Secure Electronic Marketplace for Europe*, Lect. Notes Comput. Sci., 1854, Springer-Verlag, Heidelberg, 2000.

Lai, X. and Massey, J., Markov ciphers and differential cryptanalysis, in *Proc. Eurocrypt'91*, Lect. Notes Comput. Sci., 547, Springer-Verlag, Heidelberg, 1991a.

Lai, X. and Massey, J., A proposal for a new block encryption standard, in *Proc. Eurocrypt'90*, Lect. Notes Comput. Sci., 473, Springer-Verlag, Heidelberg, 1991b.

Lambe, G., E-cash fails to generate enthusiasm, *Banking Technol.*, 15, 10, 6, 1998/1999.

Landais, Y., Les produits et services EDI offerts par un opérateur, Transpac, *Télécom Interview*, 33, 60–68, 1997.

Lang, B., Contre la mainmise sur la propriété intellectuelle: des logiciels libres à la disposition de tous, *Le Monde diplomatique*, January 26, 1998.

Lash, C., The revolt of the elites: have they canceled their allegiance to America? *Harper's Magazine*, November 39–77, 1994.

Lefebvre, L.A. and Lefebvre, E., Moving towards the virtual economy: a major paradigm shift, in *Management of Technology, Sustainable Development and Eco-Efficiency*, Lefebvre, L.A., Mason, R.M., and Khalil, T., Eds., 7th Int. Conf. Manage. Technol., Orlando, FL, Elsevier, Amsterdam; New York, 1998, pp. 13–23.

Les Échos, Des internautes plus curieux que la moyenne, November 4, 56, 1998.

Lindley, R., *Smart Card Innovation*, Saim Pty, Ltd., Australia, 1997.

Liu, M., SDML & XML, April 1998. Available at http://www.fstc.org/projects/sdml/sdml_comp.html.

Loeb, L., *Secure Electronic Transactions: Introduction and Technical Reference*, Artech House, Norwood, MA, 1998.

Lorentz, F., Commerce électronique: une nouvelle donne pour les consommateurs, les entreprises, les citoyens et les pouvoirs publics, January 1998. Available at http://www.telecom.gouv.fr/francais.

Lorenz, G.W., Electronic stored value payment systems, market position, and regulatory issues, *Am. Univ. Law Rev.*, 46, 1177–1206, 1997.

Mamais, G., Markaki, M., Sherif, M.H., and Stassinopoulos, G., Evaluation of the Casner-Jacobson algorithm for compressing the RTP/UDP/IP headers, in *Proc. ISCC98*, Athens, Greece, IEEE Computer Society Press, Washington, 1998, pp. 543–548.

Manasse, M.S., The Millicent protocols for electronic commerce, in *Proc. 1st USENIX Workshop Electron. Commerce*, New York, 1995, pp. 117–123.

Manchester, P., Why collaboration will be crucial, *Financial Times, FI-IT Review*, April 3, 2, 2002.

Martres, D. and Sabatier, G., *La Monnaie Électronique, Que sais-je?* No. 2370, Presses Universitaires de France, Paris, 1987.

Le Matin (Lausanne), *Carte Cash: vous rechignez*, 1, February 6, 1998.

Matsunaga, M., I-mode media concept, *NTT DoCoMo Tech. J.*, 1, 1, 10–19, 1999.

Mayer, M., *The Bankers: The Next Generation*, Truman Talley Books/Dutton, New York, 1997.

McCarthy, J.C., The social impact of electronic commerce, *IEEE Commun. Mag.*, 37, 9, 58–60, 1999.

McCrindle, J., *Smart Cards*, Springer-Verlag, Heidelberg, 1990.

Menezes, A., *Elliptic Curve Public Key Cryptosystems*, Kluwer, Dordrecht, 1993.

Menezes, A.J., van Oorschot, P.C., and Vanstone, S.A., *Handbook of Applied Cryptography*, CRC Press LLC, Boca Raton, FL, 1997.

MENTIS, *Defining e-commerce*, GartnerGroup, Durham, NC, 1998.

Messerges, T.S., Dabbish, E.A., and Sloan, R.H., Investigations of power analysis attacks on smartcard, in *Proc. USENIX Workshop on Smartcard Technol.*, Chicago, IL, 1999, pp. 151–161.

Mevinsky, G. and Neuman, B.C., NetCash: a design for practical electronic currency on the Internet, in *Proc. 1st ACM Conf. Comput. Commun.* Security, 1993. Available at http://nii.isi.edu/info/netcheque/documentation.html.

Michard, A., *XML Langage et Applications*, Eyrolles, Paris, 1999.

Millicent, Millicent-Specific Elements for an HTTP Payment Protocol, V.1.3, 1995. Available at http://www.millicent.com/htlm/specs/mcproto.htm.

Mitchell, J.C., Shmatikov, V., and Stern, U., Finite-state analysis of SSL 3.0, in *Proc. 7th USENIX Security Symp.*, San Antonio, TX, 1998, pp. 26–29.

Le Monde, Les fraudes sur les cartes bancaires en hausse, November 25, 2000.

Montgomery, M. and Krishna, K., Secure object sharing in Java Card, in *Proc. USENIX Workshop Smartcard Technol.*, Chicago, IL, 1999, pp. 119–127.

Moore, D., Voelker, G.M., and Savage, S., Inferring Internet denial-of-service activity, *Proc. 10th USENIX Security Symp.*, August 13–17, Washington, D.C., 2001.

Morrison, S., California hopes to download revenue from cyberspace, *Financial Times*, May 8, 2, 2003.

NTT DoCoMo, *Financial Times*, October 7–19, 2002.

Nakamoto, M., Sony chief in warning to US, *Financial Times*, June 13, 19, 2002.

Nalwa, V.S., Automatic on-line signature verification, in *Biometrics: Personal Identification in Networked Society*, Jain, A., Bolle, R., and Pankati, S., Eds., Kluwer, Dordrecht, 1999, pp. 143–163.

Nanavati, S., Thieme, M., and Nanavati, R., *Biometrics: Identity Verification in a Network World*, John Wiley & Sons, New York, 2002.

National Institute of Standards and Technology (NIST), FIPS Publication 180, Secure Hash Standard (SHS), May, 1993.

National Institute of Standards and Technology (NIST), FIPS Publication 180-1, Secure Hash Standard (SHS), April 1995.

National Security Agency (NSA), The Mosaic Program Office, Key Management Concepts, Version 2.52, February 1994. Available at http://www.rbo.com/PROD/rmadillo/f.

Nechvatal, J., Baker, E., Bassham, L., Burr, W., Dworkin, M., Foti, J., and Roback, E., Report on the development of the Advanced Encryption Standard (AES), National Institute of Standards and Technology, October 2, 2000.

Neuman, B.C., Proxy-based authorization and accounting for distributed systems, in *Proc. 13th Int. Conf. on Distributed Comput. Syst.*, 1993, pp. 283–291.

Neuman, B.C. and Mevinsky, G., Requirements for network payement: the NetCheque™ perspective, *COMPCON'95*, 1995, pp. 32–36.

Neuman, B.C and Ts'o, T., Kerberos: an authentication service for computer networks, *IEEE Commun. Mag.*, 32, 9, 33–38, 1994.

New, D., Internet information commerce: the First Virtual (TM) approach, in *Proc. 1st USENIX Workshop Electron. Commerce*, New York, 1995, pp. 33–68.

Norton, M., Why US lags in wireless, *Banking Technol.*, 18, 3, 22–23, April, 2001.

Norris, F., Software maker report counterfeit shares, *New York Times*, February 7, 2001.

Nyström, M., PKCS #15: A cryptographic token information format standard, in *Proc. USENIX Workshop Smartcard Technol.*, Chicago, IL, 1999, pp. 37–42.

OASIS, Assertions and Protocol for the OASIS Security Assertion Markup Language (SAML), Committee specification 01, cs-sstc-core-01, May 31, 2002.

Obaidat, M.S. and Sadoun, N., Keystroke dynamics based authentication, in *Biometrics: Personal Identification in Networked Society*, Jain, A., Bolle, R., and Pankati, S., Eds., Kluwer, Dordrecht, 1999, pp. 213–225.

O'Callaghan, R. and Turner, J.A., Electronic data interchange — concept and issues, in *EDI in Europe: How it Works in Practice*, Krcmar, H., Bjørn-Andersen, N., and O'Callaghan, R., Eds., John Wiley & Sons, New York, 1995, pp. 1–19.

Oestreicher, M. and Krishna, K., Object lifetimes in Java Card, in *Proc. USENIX Workshop on Smartcard Technol.*, Chicago, IL, 1999, pp. 129–137.

Okamoto, T., An efficient divisible electronic cash scheme, in *Adv. Cryptology — Crypto'95, Proc.*, Springer-Verlag, Heidelberg, 1995, pp. 438–451.

O'Mahony, D., Pierce, M., and Tewari, H., *Electronic Payment Systems*, Artech House, Norwood, MA, 1997 (2nd ed., 2001).

Omnès, S., Le système interbancaire de télécompensations, *Banque Stratégie*, 131, 22–24, October, 1996.

Oppliger, R., *Internet and Intranet Security*, Artech House, Norwood, MA, 1998.

Palme, J., *Electronic Mail*, Artech House, Norwood, MA, 1995.

Parsons, M., The accidental rise of smart cards, *Red Herring*, 109, 72–73, January, 2002.

Pasini, S. and Chaloux, J., EDI and activity based management for business — the case of Whirlpool in Italy, in *EDI in Europe: How it Works in Practice*, Krcmar, H., Bjørn-Andersen, N., and O'Callaghan, R., Eds., John Wiley & Sons, 1995, pp. 85–112.

Pays, P.-A., Systèmes de paiement pour le commerce électronique, in *le 6ème Colloque Francophone sur l'Ingénierie des Protocoles*, Liège, Belgium, 1997.

Pays, P.-A. and de Comarmont, F., An intermediation and payment system technology, *Comput. Networks ISDN Syst.*, 28, 7–11, 1197–1206, 1996.

PC/SC Workgroup, Interoperability specification for ICCs and personal computer systems, December, 1997. Available at http://www.smartcardsys.com.

Pedersen, T., Electronic payments of small amounts, in *Proc. Int. Workshop on Security Protocols*, Cambridge, U.K., April, 1996, Lomas, M., Ed., Lect. Notes Comput. Sci., No. 1189, Springer-Verlag, Heidelberg, 1997, pp. 56–68.

Penrose, P., Upwardly mobile, *Banking Technol.*, 15, 1, 23–36, 1998a.

Penrose, P., Share and share alike, *Banking Technol.*, 15, 1, 50, 1998b.

Pentland, A. and Choudhury, T., Face recognition for smart environments, *Computer*, 3, 2, 50–65, 2000.

Perkins, C.E., *Mobile IP: Design Principles and Practices*, Addison-Wesley, Reading, MA, 1998.

Phillips, C. and Meeker, M., *Collaborative Commerce*, Morgan Stanley Dean Witter, Equity Research North America, April 2000.

Phillips, P.J., Martin, A., Wilson, C.L., and Przybocki, M., An introduction to evaluating biometric systems, *Computer*, 3, 2, 56–63, 2000.

Pimont, T., i-mode, ou le succès d'un service distant made in Japan, *Décision Micro & Réseaux*, 444, 20, November 13, 2000.

Plassard, F., Une économie de don et de réciprocité, *Manière de Voir*, 41, 14–16, 1998.

Porter, M.E., *Competitive Strategy*, Free Press, New York, 1980.

Praca, D. and Barral, C., From smart cards to smart objects: the road to new smart technologies, *Comput. Networks*, 36, 381–389, 2001.

Presidential Executive Memorandum, Streamlining Procurement through Electronic Commerce, October 26, 1993.

Rabin, M.O., Digital Signatures and Public-Key Functions as Intractable as Factorization, Technical Report MIT/LCS/TR-212, MIT Laboratory for Computer Science, 1979.

Remacle, F., Swift, actif stratégique du monde bancaire, *Banque Stratégie*, 131, 18–21, October 1996.

Remery, P., A system of payment using coin purse cards, in *Smart Card 2000, Proc. IFIP WG 11.6 Int. Conf. Smart Card 2000: The Future of IC Cards*, Laxenburg, Austria, October 1987, Chaum, D. and Schaumüller-Bichl, I., Eds., Elsevier/ North Holland, Amsterdam, 1989, pp. 49–55.

Renfro, S.G., VeriSign CZAG: privacy leak in X.509 certificates, *Proc. 11th USENIX Security Symp.*, San Francisco, CA, 2002.

Rescorla, E., *SSL and TLS: Designing and Building Secure Systems*, Addison-Wesley, Reading, MA, 2001.

Rescorla, E., Cain, A., and Korver, B., SSLACC: a clustered SSL accelerator, *Proc. 11th USENIX Security Symp.*, San Francisco, CA, August 5–9, 2002.

Richmond, R., Scammed! Web merchants use new tools to keep buyers from ripping them off, *The Wall Street Journal*, January 27, R6, 2003.

Rila, L. and Mitchell, C.J., Security analysis of smartcard to card reader communications for biometric cardholder authentication, in *Proc. 5th Smart Card Res. Adv. Appl. Conf.*, San Jose, CA, 2002.

Rivest, R.L., The RC5 encryption algorithm, *CryptoBytes*, 1, 1, 9–11, 1995.

Rivest, R.L. and Shamir, A., PayWord and MicroMint: two simple micropayment schemes, in *Security Protocols, Proc. Int. Workshop*, Lomas, M., Ed., Cambridge, U.K., April, 1996, Lect. Notes Comput. Sci., 1189, Springer-Verlag, Heidelberg, 1997, pp. 69–87.

Rivest, R.L., Shamir, A., and Adleman, L.M., A method for obtaining digital signatures and public key cryptosystems, *Commun. ACM*, 21, 2, 120–126, 1978.

Rolin, P., La sécurité dans les réseaux, in *Réseaux de Communication et Conception de Protocoles*, Juanole, G., Sehrouchni, A., and Seret, D., Eds., Hermès, Paris, 1995, pp. 80–103.

Romain, H., Kléline: un système français de paiement électronique on-line, in *l'Internet et la Vente*, Aimetti, J.-P., Ed., Les Éditions d'Organisation, Paris, 1997, pp. 96–99.

Rubin, A.D., Geer, D., and Ranum, M.J., *Web Security Sourcebook*, John Wiley & Sons, New York, 1997.

Saarinen, M.-J., Attacks against the WAP WTLS protocol, 2000. Available at http:// www.jyu.fi/~mjos/wtls.pdf.

Sabatier, G., *Le porte-monnaie électronique et le porte-monnaie virtuel*, *Que sais-je*? No. 3261, Presses Universitaires de France, Paris, 1997.

Salehi, N., Obraczka, K., and Neuman, C., The performance of a reliable, request-response transport protocol, in *Proc. 4th IEEE Symp. Comput. Commun. ISCC'99*, Red Sea, Egypt, 1999, IEEE Computer Society Press, Washington, pp. 102–108.

Sandoval, V., *Technologie de l'EDI*, Hermès, Paris, 1990.

Schneider, I., Push and pay, *Bank Systems & Technology*, 39, 1, 28–30, 2002.

Schneier, B., *Applied Cryptography*, 2nd ed., John Wiley & Sons, New York, 1996a.

Schneier, B., Why cryptography is harder than it looks, 1996b. Available at http://www.counterpane.com/whycrypt.html.

Schneier, B., Security piftalls in cryptology, 1998a. Available at http://www.counterpane.com/pitfalls.html.

Schneier, B., Cryptography for the Internet, *3rd USENIX Workshop Electron. Commerce*, Boston, MA, Tutorial, 1998b.

Schuba, C.L., Krsul, I.V., Kuhn, M.G., Spafford, E.H., Sundaram, A. and Zamboni, D., Analysis of a denial of service attack on TCP, *IEEE Symp. Security and Privacy*, 1997, pp. 208–223.

Segev, A., Porra, A.J., and Roldan, M., Financial EDI over the Internet, Case study II, The Bank of America and Lawrence Livermore National Laboratory pilot, in *Proc. 2nd USENIX Workshop Electron. Commerce*, Oakland, CA, 1996, 173–190.

SEIS (Secured Electronic Information in Society), *Specifications of SEIS S1, SEIS Cards, Electronic ID Application*, February 19. 1998. Available at http://www.seis.se.

SEMPER Consortium, Architecture of Payment Gateway, Deliverable D14 of ACTS Project AC026, November 22, 1996. Available at http://www.semper.org.

Service Central de la Sécurité des Systèmes d'Information, Common Criteria for IT Security Evaluation Protection Profile — Smartcard Integrated Circuit Protection Profile, October, registered at the French Certification Body, 1997.

SET (Secure Electronic Transaction) Specification, *Book 1: Business Description; Book 2: Programmer's Guide; Book 3: Format Protocol Definition*, Version 1.0, May 31, 1997. Available at http://www.mastercard.com/set or http://www.visa.com/cgi-bin/vee/sf/set.

SETCo, *Online PIN Extensions to SET Secure Electronic Transaction™ Version 1.0*, May 25, 1999a. Available at http://www.setco.org.

SETCo, *Common Chip Extension SET™ 1.0*, September 29, 1999b. Available at http://www.setco.org.

Sherif, M.H., L'Internet, la Société de Spectacles et la Contestation, Institut de Recherches Internationales, Groupe de Recherche Économique et Sociale, Université Paris-Dauphine, February 1997.

Sherif, M.H., Sehrouchni, A., Gaid, A.Y., and Farazmandnia, F., SET et SSL: échanges sécurisés sur l'Internet, *Document Numérique*, 1, 4, 421–440, 1997.

Sherif, M.H., Sehrouchni, A., Gaid, A.Y., and Farazmandnia, F. SET and SSL Electronic payments on the Internet, in *Proc. ISCC98*, Athens, Greece, 1998, pp. 353–358.

Sherif, M.H, Ash, G.R., and Han, J., Transparent processing of Resource Management (RM) cells that are not defined in ATM-TM-011-21.00, I.371, or I.361, ATM Forum contribution 00-479, January 22–26, Phoenix, AZ, 2001.

Sherman, S.A., Skibom, R., and Murray, E.A., Secure network access using multiple applications of AT&T's Smart Card, *AT&T Tech. J.*, 61–72, September/October, 1994.

Simon, D.R., Anonymous communication and anonymous cash, in *Adv. Crytpotology — CRYPTO'96, Proc. 16th Annu. Int. Cryptology Conf.*, Santa Barbara, CA, Springer-Verlag, New York, 1996, pp. 61–73.

Simpson, W.A., IKE/ISAMP is considered harmful, *login*, 24, 6, 48–58, 1999.

Sirbu, M.A., Credits and debits on the Internet, *IEEE Spectrum*, 34, 2, 23–29, 1997.

Sirbu, M.A. and Chuang, J.C.-I. [online], Distributed authentication in Kerberos using public key cryptography, 1996. Available at http://www.cs.cmu.edut/afs/andrew.cmu.edu/inst/ini.

Sirbu, M.A. and Tygar, J.D., NetBill: an Internet commerce system optimized for network delivered services, *IEEE Personal Commun.*, 2, 4, 6–11, 1995. Available at http://www.ini.cmu.edu/netbill/pubs/CompCon.html.

Sivori, J.R., Evaluate receipts and settlements at Bell Atlantic, *Commun. ACM*, 39, 6, 24–28, 1996.

Steedman, D., *Abstract Syntax Notation One ASN.1: The Tutorial and Reference*, Technology Appraisals, Twickenham, U.K., 1993.

Sun Microsystems, Inc., The JavaCard 2.1 Platform Specifications, 1999. Available at http://java.sun.com/products/javacard.

Takeda, M., Uchida, S., Hiramatsu, K., and Matsunami, T., Finger image identification method for personal verification, *Proc. 10th Int. Conf. Pattern Recog.*, Vol. 1, pp. 761–766, 1990.

Télécommunications, No. 4529, citing a report of the Observatoire du commerce et des échanges électroniques (OC2E), March 18, 5, 1998.

Tenenbaum, J.M., Medich, C., Schiffman, A.M., and Wong, W.T., CommerceNet: spontaneous electronic commerce on the Internet, in *COMPCON'95*, 1995, pp. 38–43.

Thierauf, R.J., *Electronic Data Interchange in Finance and Accounting*, Quorum Books, New York, 1990.

Thorigné, Y. and Reiter, R., Nouvelle technologie de la carte à mémoire: la carte sans contact, *L'Echo des Recherches*, 43–48, 4th Quarter, 1994.

Troulet-Lambert, O., L'évolution de l'EDI et l'EDI-ouvert: les travaux du SC 30, *Télécom Interview*, 33, 41–46, 1997.

Tung, B., *Kerberos: A Network Authentication System*, Addison-Wesley Longman, Reading, MA, 1999.

Turner, S., Net effect, *Banking Technol.*, 15, 8, 49–52, 1998.

Tweddle, D., EDI in international trade: a customs view, in *The EDI Handbook*, Gikins, M. and Hitchcock, D., Eds., Blenheim Online, London, 1988, pp. 139–146.

Tyson-Davies, R., The function of APACS, in *Electronic Banking and Security*, Welch, B., Ed., Blackwell Scientific, Oxford, 1994, pp. 11–30.

Ugon, M. [online], L'odyssée de la carte à puce. Available at http://www/cardshow.com/guide/carte/odyssee.html.

Ugon, M., The microcomputer smart card: a multi-application device which secures telecommunications, in *EuroComm 88, Proc. Int. Congr. Bus., Public and Home Commun.*, Amsterdam, Schuringa, T.M., Ed., Elsevier/North Holland, Amsterdam, 1989, pp. 265–282.

UNCITRAL Model Law on Electronic Commerce, 1996. Available at http://www.unicitral.org.

U.S. Department of Treasury FAQ. Available at http://www.echeck.org/kitprint/ustfaq/ustfaq.htm.

Van Herreweghen, E. and Wille, U., Risks and potentials of using EMV for Internet payments, in *Proc. USENIX Workshop Smartcard Technol.*, Chicago, IL, 1999, pp. 163–173.

Van Hove, L., The New York City Smart Card trial in perspective: a research note, *Int. J. Electron. Commerce*, 5, 2,119–131, 2001.

Van Oorschot, P. and Wiener, M., Parallel collision search with applications to hash functions and discrete logarithms, in *Proc. 2nd ACM Conf. on Comput. and Commun. Security*, 1994, pp. 210–128.

Varadharajan, V. and Mu, Y., On the design of secure electronic payment schemes for Internet, in *Proc. 12th Annu. Comput. Security Appl. Conf.*, San Diego, CA, IEEE Computer Society Press, Washington, 1996, pp. 78–87.

VeriSign, VeriSign Certification Practice Statement, Version 1.2, May 15, 1997.

Wagner, D. and Schneier, B., Analysis of the SSL 3.0 protocol, in *Proc. 2nd USENIX Workshop Electron. Commerce*, Oakland, CA, 1996, pp. 29–40.

Waidner, M., Development of a secure electronic marketplace for Europe, in *Proc. ESORICS 96, 4th European Symp. Res. Comput. Security*, Rome, Lect. Notes Comput. Sci., 1146, Springer-Verlag, Heidelberg, 1996, pp. 1–14.

Waldmeir, P., Freedom under attack: civil liberties in the workplace are the latest casualties of the US reaction to terrorism, *Financial Times*, December 13, 10, 2001.

Walker, R., 1992: Maintaining the UK's competitive edge in EDI, in *The EDI Handbook*, Gikins, M. and Hitchcock, D., Eds., Blenheim Online, London, 1988, pp. 3–10.

Waters, R., How fraudsters set traps and take the credit, FT-IT Review, *Financial Times*, May 21, 2003.

Wayner, P., *Digital Cash: Commerce on the Net*, 2nd ed., Academic Press Professional, New York, 1997.

Westland, J.C., Kwok, M., Shu, J., Kwok, T., and Ho, H., Customer and merchant acceptance of electronic cash: evidence from Mondex in Hong Kong, *Int. J. Electron. Commerce*, 2, 4, 5–26, 1998.

Wiener, M.J., Performance comparison of public-key cryptosystems, *CryptoBytes*, 4, 1, 1–5, 1998. Available at http://www.rsa.com/rsalabs/pubs/cryptobytes.html.

Wildes, R.P., Iris recognition: an emerging biometric technology, *Proc. IEEE*, 85, 9, 1348–1363, 1997.

Wireless Application Forum, *Wireless Application Protocol — Wireless Transport Layer Security Specification WAP WTLS*, Version 05-Nov., 1999.

Yamada, H., Masaki, H., Yagi, H., and Oda, T., A new integrated access scheme for internet and telephony services, in *Proc. Int. Symp. Services Local Access* (ISSLS), Venice, Italy, 1998, pp. 259–265.

Ylönen, T., The SSH (Secure Shell) Remote Login Protocol, November 1995. Available at http://www.tigerlair.com/ssh/faq/ssh1-draft.txt.

Ylönen, T., SSH — secure login connections over the Internet, in *Proc. Sixth USENIX Security Symp.*, 1996, pp. 37–42.

Zaba, S., E-commerce payment protocols: requirements and analysis, in *Proc. 9th IEEE Comput. Security Found. Workshop*, 1996, pp. 78–80.

Index

A

Access control, 104–106, 142, 147
ACH (Automated Clearing House), 4
Active attacks, 73–74, 146–147
ADSL (Asymmetrical Digital Subscriber
    Line), 16
Aeronautical industry, 178–179
AES (Advanced encryption standard), 153
AH (Authentication Header) protocol, 79
Alert messages, 288–289
    TLS, 288–289
    WTLS, 299
Alphanumeric data, structured, 187–188
    definitions, 188–189
    EDIFACT, 5, 188, 190–191, 198–200
    UNB/UNZ and UIB/UIZ segments,
        191–192
    UNG/UNE segments, 193
    UNH/UNT segments, 191–193
    UNO/UNP segments, 193–194
    UNS segment, 193
    X12, 189–190
Amazon.com, 9
American Express card, 42, 50, 384
Anonymity
in transactions, 54–55
ANSI X12, 188, 189–190, 195
    funds transfer with, 228
    security, 207–208
Application layer security services, 82–83
Archives dematerialization, 545–546
ASN.1 (Abstract Syntax Notation 1), 309–310
ATM (automated teller machine), 373
Atomicity, 53
Attribute certificates, 141–142
Audits, 143
Australia, 7
Austria, 376
Authentication
    electronic purse, 387
    EMV card, 498–503
    integrated-circuit card, 493–496
    memory card reader for Minitel, 496–497
    one-way, 138–139
    of participants, 102–103
    SSL, 252–253
    three-way, 139–140
    two-way, 139
Automotive industry, 179–180

B

BACS (Banker's Automated Clearing
        Service), 4, 67–68
Banking, 24
    business-to-business electronic commerce
        in, 178
    certificate management for, 133–134
    clearance and settlement, 65–69
Banksys, 384
Belgium
    electronic purses in, 51
    Internet use in, 11
    Proton use in, 384
    criptural money in, 31
    smart card use in, 482
Billing, electronic, 228–229
Bills of exchange, 42
Biometric identification, 94–95, 100–102
BIPS (Bank Internet Payment System),
        466–469
BIS (Bank for International Settlements), 457
BizTalk, 200
Blind signatures, 90
Block encryption, 147–155
BSP (Bank Settlement Payment), 4
BTX (Bildschirmtext) system, 5
Business-to-business commerce, 4–5
    aeronautical applications, 178–179
    automotive applications, 179–180
    banking applications, 178
    by check, 36
    effect of the Internet on, 180–181
    electronic platforms, 181–182
    history of, 177–178
    obstacles facing electronic, 182–184
    overview of, 174–177

standardization of the exchanges of,
        230–235
    systems architecture, 184–187
Business-to-consumer commerce, 5–7

C

CAFE (Conditional Access for Europe), 48,
        523–526
CALS (Continuous Acquisition and Life-
        cycle Support), 5
Canada
    business-to-business electronic commerce
        in, 177
    check use in, 33
    credit transfers in, 37
    cross-certification in, 132–133
    Internet privacy in, 541
    scriptural money in, 29
Cash
    as instrument of payment, 31–33, 55–57
    registers, 528
Categories of electronic commerce
    business-to-business, 4–5
    objectives and, 3–4
CCD (Cash Concentration and
        Disbursement), 66–67
CCS (Common Channel Signaling), 107
CD-ROM, payments with, 369–370
CEFACT, 234
CEFIC (Conseil Européen des Fédérations de
        l'Industrie Chimique), 180
Censorship, 543–544
CertCo & Digital Trust Co., 316–317
Certificate management, 122–125
    attribute certificates and, 141–142
    banking applications, 133–134
    basic operation, 125
    certificate revocation and, 140–141
    certification path, 128
    cross-certification, 131–133
    hierarchical certification path, 128–131
    nonhierarchical certification path, 131
    online, 133
    SET, 316–318
    SSL, 264–265
    VeriSign architecture, 134–137
    X.509 certificate description and, 126–128
CHAPS (Clearing House Automated
        Payment System), 67–68
Charging protocol, 59
Chaum, David, 445
CheckFree, 229
Checks, *see also* Electronic check presentment
        (ECP)

classical processing of paper, 458–459
dematerialized processing of paper-
        based, 459–462
electronic presentment of (ECP), 36–37
as instrument of payment, 33–37, 55–57
Cheque & Credit Clearing Company Ltd.,
        67–68
Chipper®, 371, 374–376
CHIPS (Clearing House Interbank Payment
        System), 66
Cipher suites, 249–252, 265–267, 286
Clearance and settlement
    banking, 65–66
    DigiCash, 448
    digital money, 436, 444
    3-D Secure, 367–368
    First Virtual, 394
    in France, 68–69
    KLELine, 406
    MicroMint, 422
    NetBill, 401
    PayWord, 419
    in the United Kingdom, 67–68
    in the United States, 66–67
ClickShare, 7, 430
ClientHello, 263–264, 280–281, 295–296, 302
Clinton, Bill, 5, 10, 101
CommerceNet, 234
Commerce XML, 200
Commercial transactions, security of, 71–72
Communication infrastructure for electronic
        commerce, 534–536
Confidentiality, message
    public key cryptography for, 84–86, 88–90,
        146
    SSL, 242–243
    symmetric cryptography for, 83–84,
        91–94, 146
Connection establishment, SSL, 271–274
Consequences of electronic commerce, 21–25
Consumers
    business-to-, 5–7
COPPA (Children Online Privacy Protection
        Act), 541
Corporate cards, 42
Counterfeit money, 442–445
Credit transfers, 37–39
Creic (Centre Régionaux d'Échanges
        d'Images-Chèques), 68
CRI (Centrale des Règlements
        Interbancaires), 69
Cross-certification, 131–133
C-SET, 343–344
    architecture, 344–346
    cardholder registration, 346–348

encryption algorithm, 351–352
interoperability with SET, 352–353
payment software distribution, 348
purchase and payment transactions,
  348–351
CTP (Corporate Trade Payments), 66–67
CTX (Corporate Trade Exchange), 66–67
Cuba, 487
Cyber-COMM, 343–344

D

Data integrity, 86–87
blind signatures and, 90–91
SSL, 243
verification with one-way hash function,
  87–88
verification with public key
  cryptography, 88–90, 112–113
verification with symmetric
  cryptography, 91–94, 112–113
Definition of electronic commerce, 1–2
Dematerialized money, 46–49
business-to-business commerce and,
  174–177
direct payments to the merchant via, 62
harmonization and standardization of,
  536–537
payment via intermediaries, 62–65
protocols of systems of, 57–61
transactional properties of, 53–55
Denial of service, 106–108
DER (Distinguished Encoding Rules), 309
DES (Data Encryption Standard), 154
Diffie-Hellman exchange, 118–119, 252–253
DigiCash, 48, 445
delivery, 448
evaluation, 449
financial settlement, 448
loading of value, 446–447
purchase protocol, 447–448
registration, 446
Digital Millennium Copyright Act of 1998,
  539
Digital money, 48–49, 433, *see also* Money
building blocks for systems of, 434–445
counterfeit, 442–445
creditor untraceability and, 438
debtor untraceability and, 434–437
denominations, 439–442
deposit and settlement, 436
DigiCash, 48, 445–449
double spending proof, 444–445
financial settlement and verification, 444
improvement of protection, 436–437

loading of value, 435–436
mutual untraceability and, 438–439
NetCash, 450–455
purchase protocol, 436, 444
Diner's Card, 42
Direct debit, 40–41
Documents or forms, structured, 195–197
BizTalk, 200
Commerce XML, 200
electronic business XML, 201
SAML, 201–202
SGML, 197–198
SOAP, 202
UDDI, 202
WSDL, 203
XML, 198–200
DSA (Digital Signature Algorithm), 161–162
3-D Secure, 362–363
clearance and settlement, 367–368
enrollment, 364–365
purchase and payment protocol, 365–367
security, 368–369
DSL (Digital Subscriber Line), 16
Dual signature, 314–316

E

Easycode cards, 427
EBay, 8, 428
Ecash, *see* DigiCash
ECC (Elliptic Curve Cryptography), 159–160
ECheck, 470–474
ECoin, 424–425
EDI (Electronic Data Interchange), 1, 2, 5
business-to-business commerce and,
  181–184
integration with business processes,
  229–230
integration with XML, 198–200
interoperability of S/MIME and secured,
  221–222
messaging, 203–206, 220–221
relationship with electronic funds
  transfer, 223–228
security of, 206–223
standardization, 230–233
structured alphanumeric data and,
  187–188
XML integration, 234
EDIFACT, 5, 188, 190–195
funds transfer with, 226–227
security, 208–215
standardization, 230–233
Electronic billing, 228–229
Electronic business XML, 201

Electronic check presentment (ECP), 36–37,
        459–462, *see also* Checks
    BIPS, 466–469
    check imaging, 461–462
    compared to bankcards, 473–476
    eCheck, 470–474
    NetCheque, 462–465
    point-of-sale approval, 461
Electronic commerce
    archives dematerialization, 545–546
    banking and, 24
    business-to-business, 4–5
    business-to-consumer, 5–7
    categories of, 3–8
    clients, 21–22
    consequences of, 21–25
    definition of, 1–2
    fraud prevention and, 545
    harmonization and standardization for,
        536–537
    infrastructure for, 13–15, 534–536
    new entrants into, 23
    peer-to-peer, 8
    prepaid cards in, 7
    privacy and, 539–543
    role of governments in, 24–25
    substitution of paper money with, 22–23
    suppliers, 22
    taxation of, 544–545
Electronic funds transfer (EFT), 29
    with EDIFACT, 226–227
    relationship of EDI with, 223–228
    with X12, 228
Electronic mail systems, 427–428
Electronic money, 46–47
    CAFE, 523–526
    issuance of, 537–538
Electronic purses and token holders, 49–50
    diffusion of, 51–52
    harmonization of, 386–389
Electronic surveillance and privacy, 539–543
EMV specifications, 487–489
    authentication, 498–503
    evaluation, 503–504
Encryption
    block, 147–155
    cracks, 143–146
    differences between SSL and tLS, 286
EPC (Every Penny Counts, Inc.), 528–529
EPO (Electronic Payment Order), 399–401
ESP (Encapsulating Security Payload), 79–81
ETEBAC5, 68
ETSI (European Telecommunications
        Standards Institute), 17

European Union, the, 479–480, 543, *see also*
        individual countries
Evaluation
    DigiCash, 449
    EMV specifications, 503–504
    KLELine, 406–407
    MicroMint, 424
    Millicent, 414–415
    NetBill, 401–402
    NetCash, 454–455
    PayWord, 421
    SET, 339–340
EyeDentify, Inc., 97

F

Face recognition, 98–99
FACNET (Federal Acquisition Computer
        Network), 5
Fedwire, 66
Fidelity cards, 528–529
Fiduciary money, 28, 31–33
Filtering, 543–544
Financial agents certificates, 319
Financial settlement, *see* Clearance and
        settlement
Fingerprint recognition, 99–100
Finland, 11–12
First virtual, 391
    acquisition and financial settlement, 394
    buyer's subscription, 392
    evaluation, 395
    purchase protocol, 392–394
    security, 394
Forgery, 423
France
    banking clearance and settlement in,
        68–69
    business-to-business electronic commerce
        standardization in, 233
    cash use in, 33
    check use in, 33, 457, 458
    credit transfers in, 37
    direct debit in, 40
    electronic check presentment in, 460–461
    EMV specifications in, 489
    fraud in, 362
    government role in electronic commerce,
        24–25
    Internet privacy in, 542
    Internet use in, 11
    Minitel system, 1, 2, 5–7, 8, 9, 46, 72, 228,
        458, 496–497, 533
    repaid telephone cards in, 7
    scriptural money in, 29

smart card use in, 479, 482, 497–498
Télécom, 6
Fraud prevention, 362–369, 545
Free Software Association, 2
French Association for Commerce and
  Electronic Interchange, 1
FSML (Financial Services Markup
  Language), 473

G

GeldKarte, 371, 376–381
Gemplus, 376
General public license software, 2
Germany
  BTX (Bildschirmtext) system, 5
  business-to-business electronic commerce
    standardization in, 233
  cash use in, 33
  direct debit in, 41
  GeldKarte use in, 376
  Internet use in, 11
  scriptural money in, 31
  smart card use in, 482
Giesecke & Devrient, 376
Gore, Al, 10
Government roles in electronic commerce,
  24–25
GPRS (General Packet Radio Service), 16, 292
Greece, 11
GSM (Global System for Mobile
  Communication), 291
GSM (Groupe Spécial Mobile), 16

H

Hand geometry recognition, 100
Handwritten recognition, 96
Harmonization and standardization, 536–537
Hash operations, 87–88, 318–319, 436–437
Hierarchical certification path, 128–131
High-context societies, 11–12, 538
Hong Kong, 29
  cash use in, 33
  check use in, 33, 36
  credit transfers in, 37–39
  Mondex use in, 384
HTML (HyperText Markup Language), 15
Human-Authentication Application Program
  Interface (HA-API), 101
Humpich, Serge, 498
HyTime (Hypermedia/Time-Based
  Document Structuring
  Language), 198

I

IANA (Internet Assigned Numbers
  Authority), 241
IDEA (International Data Encryption
  Algorithm), 154
Identification of participants
  authentication and, 102–103, 147
  biometric identification, 94–95, 100–102
  face recognition, 98–99
  fingerprint recognition, 99–100
  hand geometry recognition, 100
  handwritten recognition, 96
  iris recognition, 97–98
  keystroke recognition, 96–97
  retinal recognition, 97
  voice recognition, 95–96
Identity-based access control, 104–106
IETF (Internet Engineering Task Force)
  PGP/MIME encrypted and signed,
    217–218
  security proposals, 216–217
  S/MIME message encrypted and signed,
    219–220
  standardization, 234
Infrastructure for electronic commerce,
  13–15, 534–536
Integrated-circuit cards, *see also* Smart cards
  attacks against chip-reader
    communication channel, 514–515
  attacks due to negligence, 513–514
  contactless, 482–483
  with contacts, 482
  EMV specifications, 487–489, 498–503
  integration with computer systems,
    509–512
  ISO standards, 486–487
  logical security during usage, 493–496,
    512–513
  memory types, 484–485
  operating systems, 485
  physical security during usage, 492–493,
    513
  security during production, 489–492
  standards for, 486–489
Intellectual property, protection of, 538–539
Interbank transfers, 41–42
Interchange structure, 194
Intermediaries, payment via, 62–65
International Organization for
  Standardization (ISO), 74–75
Internet, the, 1
  effect on business-to-business commerce,
    180–181

electronic surveillance and privacy on, 539–543
filtering and censorship of, 543–544
influence of, 8–12
messaging, 204–206
privatization of, 10
scope and penetration of, 11–12
successful use of, 8–9
traffic multiplexing and, 17–20
transactional security and, 9–10
IPIN, 7
Iran, 488
Iraq, 488
Iridian Technologies, 97
Iris recognition, 97–98
ISAKMP (Internet Security Association and Key Management Protocol), 119–121
ISDN (Integrated Services Digital Network), 16, 17, 20
ISO/IEC 7816-4, 504–505
Italy
    Internet use in, 11
    scriptural money in, 31
    smart card use in, 374
ITLS, 301–302

J

Japan
    NTT DoCoMo, 7, 16, 72
    prepaid cards in, 7
    scriptural money in, 29
    smart card use in, 479
JEPI (Joint Electronic Payment Initiative), 526
Jeton holders, 49–51, 424–425

K

Kerberos, 113–117
Keys, cryptographic
    comparison between symmetric and public, 112–113
    deletion, backup, and archiving, 112
    distribution, 111
    Exchange Algorithm (KEA), 121–122
    exchange of secret, 113–117
    production and storage, 110–111
    revocation, 112
    utilization, withdrawal, and replacement, 111–112
Keystroke recognition, 96–97
KLELine, 391, 402–403
    evaluation, 406–407
    financial settlement, 406

purchase and payment protocols, 403–405
registration, 403

L

LETS (Local Exchange Trading System), 2
Libya, 488
Link-layer protocol PPP (Point-to-Point Protocol), 78–79
Linux, 81
Low-context societies, 11–12, 538

M

MasterSecret, 286–288, 297–299
Memory cards, 481, 484–485
Merchants
    GeldKarte use by, 375–381
    hybrid SSL/SET notification of, 360
    SET certificates, 319
    SET registration, 325–326
Message Authentication Code (MAC), 91–94
MicroMint, 391, 421
    evaluation, 424
    financial settlement, 422
    purchase protocol, 422
    registration and loading of value, 422
    security, 422–424
Micropayment systems, 372–374, *see also* Payments; Remote micropayments
Microprocessor cards, 481
    logical security during usage, 493–496
    physical security during usage, 492–493
    security during production, 489–491
Middleware, 15
Millicent, 391, 408
    evaluation, 414–415
    purchase protocol, 412–413
    registration and loading of value, 411–412
    scrip description, 409–411
    security, 409
Minitel, 1, 2, 5–7, 8, 9, 46, 72, 228, 458, 533
    like systems, 430
    memory card reader, 496–497
Mondex, 8, 371, 381–382
    payments, 382–383
    pilot experiments, 384
    security, 383–384
Money
    counterfeit, 442–445
    definition and mechanisms of, 27–29
    dematerialized, 46–49, 53–55, 57–65, 536–537
    digital, 48–49, 433–455

electronic, 46–47, 523–526, 537–538
fiduciary, 28, 31–33
scriptural, 28, 33–37
virtual, 47–48
Motorola, 376
Multiapplication smart cards, 504–509

N

NACHA (National Automated Clearing
House Association), 4, 29, 66–67,
133–134
NAETEA, 302–305
Napster, 8
Neighborhood commerce, 7
Neopost, 9
NetBill, 391
delivery phase, 398–399
evaluation, 401–402
financial settlement, 401
negotiation phase, 398
order phase, 398
payment phase, 399–401
registration and loading of value, 395
NetCash, 449
evaluation, 454–455
extensions, 451–453
purchase protocol, 450–451
registration and value purchase, 450
NetCheque, 462–465
Netherlands, the
bank card use in, 45
check use in, 36
Chipper® use in, 371, 374–376
direct debit in, 41
GeldKarte use in, 376
Internet use in, 11
scriptural money in, 31
Net Nanny Software International, 97
Network access, 15–20
Network layer security services, 79–81
Nonhierarchical certification path, 131
Nonrepudiation, 108–110, 147
North Korea, 488
Norway, 12
NTT DoCoMo, 7, 16, 72
NYCH (New York Clearing House), 66

O

OASIS (Organization for the Advancement of
Structured Information
Standards), 235
OBI (Open Buying on the Internet), 234
Odysseo, 407

One-way authentication, 138–139
One-way hash function, 87–88
OpenCard Framework, 509, 510
Open financial networks, security of, 72–73
Order information, SET, 330–333
OSI (Open Systems Interconnection), 75–78
OTP (Open Trading Protocol) Consortium,
234

P

Paris Net Settlement (PNS), 69
Participants
authentication, 102–103, 138–139, 252–253
hybrid SSL/SET notification of, 360
SET registration, 320–326
Passive attacks, 73, 146–147, 512–513
Payments, *see also* Micropayment systems;
Remote micropayments
cards, 42–45
with CD-ROM, 369–370
comparison of means of, 55–57
direct merchant, 62
3-D Secure protocol for, 365–367
eCheck, 470–472
GeldKarte, 377–380
hybrid SSL/SET, 360
instructions, SET, 330–333
instruments of, 29–31
bills of exchange as, 42
cash as, 31–33, 55–57
checks as, 33–37, 55–57
credit transfers as, 37–39
direct debit as, 40–41
interbank transfers as, 41–42
payment cards as, 42–45
KLELine, 403–405
Mondex, 382–383
NetBill, 399–401
NetCheque, 464–465
protocol, 59
Proton, 385
SEMPER, 522–523
via intermediaries, 62–65
PayPal, 428–430
PayWord, 391, 415–416
commitment, 417–418
computational load, 419–420
delivery, 418–419
evaluation, 421
financial settlement, 419
purchase protocol, 417–419
registration and loading of value, 416–417
PC/SC smart cards, 509, 511–512
Peer-to-peer commerce, 8

Performance acceleration, SSL, 274–276
Personal identification number (PIN), 54
PESIT, 68–69
PGP (Pretty Good Privacy), 159, 206, 216–218
PICS (Platform for Internet Content
    Selection), 526–527
PIN (personal identification number), *see*
    C-SET
POSTs (point-of-sale terminals), 388–389, 528
  check approval, 461
P3P (Platform for Privacy Preference),
    526–527
Prepaid cards, 7, 427
Privacy and electronic surveillance, 539–543
Procurement cards, 43
Proton, 371, 384–386
PSTN (Public Switched Telephone Network),
    5, 17, 20, 392
Public key cryptography, 84–86, 88–91,
    112–113, 146, 155–161
  Diffie-Hellman exchange of, 118–119
  Kerberos and, 117–118
  standards (PKCS), 157–159
Purchasing protocol, 59
  cards, 43
  DigiCash, 447–448
  digital money, 436, 444
  3-D Secure, 365–367
  First Virtual, 392–394
  KLELine, 403–405
  MicroMint, 422
  Millicent, 412–413
  NetBill, 396–397
  NetCash, 450–451
  PayWord, 417–419
  SET, 326–337
Purses and holders
  diffusion of electronic, 51–52
  electronic, 49–50
  virtual, 50–51

Q

Quality of service considerations, 529–530

R

RADSL (Rate Adaptive Digital Subscriber
    Line), 16
Remote micropayments, *see also* Payments
  eCoin, 424–425
  electronic mail based, 427–428
  first-generation, 425–426
  First Virtual, 391, 392–395
  history of, 391–392

KLELine, 391, 402–407
MicroMint, 391, 421–424
Millicent, 408–415
NetBill, 391, 395–402
PayPal, 428–430
PayWord, 391, 415–421
prepaid card, 427
purchase protocol, 396–397
second-generation, 427–430
security, 392–395
Réseaux Réciproques d'Échange de Savoirs
    (RRES), 2
Retinal recognition, 97
Role-based access control, 104–106, 142
Root authority certificates, 319
RosettaNet, 235
RTP (Real-Time Protocol), 18

S

SABRE, 4
SAGITTAIRE, 4, 69
SAML (Security Assertion Markup
    Language), 201–202
Scriptural money, 28, 33–37
Secure Shell, 82
Security
  access control and, 104–106
  business-to-business commerce, 187
  of commercial transactions, 71–72
  data integrity and, 86–94
  denial of service and, 106–108
  digital money, 436–437
  3-D Secure, 368–369
  EDI, 206–223
  EDIFACT, 208–215
  First Virtual, 394
  identification of participants and, 94–103
  in-band segments, 209–213
  Kerberos and, 113–117
  MicroMint, 422–424
  microprocessor card, 489–504
  Millicent, 409
  Mondex, 383–384
  nonrepudiation and, 108–110
  objectives, 73–75
  of open financial networks, 72–73
  OSI model for cryptographic, 75–78
  out-of-band segments, 213–215
  services at the application layer, 82–83
  services at the link layer, 78–79
  services at the network layer, 79–81
  services definitions and locations, 75–78
  SET, 311–316
  smart card, 512–515

SSL, 241–243
XML exchange, 223
Seignorage, 537
SEL (Systèmes d'Échange Locaux), 2
SEMPER (Secure Electronic Marketplace for Europe)
  architecture, 520–522
  history of, 519–520
  payment manager, 523
  payment terminology in, 522–523
Sequence numbers, 109–110
ServerHello, 264, 302
SET (Secure Electronic Transaction)
  architecture, 308–311
  authorization requests, 334
  capture, 336–337
  cardholder certificate, 318–319
  cardholder registration, 320
  certificate durations, 320
  certificate management, 316–318
  certification, 316–326
  completing and sending forms, 322–325
  cryptographic algorithms, 312–314
  evaluation, 339–340
  financial agents certificate, 319
  granting authorization in, 334–336
  history of, 307
  hybrid SSL/, 353–362
  implementations, 338–339
  initialization, 321–322, 329–330
  interoperability with C-SET, 352–353
  merchant certificates, 319
  merchant's registration, 325–326
  method of dual signature, 314–316
  optional procedures in, 337
  order information and payment instruction, 330–333
  participant registration, 320–326
  payment messages, 327–329
  purchasing transaction, 326–337
  request for registration form, 322
  root authority certificates, 319
  security services of, 311–316
  transaction progress, 329–330
Settlement, *see* Clearance and settlement
SGML, 197–198
S-HTTP (Secure HyperText Transfer Protocol), 206, 239
SIC (Swiss Interbank Clearing), 4
SIngapore
  check use in, 33
  electronic purses in, 51–52
SITA (Société Internationale de Télécommunications Aéronautiques), 4, 178

SIT (Système Interbancaire de Télécompensation), 4, 68
SKIPJACK, 154–155
SKIP (Simple Key Management for Internet Protocols), 121
Smart cards, 12, 15, 29, 373–374, 427, 479–480, 515–517, *see also* Integrated-circuit cards
  applications, 480–482, 506–509
  authentication, 497–498
  classification of, 480–482
  contactless, 482–483
  with contacts, 482
  integration with computer systems, 509–512
  ISO/IEC 7816-4, 504–505
  multiapplication, 504–509
  OpenCard Framework, 509, 510
  PC/SC, 509, 511–512
  security, 512–515
  Swedish electronic identity (EID), 506
S/Mime, 219–222
SMS (Short Message System), 291
SMTP/MIME, 204–206
SOAP (Simple Object Access Protocol), 202
SOA (Source of Authority), 142
Software
  free, 9
  general public license, 2
South Africa, 7
Spain, 11
Spam, 544
SSL (Secure Socket Layer), 77
  Alert protocol, 243, 259–260
  certificate message, 264–265, 284
  ChangeCipherSpec (CCS) protocol, 243, 258
  cipher suite calculation, 265–267
  ClientHello message, 263–264, 280–281
  compared to TLS, 285–290
  connection establishment, 271–274
  connection state variables, 246–247
  decryption and verification of data, 270
  definition, 239
  differences between TLS and, 285–290
  exchanges, 244–247
  finished message, 267–268
  functional architecture, 240–241
  Handshake protocol, 243, 249–257, 279–284
  hybrid SET/, 353–362
  implementations, 276–277
  MAC computation and encryption, 270
  new session establishment, 263–269
  parameters computation, 247–248

performance acceleration, 274–276
processing assumptions, 262
processing at the record layer, 268–269
processing of application data, 270
protocol general presentation, 239–243
Record protocol, 243, 258–259
security services, 241–243
ServerHello message, 264
sessions state variables, 245–246
subprotocols, 243–261
Stallman, Richard, 2
Stamps.com, 9
Sudan, the, 488
Sweden
    bank card use in, 45
    cash use in, 31, 33
    check use in, 36, 457
    electronic identity card (EID), 506
    Internet use in, 11
    scriptural money in, 31
SWIFT (Society for Worldwide Interbank
        Financial Telecommunications),
        4, 69, 178
Switzerland
    bank card use in, 45
    check use in, 36, 457
    GeldKarte use in, 376
    scriptural money in, 31
Symmetric block encryption algorithms,
        153–155
Symmetric cryptography, 83–84, 91–94,
        112–113, 146
Syria, 488

T

TARGET (Trans-European Automated Real-
        Time Gross Settlement Express
        Transfer system), 69
Taxation of electronic commerce, 544–545
TBF (Transfers Banque de France), 69
Telephone cards, 47–48, 481
Téletel, 2
Telstra, 7
Theft, 423–424
Three-way authentication, 139–140
Time-stamping, 109–110
Tip (Titre Interbancaire de Paiement), 41–42
TLS (Transport Layer Security), 77
    alert messages, 288–289
    available cipher suite for, 286
    compared to SSL, 285–290
    computation of MasterSecret and the
            derivation of keys for, 286–288
    differences between SSL and, 285–290

modifications to WTLS, 292–293
responses to record blocks of unknown
        type, 289–290
Town Clearing Company Ltd., 67–68
Traceability, 55
Traffic multiplexing, 17–20
Transactional security and the Internet, 9–10
Triple DES, 154
Two-way authentication, 139

U

UDDI (Universal Description, Discovery, and
        Integration), 202
UDEF (Universal Data Element Framework),
        5
UIB/UIZ segments, 191–192
UMTS (Universal Mobile Telecommunication
        System), 292
UNB/UNZ segments, 191–192
UNCITRAL (United Nations Commission on
        International Trade Law), 5
UNG/UNE segments, 193
UNH/UNT segments, 191–193
United Kingdom, the
    banking clearance and settlement in,
            67–68
    business-to-business electronic commerce
            in, 177
    business-to-business electronic commerce
            standardization in, 233
    cash use in, 31, 33
    check use in, 33
    EMV specifications in, 489
    fraud in, 362
    internet use in, 11
    Mondex use in, 381–382, 384
    scriptural money in, 29
United States, the
    banking clearance and settlement in,
            66–67
    business-to-business electronic commerce
            in, 177, 233
    cash use in, 33
    check use in, 33, 36
    credit transfers in, 37–39
    Internet use in, 11
    Mondex use in, 384
    scriptural money in, 29
UNO/UNP segments, 193–194
UNS segment, 193
User behavior analysis, 527–528

V

VDSL (Very High Bit Rate Digital Subscriber Line), 16
VeriSign, 134–137, 143, 307
Virtual money, 47–48, 424–425
Virtual purses and jeton holders, 50–51
Voice recognition, 95–96

W

WAP (Wireless Application Protocol), 16
Wired-logic cards, 481
Wireless access, 16
Wireline access, 16
WISP, 7, 430
WML (Wireless Markup Language), 291
World Wide Web, the, 10
WSDL (Web Services Description Language), 203
WTLS (Wireless Transport Layer Security), 285
  alert messages, 299
  architecture, 290–292
  calculation of secrets, 297–299
  cryptographic algorithms, 294–295
  exchange protocol during handshake, 296–297
  identifier and certificate formats, 293–294
  ITLS and, 301–302
  modifications from TLS, 292–293
  NAETEA and, 302–305
  parameter sizes, 299
  possible location of the WAP/Web Gateway, 300–301
  protocols of, 292–293
  record, 299
  service constraints, 299–300

X

X12, 188, 189–190, 195
  funds transfer with, 228
  security, 207–208
X.509 certificate, 126–128, 242
XHTML (eXtensible HyperText Markup Language), 292
X.400 messaging, 203–204
XML (Extensible Markup Language), 10, 183–187, 198–202
  Commerce, 200
  EDI integration, 234
  electronic business, 201
  exchange security, 223
  standardization, 235

Z

Zhone Technologies, 538

T - #0216 - 101024 - C0 - 234/156/34 [36] - CB - 9780849315091 - Gloss Lamination